DIFFERENTIAL GEOMETRIC STRUCTURES

Walter A. Poor

Center for Naval Analyses

DOVER PUBLICATIONS, INC.
Mineola, New York

Bibliographical Note

This Dover edition, first published in 2007, is an unabridged, slightly corrected republication of the work originally published by McGraw-Hill, Inc., New York, in 1981.

Library of Congress Cataloging-in-Publication Data

Poor, Walter A.
Differential geometric structures / Walter A. Poor.
p. cm.
"This Dover edition, first published in 2007, is an unabridged, slightly corrected republication of the work originally published by McGraw-Hill, Inc., New York, in 1981."
Includes bibliographical references and index.
ISBN 0-486-45844-X (pbk.)
1. Geometry, Differential. I. Title.

QA641.P616 2007
516.3'6—dc22

2006102449

Manufactured in the United States of America
Dover Publications, Inc., 31 East 2nd Street, Mineola, N.Y. 11501

To my father
a gifted applied geometer

CONTENTS

PREFACE

This text is intended to serve as an introduction to differential geometry. The prerequisites are the basic facts about manifolds as presented in the books by Auslander and Mackenzie, Boothby, Brickell and Clark, Hu, Lang, Matsushima, Milnor, Singer and Thorpe, or Warner; with extra work, the material from some modern advanced calculus books (for example, Goffman, Loomis and Sternberg, or Spivak) would also be sufficient. For the sake of uniformity almost all references to the elementary concepts of manifolds will be to Warner.

For example, a sample first course in differential geometry could cover the first two chapters of Warner and the first part of this book.

Later sections of the book make heavier demands on manifold theory, for which the reader is again usually referred to Warner. For example, a second geometry course could proceed through the third chapter of this book, Chapters 3 and 5 from Warner, and Chapter 4 of this book (with Warner's last chapter for reference). Similarly, a course on the basic geometry of Lie groups and symmetric spaces could use Chapter 4 of Warner and Chapters 6 and 7 of this book. Other courses are also possible.

The book should also be useful for independent study and as a reference work for mathematicians and physicists who need the basic ideas from differential geometry.

The book is in no way intended to be encyclopedic; for example, variational theory and submanifolds are not included, although they are fairly standard topics in first-year courses. For such topics the reader is referred to [BC], [CE], [GKM], [Hi], [KN], [Spi], or [Sb] in the bibliography.

Some comments on the more unusual features of the book follow.

In 1917 Levi-Civita interpreted the classical Ricci tensor calculus as an analytic description of a geometric concept which he called parallel transport. É. Cartan reversed the process in his study of affine, conformal, and projective differential geometry by postulating the existence of some system of parallel transport between nearby fibers of an appropriate fiber bundle, and then deriving the differential equations for parallel transport; these were written in terms of local connection forms. Since fiber bundles were still to be defined, Cartan's initial description of parallel transport was necessarily heuristic, and seems to have been taken by most people as an intuitive guideline, rather than as a rigorous starting point. For this reason, subsequent research either continued with the classical tensor calculus, or else emphasized the connection forms as the fundamental concept.

Once a satisfactory theory of fiber bundles had been developed, everything was reinterpreted in terms of connections on fiber bundles, global connection forms on principal fiber bundles, index-free covariant differentiation, and sprays.

In this book Cartan's viewpoint is taken: a geometric structure is defined by specifying the parallel transport in an appropriate fiber bundle (for a comparison of this approach with that of Riemann and Klein, see [Ca5]). Parallel transport is defined axiomatically, and everything else is then derived from it. Most of the book is devoted to the simplest case—linear parallel transport in a vector bundle. Only after the reader has come to grips with this case is parallel transport considered in more general fiber bundles.

The axioms for parallel transport in this book are my attempt to state carefully the heuristic comments in Section 2.1 of the book *Riemannsche Geometrie im Grossen*, by Gromoll, Klingenberg, and Meyer.†

The book begins with a study of fiber bundles. In Chapter 2, parallel transport in vector bundles is considered as an abstraction of parallelism of vectors in Euclidean space. After the statement of the axioms and some examples, holonomy is looked at briefly. The connection (as a horizontal distribution) is then defined, and connections and parallel transport are shown to be equivalent concepts. More study of holonomy leads to the curvature tensor. Only then is parallel transport rephrased in terms of covariant differentiation, which is the main computational tool used throughout the book; it is used to study the space of connections on a vector bundle, and for an introduction to characteristic classes using the exterior covariant derivative operator. The

† Gromoll later told me that those comments were based on axioms developed by Dombrowski, who in turn kindly informed me of prior axioms given by Rinow.

chapter closes with the special case of the tangent bundle and Cartan's generalization, the affine tangent bundle.

Chapter 3 covers Riemannian vector bundles, the Levi-Civita connection, curvature, and the metric space structure of a Riemannian manifold; it concludes with Chern's proof of the Gauss-Bonnet theorem.

The fourth chapter covers the Laplacian and basic harmonic theory. Weitzenböck's formula for the Laplacian is followed by Chern's much less familiar formula; so far as I know this has not appeared in any other differential geometry book (in fact the only references I know for it are Chern's original paper, Weil's *Séminaire Bourbaki* lecture notes, and my *Proceedings* note). Chern's formula is used to give a simple proof of D. Meyer's positive curvature operator theorem.

Chapter 5 covers harmonic, Killing, conformal, affine, and projective vector fields; the conformal vector fields are described on the sphere, thought of as the so-called Möbius space.

Lie groups appear often in the early chapters as examples; their basic geometry is explored further in Chapter 6. The difference between positive and negative curvature in the study of Lie groups is exemplified by the fact that although S^1 and S^3 are the only spheres which are Lie groups (a geometric proof is presented), hyperbolic space is a Lie group in each dimension. The classical simple Lie groups are studied in some detail, as are the spinor groups. The chapter concludes with the basic geometry of homogeneous spaces.

Chapter 7 introduces affine symmetric spaces, which are then described via symmetric pairs and Loos' axioms for the affine symmetries; the geodesic spray is derived directly from these axioms. This is followed by a discussion of locally affine symmetric spaces, symmetric Lie algebras, and Riemannian symmetric spaces.

Chapter 8 starts with symplectic vector bundles, shows that they are related to Hermitian vector bundles, and then covers enough complex manifold theory for an introduction to Kähler manifolds.

The last chapter deals with parallel transport, first in principal fiber bundles, and then in associated bundles for which the fiber has a prescribed geometric structure. Cartan connections are used to reinterpret affine and conformal differential geometry. The final topic is spin structures, culminating in Lichnerowicz' harmonic spinors theorem.

There are many examples and exercises scattered throughout the text; the reader is encouraged at least to read through each exercise before proceeding further. Because of space considerations, some of the examples must also be considered as exercises.

The book is based in part on a course which I taught at the University of Bonn in 1975 to 1976; I would like to thank the Sonderforschungsbereich 40 at the University of Bonn for its support. Thanks are also due to Professors Hirzebruch and Klingenberg for inviting me to Bonn, and to the students at Bonn; among other things, the students showed me the geometric proof given in Chapter 5 that S^1 and S^3 are the only spheres which are Lie groups.

The book was begun while I was supported by grant NSF No. MCS72 05055 A04 at the Institute for Advanced Study; my thanks to the National Science Foundation for its support, and to Professor Milnor for inviting me to the Institute. Further support was provided by a Mellon grant while I was at Skidmore College; for this I thank the Mellon Foundation and Dean Weller. The book was completed during a visit to Rensselaer Polytechnic Institute; I would like to thank Professor DiPrima for the invitation, and Rensselaer for the congenial working atmosphere.

It is a pleasure to express my deep gratitude to Jean-Pierre Bourguignon and Wolfgang Ziller, who read preliminary versions of the typescript and made innumerable penetrating, helpful suggestions (and corrections).

Others who are to be thanked for their comments on parts of the typescript are Glen Castore, Jeff Cheeger, David Elliot, Robert Greene, I. M. James, Michio Kuga, Robert Maltz, Charles Marshall, John Milnor, Katsumi Nomizu, Phillip Parker, John Thorpe, Frank Warner, and Steve Wilson.

Fred Cohen and Ravindra Kulkarni deserve special thanks; the former convinced me to write the book, and the latter read the finished product.

It is appropriate to thank Leonard Charlap for telling me to study parallel transport rather than covariant derivatives, Takushiro Ochiai and Tsunero Takahashi for explaining to me what I was doing on principal bundles, Ernst Heintze for helping me with Lie algebras, Seiki Nishikawa for explaining Cartan connections to me, and Nigel Hitchin and Jacques Tits for teaching me about spinors. Allen Adler and Edward Spitznagel have answered many questions on all sorts of topics.

Especially important to me has been the general encouragement I have received from Marcel Berger and Shiing-Shen Chern, and also from my wife Ellen Sara.

Finally I would like to thank my teachers—John Brillhart, Jeff Cheeger, David Ebin, Detlef Gromoll, Irwin Kra, Jim Simons, and John Thorpe.

Irene Abaganale and Peggy Murray typed the first draft for me, Rosanne Hammond and JJ Williams helped on the second draft, and my daughter Nureet helped me type the final draft of the book. Carolyn Boyce proofread it with me. To all these people I say thank you.

Most of the basic notation is the same as in Warner's book; one difference the reader should be aware of at the beginning has to do with the differential of a C^∞ map. If $f\colon M \to N$ is a C^∞ map of manifolds, then $f_*\colon TM \to TN$ denotes the induced tangent map (or differential), and $f^*\colon T^*N \to T^*M$ denotes the induced cotangent map. For each $p \in M$, the tangent space at p is denoted by M_p, and the tangent map induced by f at p is $f_*|_p\colon M_p \to N_{f(p)}$.

The vector field $d/dt \in \mathfrak{X}\mathbb{R}$ on the real line is often denoted by $D\colon f' = Df$ for $f \in C^\infty\mathbb{R}$.

Elements of $\mathbb{R}^n$ are usually written as row vectors except when they must be written as column vectors for matrix multiplication. The transpose of a matrix A is denoted by ${}^tA$.

Walter A. Poor

CHAPTER
ONE

AN INTRODUCTION TO FIBER BUNDLES

Fiber bundles constitute an important generalization of the product of two topological spaces; locally a fiber bundle is the product of two given topological spaces. For differential geometric purposes, the appropriate definition must be in the category of C^∞ manifolds.

THE DEFINITION OF A FIBER BUNDLE

1.1 **Notation** Given C^∞ manifolds M_1 and M_2, the topological product $M_1 \times M_2$ is naturally a C^∞ manifold [W: 1.5(g); 1, exercise 24]. Denote the C^∞ projection maps of $M_1 \times M_2$ onto M_1 and M_2 by pr_1 and pr_2, respectively.

$$\begin{array}{ccc} & M_1 \times M_2 & \\ {\scriptstyle \mathrm{pr}_1} \swarrow & & \searrow {\scriptstyle \mathrm{pr}_2} \\ M_1 & & M_2 \end{array}$$

1.2 **Definition** A C^∞ *fiber bundle* consists of:

(i) three C^∞ manifolds,
E, called the *total space* of the bundle,
M, called the *base space* of the bundle, and
F, called the *standard fiber* of the bundle;

(ii) a surjective C^∞ map $\pi: E \to M$, called the *projection*;

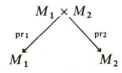

(iii) an open covering $\mathscr{U}$ of M, and for each U in $\mathscr{U}$ a C^∞ map $\varphi\colon \pi^{-1}U \to F$ such that the map $(\pi, \varphi)\colon \pi^{-1}U \to U \times F$ is a diffeomorphism; the map (π, φ) is called a *bundle chart* on E (or a *local trivialization* of π) over U, and φ is often called the *principal part* of the bundle chart.

1.3 Convention A precisely specified collection of bundle charts on a fiber bundle is not nearly so important as the fact that bundle charts exist relative to some open covering of M. For example, suppose that V is an open subset of M, and that α is a diffeomorphism from $\pi^{-1}V$ to $V \times F$ such that the diagram

$$\begin{array}{ccc} \pi^{-1}V & \xrightarrow{\ \alpha\ } & V \times F \\ & {\scriptstyle \pi}\searrow \qquad \swarrow{\scriptstyle \mathrm{pr}_1} & \\ & V & \end{array}$$

commutes; set $\psi := \mathrm{pr}_2 \circ \alpha$. For all practical purposes, the pair $(V, (\pi, \psi))$ might as well be a bundle chart of E over M. Any collection of bundle charts on E over M relative to an open covering of M will be called a *bundle atlas* on E. In particular, the union of two bundle atlases on E over M is a new bundle atlas on E over M; in this case, the two original fiber bundles on E over M will be identified with the fiber bundle determined by the union of the two original bundle atlases.

Alternatively, we could require that a bundle atlas be maximal, but this is not necessary for most purposes.

1.4 Definition For each p in M, the subset $E_p := \pi^{-1}(p)$ of the total space E of a fiber bundle over M is called the *fiber* of π (or the *fiber* of E) over p; the projection π is sometimes called a C^∞ *fibration* of E. Very often the adjective C^∞ will be deleted.

1.5 Proposition The projection map $\pi\colon E \to M$ of a C^∞ fiber bundle is a *submersion*, that is, for each point ξ in E, the induced tangent map $\pi_*|_\xi\colon E_\xi \to M_{\pi(\xi)}$ is surjective; furthermore, for each $p \in M$, the fiber $E_p = \pi^{-1}p$ in E over p is an embedded submanifold diffeomorphic to the standard fiber F of the bundle.

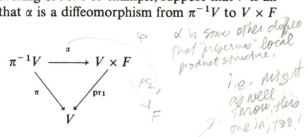

PROOF Let (π, ψ) be a bundle chart on E over an open set U in M. For each $(p, v) \in U \times F$, $\mathrm{pr}_{1*}|_{(p,v)}$ maps the tangent space $(U \times F)_{(p,v)}$ onto $M_p = U_p$, so pr_1 is a submersion of $U \times F$ onto U. Since $(\pi, \psi)_*$ maps $T(\pi^{-1}U)$ diffeomorphically onto $T(U \times F)$, $\pi_* = \mathrm{pr}_{1*} \circ (\pi, \psi)_*$ is then surjective at each $\xi \in \pi^{-1}U$; thus π is a submersion. By the implicit function theorem [W: 1.38], each fiber of π is naturally an embedded submanifold of E. For each $p \in U$, (π, ψ) maps E_p diffeomorphically onto the embedded submanifold $\{p\} \times F$ of $U \times F$, and pr_2 is a diffeomorphism from this onto F.

If (π, ψ) is a bundle chart on E over U in M, and if $p \in U$, then $(\psi|_{E_p})^{-1}$ is a diffeomorphism from F to E_p in E; therefore the bundle E over M with fiber F is occasionally denoted, by abuse of notation, by the symbol $F \to E \xrightarrow{\pi} M$. More often, E is used by itself.

1.6 Let $\mathrm{Diff}(F)$ denote the diffeomorphism group of F under composition.

Assume that (π, φ) and (π, ψ) are bundle charts on E over overlapping open sets U and V in M, respectively. The restrictions of (π, φ) and (π, ψ) to $\pi^{-1}(U \cap V)$ are also bundle charts on E by 1.3; a priori there is no reason for them to agree, but their difference is easily measured in $\mathrm{Diff}(F)$. For each $p \in U \cap V$, φ and ψ map the fiber E_p diffeomorphically to F; thus $\varphi \circ (\psi|_{E_p})^{-1}$ is a diffeomorphism of F; the result is a map

$$f_{\varphi,\psi}\colon U \cap V \to \mathrm{Diff}(F) \qquad p \to \varphi \circ (\psi|_{E_p})^{-1}$$

satisfying the relations

$$\varphi = (f_{\varphi,\psi} \circ \pi) \cdot \psi \qquad f_{\varphi,\psi}(p)^{-1} = f_{\psi,\varphi}(p)$$

where $\cdot$ denotes the action of $\mathrm{Diff}(F)$ on F. Furthermore, if (π, η) is a bundle chart on E over W in M with $U \cap V \cap W$ nonempty, then

$$f_{\varphi,\psi}(p) = f_{\varphi,\eta}(p) \cdot f_{\eta,\psi}(p) \qquad p \in U \cap V \cap W$$

Definition Given bundle charts (π, φ) and (π, ψ) over open sets U and V in M with $U \cap V \neq \varnothing$, the map $f_{\varphi,\psi}\colon U \cap V \to \mathrm{Diff}(F)$ is called the *transition function* from ψ to φ.

1.7 **Comments** Let (π, ψ) be a bundle chart on E over U in M. For each C^∞ manifold chart x on $V \subset U$ in M and each C^∞ chart y on W in F, $(x \circ \pi, y \circ \psi)$ is a C^∞ manifold chart on an open set in E; in fact, by using charts on M and F this way, one obtains an atlas of manifold charts on E. Thus a bundle atlas on E generates the C^∞ structure on the manifold E. In practice, this means that one often starts with a surjective map π from a set E to a manifold M, then defines a bundle atlas on E, and finally defines a C^∞ structure on E by checking that the bundle charts on E are C^∞-compatible; only then is E a manifold, and therefore eligible for use as the total space of a C^∞ fiber bundle over M (see also 1.12*i* and *j*).

1.8 A fiber bundle is locally a product manifold; a bundle chart on E exhibits this property by inducing a local C^∞ product structure on E over an open subset of M. The local product structures induced on E over an open set in M by different bundle charts can differ greatly because the group Diff(F) is so big. For a fiber bundle to be of use in elementary differential geometry, one usually restricts severely the choice of possible transition functions; these restrictions are the content of the next three definitions. Not all fiber bundles allow these restrictions [Om].

1.9 **Definition** A Lie group G is said to be a *Lie transformation group* on a manifold F if there is given a C^∞ *left action* of G on F, that is [W: 3.44], a C^∞ map $\mu: G \times F \to F$ such that

$$\mu(gh, \xi) = \mu(g, \mu(h, \xi)) \qquad \text{and} \qquad \mu(e, \xi) = \xi \qquad \xi \in F$$

for all $g, h \in G$, where e is the identity element of G. Such an action μ can be interpreted as a homomorphism from G to Diff(F): for each $g \in G$, μ_g is the element of Diff(F) such that $\mu_g \xi := \mu(g, \xi)$ for all $\xi \in F$.

1.10 **Definition** Let G be a Lie group acting on F on the left. Bundle charts (π, φ) and (π, ψ) on E over U and V in M, respectively, will be said to be *G-compatible* if $U \cap V = \varnothing$, or if $U \cap V \neq \varnothing$ and there exists a C^∞ map g from $U \cap V$ to G such that $f_{\varphi,\psi}(p) = \mu_{g(p)} \in \text{Diff}(F)$ for all $p \in U \cap V$. In this case we shall identify $f_{\varphi,\psi}$ with g, and consider $f_{\varphi,\psi}$ as a C^∞ map from $U \cap V$ to G; this yields the identity $f_{\varphi,\psi}(p) \cdot \xi = \mu(f_{\varphi,\psi}(p), \xi)$, $p \in U \cap V$, $\xi \in F$, where μ is the action of G on F.

1.11 **Definition** Let G be a Lie group. A C^∞ fiber bundle $F \to E \xrightarrow{\pi} M$ is called a *C^∞ fiber bundle with structure group* G if G is a Lie transformation group on F, and there is given a bundle atlas $\mathscr{A}$ on E such that any two bundle charts in $\mathscr{A}$ are G-compatible. A C^∞ fiber bundle chart is *admissible* as a bundle chart on the bundle E with structure group G if and only if it is G-compatible with every element of the given G-bundle atlas.

Although G will be referred to as *the* structure group of the bundle, it must be emphasized that G is not the only possible choice of structure group. For example, let a Lie group H act trivially on F; the Lie group $G \times H$ is then a structure group for E in a natural way. More important, often a subgroup of G can be chosen as structure group of E, just as G itself is homomorphic to a subgroup of Diff(F); there will be many examples of this phenomenon throughout the book. If a subgroup of G is specified as the structure group of E, the choice of admissible bundle charts is reduced because of the new restrictions on the transition functions. This process is called *reducing the group* of the bundle.

1.12 Examples

(*a*) ***Trivial bundles*** The product $M \times N$ of manifolds M and N is naturally the total space of the fiber bundle $N \longrightarrow M \times N \xrightarrow{\text{pr}_1} M$. The trivial group $G = \{\text{id}_N\}$ can be chosen as structure group, so the bundle is called *trivial.* Similarly, $M \times N$ is the total space of the trivial fiber bundle $M \longrightarrow M \times N \xrightarrow{\text{pr}_2} N$. These bundles are naturally trivialized. A bundle chart on an arbitrary fiber bundle trivializes the bundle locally; in general, there is no natural local trivialization.

(*b*) ***The Möbius strip*** The circle S^1 is the quotient $\mathbb{R}/\sim$, where for $x \in \mathbb{R}$ and $n \in \mathbb{Z}$, $x \sim x + 2\pi n$. Define an equivalence relation $\simeq$ on $\mathbb{R} \times (-1, 1)$ by $(x, t) \simeq (x + 2\pi n, (-1)^n t)$, $n \in \mathbb{Z}$, and set $E := (\mathbb{R} \times (-1, 1))/\simeq$. The result is the familiar Möbius strip. Define $\Pi: E \to S^1$ by $\Pi[x, t]_\simeq := [x]_\sim$, where the brackets denote the equivalence classes. Bundle charts (Π, φ) and (Π, ψ) are defined on E by $\varphi[x, t]_\simeq := t$, $|t| < \pi$, and $\psi[x, t]_\simeq := -t$, $0 < t < 2\pi$. The group of E is $\mathbb{Z}_2$. *Exercise:* Exhibit the Klein bottle as a bundle over S^1 with fiber S^1.

(*c*) ***The tangent bundle of a manifold*** [W: 1.25] Let M be a C^∞ manifold. The union over all $p \in M$ of the tangent space M_p on M at p is the total space TM of the *tangent bundle* of M; the projection map sends M_p to p. It follows from the definition of tangent vectors that if U in M is open, then U_p can be canonically identified with M_p for all $p \in U$. Hence if π is the projection of TM to M, then $\pi^{-1}U$ and TU are canonically identified.

The abbreviation M_p will be used for the fiber in TM at p, rather than the symbol TM_p. Caution: if E is the total space of a fiber bundle over M, then the symbol E_ξ must be read carefully. If $\xi \in M$, then E_ξ is the fiber in E over ξ, while if $\xi \in E$, then E_ξ is the tangent space on E at ξ.

The standard fiber of TM is $\mathbb{R}^n$, $n = \dim M$.

There is a slight discrepancy between the usual bundle charts on TM [W: 1.25] and bundle charts in general, as defined in 1.2. Given a chart $x = (x^1, \ldots, x^n): U \to \mathbb{R}^n$ on M, the associated tangent bundle chart on $\pi^{-1}U = TU$ is the induced tangent map of x:

$$x_* = (x \circ \pi, dx): TU \to x(U) \times \mathbb{R}^n \subset \mathbb{R}^{2n} = T\mathbb{R}^n$$

$$v \mapsto (x \circ \pi(v), dx^1(v), \ldots, dx^n(v))$$

Replacing x_* by (π, dx): $TU \to U \times \mathbb{R}^n$ we obtain a bundle chart according to the definition in 1.2. Since x is a diffeomorphism between U and $x(U) \subset \mathbb{R}^n$, the difference between the two versions of the bundle chart is purely formal.

As the structure group of TM we can choose $GL(n, \mathbb{R})$. *Proof:* Given charts x and y near $p \in M$, the value at p of the transition function from y_* to x_* is $f_{x,y}(p) = dx \circ (dy|_{M_p})^{-1} = [\partial x^i/\partial y^j](p)$, which is the usual Jacobian matrix of x and y at p. Therefore $f_{x,y}$ is a C^∞ map from a neighborhood of p to $GL(n, \mathbb{R})$. *QED.*

(*d*) ***Subbundles*** Let M and N be manifolds; call (M, N) a *manifold pair* if N is a submanifold of M. Suppose that (F_1, F_2), (E_1, E_2), and (M_1, M_2) are manifold pairs such that $F_i \longrightarrow E_i \xrightarrow{\pi_i} M_i$ is a fiber bundle, $i = 1, 2$. Call E_2 a *subbundle* of E_1 if the following condition is satisfied for each bundle chart (π_2, φ) on E_2 over an open set U in M_2: given $p \in U$, there exists an open neighborhood V of p in M_1 and a bundle chart (π_1, ψ) on E_1 over V such that

$$(\pi_1, \psi)|_{\pi_1^{-1}(U \cap V)} = (\pi_2, \varphi|_{\pi_2^{-1}(U \cap V)})$$

Not all bundle charts on E_1 restrict to bundle charts on E_2 if $F_1 \neq F_2$, for there are diffeomorphisms of F_1 which do not map F_2 to F_2.

As an example of a subbundle, $F \to \pi^{-1}X \xrightarrow{\pi} X$ is a subbundle of $F \to E \xrightarrow{\pi} M$ for each submanifold X of M; if E is trivial over a neighborhood of X, then $\pi^{-1}X$ is a trivial bundle.

(*e*) ***The universal line bundle over a projective space*** Let $\mathbb{F}$ denote one of the fields $\mathbb{R}$ or $\mathbb{C}$, and fix $n > 0$. The *projective space* $\mathbb{F}P^n$ of dimension n over $\mathbb{F}$ is the set $\mathbb{F}^{n+1} - \{0\}$ modulo the equivalence relation $\sim$ such that $u \sim v$ if and only if there exists $c \in \mathbb{F} - \{0\}$ such that $u = cv$ [W: 3.65(c, d)]. Denote the projection $\mathbb{F}^{n+1} - \{0\} \to \mathbb{F}P^n$ by $\not{p}$. For $j = 0, \ldots, n$, define a chart z_j on the open set $U_j := \{\not{p}(v^0, \ldots, v^n) \in \mathbb{F}P^n \mid v^j \neq 0\}$ by

$$z_j \circ \not{p}(v^0, \ldots, v^n) := \left(\frac{v^0}{v^j}, \ldots, \widehat{\frac{v^j}{v^j}}, \ldots, \frac{v^n}{v^j}\right)$$

where $\hat{}$ indicates an entry to be deleted. The charts (z_j, U_j) generate the C^∞ structure on $\mathbb{F}P^n$ (see [W: 1.5(c)] for the C^∞ structure on C^n); the differentiability of the transition function from a chart z_k to a chart z_j must be checked. If $\xi \in z_k(U_j \cap U_k) \subset \mathbb{F}^n$,

$$z_j \circ z_k^{-1}(\xi) = \begin{cases} \dfrac{1}{\xi^{j+1}}(\xi^1, \ldots, \widehat{\xi^{j+1}}, \ldots, \xi^k, 1, \xi^{k+1}, \ldots, \xi^n) & \text{if } j < k \\[2ex] \dfrac{1}{\xi^j}(\xi^1, \ldots, \xi^k, 1, \xi^{k+1}, \ldots, \widehat{\xi^j}, \ldots, \xi^n) & \text{if } j > k \end{cases}$$

The denominators are nonzero because $\xi \in z_k(U_j)$. This map is obviously C^∞.

Now we make $\mathbb{F}^{n+1} - \{0\}$ into the total space of a fiber bundle over $\mathbb{F}P^n$ with standard fiber $\mathbb{F} - \{0\}$. Define $\psi_j : \pi^{-1}U_j \to \mathbb{F} - \{0\}$ by $\psi_j(v^0, \ldots, v^n) := v^j$; the maps $(\not{p}, \psi_j)$ will be bundle charts once it is checked that they generate a C^∞ structure on $\mathbb{F}^{n+1} - \{0\}$.

Let $p \in U_j \cap U_k$, $j \neq k$, and fix $v \in \not{p}^{-1}(p)$; by the definition of U_k, $v^k \neq 0$. For all $\zeta \in \mathbb{F} - \{0\}$, $\psi_k|_{\not{p}^{-1}(p)}^{-1}(\zeta) = (\zeta/v^k)v$, so

$$f_{\psi_j, \psi_k}(p)\zeta = \psi_j \circ \psi_k|_{\not{p}^{-1}(p)}^{-1}(\zeta) = \psi_j\left(\frac{\zeta}{v^k} v\right) = \frac{\zeta v^j}{v^k} = \frac{z_j(v)}{z_k(v)} \cdot \zeta$$

Thus, $f_{\psi_j, \psi_k}(p)$ is a diffeomorphism of $\mathbb{F} - \{0\}$; furthermore, the map

$$(p, \zeta) \mapsto (\not{p}, \psi_j) \circ (\not{p}, \psi_k)^{-1}(p, \zeta) = (p, f_{\psi_j, \psi_k}(p) \cdot \zeta)$$

is a diffeomorphism of $(U_j \cap U_k) \times (\mathbb{F} - \{0\})$. Thus the maps $(\not{p}, \psi_j)$ determine a C^∞ structure on $\mathbb{F}^{n+1} - \{0\}$ (actually, this is just the usual structure of $\mathbb{F}^{n+1} - \{0\}$ as an open submanifold of $\mathbb{F}^{n+1}$).

With respect to the local coordinates on $\mathbb{F}^{n+1} - \{0\}$,

$$z_j \circ \not{p} \circ (\not{p}, \psi_j)^{-1} = z_j \circ \mathrm{pr}_1 : U_j \times (\mathbb{F} - \{0\}) \xrightarrow{C^\infty} \mathbb{F}^n$$

so $\not{p}$ is a C^∞ map from $\mathbb{F}^{n+1} - \{0\}$ to $\mathbb{F}P^n$ [W: 1.6].

Now that the C^∞ structure of $\mathbb{F}^{n+1} - \{0\}$ has been established and $\not{p}$ has been shown to be C^∞, the maps $(\not{p}, \psi_j)$ are bundle charts on $\mathbb{F}^{n+1} - \{0\}$. Since each transition function f_{ψ_j, ψ_k} maps $U_j \cap U_k$ differentiably to $GL(1, \mathbb{F})$, the structure group of the bundle is $GL(1, \mathbb{F})$.

The bundle $\mathbb{F}^{n+1} - \{0\}$ over $\mathbb{F}P^n$ is canonically a subbundle of a bundle E over $\mathbb{F}P^n$ with fiber $\mathbb{F}$. For each $p \in \mathbb{F}P^n$, define $E_p := \not{p}^{-1}(p) \cup \{(0, p)\}$. All that has really been added to $\not{p}^{-1}(p)$ is the point 0 in $\mathbb{F}$; calling the new point $(0, p)$ instead of just 0 is a technical device to guarantee that distinct fibers in the new bundle have distinct zero points. Set E equal to the union over all $p \in \mathbb{F}P^n$ of E_p. Extend the maps $\not{p}$ and ψ_j from $\mathbb{F}^{n+1} - \{0\}$ to E by defining $\not{p}(0, p) := p \in \mathbb{F}P^n$, and $\psi_j(0, p) := 0 \in \mathbb{F}$, $p \in U_j$.

Exercise The extended maps $(\not{p}, \psi_j)$ are bundle charts on E over the sets U_j in $\mathbb{F}P^n$; $\mathbb{F}^{n+1} - \{0\}$ is a subbundle of E.

The fiber bundle $\mathbb{F} \to E \xrightarrow{\not{p}} \mathbb{F}P^n$ is called the *universal line bundle* over $\mathbb{F}P^n$ (see [Ch5: 6, exercise 2], [Ws: 1.2.6]). Topologists call this the canonical line bundle, but in geometry that term usually refers to something else.

Exercises Define a map from $\mathbb{F}P^n$ to $\mathbb{F}P^{n+1}$ by sending $\not p(v^0, \ldots, v^n)$ to $\not p(v^0, \ldots, v^n, 0)$; show that this map is an embedding. Show that the universal line bundle over $\mathbb{F}P^n$ is naturally a subbundle of the universal line bundle over $\mathbb{F}P^{n+1}$. The Grassmann manifold [W: 3.65(f)] of 1-planes in $\mathbb{F}^n$ is $\mathbb{F}P^{n+1}$; generalize the universal line bundle over $\mathbb{F}P^{n-1}$ to the universal k-plane bundle over the Grassmann manifold of k-planes in $\mathbb{F}^n$.

(f) ***Product bundles*** If $F_j \to E_j \to M_j$ is a fiber bundle, $j = 1, 2$, then $F_1 \times F_2 \to E_1 \times E_2 \to M_1 \times M_2$ is also a fiber bundle. For example, $TM_1 \times TM_2$ is the tangent bundle of $M_1 \times M_2$. The structure group of $E_1 \times E_2$ is the product of the structure groups of the E_j.

(g) ***The pullback of a bundle*** Suppose that $h\colon N \to M$ is a C^∞ map, and let $F \to E \xrightarrow{\pi} M$ be a fiber bundle. By example (a), $N \times E$ is a fiber bundle over N with fiber E. The map h now determines a subbundle of $N \times E$, which in most cases is far more interesting than the original bundle $N \times E$. Set

$$h^*E := \{(p, \xi) \in N \times E \mid h(p) = \pi(\xi)\}; \qquad \begin{array}{ccc} h^*E & \xrightarrow{\mathrm{pr}_2} & E \\ {\scriptstyle \mathrm{pr}_1}\downarrow & & \downarrow{\scriptstyle \pi} \\ N & \xrightarrow[h]{} & M \end{array}$$

project h^*E onto N by the map $\mathrm{pr}_1|_{h^*E}$, which will also be denoted by pr_1. Similarly, denote the restriction of pr_2 to h^*E by pr_2.

The fiber of h^*E at $p \in N$ is $(h^*E)_p = \{p\} \times E_{h(p)}$, which is diffeomorphic to $E_{h(p)}$ under pr_2; thus the standard fiber of h^*E is F.

Assume that (π, ψ) is a local trivialization of E over an open set U in M; the subset $(h \circ \mathrm{pr}_1)^{-1}U$ of h^*E is a trivial bundle over the open set $h^{-1}U$ in N; in fact, a bundle chart on h^*E over $h^{-1}U$ is the map $(\mathrm{pr}_1, \psi \circ \mathrm{pr}_2)$. The bundle h^*E is called the *pullback of E by the map h*. It will appear frequently throughout the book.

Exercises Prove that h^*E over N is a subbundle of $N \times E$ over N. If the structure group of E is a Lie group G, show that the structure group of h^*E is a Lie subgroup of G. Since h^*E is a submanifold of $N \times E$, its tangent bundle is a submanifold of $T(N \times E)$; prove that $T(h^*E) = \{(u, v) \in TN \times TE \mid h_* u = \pi_* v\}$.

(*h*) ***Composite bundles*** Assume that $F \to E \xrightarrow{\pi} M$ and $F_1 \to E_1 \xrightarrow{\pi_1} E$ are fiber bundles. Composition of the projection maps yields the *composite fiber bundle* $F \times F_1 \longrightarrow E_1 \xrightarrow{\pi \circ \pi_1} M$. If (π, φ) is a bundle chart on E over $U \subset M$ and (π_1, ψ) is a bundle chart on E_1 over $V \subset E$, with $W := \pi(V) \cap U \neq \varnothing$, then the map $(\pi \circ \pi_1, \varphi \circ \pi_1, \psi)$ is a bundle chart on E_1 over W. The transition function from the map $\beta = (\varphi \circ \pi_1, \psi)$ to the map $\alpha = (\eta \circ \pi_1, \mu)$ satisfies the rule

$$f_{\alpha, \beta}(p) \cdot (\zeta, \xi) = (f_{\eta, \varphi}(p) \cdot \zeta, f_{\mu, \psi} \circ (\pi, \varphi)^{-1}(p, \zeta) \cdot \xi)$$

for $(\zeta, \xi) \in F \times F_1$ and appropriate $p \in M$.

For example, the composite of TM and TTM is the bundle TTM over M with standard fiber isomorphic to $\mathbb{R}^{3n}$. The composite of the universal line bundle E over $\mathbb{R}P^n$ and the tangent bundle TE is the bundle TE over $\mathbb{R}P^n$ with fiber $\mathbb{R}^{n+2}$.

(*i*) Let $\mathscr{U}$ be an open covering of a manifold M, and suppose that for each pair $U, V \in \mathscr{U}$ with $U \cap V \neq \varnothing$ there is given a C^∞ map $f_{U,V}$ from $U \cap V$ to a fixed Lie group G. Suppose given a C^∞ action of G on a manifold F. If the maps $f_{U,V}$ satisfy the relations stated in 1.6, that is, $f_{U,V}(p)^{-1} = f_{V,U}(p)$ for all $p \in U \cap V$, and $f_{U,V}(p) = f_{U,W}(p) \cdot f_{W,V}(p)$ for $p \in U \cap V \cap W$ whenever this intersection is nonempty, then the open covering $\mathscr{U}$ and the given maps $f_{U,V}$ can be used to construct a fiber bundle over M with fiber F and group G. This procedure will be outlined first, and then applied to a special famous example in (*j*).

Suppose $U \cap V \neq \varnothing$. Over the open sets U and V in M there are the product bundles $U \times F$ and $V \times F$. If $p \in U \cap V$, identify the point (p, ξ) in $U \times F$ with the point $(p, f_{U,V}(p) \cdot \xi)$ in $V \times F$. By the first of the two relations stated above, it follows that $(p, f_{U,V}(p) \cdot \xi)$ in $V \times F$ is identified with $(p, f_{V,U}(p) f_{U,V}(p)\xi) = (p, \xi)$ in $U \times F$. Let E be the disjoint union of $U \times F$ and $V \times F$ modulo the identification just described. Define π from E to $U \cup V$ to be projection onto the first factor. (See p. 10.) There are natural bundle charts on E over U and V; if $v \in E$ comes from a point $(p, \zeta) \in U \times F$, define $(\pi, \varphi)(v) := (p, \zeta) \in U \times F$, while if $v \in E$ comes from a point $(p, \xi) \in V \times F$, define $(\pi, \psi)(v) := (p, \xi) \in V \times F$. The maps $f_{U,V}$ and $f_{V,U}$ are the transition functions between the bundle charts.

The second relation stated above for the collection of $f_{U,V}$ is the necessary consistency requirement if E is extended to $U \cup V \cup W$, where $U \cap V \cap W \neq \varnothing$.

(*j*) ***Milnor's exotic spheres*** Let N be the "north pole" $(1, 0, \ldots, 0)$ in $S^n \subset \mathbb{R}^{n+1}$, and S the "south pole" $(-1, 0, \ldots, 0) \in S^n$. Set $U := S^n - \{N\}$ and $V := S^n - \{S\}$. The *stereographic projection of* S^n

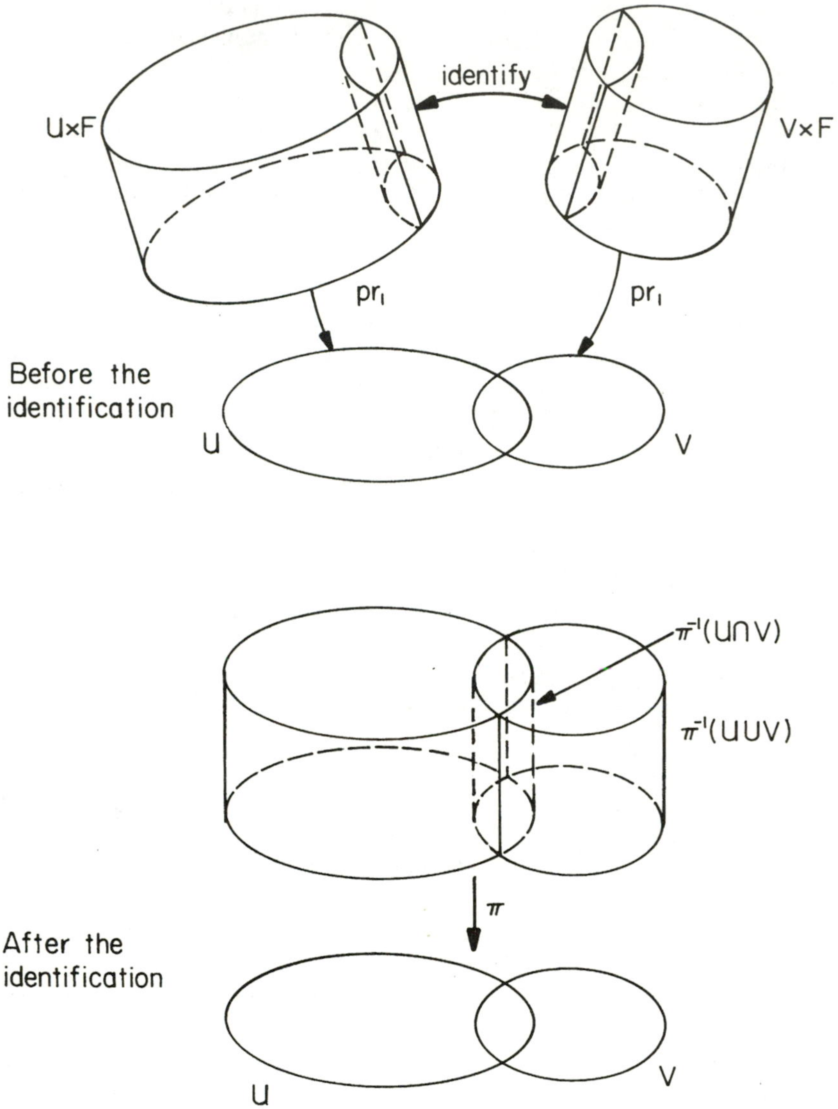
identify
U×F
V×F
pr₁
pr₁
Before the identification
U
V
π⁻¹(U∩V)
π⁻¹(U∪V)
π
After the identification
U
V

onto the equatorial plane *from N* is the map

$$x\colon U \to \mathbb{R}^n$$

$$(u^0, \ldots, u^n) \mapsto \frac{1}{1-u^0}(u^1, \ldots, u^n)$$

The *stereographic projection of S^n from S* is the map

$$y\colon V \to \mathbb{R}^n$$

$$(u^0, \ldots, u^n) \mapsto \frac{1}{1+u^0}(u^1, \ldots, u^n)$$

For each $u \in U \cap V$, $y(u) = x(u)/\|x(u)\|^2$, and $x(u) = y(u)/\|y(u)\|^2$, where $\|v\|^2$ denotes the square of the norm of $v \in \mathbb{R}^{n+1}$ with respect to the standard inner product on $\mathbb{R}^{n+1}$. The maps x and y are charts on S^n which generate the C^∞ structure.

The result is that S^n can be constructed as the disjoint union of two copies of $\mathbb{R}^n$ modulo the identification of each nonzero vector u in the first copy with the vector $u/\|u\|^2$ in the second copy; symbolically, $S^n = (\mathbb{R}^n \coprod \mathbb{R}^n)/\sim$, where $u \sim u/\|u\|^2$, $u \neq 0$.

The *quaternions* are the 4-dimensional real division algebra $\mathbb{H}$ with real basis $\{1, i, j, k\}$ such that multiplication satisfies $i^2 = j^2 = k^2 = ijk = jki = kij = -1$. *Conjugation* in $\mathbb{H}$ is the real involution of $\mathbb{H}$ defined by $\overline{a + bi + cj + dk} := a - bi - cj - dk$, $a, b, c, d \in \mathbb{R}$. It follows that given $z_1, z_2 \in \mathbb{H}$, $\overline{z_1 z_2} = \bar{z}_2 \bar{z}_1$, and that the reciprocal of a nonzero element $z \in \mathbb{H}$ is $\bar{z}/(z\bar{z})$. The unit sphere $S^3 \subset \mathbb{R}^4$ is thus identified with the set of all $z \in \mathbb{H}$ such that $z\bar{z} = 1$.

Let a be an odd integer; set $b := (1 + a)/2$ and $c := (1 - a)/2$. Define

$$E_a := [(\mathbb{R}^4 \times S^3) \coprod (\mathbb{R}^4 \times S^3)]/\simeq$$

where if (u, v) belongs to the first copy of $\mathbb{R}^4 \times S^3$ and $u \neq 0$, then

$$(u, v) \simeq \left(\frac{u}{\|u\|^2}, \frac{u^b v u^c}{\|u\|}\right)$$

The symbol $u^b v u^c$ indicates quaternionic multiplication in $\mathbb{R}^4 \cong \mathbb{H}$.

Define $\pi\colon E_a \to S^4$ such that $\pi[u, v] := x^{-1}(u)$ for (u, v) in the first copy of $\mathbb{R}^4 \times S^3$, and $\pi[u, v] := y^{-1}(u)$ for (u, v) in the second copy of $\mathbb{R}^4 \times S^3$, where brackets indicate the coset in E_a. The map π is well-defined by the properties of stereographic projection.

To define bundle charts on E_a, set $\varphi[u, v] := v \in S^3$ for (u, v) in the first copy of $\mathbb{R}^4 \times S^3$, and $\psi[u, v] := v$ for (u, v) in the second copy for $\mathbb{R}^4 \times S^3$. It follows that given $(u, v) \in (\mathbb{R}^4 - \{0\}) \times S^3$,

$(\pi, \psi) \circ (\pi, \varphi)^{-1}(u, v) \simeq (u, v)$; this implies that (π, ψ) and (π, φ) are C^∞-compatible, and therefore define a C^∞ structure on E_a. Simultaneously they determine a fiber bundle structure on E_a over S^4.

The manifolds E_a were constructed by Milnor [Mi*1*], who proved that they are homeomorphic to the standard 7-sphere S^7, but that E_a is not diffeomorphic to S^7 if a^2 is not congruent to 1 mod 7; these manifolds are called *exotic 7-spheres*. In all there are 28 diffeomorphism classes of 7-spheres, 16 of which are of the Milnor type, that is, are S^3-bundles over S^4 [EK]. A similar construction using Cayley numbers yields exotic 15-spheres [Sh].

Exercise Show that E_1 is the standard 7-sphere S^7.

VECTOR BUNDLES

1.13 Suppose that $E \xrightarrow{\pi} M$ is a C^∞ fiber bundle with fiber F. Recall that by the definition of E as the total space of the bundle, certain manifold charts on the C^∞ manifold E have a preferred status as bundle charts on E; the rest of the C^∞ charts on E are ignored because they do not respect the fibered structure defined on E by the projection π. If E has a Lie group G as a structure group, the bundle atlas is reduced still further (1.10, 1.11) so that the transition functions are G-valued, that is, the admissible bundle charts are G-compatible. Now we restrict our attention to groups which act linearly on a vector space.

Definition Let $\mathbb{F}$ be one of the fields $\mathbb{R}$ or $\mathbb{C}$, and let $\mathscr{V}$ be an m-dimensional vector space over $\mathbb{F}$. A fiber bundle E over M with standard fiber $\mathscr{V}$ is called an $\mathbb{F}$-*vector bundle* if each fiber of E is endowed with a vector space structure over $\mathbb{F}$, and if there is given a bundle atlas $\mathscr{A}$, called a *vector bundle atlas*, such that each bundle chart in $\mathscr{A}$ is fiberwise linear: if $(\pi, \psi)\colon \pi^{-1}U \to U \times \mathscr{V}$ belongs to $\mathscr{A}$, then the map $\psi|_{E_p}\colon E_p \to \mathscr{V}$ must be a vector space isomorphism for all $p \in U$. The bundle E is also called a *vector bundle of rank m over* $\mathbb{F}$.

Quaternionic vector bundles are also defined, but will not be used in this book.

1.14 **Proposition** Let $\mathscr{V}$ be a vector space over $\mathbb{F}$. A fiber bundle E over M with standard fiber $\mathscr{V}$ is an $\mathbb{F}$-vector bundle if and only if E admits $GL(\mathscr{V})$ as structure group, where $GL(\mathscr{V})$ is the group of all invertible linear transformations of the vector space $\mathscr{V}$.

PROOF Let (π, φ) and (π, ψ) be admissible bundle charts over open sets U and V in M. Fix $p \in U \cap V$. For all $\zeta, \xi \in \mathscr{V}$ and $c \in \mathbb{F}$,

$$f_{\varphi,\psi}(p) \cdot (c\zeta + \xi) = \varphi \circ \psi\,|_{E_p}^{-1}(c\zeta + \xi) = \varphi(c\psi\,|_{E_p}^{-1}\zeta + \psi\,|_{E_p}^{-1}\xi)$$
$$= cf_{\varphi,\psi}(p)\zeta + f_{\varphi,\psi}(p)\xi$$

so $f_{\varphi,\psi}(p)$ acts linearly on $\mathscr{V}$; since it is invertible, it is in $GL(\mathscr{V})$.

A basis $\{\epsilon_j\}$ for $\mathscr{V}$ fixes a Lie group isomorphism (which is a diffeomorphism by definition [W: 3.13]) from $GL(\mathscr{V})$ to $GL(m, \mathbb{F})$; if $\{\omega^i\}$ is the dual basis for $\mathscr{V}^*$, then with this identification,

$$f_{\varphi,\psi}(p) = [\omega^i \circ \varphi \circ (\pi, \psi)^{-1}(p, \epsilon_j)] \in GL(m, \mathbb{F}) \qquad p \in U \cap V$$

Thus $f_{\varphi,\psi}$ is a matrix of C^∞ functions on $U \cap V$, and is therefore a C^∞ map from $U \cap V$ to $GL(m, \mathbb{F})$.

Conversely, for $p \in U$ define a vector space structure on E_p by

$$cu + v := (\pi, \psi)^{-1}(c\psi u + \psi v) \qquad c \in \mathbb{F},\ u, v \in E_p$$

By definition, ψ maps E_p isomorphically to $\mathscr{V}$. If $p \in U \cap V$, then φ is also an isomorphism from E_p to $\mathscr{V}$ since $f_{\varphi,\psi}(p) \in GL(\mathscr{V})$; thus the vector space structure of E_p is well-defined.

1.15 Examples

(*a*) The trivial bundle $M \times \mathscr{V}$ admits a canonical vector bundle structure. Notice that for the vector bundle structure on $\mathrm{pr}_1 : \mathbb{F}^n \times \mathscr{V} \to \mathbb{F}^n$,

$$(u, v) + c(u, w) = (u, cv + w) \qquad (u, v), (u, w) \in \mathbb{F}^n \times \mathscr{V}, c \in \mathbb{F}$$

(*b*) The tangent and cotangent bundles of M [W: 1.25] are real vector bundles, as are the tensor bundle and the exterior bundle of M [W: 1.14]. These will be dealt with further in 1.41.

(*c*) The universal line bundle (1.12*e*) over $\mathbb{F}P^n$ is a vector bundle of rank 1 over $\mathbb{F}$.

1.16 **Definition** Let $E_j \xrightarrow{\pi_j} M$ be an $\mathbb{F}$-vector bundle, $j = 1, 2$. A C^∞ map f from E_1 to E_2 is called a *vector bundle homomorphism* if f preserves the fibers (that is, if $\pi_2 \circ f = \pi_1$) and if the restriction of f to each fiber of E_1 is linear. If f maps $E_1|_p$ isomorphically to $E_2|_p$ for all $p \in M$, f is called a *vector bundle isomorphism*.

A vector bundle homomorphism from E_1 to E_2 is a vector bundle isomorphism if and only if it is a diffeomorphism.

1.17 **Definition** If $g: M_1 \to M_2$ is C^∞, and if $E_j \overset{\pi_j}{\to} M_j$ is a vector bundle over $\mathbb{F}$, $j = 1, 2$, then a C^∞ map $f: E_1 \to E_2$ is called a *vector bundle homomorphism along* g if $\pi_2 \circ f = g \circ \pi_1$ and if for each $p \in M$, f maps $E_1|_p$ linearly to $E_2|_{g(p)}$. If f maps $E_1|_p$ isomorphically to $E_2|_{g(p)}$ for all $p \in M$, f is called an *isomorphism along* g.

1.18 **Proposition** Let $h: N \to M$ be C^∞. If E is a vector bundle over M, then h^*E is a vector bundle over N which is isomorphic to E along h. Furthermore, h^*E is unique up to isomorphism: a vector bundle E_1 over N is isomorphic to h^*E if and only if it is isomorphic to E along h.

PROOF Let $p \in N$; by 1.12*g*, the fiber in h^*E at p is $\{p\} \times E_{h(p)}$; this is canonically a vector space isomorphic to $E_{h(p)}$ by pr_2.

If (π, ψ) is a vector bundle chart on E over $U \subset M$, then $(\mathrm{pr}_1, \psi \circ \mathrm{pr}_2)$ is a vector bundle chart on h^*E over $h^{-1}U$.

The C^∞ map pr_2 is an isomorphism along h because $\pi \circ \mathrm{pr}_2 = h \circ \mathrm{pr}_1$ and pr_2 is an isomorphism on each fiber. Uniqueness is trivial.

Exercises Let (π, φ) and (π, ψ) be vector bundle charts on E over U and V, with $U \cap V \neq \varnothing$. Calculate the transition function from $\psi \circ \mathrm{pr}_2$ to $\varphi \circ \mathrm{pr}_2$ on $h^{-1}(U \cap V)$; prove that h^*E has $GL(m, \mathbb{F})$ as structure group. If E is a vector bundle over M, and if $h: N \to M$ and $k: L \to N$ are C^∞, prove that k^*h^*E and $(h \circ k)^*E$ are isomorphic.

1.19 **Definition** Let $E_2 \to M_2$ be a subbundle of $E_1 \to M_1$; assume that E_1 and E_2 are $\mathbb{F}$-vector bundles. If the inclusion map of E_2 into E_1 is a homomorphism along the inclusion map of M_2 into M_1, then E_2 is called a *vector subbundle* of E_1.

1.20 An n-dimensional complex vector space is canonically a real vector space of dimension $2n$—simply forget about multiplication by $i = \sqrt{-1}$. For example, the real vector space underlying $\mathbb{C}^n$ is isomorphic to $\mathbb{R}^{2n}$

under the *standard identification*

$$\mathbb{C}^n \cong \mathbb{R}^{2n}$$

$$u + iv \leftrightarrow (u, v) \qquad u, v \in \mathbb{R}^n$$

that is, $(u^1 + iv^1, \ldots, u^n + iv^n) \leftrightarrow (u^1, \ldots, u^n, v^1, \ldots, v^n)$. Notice that $i(u + iv) = -v + iu \leftrightarrow (-v, u)$, which will be denoted by $J_o(u, v)$.

A complex linear map of complex vector spaces canonically induces a real linear map of the underlying real vector spaces.

> **Definition** Let $E \xrightarrow{\pi} M$ be a $\mathbb{C}$-vector bundle. The *realification* of E is the bundle $\pi\colon E \to M$ with each fiber thought of in the natural way as a real vector space. Each complex vector bundle chart on E induces a real vector bundle chart on E which, fiberwise, is the induced real linear map on the underlying real vector spaces.

1.21 Examples

(*a*) The realification of $\mathbb{C}^n \times \mathbb{C}^m \xrightarrow{\text{pr}_1} \mathbb{C}^n$ is $\mathbb{C}^n \times \mathbb{R}^{2m} \xrightarrow{\text{pr}_1} \mathbb{C}^n$. As a special case, the realification of $\mathbb{C}^n \times \mathbb{C}^n \xrightarrow{\text{pr}_1} \mathbb{C}^n$ is isomorphic to the tangent bundle $T\mathbb{C}^n$ of $\mathbb{C}^n$; conversely, $T\mathbb{C}^n$ admits a complex vector bundle structure isomorphic to that of the trivial bundle $\mathbb{C}^{2n}$ over $\mathbb{C}^n$.

(*b*) The tangent bundle $T\mathbb{C}P^n$ of n-dimensional complex projective space $\mathbb{C}P^n$ has a canonical complex vector bundle structure. *Proof:* It suffices to define a complex vector space structure on each fiber, and then to construct complex vector bundle charts. Let v be tangent to $\mathbb{C}P^n$ at $p \in U_j$, where (z_j, U_j) is one of the charts on $\mathbb{C}P^n$ from 1.12*e*. Define $iv := z_{j*}|_p^{-1}(iz_{j*}v)$; here $z_{j*}v \in T\mathbb{C}^n$ [which is a complex vector bundle by (*a*)], so the product of i and $z_{j*}v$ is well-defined and has the required properties. The tangent bundle chart z_{j*} is now fiberwise complex linear, but it must be checked that z_{k*} is also complex linear on fibers over $U_j \cap U_k$; a direct proof would be too much of a digression now (for a more general proof, see 8.35), so an indirect check will be presented instead. In 1.12*e*, $\mathbb{C}^{n+1} - \{0\}$ was fibered over $\mathbb{C}P^n$ by a C^∞ map $\wp$. Fix $q \in \mathbb{C}^{n+1} - \{0\}$ such that $\wp(q) = p$. Since $\wp$ is a submersion by 1.5, there exists a tangent vector ξ on $\mathbb{C}^{n+1} - \{0\}$ at q such that $\wp_*\xi = v$. The claim now is that $\wp_*(i\xi) = iv$; since z_{j*} does not appear in the definition of $\wp_*(i\xi)$, iv must then be independent of the choice of tangent bundle chart used in the definition. To prove this equality, consider ξ as the point $(q, \tilde{v}) = (q, (v^0, \ldots, v^n)) = d/dt|_0(q + t\tilde{v})$ in $(\mathbb{C}^{n+1} - \{0\}) \times \mathbb{C}^{n+1}$; then $i\xi = d/dt|_0(q + ti\tilde{v})$, and so

$$\begin{aligned} z_{j*}\not{p}_*(i\xi) &= \frac{d}{dt}\bigg|_0 [z_j \circ \not{p}(q + ti\tilde{v})] \\ &= \frac{d}{dt}\bigg|_0 \left(\frac{1}{q^j + tiv^j}(q^0 + tiv^0, \ldots, \widehat{q^j + tiv^j}, \ldots, q^n + tiv^n)\right) \\ &= \left(q, \frac{iv^j}{q^j}(v^0, \ldots, \widehat{v^j}, \ldots, v^n) - \frac{iv^j}{q^{j2}}(q^0, \ldots, \widehat{q^j}, \ldots, q^n)\right) \\ &= i\left(q, \frac{v^j}{q^j}(v^0, \ldots, \widehat{v^j}, \ldots, v^n) - \frac{v^j}{q^{j2}}(q^0, \ldots, \widehat{q^j}, \ldots, q^n)\right) \\ &= i \cdot z_{j*}\not{p}_*\xi = i \cdot z_{j*}v \end{aligned}$$

Thus $\not{p}_*(i\xi) = z_{j*}|_p^{-1}(iz_{j*}v) = iv$, as required. Since the sets U_j form an open covering of $\mathbb{C}P^n$, the tangent bundle charts z_{j*} determine the complex vector bundle structure of $T\mathbb{C}P^n$. *QED.*

The projective complex line $\mathbb{C}P^1$ is diffeomorphic to the 2-sphere S^2; thus $TS^2 \cong T\mathbb{C}P^1$ admits a complex vector bundle structure. An obvious problem is to find all even $n \geq 2$ such that TS^n admits a complex vector bundle structure. It can be shown [KN: IX, example 2.6] that TS^6 can be made into a complex vector bundle; Borel and Serre [BS] proved that TS^n is not a complex vector bundle for $n \neq 2, 6$.

1.22 **Definition** The *complexification* $\mathscr{V}^{\mathbb{C}}$ of a real vector space $\mathscr{V}$ is the set of all formal symbols $u + iv$, where $u, v \in \mathscr{V}$, with *addition* and *scalar multiplication* defined by

$$(u + iv) + (w + ix) := (u + w) + i(v + x)$$

$$(a + ib) \cdot (u + iv) := (au - bv) + i(av + bu) \qquad a, b \in \mathbb{R}$$

Complex conjugation in $\mathscr{V}^{\mathbb{C}}$ is defined by $\overline{u + iv} := u - iv$, $u, v \in \mathscr{V}$. A *real subspace* of $\mathscr{V}$ is a real vector space $\mathscr{W}$ contained in $\mathscr{V}^{\mathbb{C}}$ such that $\bar{w} = w$ for all $w \in \mathscr{W}$.

It is easily checked that $\mathscr{V}^{\mathbb{C}}$ is a complex vector space of complex dimension equal to the real dimension of $\mathscr{V}$. The set of all $u + i0$, $u \in \mathscr{V}$, is a real subspace of $\mathscr{V}^{\mathbb{C}}$ naturally isomorphic to $\mathscr{V}$. As a real vector space, $\mathscr{V}^{\mathbb{C}}$ is isomorphic to the tensor product $\mathbb{C} \otimes_{\mathbb{R}} \mathscr{V}$.

1.23 **Definition** Let $\mathscr{V} \to E \xrightarrow{\pi} M$ be a real vector bundle. The *complexification of E* is the complex vector bundle $\pi\colon E^{\mathbb{C}} \to M$ with standard fiber $\mathscr{V}^{\mathbb{C}}$ such that the fiber $(E^{\mathbb{C}})_p$ over $p \in M$ is the complexification $(E_p)^{\mathbb{C}}$ of E_p. If (π, ψ) is a vector bundle chart on E over U in M, then $(\pi, \psi^{\mathbb{C}})$ is a complex vector bundle chart on $E^{\mathbb{C}}$ over U, where for $p \in U$ and $u, v \in E_p$,

$$\psi^{\mathbb{C}}(u + iv) := \psi(u) + i\psi(v) \in \mathscr{V}^{\mathbb{C}}$$

Complex conjugation in $E^{\mathbb{C}}$ is defined fiberwise: $\overline{u + iv} := u - iv$ for u, $v \in E_p$, $p \in M$. A *real subbundle* of $E^{\mathbb{C}}$ is a real vector bundle E_1 contained in E such that $\bar{v} = v$ for all $v \in E_1$.

The original vector bundle E is isomorphic to the real subbundle of $E^{\mathbb{C}}$ consisting of all elements of the form $u + i0$, $u \in E$.

THE VERTICAL BUNDLE

Let $E \xrightarrow{\pi} M$ be a fiber bundle; recall (1.12*c*) that if $p \in M$, then E_p is the fiber $\pi^{-1}p \subset E$ (which is an embedded submanifold by 1.5), while if $\xi \in E$, then E_ξ is the tangent space at ξ.

1.24 **Lemma** Let $F \to E \xrightarrow{\pi} M$ be a fiber bundle; fix $p \in M$. Set $N := E_p$, and let ι: $N \to E$ be the inclusion. For all $\xi \in N$,

$$\iota_* N_\xi = \ker[\pi_*|_{\iota\xi}\colon E_{\iota\xi} \to M_p] = \pi_*|_{\iota\xi}^{-1}(0_p) \subset E_{\iota\xi}$$

where $0_p \in M_p$ is the zero vector at p. If ψ: $N \to F$ is a diffeomorphism and if x is a chart on an open set V in F, then for all $\xi \in \psi^{-1}V$, $dx \circ \psi_*|_\xi$ maps N_ξ isomorphically onto $\mathbb{R}^m$, $m := \dim F$.

Proof For each C^∞ curve γ in N, $\pi \circ \iota \circ \gamma$ is constant, so $\pi_* \iota_* \dot{\gamma}(0) = 0_p$; thus $\iota_* N_\xi \subset \pi_*|_{\iota\xi}^{-1}(0_p)$. But

$$\dim(\pi_*|_{\iota\xi}^{-1}(0_p)) = \dim E - \dim M = \dim F = \dim N_\xi$$

so $\iota_* N = \pi_*|_{\iota\xi}^{-1}(0_p)$ since $\iota_*|_\xi$ is nonsingular. The rest of the proposition follows since dx is the principal part $\mathrm{pr}_2 \circ x_*$ of the bundle chart induced on TF by x on F (1.2 and 1.12*c*).

From now on, suppress the inclusion ι, and simply write

$$(E_p)_\xi = \pi_*|_\xi^{-1}(0_p) \subset E_\xi$$

1.25 **Definition** Let $F \to E \xrightarrow{\pi} M$ be a fiber bundle, with $\dim F =: m$. The *vertical bundle on E* is the real vector bundle with total space

$$\mathscr{V}E := \pi_*^{-1}(0) = \bigcup_{\xi \in E} \ker[\pi_*|_\xi\colon E_\xi \to M_{\pi\xi}] \subset TE$$

Denote the projection map by $\pi_{\mathscr{V}}: \mathscr{V}E \to E$. The bundle atlas on $\mathscr{V}E$ consists of the vector bundle charts of the form

$$(\pi_{\mathscr{V}}, dx \circ \psi_*): \pi_{\mathscr{V}}^{-1}(\pi^{-1}(U) \cap \psi^{-1}V) \to (\pi^{-1}(U) \cap \psi^{-1}V) \times \mathbb{R}^m$$

where (π, ψ) is a bundle chart on E over U in M, and x is a chart on V in F.

1.26 The next proposition says that the vertical bundle is functorial.

Proposition If f^*E is the pullback of a bundle E over M by a map $f: N \to M$, then the vector bundles $\mathscr{V}f^*E$ and $\mathrm{pr}_2^*\mathscr{V}E$ are isomorphic.

PROOF By 1.18 it is sufficient to prove that $\mathscr{V}f^*E$ is isomorphic to $\mathscr{V}E$ along the map $\mathrm{pr}_2: f^*E \to E$. By an exercise in 1.12*g*, $Tf^*E = \{(u, v) \in TN \times TE \mid f_* u = \pi_* v\}$. Thus if $(u, v) \in Tf^*E$, $\mathrm{pr}_{1*}(u, v) = u$ is zero if and only if $v \in \mathscr{V}E$. The map $(u, v) \mapsto v$ is an isomorphism from $\mathscr{V}f^*E$ to $\mathscr{V}E$ along pr_2.

1.27 **Proposition** If $E \xrightarrow{\pi} M$ is a vector bundle, then the vertical vector bundle $\mathscr{V}E$ is isomorphic to the vector bundle π^*E on E, that is, $\mathscr{V}E$ is isomorphic to E along π. In particular, if E is a complex vector bundle, then so is $\mathscr{V}E$.

PROOF If $(\zeta, \xi) \in \pi^*E$, then $\pi(\zeta + t\xi)$ is constant in t. Thus the map from π^*E to TE such that $(\zeta, \xi) \mapsto d/dt|_0(\zeta + t\xi)$ maps into $\mathscr{V}E$; this map is clearly a real vector bundle isomorphism. If E is a complex vector bundle, define the complex vector bundle structure on $\mathscr{V}E$ so that the given map is complex linear on fibers.

In the real case, $\mathscr{V}E$ is a vector subbundle of TE; this is also true for the realification of $\mathscr{V}E$ if E is a complex vector bundle.

1.28 **Notation** Let $\mathscr{V}E$ be the vertical bundle on a vector bundle E over M. Denote the isomorphism from π^*E to $\mathscr{V}E$ by $\mathscr{J}$:

$$\mathscr{J}: \pi^*E \to \mathscr{V}E \qquad (\zeta, \xi) \mapsto \left.\frac{d}{dt}\right|_0 (\zeta + t\xi) =: \mathscr{J}_\zeta \xi$$

Denote the vector bundle isomorphism from $\mathscr{V}E$ to E along π by pr_2:

$$\mathrm{pr}_2: \mathscr{V}E \to E \qquad \mathscr{J}_\zeta \xi \mapsto \xi$$

The map $\mathrm{pr}_2: \mathscr{V}E \to E$ picks off the second "component" ξ of $\mathscr{J}_\zeta \xi$, whereas the projection map $\pi_{\mathscr{V}}: \mathscr{V}E \to E$ yields the first component ζ. In fact the isomorphism $\mathscr{J}$ makes the following bundle diagrams equivalent:

$$\begin{array}{ccc} \pi^*E & \xrightarrow{\text{pr}_2} & E \\ {\scriptstyle \text{pr}_1}\downarrow & & \downarrow{\scriptstyle \pi} \\ E & \xrightarrow[\pi]{} & M \end{array} \quad \cong \quad \begin{array}{ccc} \mathscr{V}E & \xrightarrow{\text{pr}_2} & E \\ {\scriptstyle \pi_{\mathscr{V}}}\downarrow & & \downarrow{\scriptstyle \pi} \\ E & \xrightarrow[\pi]{} & M \end{array}$$

1.29 Now we will study briefly the composite bundles π^*E and $\mathscr{V}E$ over M; for π^*E the terminology will be different in anticipation of vector bundle operations.

Definition Let E_1 and E_2 be $\mathbb{F}$-vector bundles over M. The *Whitney sum* of E_1 and E_2 is the pullback bundle $E_1 \oplus E_2 := \Delta^*(E_1 \times E_2) \to M$ of the product bundle $E_1 \times E_2 \to M \times M$ by the so-called diagonal embedding map $\Delta\colon M \to M \times M, p \mapsto (p, p)$.

The standard fiber of $E_1 \oplus E_2$ is the direct sum of the standard fibers of E_1 and E_2 by 1.12*f* and *g*.

1.30 Proposition If $E \xrightarrow{\pi} M$ is a vector bundle, the composite bundle π^*E over M is a vector bundle isomorphic to the Whitney sum $E \oplus E$.

Proof The "identity map" $(\zeta, \xi) \leftrightarrow (\zeta, \xi)$ is an isomorphism

$$\pi^*E = \{(\zeta, \xi) \in E \times E \mid \pi\zeta = \pi\xi\} \cong \Delta^*(E \times E) = E \oplus E$$

1.31 Corollary If E is a vector bundle, the composite bundle (1.12*h*) $\pi \circ \pi_{\mathscr{V}}\colon \mathscr{V}E \to M$ is a vector bundle isomorphic to the Whitney sum $E \oplus E$.

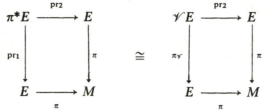

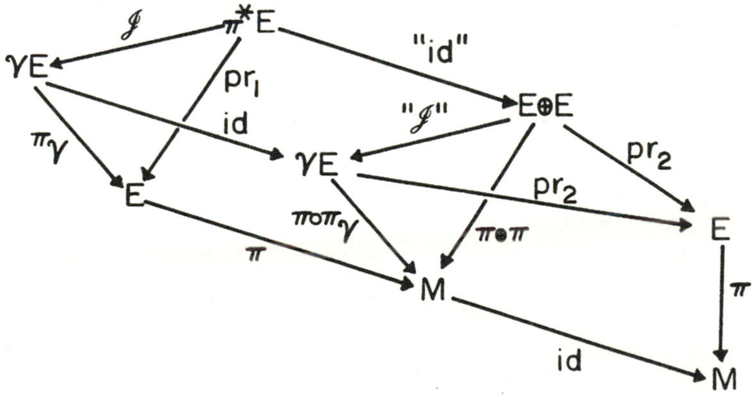

In summary, the bundles E, π^*E, $E \oplus E$, $\mathscr{V}E$ over E, and $\mathscr{V}E$ over M are related by the commutative diagram above of bundle isomorphisms along maps.

Exercise Study this diagram in the special case of the universal line bundle over the projective space $\mathbb{F}P^n$.

1.32 Assume that $E \xrightarrow{\pi} M$ is an $\mathbb{F}$-vector bundle. The tangent bundle of E is naturally an $\mathbb{F}$-vector bundle over TM.

Definition If $E \xrightarrow{\pi} M$ is an $\mathbb{F}$-vector bundle, define a vector bundle structure on $\pi_*: TE \to TM$ by

$$u + v := A_*(u, v) \qquad u, v \in TE, \ \pi_* u = \pi_* v \in TM$$

$$c \cdot v := \mu_{c*} v \qquad c \in \mathbb{F}, \ v \in TE$$

where $A(\zeta, \xi) := \zeta + \xi$ for $(\zeta, \xi) \in E \oplus E$, and $\mu_c \xi := c\xi$, $c \in \mathbb{F}$, $\xi \in E$. Each vector bundle chart (π, ψ) on E generates a vector bundle chart $(\pi, \psi)_*$ on TE over TM.

Exercise Check that TE is an $\mathbb{F}$-vector bundle over TM.

1.33 **Proposition** Let $E \to M$ be a vector bundle. The composite bundle $\pi \circ \pi_{\mathscr{V}}: \mathscr{V}E \to M$ is a vector subbundle of $\pi_*: TE \to TM$.

PROOF Since $\mathscr{V}E$ is a subset of TE, by 1.12*d* all we need is to have M as a submanifold of TM in such a way that the bundle charts on $\mathscr{V}E$ extend to bundle charts on TE.

Embed M into TM by the map 0_M which sends $p \in M$ to the zero vector $0_p \in M_p$.

$$\begin{array}{ccc} \mathscr{V}E & \longrightarrow & TE \\ {\scriptstyle \pi \circ \pi_{\mathscr{V}}} \downarrow & & \downarrow {\scriptstyle \pi_*} \\ M & \underset{0_M}{\longrightarrow} & TM \end{array}$$

Given a vector bundle chart (π, ψ) on E over U in M, the associated vector bundle chart $(\pi \circ \pi_{\mathscr{V}}, \psi_*)$ on $\mathscr{V}E$ over U is the restriction of the vector bundle chart (π_*, ψ_*) on TE over TU.

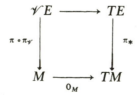

OPERATIONS ON VECTOR BUNDLES

Let E_1 and E_2 be $\mathbb{F}$-vector bundles over M; by 1.29, the total space of the Whitney sum of E_1 and E_2 is the union over all $p \in M$ of the direct sum of the fibers of E_1 and E_2 at p. The topology, C^∞ structure, and vector bundle structure of $E_1 \oplus E_2$ were defined by the trick of considering $E_1 \oplus E_2$ as the pullback of the vector bundle $E_1 \times E_2 \to M \times M$ by the diagonal embedding Δ of M into $M \times M$.

Given a vector bundle E over M, define the set $E^* := \bigcup_{p \in M} E_p^*$, where E_p^* is the dual of the vector space E_p; by analogy with the process for defining the cotangent bundle of a manifold, we could now define a topology, C^∞ structure, and vector bundle structure on E^* to obtain a vector bundle over M for which the standard fiber is the dual of the standard fiber of E.

Similarly, given vector bundles E_1 and E_2 over M, we could define the tensor product $E_1 \otimes E_2$, the homomorphism bundle $\mathrm{Hom}(E_1, E_2)$, and so forth, all by applying the stepwise process outlined above to each individual case. To avoid the ad hoc constructions this would entail, it is expedient to unify everything by the use of functors.

1.34 Fix $\mathbb{F} = \mathbb{R}$ or $\mathbb{C}$. Recall that if $\mathscr{V}$ and $\mathscr{W}$ are finite-dimensional vector spaces over $\mathbb{F}$, then so is the space $\mathrm{Hom}(\mathscr{V}, \mathscr{W})$ of linear maps from $\mathscr{V}$ to $\mathscr{W}$; in particular, $\mathrm{Hom}(\mathscr{V}, \mathscr{W})$ is a manifold [*W*: 1.5(b, c)].

Let $\mathscr{C}$ be the category of finite-dimensional vector spaces over $\mathbb{F}$ and homomorphisms between these vector spaces.

Definition A *covariant C^∞ functor L of one variable on $\mathscr{C}$* consists of
(i) a map $\mathscr{C} \to \mathscr{C}$, denoted by L, and
(ii) (for each $\mathscr{V}$ and $\mathscr{W}$ in $\mathscr{C}$) a map $\mathrm{Hom}(\mathscr{V}, \mathscr{W}) \xrightarrow{C^\infty} \mathrm{Hom}(L\mathscr{V}, L\mathscr{W})$, also denoted by L, such that
(iii) $L(\mathrm{id}_{\mathscr{V}}) = \mathrm{id}_{L\mathscr{V}}$ for all $\mathscr{V} \in \mathscr{C}$, and
(iv) $L(f \circ g) = L(f) \circ L(g)$ for $g: \mathscr{U} \to \mathscr{V}$ and $f: \mathscr{V} \to \mathscr{W}$.
It follows that if f is an isomorphism, then so is $L(f)$.

1.35 Examples Let k be a positive integer.
(*a*) Assign to $\mathscr{V} \in \mathscr{C}$ the direct sum $\mathscr{V} \oplus \cdots \oplus \mathscr{V}$ of k copies of $\mathscr{V}$, and for $f \in \mathrm{Hom}(\mathscr{V}, \mathscr{W})$, define $f \oplus \cdots \oplus f \in \mathrm{Hom}(\mathscr{V} \oplus \cdots \oplus \mathscr{V}, \mathscr{W} \oplus \cdots \oplus \mathscr{W})$ by $(f \oplus \cdots \oplus f)(v_1, \ldots, v_k) := (fv_1, \ldots, fv_k)$.
(*b*) Assign to $\mathscr{V} \in \mathscr{C}$ the tensor product $\mathscr{V} \otimes \cdots \otimes \mathscr{V}$ (k factors), and for $f \in \mathrm{Hom}(\mathscr{V}, \mathscr{W})$, define $f \otimes \cdots \otimes f \in \mathrm{Hom}(\mathscr{V} \otimes \cdots \otimes \mathscr{V}, \mathscr{W} \otimes \cdots \otimes \mathscr{W})$ by $(f \otimes \cdots \otimes f)(v_1 \otimes \cdots \otimes v_k) := f(v_1) \otimes \cdots \otimes f(v_k)$.
(*c*) Assign to $\mathscr{V} \in \mathscr{C}$ the exterior algebra $\Lambda(\mathscr{V})$, and for $f \in$

$\text{Hom}(\mathscr{V}, \mathscr{W})$, define $\Lambda(f) \in \text{Hom}(\Lambda(\mathscr{V}), \Lambda(\mathscr{W}))$ by $\Lambda(f)(v_1 \wedge \cdots \wedge v_k) := f(v_1) \wedge \cdots \wedge f(v_k)$, and $f(1) := 1$.

1.36 Similarly, we can consider C^∞ covariant functors on $\mathscr{C}$ of more than one variable; for example, the map $\otimes: \mathscr{C} \times \mathscr{C} \to \mathscr{C}$, $(\mathscr{V}, \mathscr{W}) \mapsto \mathscr{V} \otimes \mathscr{W}$, and the map $\otimes: \text{Hom}(\mathscr{V}, \mathscr{W}) \times \text{Hom}(\mathscr{V}', \mathscr{W}') \to \text{Hom}(\mathscr{V} \otimes \mathscr{V}', \mathscr{W} \otimes \mathscr{W}')$, $(f, g) \mapsto f \otimes g$, together make up a covariant functor of two variables on $\mathscr{C}$.

1.37 **Definition** *A contravariant C^∞ functor L of one variable on $\mathscr{C}$* consists of

(i) a map $L: \mathscr{C} \to \mathscr{C}$, and
(ii) a map $L: \text{Hom}(\mathscr{V}, \mathscr{W}) \overset{C^\infty}{\to} \text{Hom}(L\mathscr{W}, L\mathscr{V})$, $\mathscr{V}, \mathscr{W} \in \mathscr{C}$, such that
(iii) $L(\text{id}_{\mathscr{V}}) = \text{id}_{L\mathscr{V}}$ for all $\mathscr{V} \in \mathscr{C}$, and
(iv) $L(f \circ g) = L(g) \circ L(f)$ for $g \in \text{Hom}(\mathscr{U}, \mathscr{V})$ and $f \in \text{Hom}(\mathscr{V}, \mathscr{W})$.

The most important example for geometric purposes is the "dual" functor: assign to $\mathscr{V} \in \mathscr{C}$ the dual space $\mathscr{V}^* \in \mathscr{C}$, and to $f \in \text{Hom}(\mathscr{V}, \mathscr{W})$ the transpose map $f^* \in \text{Hom}(\mathscr{W}^*, \mathscr{V}^*)$, where $(f^*\mu)(v) := \mu(fv)$.

Similarly, we can consider C^∞ contravariant functors of more than one variable. More generally, there are C^∞ functors of several variables on $\mathscr{C}$ which are contravariant in some arguments and covariant in the remaining arguments; for example, the functor

$$(\mathscr{V}, \mathscr{W}) \mapsto \mathscr{V}^* \otimes \mathscr{W} \qquad (f, g) \mapsto f^* \otimes g$$

is contravariant in the first variable and covariant in the second variable. Because of the canonical isomorphism between $\mathscr{V}^* \otimes \mathscr{W}$ and $\text{Hom}(\mathscr{V}, \mathscr{W})$ for $\mathscr{V}, \mathscr{W} \in \mathscr{C}$ [W: 2.2(d)], this functor is usually written Hom.

1.38 **Proposition** Let L be a C^∞ functor of k variables on $\mathscr{C}$, and let $E_1, \ldots, E_k$ be $\mathbb{F}$-vector bundles on M. The set

$$E = L(E_1, \ldots, E_k) := \bigcup_{p \in M} L(E_1|_p, \ldots, E_k|_p)$$

is naturally an $\mathbb{F}$-vector bundle over M.

Proof For the sake of definiteness, let $k = 2$, and assume L is contravariant in the first variable and covariant in the second; the general case is only a matter of more complicated notation. If the standard fiber of E_i is $\mathscr{V}_i$, then the standard fiber of E will be $L(\mathscr{V}_1, \mathscr{V}_2)$.

Define $\pi\colon E \to M$ by $\pi L(E_1|_p, E_2|_p) := p \in M$.

Each fiber of E is given as a vector space, so we only need to define the vector bundle charts. Let U be an open set in M over which both E_i are trivial, and let (π_i, ψ_i) be a vector bundle chart on E_i over U. Write $\psi_i|_p$ for the restriction to $E_i|_p$ of the map ψ_i; thus $\psi_i|_p\colon E_i|_p \to \mathscr{V}_i$. For each $p \in U$, define

$$\psi|_p := L(\psi_1|_p^{-1}, \psi_2|_p) \in \mathrm{Hom}(E_p, L(\mathscr{V}_1, \mathscr{V}_2))$$

By functoriality, $\psi|_p$ is an isomorphism.

Define the vector bundle structure on E over U by requiring that (π, ψ) be a vector bundle chart, that is, a vector bundle isomorphism from $\pi^{-1}U$ to $U \times L(\mathscr{V}_1, \mathscr{V}_2)$; notice that this definition simultaneously fixes the underlying topological and C^∞ structure of E over U. It must be checked that this structure is independent of the choice of vector bundle charts (π_i, ψ_i) over U.

Let (π_i, φ_i) be another vector bundle chart on E_i over U, and let (π, φ) be the associated candidate for a vector bundle chart on E over U. For each $p \in U$,

$$\begin{aligned} f_{\varphi,\psi}(p) &= \varphi \circ \psi|_p^{-1} = L(\varphi_1|_p^{-1}, \varphi_2|_p) \circ L(\psi_1|_p, \psi_2|_p^{-1}) \\ &= L(\psi_1|_p \circ \varphi_1|_p^{-1}, \varphi_2|_p \circ \psi_2|_p^{-1}) = L(f_{\psi_1,\varphi_1}(p), f_{\varphi_2,\psi_2}(p)) \end{aligned}$$

By the properties of f_{ψ_1,φ_1}, f_{φ_2,ψ_2}, and L, this is an isomorphism of $L(\mathscr{V}_1, \mathscr{V}_2)$. Finally, $f_{\varphi,\psi}$ is a C^∞ map on U.

Exercises What is the structure group of $L(E_1, E_2)$ (given structure groups G_j for the E_j)? If L is the direct sum functor $\oplus\colon \mathscr{C} \times \mathscr{C} \to \mathscr{C}$, check that $L(E_1, E_2)$ is the Whitney sum $E_1 \oplus E_1$ defined in 1.29.

1.39 Examples

(a) The *dual of a vector bundle E* is the vector bundle E^*; the fiber in E^* at p is the vector space E_p^* of linear functionals on E_p.

(b) The *tensor product of vector bundles* E_1 *and* E_2 is the vector bundle $E_1 \otimes E_2$; the fiber at $p \in M$ is the tensor product $E_1|_p \otimes E_2|_p$. The canonical isomorphism $\mathscr{V}_1 \otimes \mathscr{V}_2 \cong \mathscr{V}_2 \otimes \mathscr{V}_1$ for all $\mathscr{V}_1, \mathscr{V}_2 \in \mathscr{C}$ induces a canonical vector bundle isomorphism $E_1 \otimes E_2 \cong E_2 \otimes E_1$. Similarly, $E_1 \otimes (E_2 \otimes E_3) \cong (E_1 \otimes E_2) \otimes E_3$, and $E_1^* \otimes E_2^* \cong (E_1 \otimes E_2)^*$. If E is an $\mathbb{F}$-vector bundle, $E \otimes (M \times \mathbb{F}) \cong E$; if $\mathbb{F} = \mathbb{R}$, then $E \otimes (M \times \mathbb{C})$ is naturally a complex vector bundle isomorphic to $E^{\mathbb{C}}$ (1.23).

Denote by $\otimes^k E$ the bundle $E \otimes \cdots \otimes E$ (k factors).

(c) The *homomorphism bundle from* E_1 *to* E_2 is the vector bundle $\mathrm{Hom}(E_1, E_2)$; the fiber at $p \in M$ is the vector space $\mathrm{Hom}(E_1|_p, E_2|_p)$. The isomorphism $\mathrm{Hom}(\mathscr{V}_1, \mathscr{V}_2) \cong \mathscr{V}_1^* \otimes \mathscr{V}_2$ for all $\mathscr{V}_1$,

$\mathscr{V}_2 \in \mathscr{C}$ induces a vector bundle isomorphism $\mathrm{Hom}(E_1, E_2) \cong E_1^* \otimes E_2$. For example,

$$\mathrm{Hom}(E, M \times \mathbb{F}) \cong E^* \otimes (M \times \mathbb{F}) \cong E^*$$

1.40 Recall that a vector space $\mathscr{V}$ is called an *algebra* if there is given a linear map $\mathscr{V} \otimes \mathscr{V} \to \mathscr{V}$, called *multiplication* in $\mathscr{V}$.

Definition Let $\mathscr{V}$ be an algebra over $\mathbb{F}$. An $\mathbb{F}$-vector bundle E over M is called an *algebra bundle with standard fiber* $\mathscr{V}$ if there is given a vector bundle homomorphism $E \otimes E \to E$ and an *algebra bundle atlas* on E, that is, a vector bundle atlas $\mathscr{A}$ on E such that if $(\pi, \psi) \in \mathscr{A}$ is a chart on E over $U \subset M$, then for all $p \in U$, ψ maps the algebra E_p isomorphically to the algebra $\mathscr{V}$.

1.41 Examples

(*a*) The *tensor algebra bundle of a vector bundle* E is the vector bundle with standard fiber $\otimes(\mathscr{V})$ ($\mathscr{V}$ being the standard fiber of E)

$$\begin{aligned}\otimes(E) &:= \bigoplus_{k,\, l \geq 0} (\otimes^k E) \otimes (\otimes^l E^*) \\ &= \bigoplus_{k,\, l \geq 0} (\underbrace{E \otimes \cdots \otimes E}_{k \text{ factors}} \otimes \underbrace{E^* \otimes \cdots \otimes E^*}_{l \text{ factors}})\end{aligned}$$

with the natural tensor algebra multiplication on each fiber; this can be defined functorially: the multiplication homomorphism $[\otimes(\mathscr{V})] \otimes [\otimes(\mathscr{V})] \to \otimes(\mathscr{V})$ on the standard fiber of $\otimes(E)$ induces the multiplication on each fiber by 1.34ii and 1.37ii.

(*b*) The *exterior algebra bundle of* E is the vector bundle

$$\Lambda(E) := \bigoplus_{k \geq 0} \Lambda^k E = \bigoplus_{k \geq 0} (\underbrace{E \wedge \cdots \wedge E}_{k \text{ factors}})$$

with the natural multiplication on the fibers.

For each k, $\Lambda^k(E^*)$ is isomorphic to $(\Lambda^k E)^*$ by the nonsingular pairing $(\omega^1 \wedge \cdots \wedge \omega^k)(\xi_1, \ldots, \xi_k) := \det[\omega^i(\xi_j)]$ on the decomposable elements. The isomorphisms $\Lambda^k(E^*) \cong (\Lambda^k E)^*$ extend naturally to an isomorphism $\Lambda(E^*) = \Lambda(E)^*$. If $\omega \in \Lambda^k(E^*)$, $\mu \in \Lambda^l(E^*)$, and $\xi_1, \ldots, \xi_{k+l} \in E$ all lie over the same point in M, then

$$\omega \wedge \mu(\xi_1, \ldots, \xi_{k+l}) = \frac{1}{k!\,l!} \sum_{\sigma} \mathrm{sgn}(\sigma)\omega(\xi_{\sigma 1}, \ldots, \xi_{\sigma k}) \cdot \mu(\xi_{\sigma k+1}, \ldots, \xi_{\sigma k+l})$$

where σ ranges over all permutations of $1, \ldots, k + l$.

(*c*) The *endomorphism algebra bundle* E is the vector bundle $\mathrm{End}(E) :=$

Hom(E, E), with the natural fiberwise associative multiplication by composition: $f \cdot g := f \circ g, f, g \in \operatorname{End}(E)_p = \operatorname{End}(E_p)$.

(*d*) The *general linear algebra bundle of* E is the vector bundle $\mathfrak{gl}(E) :=$ Hom(E, E), with commutation as multiplication on the fibers: $[f, g] := f \circ g - g \circ f, f, g \in \mathfrak{gl}(E)_p = \mathfrak{gl}(E_p)$.

Emphasis: The algebra bundles End(E) and $\mathfrak{gl}(E)$ have the same underlying vector bundle—Hom(E, E) $\cong E^* \otimes E$—but differ in the algebra structure of the fibers and in the admissible algebra bundle charts.

PRINCIPAL AND ASSOCIATED BUNDLES

1.42 **Definition** A C^∞ *right action of a Lie group G on a manifold P* is a C^∞ map $\mu: P \times G \to P$ such that $\mu(p, gh) = \mu(\mu(p, g), h)$ and $\mu(p, e) = p$, where e is the identity element of G. An action of G on P is called *free* if the identity element e of G is the only element of G which fixes a point in P: $\mu(p, g) = p$ for some $p \in P$ implies $g = e$.

The reader should compare this with a left action of G (1.9).

1.43 **Definition** Let $\not{p}: P \to M$ be a C^∞ fiber bundle for which the standard fiber is a Lie group G. If there is given a free C^∞ right action of G on P, and if there is given a bundle atlas $\mathscr{A}$ on E such that for each bundle chart $(\not{p}, \psi) \in \mathscr{A}$ over $U \subset M$,

$$\psi(bg) = \psi(b) \cdot g \in G \qquad b \in \not{p}^{-1}U, g \in G$$

then P is called a *principal fiber bundle over M with group G* (or a *principal G-bundle over M*), and $\mathscr{A}$ is called a *principal bundle atlas.*

It follows immediately that as structure group for P we may take G itself, for if $(\not{p}, \varphi)$ and $(\not{p}, \psi) \in \mathscr{A}$ are bundle charts over $U \cap V$ in M, then the transition function satisfies $f_{\varphi,\psi}(p) = \varphi(b)\psi(b)^{-1} \in G$, where $\not{p}b = p \in U \cap V$; the choice of $b \in \not{p}^{-1}p$ is irrelevant because

$$\varphi(bg)\psi(bg)^{-1} = \varphi(b)gg^{-1}\psi(b)^{-1} = \varphi(b)\psi(b)^{-1} \qquad g \in G$$

If P is a principal G-bundle over M, then M is the quotient space P/G and $\not{p}$ is the canonical projection of P onto P/G. In fact, let b and $\tilde{b} \in \not{p}^{-1}p \subset P$, $p \in M$, and let $(\not{p}, \psi)$ be a principal bundle chart on P over a neighborhood U of p. Set $g := \psi(b)$ and $h := \psi(\tilde{b})$. By the homomorphism property for ψ in the definition,

$$(\not{p}, \psi)(bg^{-1}h) = (p, \psi(b)g^{-1}h) = (\not{p}\tilde{b}, h) = (\not{p}, \psi)(\tilde{b})$$

Since $(\not{p}, \psi)$ is injective, $\tilde{b} = bg^{-1}h$.

1.44 Alternatively, a principal fiber bundle over M could be defined to be a fiber bundle $P \xrightarrow{\not{p}} M$ with a left free action (1.9) of G on P and an atlas of bundle charts $(\not{p}, \psi)$ such that $\psi(gb) = g\psi(b)$.

1.45 Examples

(*a*) The trivial bundle $M \times G \xrightarrow{\text{pr}_1} M$ is a trivial principal bundle, where G acts on $M \times G$ on the right by $(p, g) \cdot h := (p, gh)$. This is essentially the local product structure of any principal G-bundle P over M: a principal bundle chart $(\not{p}, \psi)$ on P over $U \subset M$ is an *isomorphism* from the principal subbundle $\not{p}^{-1}U$ of P to $U \times G$.

(*b*) The nonzero elements of $\mathbb{F}$ form a group under multiplication, and this group acts on $\mathbb{F}^{n+1} - \{0\}$ by scalar multiplication in the vector space $\mathbb{F}^{n+1}$. Thus we have an action of $\mathbb{F} - \{0\}$ on the total space $\mathbb{F}^{n+1} - \{0\}$ of the fiber bundle over $\mathbb{F}P^n$ from 1.12*e*. The bundle chart $(\not{p}, \psi_j)$ over $U_j \subset \mathbb{F}P^n$ was defined by $\psi_j(v^0, \ldots, v^n) := v^j$; the maps ψ_j satisfy the rule $\psi_j(v \cdot c) = \psi_j(v)c$ for $v \in \not{p}^{-1}U_j$ and $c \in \mathbb{F} - \{0\}$. Thus $\not{p}$: $\mathbb{F}^{n+1} - \{0\} \to \mathbb{F}P^n$ is a principal fiber bundle with group $\mathbb{F} - \{0\}$.

(*c*) If $\not{p}$: $P \to M$ is a principal G-bundle, then the pullback f^*P of P by a C^∞ map f: $N \to M$ is a principal G-bundle. The action of G on f^*P is given by $(p, b) \cdot g := (p, bg)$, $(p, b) \in f^*P$, $g \in G$.

(*d*) Suppose H is a closed subgroup of a Lie group G. Let M be the homogeneous space G/H, and let $\not{p}$: $G \to G/H = M$ be the projection map; M is a manifold, $\not{p}$ is C^∞, and there exists an open covering $\mathscr{U}$ of M such that for each $U \in \mathscr{U}$ there is a C^∞ map σ: $U \to G$ for which $\not{p} \circ \sigma = \text{id}$ [W: 3.58]. The subgroup H acts naturally on G by multiplication, $(g, h) \mapsto gh$, $g \in G$, $h \in H \subset G$. Fix $U \in \mathscr{U}$, and define a C^∞ map f: $U \times H \to G$ by $f(p, h) := \sigma(p)h$; the inverse map f^{-1} is also C^∞, and is therefore a diffeomorphism. Define $\psi := \text{pr}_2 \circ f^{-1}$: $\not{p}^{-1}U \to H$. Fix $g \in \not{p}^{-1}U$; since $M = G/H$, there exists $h \in H$ such that $g = \sigma(q)h$, where $q := \not{p}(g)$. For all $h_1 \in H$,

$$\psi(gh_1) = \psi(\sigma(q)hh_1) = hh_1 = \psi(\sigma(q)h)h_1 = \psi(g)h_1$$

Thus the maps $(\not{p}, \psi)$ associated with the open covering $\mathscr{U}$ of M by this process form a principal bundle atlas on G over M. Hence, if H is a closed subgroup of G, then G is a principal H-bundle over G/H.

(*e*) Let E be a vector bundle over M with standard fiber $\mathscr{V}$. Fix $p \in M$. Any two isomorphisms b and b_1 from $\mathscr{V}$ to E_p differ by an element of $GL(\mathscr{V})$, that is, there exists $g \in GL(\mathscr{V})$ such that $b_1 = b \circ g$. In fact, for a fixed isomorphism b from $\mathscr{V}$ to E_p, the map $g \mapsto b \circ g$ is a bijective map from $GL(\mathscr{V})$ to the set $B(E_p)$ of all isomorphisms from $\mathscr{V}$ to E_p. Injectivity: if $b \circ g = b$, then g is the identity element of

$GL(\mathscr{V})$. Surjectivity: if b_1 is another isomorphism, then $b_1 = b \circ (b^{-1} \circ b_1)$, and $b^{-1} \circ b_1 \in GL(\mathscr{V})$.

Fix a basis $\{\epsilon_j\}$ for $\mathscr{V}$; each isomorphism $b\colon \mathscr{V} \to E_p$ determines a basis $\{b(\epsilon_j) \in E_p\}$ for E_p, and conversely. Thus the set of isomorphisms from $\mathscr{V}$ to E_p is in one-to-one correspondence with the *Stiefel manifold* [W: 3.65(e)] *of bases* (or *frames*) *of* E_p.

For each $p \in M$, let $B(E_p)$ be the set of isomorphisms from $\mathscr{V}$ to E_p; define the total space BE of the *bundle of bases of* E by

$$BE := \bigcup_{p \in M} B(E_p)$$

Let $\not{p}\colon BE \to M$ such that $\not{p}B(E_p) := p$; thus $BE_p = \not{p}^{-1}p := B(E_p)$.

Given a vector bundle chart (π, ψ) on E over U in M, $\psi|_{E_p}$ is an isomorphism from E_p to $\mathscr{V}$, so $b_p := \psi|_{E_p}^{-1} \in BE_p$. Define $\tilde{\psi}$ from $\not{p}^{-1}U$ to $GL(\mathscr{V})$ by $\tilde{\psi}(b_p \circ g) := g$, $p \in U$, $g \in GL(\mathscr{V})$. The map $(\not{p}, \tilde{\psi})$ from $\not{p}^{-1}U$ to $U \times GL(\mathscr{V})$ will be a principal bundle chart on BE.

If $(\not{p}, \tilde{\varphi})$ is the map associated with another vector bundle chart (π, φ) on E over U, then for $(p, g) \in U \times GL(\mathscr{V})$,

$$(\not{p}, \tilde{\varphi}) \circ (\not{p}, \tilde{\psi})^{-1}(p, g) = (\not{p}, \tilde{\varphi})(\psi|_{E_p}^{-1} \circ g) = (p, \varphi \circ \psi|_{E_p}^{-1} \circ g)$$

Thus the transition function from $\tilde{\psi}$ to $\tilde{\varphi}$ is just the original transition function $f_{\varphi,\psi}\colon U \to GL(\mathscr{V})$ from ψ to φ in E; this is C^∞ since E is a C^∞ bundle. Thus $(\not{p}, \tilde{\varphi})$ and $(\not{p}, \tilde{\psi})$ define the same C^∞ structure on $\not{p}^{-1}U \subset BE$; they also determine the fiber bundle structure of BE over U. As structure group we can take $GL(\mathscr{V})$.

The map $BE \times GL(\mathscr{V}) \to BE$ such that $(b, g) \mapsto b \circ g$ is a C^∞ right action of $GL(\mathscr{V})$ on BE; it is free because $GL(\mathscr{V})$ acts freely on $B(\mathscr{V})$. Furthermore, the action preserves the fibered structure of BE: $\not{p}(bg) = \not{p}(b)$. Finally, let $(\not{p}, \tilde{\psi})$ be the bundle chart on BE associated with a vector bundle chart (π, ψ) on E over U in M. Given $p \in U$ and $b \in BE_p = B(E_p)$, there exists $g \in GL(\mathscr{V})$ such that $b = \psi|_{E_p}^{-1} \circ g =: b_p \circ g$ as before; hence for all $g_1 \in GL(\mathscr{V})$,

$$\tilde{\psi}(bg_1) = \tilde{\psi}(b_p g g_1) = g g_1 = \tilde{\psi}(b_p g) g_1 = \psi(b) g_1$$

so $(\not{p}, \tilde{\psi})$ is a principal bundle chart on BE over U in M.

Thus BE is a principal $GL(\mathscr{V})$-bundle over M.

Exercises Fix a basis for $\mathscr{V}$, and carry out the work above in terms of matrices [Hi: 1, exercise 22]. Note: Elements of $\mathbb{F}^m$ should be written as column vectors, and $GL(m, \mathbb{F})$ acts on $\mathbb{F}^m$ on the left.

Let f^*E be the pullback of the vector bundle E by a C^∞ map $f\colon N \to M$; prove that f^*BE and Bf^*E are isomorphic principal bundles over N, that is, there is a bundle diffeomorphism from f^*BE to Bf^*E which commutes with the actions of $GL(\mathscr{V})$ on the bundles.

1.46 In 1.45*e* we obtained BE as the bundle of bases for the fibers of E. Conversely we can recover E from BE. Intuitively, each $b \in BE$ glues a copy of $\mathscr{V}$ onto M at the basepoint $p = \wp b \in M$. The properties of BE as a principal bundle guarantee that the copy $b(\mathscr{V})$ of $\mathscr{V}$ at p is independent of the choice of $b \in \wp^{-1}p$, and that the union over all $p \in M$ of the attached vector space at p is isomorphic to E.

This construction is made precise in the next definition.

Definition Let $B \xrightarrow{\wp} M$ be a principal G-bundle, and let G act on a manifold F on the left. Define $B \times_G F := (B \times F)/\sim$, where

$$(bg, \xi) \sim (b, g\xi) \qquad \text{that is} \qquad (b, \xi) \sim (bg, g^{-1}\xi)$$

for $(b, \xi) \in B \times F$, $g \in G$. Denote the $\sim$-equivalence class of (b, ξ) in $B \times F$ by $\mathscr{P}(b, \xi)$, and define $\pi\colon B \times_G F \to M$ by $\pi\mathscr{P}(b, \xi) := \wp(b)$. The set $B \times_G F$ is called the *bundle with fiber F associated with B and the given action of G on F.*

It must still be proved that $B \times_G F$ is a fiber bundle over M.

1.47 Proposition Let $B \xrightarrow{\wp} M$ be a principal G-bundle, and let G act on a manifold F on the left. The map $\pi\colon B \times_G F \to M$ is a fiber bundle. Furthermore, the map $\mathscr{P}\colon B \times F \to B \times_G F$ is a principal G-bundle, and $\mathrm{pr}_1\colon B \times F \to B$ is a principal bundle isomorphism along the map π, that is, pr_1 maps the fiber in $B \times F$ over $\mathscr{P}(b, \xi)$ diffeomorphically to $B_{\wp(b)}$, and commutes with the actions of G on $B \times F$ and B. A structure group for $B \times_G F$ is G.

$$\begin{array}{ccc} B \times F & \xrightarrow{\mathrm{pr}_1} & B \\ \downarrow{\scriptstyle \mathscr{P}} & & \downarrow{\scriptstyle \wp} \\ B \times_G F & \xrightarrow[\pi]{} & M \end{array}$$

Finally, for each $b \in B$, the map of F into $B \times_G F$ which sends ξ to $\mathscr{P}(b, \xi)$ is a diffeomorphism onto the fiber over $\wp(b)$; if F is a vector space and G acts linearly on F, then $B \times_G F$ is a vector bundle over M.

PROOF Let $(\wp, \psi)$ be a principal bundle chart on B over an open set U in M. Define $\tilde{\psi}\colon \pi^{-1}U \to F$ by $\tilde{\psi} \circ \mathscr{P}(b, \xi) := \psi(b) \cdot \xi$ for all

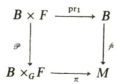

$(b, \xi) \in \not{p}^{-1}U \times F$. We must show that $(\pi, \tilde{\psi})$ is invertible. For each $p \in U$, set $\beta_p := (\not{p}, \psi)^{-1}(p, e)$, where $e \in G$ is the identity element; define $\eta\colon U \times F \to \pi^{-1}U$ by $\eta(p, \xi) := \mathscr{P}(\beta_p, \xi)$. First of all,

$$(\pi, \tilde{\psi}) \circ \eta(p, \xi) = (\pi, \tilde{\psi}) \circ \mathscr{P}(\beta_p, \xi) = (p, \psi(\beta_p) \cdot \xi) = (p, \xi)$$

For the other direction, if $p \in U$ and $b \in \not{p}^{-1}p$ in B, then

$$(\not{p}, \psi)(\beta_p \cdot \psi(b)) = (p, \psi(\beta_p) \cdot \psi(b)) = (p, \psi(b)) = (\not{p}, \psi)(b)$$

so $\beta_p \cdot \psi(b) = b \in B$; therefore,

$$\begin{aligned} \eta \circ (\pi, \tilde{\psi}) \circ \mathscr{P}(b, \xi) &= \eta(p, \psi(b) \cdot \xi) = \mathscr{P}(\beta_p, \psi(b) \cdot \xi) \\ &= \mathscr{P}(\beta_p \cdot \psi(b), \xi) = \mathscr{P}(b, \xi) \end{aligned}$$

Thus, $(\pi, \tilde{\psi})$ is invertible, and will serve as a bundle chart.

If $(\not{p}, \psi)$ and $(\not{p}, \varphi)$ are principal bundle charts on B over intersecting open sets U and V in M, then for $(p, \xi) \in (U \cap V) \times F$,

$$\begin{aligned} f_{\tilde{\varphi}, \tilde{\psi}}(p) \cdot \xi &= \tilde{\varphi} \circ (\pi, \tilde{\psi})^{-1}(p, \xi) = \varphi \circ \mathscr{P}((\not{p}, \psi)^{-1}(p, e), \xi) \\ &= (\varphi \circ (\not{p}, \psi)^{-1}(p, e)) \cdot \xi = (f_{\varphi, \psi}(p) \cdot e) \cdot \xi = f_{\varphi, \psi}(p) \cdot \xi \end{aligned}$$

so $f_{\tilde{\varphi}, \tilde{\psi}} = f_{\varphi, \psi}$ is C^∞, and the maps $(\pi, \tilde{\varphi})$ and $(\pi, \tilde{\psi})$ are C^∞-compatible. The maps $(\pi, \tilde{\psi})$ associated with principal bundle charts $(\not{p}, \psi)$ on B determine the C^∞ structure on $B \times_G F$, and also form a bundle atlas for $B \times_G F$. The choice of G for the structure group is clear.

With respect to the charts on $B \times_G F$, $(\pi, \tilde{\psi}) \circ \mathscr{P}(b, \xi) = (\not{p}(b), \psi(b)\xi)$ for $(b, \xi) \in U \times F$; this identity shows that $\mathscr{P}$ is C^∞.

The C^∞ right action of G on $B \times F$ is given by $(b, \xi) \cdot g := (bg, g^{-1}\xi)$; $\mathscr{P}(b, \xi) = \mathscr{P}(bg, g^{-1}\xi)$. The identity $\mathrm{pr}_1(bg, g^{-1}\xi) = bg = \mathrm{pr}_1(b, \xi)g$ shows that pr_1 is a principal G-bundle isomorphism along π.

Fix $b \in B$ over $p \in M$. The map from F to $\pi^{-1}p$ which sends ξ to $\mathscr{P}(b, \xi)$ is C^∞ because $\mathscr{P}$ is C^∞; its inverse is given with respect to a principal bundle chart $(\not{p}, \psi)$ and the associated bundle chart on $B \times_G F$ by $\sigma \mapsto \psi(b)^{-1} \cdot \tilde{\psi}(\sigma)$, and is therefore also C^∞.

Finally, suppose F is a vector space, and G acts linearly on F. Given $b \in B$ and $u = \mathscr{P}(b, \zeta)$, $v = \mathscr{P}(b, \xi) \in \pi^{-1}\not{p}(b)$, define $u + cv := \mathscr{P}(b, \zeta + c\xi)$ for all scalars c. This is well-defined because for all $g \in G$, $u = \mathscr{P}(bg, g^{-1}\zeta)$, $v = \mathscr{P}(bg, g^{-1}\xi)$, and

$$\mathscr{P}(bg, cg^{-1}\zeta + g^{-1}\xi) = \mathscr{P}(bg, g^{-1}(c\zeta + \xi)) = \mathscr{P}(b, c\zeta + \xi)$$

The bundle charts $(\pi, \tilde{\psi})$ are now vector bundle charts.

1.48 Examples

(*a*) If G acts trivially on F, that is, if $g\xi = \xi$ for all (g, ξ) in $G \times F$, then $B \times_G F$ is diffeomorphic as a bundle to $M \times F$.

(*b*) If G acts on itself by left multiplication, then $B \times_G G$ is a principal bundle isomorphic to B.

(*c*) If B is a left principal G-bundle (that is, G acts on B on the left—see 1.44), and if G acts on F on the right, then $F_G \times B$ can be defined; the proposition still holds after the necessary changes.

(*d*) By 1.45*b*, the canonical projection $\wp\colon \mathbb{F}^{n+1} - \{0\} \to \mathbb{F}P^n$ is a principal bundle; its group $\mathbb{F} - \{0\}$ is isomorphic to $GL(1, \mathbb{F})$ and acts naturally on $\mathbb{F}$ by multiplication. The fiber bundle with fiber $\mathbb{F}$ associated with $\mathbb{F}^{n+1} - \{0\}$ and the natural action of $GL(1, \mathbb{F})$ on $\mathbb{F}$ is just the universal line bundle over $\mathbb{F}P^n$ from 1.12*e*. Since the group is $GL(1, \mathbb{F})$, the result is a vector bundle (1.15*c*).

For the rest of the examples, let $B = BE$ be the bundle of bases for a vector bundle E over M with standard fiber $\mathscr{W}$; let $G = GL(\mathscr{W})$.

(*e*) If G acts on $\mathscr{W}^*$ by $g \cdot \omega := \omega \circ g^{-1}$, then $B \times_G (\mathscr{W}^*) \cong (B \times_G \mathscr{W})^* \cong E^*$.

(*f*) If G acts on $\mathscr{W} \otimes \mathscr{W}$ by $g(\zeta \otimes \xi) := g\zeta \otimes g\xi$, then $B \times_G (\mathscr{W} \otimes \mathscr{W}) \cong E \otimes E$.

(*g*) If G acts on $\Lambda(\mathscr{W})$ by $g(\xi_1 \wedge \cdots \wedge \xi_k) := g\xi_1 \wedge \cdots \wedge g\xi_k$ (1.35*c*), then $B \times_G \Lambda(\mathscr{W}) \cong \Lambda(E)$.

(*h*) The composite vertical bundle $\mathscr{V}E \to M$ and the isomorphic bundle $E \oplus E$ (1.31) are isomorphic to $B \times_G (\mathscr{W} \oplus \mathscr{W})$, where G acts on $\mathscr{W} \oplus \mathscr{W}$ by $g(\zeta, \xi) := (g\zeta, g\xi)$ (1.35*a*).

1.49 Exercises

1. Let $\wp\colon B \to M$ be a principal G-bundle, and let $\pi\colon E \to M$ be a fiber bundle with standard fiber F. Assume that B and E have the same transition functions, that is, there exist (1) a left action of G on F [which induces a homomorphism from G to $\mathrm{Diff}(F)$]; (2) an open covering $\mathscr{U}$ of M such that B and E are trivial over each set $U \in \mathscr{U}$; and (3) a bundle chart $(\wp, \varphi)$ for B and a bundle chart (π, ψ) for E over each $U \in \mathscr{U}$ such that if U_α and U_β in $\mathscr{U}$ have nonempty intersection, then the diagram

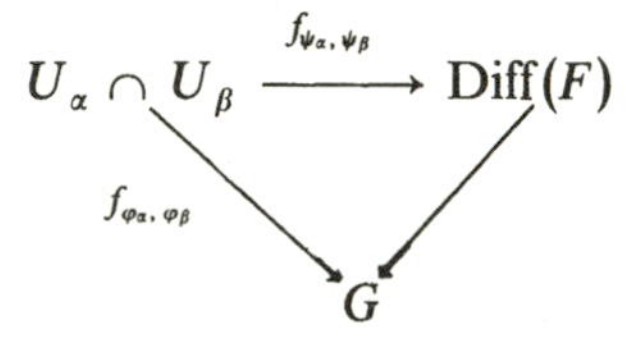

is commutative. Prove that E is diffeomorphic to $B \times_G F$ by a map which takes the fiber E_p to the fiber $(B \times_G F)_p$ for all $p \in M$.

This exercise shows that if the structure group of a fiber bundle E is a Lie group G, then E is associated to a principal G-bundle, namely the principal bundle B constructed (as in 1.12*i*) using the transition functions of E. The principal bundle B is the model for all fiber bundles over M with the given G-valued transition functions; this explains why B is called a principal G-bundle: to know all principal G-bundles over M is to know all fiber bundles over M with structure group G.

2. Construct a family of principal bundles over S^4 with group $SO(4)$ for which the associated bundles with fiber S^3 are the Milnor exotic spheres (1.12*j*).
3. Let H be a Lie subgroup of a Lie group G, let Q be a principal H-bundle over M, and let B be a principal G-bundle over M; assume that Q is a principal subbundle of B, that is, Q is a subbundle of B, and the inclusion map of Q into B commutes with the actions of H on Q and G on B. A C^∞ left action of G on a manifold F induces an action of H on F; prove that the bundles $Q \times_H F$ and $B \times_G F$ are diffeomorphic by a fiber-preserving map.

SECTIONS OF FIBER BUNDLES

A fiber bundle is a C^∞ submersion $\pi: E \to M$ of C^∞ manifolds for which the domain E of π has been endowed with some extra structure. Most C^∞ maps in the opposite direction are uninteresting because they ignore this extra structure.

1.50 **Definition** A *section of a fiber bundle* $\pi: E \to M$ is a C^∞ map $s: M \to E$ such that $\pi \circ s$ is the identity map of M. The set of sections of E is denoted by ΓE. A *local section of* E is a section of the subbundle $\pi^{-1}U \to U$ of E over an open set U in M.

1.51 **Comments** By definition, if s is a section of E, then for each $p \in M$, the value $s_p := s(p)$ belongs to the fiber E_p; thus a section of E recognizes the fibered structure of the manifold E over M.

The image $s(M) \subset E$ of a section of E is an embedded submanifold of E; if x is a chart on an open set U in M, then $x \circ \pi$ is a chart on the open set $s(U)$ in $S(M)$.

1.52 **Examples**

(*a*) Let F be a manifold. A section of the trivial bundle $M \times F \to M$ is just a C^∞ map from M to F; $\Gamma(M \times F)$ is usually denoted by $C^\infty(M, F)$.

For $\mathbb{F} = \mathbb{R}$ or $\mathbb{C}$, $C^\infty(M, \mathbb{F})$ is a ring, where the operations are

$$(f+g)(p) := f(p) + g(p) \qquad (fg)(p) := f(p)g(p) \qquad p \in M$$

The ring $C^\infty(M, \mathbb{C})$ is isomorphic to $\mathbb{C} \otimes C^\infty M$, where $C^\infty M$ is the standard abbreviation for $C^\infty(M, \mathbb{R})$.

(*b*) Let $M = G/H$, where H is a closed subgroup of a Lie group G. The local maps $\sigma: U \to G$ from 1.45*d* are local sections of the principal H-bundle G over M.

(*c*) A section X of an $\mathbb{F}$-vector bundle E over M assigns a vector $X_p \in E_p$ to each $p \in M$. The space ΓE of sections of E is a vector space over $\mathbb{F}$, where for $X, X_1, X_2 \in \Gamma E$ and $c \in \mathbb{F}$,

$$(X_1 + X_2)_p := X_1|_p + X_2|_p \qquad (cX)_p := c \cdot X_p \qquad p \in M$$

ΓE is also a module over the ring $C^\infty(M, \mathbb{F})$, where for all $p \in M$,

$$(f \cdot X)_p := f(p) \cdot X_p \qquad f \in C^\infty(M, \mathbb{F}),\ X \in \Gamma E$$

The zero element of ΓE is called the *zero section of* E: $0_E|_p = 0 \in E_p$.

A section of a vector bundle is often called a *field* on M. For example, a section of the tensor bundle $\otimes(TM)$ is called a *tensor field on* M; a section of $(\otimes^k TM) \otimes (\otimes^l T^*M)$ is called a *tensor field of type* (k, l). A $(0, 0)$-tensor field is just a C^∞ function on M, while a $(1, 0)$-tensor field is a section of TM, and is called a *vector field on* M. Denote the vector space ΓTM by $\mathfrak{X}M$.

Another example is the module $A^k M := \Gamma \Lambda^k T^*M$ of *differential k-forms on* M; the *exterior algebra of differential forms on* M is

$$A(M) := \bigoplus_{k \geq 0} A^k M = \Gamma \Lambda(T^*M)$$

The constant function $1 \in C^\infty M = A^0 M$ is the unique unit element of $A(M)$.

(*d*) If E_1 and E_2 are vector bundles over M, a bundle homomorphism from E_1 to E_2 is a section of the homomorphism bundle $\mathrm{Hom}(E_1, E_2)$; in fact, a homomorphism $f: E_1 \to E_2$ is equivalent to $s \in \Gamma\ \mathrm{Hom}(E_1, E_2)$, where s_p is f restricted to $E_1|_p$ for each $p \in M$. A section of $\mathrm{Hom}(E_1, E_2)$ is often called a *field of homomorphisms from* E_1 *to* E_2.

More generally (1.17), a homomorphism f of vector bundles E_j over M_j along a map $g: M_1 \to M_2$ is a section s of the bundle $\mathrm{Hom}(E_1, g^*E_2)$, where for each $p \in M_1$ (1.18),

$$s_p(\xi) := (p, f(\xi)) \in (g^*E_2)_p \qquad \xi \in E_1|_p$$

(*e*) Let $B \xrightarrow{\wp} M$ be a principal G-bundle. A principal bundle chart $(\wp, \psi)$ on B over $U \subset M$ is equivalent to a local section σ of B over U; if e denotes the identity element of G, then for $p \in U$,

$$\sigma_p := (\wp, \psi)^{-1}(p, e) \qquad \text{and} \qquad (\wp, \psi)(b) = (p, g) \qquad b = \sigma_p \cdot g$$

In particular, B is trivial if and only if there exists a global section σ of B over M.

Exercise The pullback $\wp^*B$ of B over itself by the projection map $\wp$ of B is trivial over B.

$$\begin{array}{ccc} \wp^*B & \xrightarrow{\text{pr2}} & B \\ {\scriptstyle \text{pr1}}\downarrow & & \downarrow{\scriptstyle \wp} \\ B & \xrightarrow[\wp]{} & M \end{array}$$

(*f*) Let E be a vector bundle over M, with standard fiber $\mathscr{V}$. A vector bundle chart (π, ψ) on E over $U \subset M$ determines an isomorphism $\psi|_{E_p}^{-1}: \mathscr{V} \to E_p$ for each $p \in U$. In particular, if we fix a basis $\{e_j\}$ for $\mathscr{V}$, then $\{\Psi_j|_p := \psi|_{E_p}^{-1} e_j\}$ is a basis for E_p; since (π, ψ) is a diffeomorphism, each Ψ_j is a section of E over U. On the other hand, for all $p \in U$ the isomorphism $\psi|_{E_p}^{-1}$ belongs to the fiber BE_p of the bundle BE of bases of E (1.45*e*); in fact, $\sigma_p := \psi|_{E_p}^{-1}$ defines a section σ of BE over U. Identify the *local basis field* $\{\Psi_j\}$ for E over U with the local section σ of BE.

Exercise By (*e*), σ determines a principal bundle chart $(\wp, \varphi)$ on BE over U. Show that $(\wp, \varphi)$ is the principal bundle chart $(\wp, \tilde{\psi})$ determined by the vector bundle chart (π, ψ) as in 1.45*e*.

A local basis field for TM is often called a *moving frame on* M. For example, given a chart x on M, $\{\partial/\partial x^j\}$ is a moving frame on M.

1.53 **Proposition** Let E_1 and E_2 be $\mathbb{F}$-vector bundles over M. The $C^\infty(M, \mathbb{F})$-modules $\Gamma\, \text{Hom}(E_1, E_2)$ and $\text{Hom}(\Gamma E_1, \Gamma E_2)$ are naturally isomorphic.

Proof Define a map $\Gamma\, \text{Hom}(E_1, E_2) \to \text{Hom}(\Gamma E_1, \Gamma E_2)$ as follows: given $s \in \Gamma\, \text{Hom}(E_1, E_2)$ and $\Xi \in \Gamma E_1$, $s\Xi$ is the section of E_2 such that $(s\Xi)_p := s_p(\Xi_p)$, $p \in M$; since s and Ξ are C^∞, so is $s\Xi$. This map is obviously a module homomorphism; now for the inverse.

Given a homomorphism L from the module ΓE_1 to the module ΓE_2, define $s \in \Gamma\ \mathrm{Hom}(E_1, E_2)$ by $s_p(\xi) := (L\Xi)_p$, $p \in M$, $\xi \in E_1|_p$, where Ξ is a section of E_1 such that $\Xi_p = \xi$. To show that s is well-defined, it suffices (by the linearity of L) to check that $\Xi_p = 0$ implies $(L\Xi)_p = 0$, $\Xi \in \Gamma E_1$. Let $\{\Phi_j\}$ be a local basis field for E_1 over a neighborhood U of p in M, and let $f \in C^\infty M$ such that the support of f is in U and $f(p) = 1$. Define $X_j \in \Gamma E_1$, $j = 1, \ldots, m$, by $X_j := f\Phi_j$ on U, and $X_j = 0$ off of U. It follows that there exist functions $g^j \in C^\infty(M, \mathbb{F})$ such that $f\Xi = \sum g^j X_j$ on M. But then

$$\begin{aligned}(L\Xi)_p = f(p)(L\Xi)_p = L(f\Xi)_p = L(\textstyle\sum g^j X_j)_p \\ = \textstyle\sum (g^j L X_j)_p = \sum g^j(p)(LX_j)_p = 0\end{aligned}$$

since $g^j(p) = 0$ for all j if $\Xi_p = 0$.

Since L is C^∞, so is s. Since the map $L \mapsto s$ and the module homomorphism above are mutually inverse, the modules are isomorphic.

For example, by 1.52c, $\Gamma\ \mathrm{Hom}(\otimes^k TM, \otimes^l TM) \cong \mathrm{Hom}(\otimes^k \mathfrak{X}M, \otimes^l \mathfrak{X}M)$; for $k = l = 1$, this reduces to $\Gamma\ \mathrm{End}\ TM \cong \mathrm{End}\ \mathfrak{X}M$ (1.41c). Similarly, $A^k M = \Gamma \Lambda^k T^*M \cong \mathrm{Hom}(\Lambda^k \mathfrak{X}M, C^\infty M)$ by 1.41b; furthermore,

$$\mu \wedge \eta(X_1, \ldots, X_{k+l}) = \frac{1}{k!\,l!} \sum_\sigma \mathrm{sgn}(\sigma)\mu(X_{\sigma 1}, \ldots, X_{\sigma k}) \cdot \eta(X_{\sigma k+1}, \ldots, X_{\sigma k+l})$$

for $\mu \in A^k M$, $\eta \in A^l M$, and $X_j \in \mathfrak{X}M$, where σ ranges over all permutations of $1, \ldots, k + l$.

1.54 **Definition** Let E be a vector bundle over M. Define $A^r(M, E) := \Gamma\ \mathrm{Hom}(\Lambda^r TM, E) \cong \Gamma(\Lambda^r T^*M \otimes E)$. Elements of $A^r(M, E)$ are called *differential r-forms on M with values in E*, or *vector-valued differential r-forms on M*. Set $A(M, E) := \bigoplus_r A^r(M, E)$.

By 1.53, $A^r(M, E) \cong \mathrm{Hom}(\Lambda^r \mathfrak{X}M, \Gamma E) \cong A^r M \otimes \Gamma E$, and $A(M, E) \cong \mathrm{Hom}(\Lambda(\mathfrak{X}M), \Gamma E) \cong A(M) \otimes \Gamma E$. Important special cases are the bundles $E = M \times \mathbb{F}^m$, $E = \Lambda(T^*M)$, $E = \otimes(TM)$, and $E = M \times \mathfrak{g}$, where $\mathfrak{g}$ is a Lie algebra. The module $A(M, M \times \mathbb{F})$ is just the ordinary module of differential forms on M with values in $\mathbb{F}$; for example, $A(M, M \times \mathbb{R}) = A(M)$.

1.55 **Definition** Suppose E is an algebra bundle over M. Denote the multiplication in E by $\zeta \cdot \xi$ for $\zeta, \xi \in E_p$, $p \in M$. Define multiplication in $A(M, E)$ such that the product $\varphi \cdot \psi \in A^{r+s}(M, E)$ of

$\varphi \in A^r(M, E)$ and $\psi \in A^s(M, E)$ is given by

$$(\varphi \cdot \psi)(X_1, \ldots, X_{r+s}) := \frac{1}{r!s!} \sum_{\sigma} \operatorname{sgn}(\sigma)\varphi(X_{\sigma 1}, \ldots, X_{\sigma r}) \cdot \psi(X_{\sigma r+1}, \ldots, X_{\sigma r+s})$$

where the sum is taken over all permutations of 1, 2, ..., $r + s$.

Any commutation relations in the algebra bundle E induce corresponding commutation relations in the module $A(M, E)$, as will be made clear in the following section.

1.56 Examples

(*a*) Let $E = M \times \mathbb{F}$. In this case, $A(M, E) = A(M, M \times \mathbb{F})$ is just the usual module of $\mathbb{F}$-valued differential forms on M; in particular, $A(M, M \times \mathbb{R}) = A(M)$. If μ is an r-form and η is an s-form, then $\mu \wedge \eta = (-1)^{rs}\eta \wedge \mu$.

(*b*) Let $E = M \times \mathfrak{g}$, where $\mathfrak{g}$ is a Lie algebra. If $\varphi \in A^r(M, E)$ and $\psi \in A^s(M, E)$, then $[\varphi, \psi] = (-1)^{rs+1}[\psi, \varphi]$. For example, if φ is a 1-form on M with values in $\mathfrak{g}$, then $[\varphi, \varphi](u, v) = [\varphi u, \varphi v] - [\varphi v, \varphi u] = 2[\varphi u, \varphi v] \neq 0$ in general, while $[\psi, \psi] = 0$ for $\psi \in A^2(M, M \times \mathfrak{g})$.

(*c*) Let E be the tensor bundle $\otimes(TM)$. If $\varphi \in A^r(M, E)$ and $\psi \in A^s(M, E)$, then there is a permutation τ of the factors of E such that $\varphi \otimes \psi = (-1)^{rs}\tau \circ (\psi \otimes \varphi)$.

(*d*) Let E be the exterior bundle $\Lambda(T^*M) = \oplus \Lambda^k T^*M$. If $\varphi \in A^r(M, \Lambda^j T^*M)$ and $\psi \in A^s(M, \Lambda^k T^*M)$, then $\varphi \wedge \psi = (-1)^{rs+jk}\psi \wedge \varphi \in A^{r+s}(M, \Lambda^{j+k}T^*M)$.

1.57 Chapter 1 will close with a number of unrelated examples of sections of bundles which will be needed later in the book.

In 1.20 and 1.21 we saw that certain real vector bundles can be given a complex vector bundle structure. This turns out to be possible if and only if the endomorphism bundle admits a section of a special type.

Proposition A real vector bundle E over M admits a complex vector bundle structure if and only if there exists a section J of the endomorphism bundle End E such that $J^2 = -\text{id}$.

PROOF $\Rightarrow$: Let E be the realification of a complex vector bundle E_1, that is, E is the real vector bundle underlying E_1, and, in particular, equals E_1 as a fiber bundle over M. For all $\xi \in E$, define $J\xi := i\xi$, where the product of $i = \sqrt{-1} \in \mathbb{C}$ and $\xi \in E$ is defined in the complex bundle E_1, and then $i\xi$ is considered as a point in E. If $X \in \Gamma E$, then as a map from M to E_1, it is C^∞; therefore iX is C^∞, which means that JX is also C^∞. Since $i^2 = -1$, $J^2 = -\text{id}$.

$\Leftarrow$: Given $J \in \Gamma$ End E such that $J^2 = -\text{id}$, define $(a + bi)\xi := a\xi + bJ\xi$, $a, b \in \mathbb{R}$, $\xi \in E$. This defines a complex vector space structure on each fiber of E.

If X is a nonvanishing section of E over an open set U in M, then so is JX, and for all $p \in U$ the vectors X_p and JX_p are linearly independent. If $Y \in \Gamma E$ over U is linearly independent of X and JX at each $p \in U$, then so is JY by the identity $J^2 = -\text{id}$. Proceeding inductively we obtain a local basis field $\{X_1, JX_1, \ldots, X_m, JX_m\}$ for E over an open set $V \subset U \subset M$. Define $\psi: \pi^{-1}U \to \mathbb{C}^m$ by $\psi \sum (f^j X_j + g^j JX_j) := (f^1 + ig^1, \ldots, f^m + ig^m)$; the map (π, ψ) is a complex vector bundle chart on E over V. *Exercise:* Check the transition functions between two such bundle charts.

For example, the Whitney sum $E \oplus E$ of a real vector bundle E with itself always admits a complex structure—define $J(X, Y) := (-Y, X)$ for $X, Y \in \Gamma E$; this should be compared with the map J_o in 1.20.

1.58 **Definition** A field J of endomorphisms of a real vector bundle E over a manifold M is called a *complex structure on E* if $J^2 = -\text{id}$. A complex structure on TM is called an *almost complex structure on M* (for reasons which will appear in chap. 8).

1.59 **Definition** Let $E \xrightarrow{\pi} M$ be a fiber bundle, and let $f: N \to M$ be C^∞. A *section of E along the map f* is a C^∞ map σ from N to E such that $\pi \circ \sigma = f$. The map σ is also called a *lift of f to E*. The set of sections of E along f will be denoted by $\Gamma_f E$. An element of $\Gamma_f TM$ is called a *vector field along f*.

$$\begin{array}{ccc} & & E \\ & \overset{\sigma}{\nearrow} & \downarrow \pi \\ N & \xrightarrow{f} & M \end{array}$$

1.60 **Proposition** Let E be a fiber bundle over M, and let f be a C^∞ map from N to M. There is a canonical one-to-one correspondence between Γf^*E and $\Gamma_f E$; if E is a vector bundle, then this correspondence is an isomorphism of modules over $C^\infty N$.

PROOF Let $s \in \Gamma f^*E$. For each $p \in N$, s_p belongs to the fiber in f^*E over p, and is therefore of the form (p, ξ), where ξ is a point in $E_{f(p)}$.

The C^∞ map $\sigma := \mathrm{pr}_2 \circ s$ from N to E has the property that $\pi \circ \sigma = f$, that is, the diagram

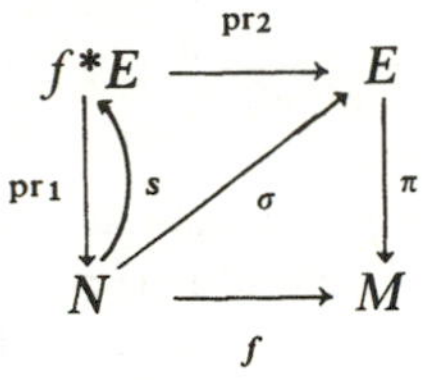

commutes. We now have a map $\Gamma f^*E \to \Gamma_f E$. Its inverse "pulls $\sigma \in \Gamma_f E$ back" to $f^*\sigma := (\mathrm{id}_N, \sigma) \in \Gamma f^*E$; for $p \in N, f^*\sigma_p = (p, \sigma_p)$.

Linearity is obvious in the vector bundle case.

1.61 If s is a section of E along $f\colon N \to M$, and if g is a C^∞ map from a manifold L to N, then $s \circ g$ is a section of E along the map $f \circ g$. In particular, sections of E can be considered as sections along the identity map of M, and each $s \in \Gamma E$ determines a section $s \circ f \in \Gamma_f E$ along each map f into M. The converse is false: if f is constant on a submanifold of N, then there exist sections of E along f which cannot be written in the form $s \circ f$, where $s \in \Gamma E$.

For $s \in \Gamma E$, denote by $f^*s \in \Gamma f^*E$ the pullback of $s \circ f \in \Gamma_f E$.

1.62 **Definition** Let $f\colon N \to M$ be C^∞, and let $X \in \mathfrak{X}N$. Define $f_* X := f_* \circ X \in \Gamma_f TM$; for all $p \in N$, $f_* X_p = f_*|_p(X_p) \in M_{f(p)}$.

1.63 Examples

(*a*) The *tangent field* $\dot{\gamma} = \gamma_* D\colon \mathbb{R} \to TM$ to a C^∞ curve γ in M is a vector field along γ, where $D := d/dt \in \mathfrak{X}\mathbb{R}$.

(*b*) A vector field X on a Lie group G is called *left-invariant* [W: 3.6] if $L_{g*}X = X \circ L_g$ for all $g \in G$, where L_g is the diffeomorphism of G such that $L_g h := gh$, $h \in G$; the *Lie algebra of* G is the Lie algebra $\mathfrak{g}$ of left-invariant vector fields on G [W: 3.8]. Similarly, a *right-invariant vector field on* G is an element $\tilde{X} \in \mathfrak{X}G$ such that $R_{g*}X = X \circ R_g$ for all $g \in G$, where $R_g h := hg$, $h \in G$.

(*c*) Let $\not{p}$ be the projection of a Lie group G onto the homogeneous space G/H, where H is a closed subgroup of G (1.45*d*). Let $\exp\colon \mathfrak{g} \to G$ be the exponential map of G [W: 3.30]. Define a map $\mathfrak{g} \to \mathfrak{X}(G/H)$ such that

$$X^*_{\not{p}(g)} := \frac{d}{dt}\bigg|_0 \not{p}(\exp(tX)g) \qquad X \in \mathfrak{g},\ g \in G$$

that is, $X^*_{\not{p}(g)} := \dot{\gamma}(0)$, where $\gamma(t) := \not{p}(\exp(tX)g) \in G/H$.

For each $X \in \mathfrak{g}$, define a right-invariant field $\tilde{X}$ on G by $\tilde{X}_g := R_{g*} X_e$, where $R_g h := hg$, $h \in G$. For all $g \in G$,

$$\wp_* \tilde{X}_g = \wp_* R_{g*} X_e = \frac{d}{dt}\bigg|_0 \wp \circ R_g \circ \exp(tX) = X^*_{\wp(g)}$$

so $\wp_* \tilde{X} = X^* \circ \wp$. Thus the right-invariant field $\tilde{X}$ on G is $\wp$-related to the field X^* on G/H determined by the left-invariant field X on G. *Exercises:* Let Ad: $G \to GL(\mathfrak{g})$ be the adjoint representation of G on its Lie algebra [W: 3.46]; prove that $g_* X^* = (\mathrm{Ad}_g X)^* \circ g$ for all $g \in G$ and $X \in \mathfrak{g}$, where the left translation g: $G/H \to G/H$ is defined by $g\wp(h) := \wp(gh)$, $h \in G$ [W: 3.63]. Prove that the map $\tilde{X} \mapsto X^*$ is a Lie algebra homomorphism from the Lie algebra of right-invariant fields on G into $\mathfrak{X}(G/H)$.

More generally, let G act on a manifold N on the left. For each $X \in \mathfrak{g}$ define $X^* \in \mathfrak{X}N$ by $X^*_p := d/dt|_0 \exp(tX) \cdot p$, $p \in N$. Prove that $\tilde{X} \mapsto X^*$ is a Lie algebra homomorphism from $\tilde{\mathfrak{g}}$ into $\mathfrak{X}N$.

1.64 **Definition** Let $E \xrightarrow{\pi} M$ be a vector bundle. The *vertical lift* of $X \in \Gamma E$ is the section $\mathscr{I}X \in \Gamma\mathscr{V}E \subset \Gamma TE = \mathfrak{X}E$ defined by

$$(\mathscr{I}X)_\xi := \mathscr{I}_\xi(X_{\pi(\xi)}) \qquad \xi \in E$$

Here $\mathscr{V}E$ is the vertical bundle defined in 1.25, and $\mathscr{I}$ is the isomorphism from π^*E to $\mathscr{V}E$ defined in 1.28.

On each fiber of E, $\mathscr{I}X$ is a constant vector field, that is, $\mathrm{pr}_2 \circ \mathscr{I}X|_{E_p}$: $E_p \to E_p$ is the constant X_p.

1.65 **Definition** A real vector bundle E over M of rank m is *orientable* if the bundle $\Lambda^m E$ admits a nonvanishing section [W: 4.1]. An *orientation* of an orientable vector bundle E is an equivalence class of nonvanishing sections of $\Lambda^m E$, where sections ω and μ are equivalent if and only if $\omega = f \cdot \mu$ for some positive $f \in C^\infty M$; this makes sense since $\Lambda^m E$ is a line bundle over M. An *oriented bundle* is an orientable bundle E together with an orientation for E; a basis $\{\eta_j\}$ for E_p is positively oriented if $\eta_1 \wedge \cdots \wedge \eta_m = \omega_p$ for some ω in the orientation of E.

A bundle is orientable if and only if its dual bundle is orientable; we will always assume that the orientations on a bundle and its dual are compatible in the sense that the basis dual to a positively oriented basis is also positively oriented. If E is an orientable bundle over a manifold with k components, then there are 2^k orientations for E.

1.66 **Definition** A manifold M is *orientable* if TM is orientable, and an orientation for TM is called an *orientation for* M. A *volume element on an oriented manifold* M is a member of the orientation for T^*M, that is, a volume element on M is a nonvanishing section $\omega \in \Gamma\Lambda^n T^*M = A^n M$ such that for each $p \in M$, $\omega_p = \omega^1 \wedge \cdots \wedge \omega^n$, where $\{\omega^i\}$ is some positively oriented basis for M_p^*.

Exercise The manifold $\mathbb{R}^n - \{0\}$ is orientable, and the map $u \mapsto u/\|u\|^2$ (1.12*j*) is orientation-reversing.

1.67 **Example** If E is a complex vector bundle over M, then the realification of E is an orientable bundle, and has a canonical orientation; if $\{\epsilon_j\}$ is a basis for E_p as a complex vector space, then $\{\epsilon_1, \ldots, \epsilon_m, J\epsilon_1, \ldots, J\epsilon_m\}$ is a basis for E_p as a real vector space. Orient E_p so that this basis is positively oriented; the result is independent of the choice of basis $\{\epsilon_j\}$ for E_p, so it extends to an orientation for E.

CHAPTER

TWO

CONNECTION THEORY FOR VECTOR BUNDLES

Let p and q be points in Euclidean space $\mathbb{R}^n$. It is classically said that tangent vectors $u \in \mathbb{R}^n_p$ and $v \in \mathbb{R}^n_q$ are *parallel* if u, v, and the line segment between p and q form three sides of a parallelogram. This happens if and only if there exists a constant vector field X on $\mathbb{R}^n$ such that $X_p = u$ and $X_q = v$. Here, "X is constant" means that $\mathrm{pr}_2 \circ X$ is constant, where pr_2 is the second factor projection of $T\mathbb{R}^n \cong \mathbb{R}^n \times \mathbb{R}^n$ onto $\mathbb{R}^n$ as in 1.1; in other words, X must be a constant-coefficient linear combination of the standard coordinate basis fields $D_j = \partial/\partial u^j \in \mathfrak{X}\mathbb{R}^n$. In this case X_p is parallel to X_0 for all $p \in \mathbb{R}^n$, so we say that X is a *parallel vector field on* $\mathbb{R}^n$.

A vector field X on $\mathbb{R}^n$ is parallel if and only if for each tangent vector v on $\mathbb{R}^n$, the usual *vector derivative* $v(X) := dX(v) = \mathrm{pr}_2\, X_* v$ of X with respect to v is zero.

If u and v are tangent to a manifold M at distinct points p and q, respectively, then direct comparison of u and v as above is meaningless unless M is an open subset of $\mathbb{R}^n$. The development of some sort of a connection between distinct tangent spaces on M is the fundamental problem in the definition of a geometric structure on M.

Several indirect procedures are possible. For example, the ordinary vector derivative on $\mathbb{R}^n$ could be generalized to the concept of a *covariant derivative on* M; the covariant derivative of a vector field $X \in \mathfrak{X}M$ on M with respect to a tangent vector $v \in M_p$ is a tangent vector $\nabla_v X$ on M at p. The covariant derivative operator ∇ satisfies the rules

$$\nabla_{au+v}X = a\nabla_u X + \nabla_v X \qquad a \in \mathbb{R},\ u, v \in M_p$$

$$\nabla_u(fX + Y) = u(f)X_p + f(p)\nabla_u X + \nabla_u Y \qquad f \in C^\infty M$$

Furthermore, $\nabla_u X$ depends differentiably on the tangent vector u. A vector field X on M is then defined to be *parallel* if $\nabla_u X = 0$ for all $u \in TM$. Covariant derivative operators will be developed in 2.52 and will be used very often throughout the book.

Another approach uses an equivalent formulation in terms of the frame bundle $LM = B(TM)$ of the tangent bundle of M; this will be treated in chap. 9. An alternative viewpoint due to Ehresmann [Eh] will be introduced in 2.26 and used often thereafter.

For now we will look directly at parallelism from an axiomatic point of view. The other formalisms will be derived as they are needed.

PARALLELISM STRUCTURES IN VECTOR BUNDLES

2.1 Suppose TM is a trivial bundle, $TM \cong M \times \mathbb{R}^n$; choose a global moving frame $\{X_1, \ldots, X_n \in \mathfrak{X}M\}$ on M, that is (1.52*f*), a family of vector fields on M such that for each $p \in M$, $\{X_j|_p\}$ is a basis for M_p. Every vector field $Z \in \mathfrak{X}M$ is a unique global linear combination $Z = \sum_j Z^j X_j$ of the X_j, $Z^j \in C^\infty M$.

Definition A manifold M is *parallelizable* if TM is trivial, and a global moving frame $\{X_j\}$ is called a *parallelization of* M. A vector field $Z = \sum Z^j X_j \in \mathfrak{X}M$ is *parallel with respect to* $\{X_j\}$ if the coefficients $Z^j \in C^\infty M$ are constant. Tangent vectors $u \in M_p$ and $v \in M_q$ are *parallel with respect to* $\{X_j\}$ if there is a parallel field $Z \in \mathfrak{X}M$ such that $Z_p = u$ and $Z_q = v$.

For example, a Lie group G is parallelizable, for the map $G \times \mathfrak{g} \to TG$ which sends (g, X) to X_g is a vector bundle isomorphism [by definition (1.63*b*), each $X \in \mathfrak{g} \subset \mathfrak{X}G$ is left-invariant, so $X_g = L_{g*}X_e$ is nonzero if X_e is nonzero]. The induced parallelization of G makes each $X \in \mathfrak{g}$ parallel.

Suppose M is parallelizable, and let $\{X_j\}$ be a parallelization of M. For all $p, q \in M$, the map $M_p \to M_q$ which sends $v \in M_p$ to the vector at q parallel to v is a vector space isomorphism. Furthermore, parallelism of vectors in TM is an equivalence relation.

Now let $\{Y_j\}$ be another global moving frame on M; the parallel vector fields on M determined by $\{Y_j\}$ are those determined by $\{X_j\}$ if and only if each Y_j is parallel with respect to $\{X_j\}$. *Proof:* If $Z \in \mathfrak{X}M$ is parallel with respect to $\{Y_j\}$, then $Z = \sum c^j Y_j$, $c^j \in \mathbb{R}$; but for each j, $Y_j = \sum_i A^i_j X_i$, with $A^i_j \in C^\infty M$, so $Z = \sum_{i,j} A^i_j c^j X_i$. If all such Z are $\{X_j\}$-parallel, then each function $\sum_j A^i_j c^j$ is constant; since the c^j are constant, so are the A^i_j. Thus each Y_j is $\{X_j\}$-parallel. The converse is obvious. *QED.*

Thus a parallelizable manifold has no unique parallelism structure. For example, let $\tilde{\mathfrak{g}}$ be the Lie algebra of right-invariant vector fields on a Lie group G (1.63*b*): given $X \in \mathfrak{X}G$, $X \in \tilde{\mathfrak{g}}$ if and only if $R_{g*}X = X \circ R_g$ for all $g \in G$, where $R_g h := hg$, $h \in G$. The tangent bundle isomorphisms $TG \cong G \times \mathfrak{g}$ and $TG \cong G \times \tilde{\mathfrak{g}}$ induce the same parallel vector fields on G if and only if each left-invariant field on G is right-invariant; this happens for a connected group G if and only if G is abelian.

2.2 A serious drawback of our definition of parallelism on a manifold is that most manifolds—for instance, even-dimensional spheres—have a nontrivial tangent bundle. To generalize the idea meaningfully it is expedient to work with arbitrary vector bundles, rather than with just the tangent bundle.

Definition If E is a trivial vector bundle over M, and if $\{X_j \in \Gamma E\}$ is a global basis field for E, that is (1.52*f*), if $\{X_j|_p\}$ is a basis for E_p for all $p \in M$, then $Y \in \Gamma E$ is *parallel with respect to* $\{X_j\}$, if Y is a constant-coefficient linear combination of the X_j.

2.3 If E is a vector bundle over M and if f: $N \to M$ is C^∞, then each $X \in \Gamma E$ determines a section of f^*E (1.61) by $(f^*X)_p = (p, X \circ f(p))$, $p \in N$. If E is trivial, then f^*E is also trivial, and a global basis field $\{X_j\}$ for E pulls back to a global basis field $\{f^*X_j\}$ for f^*E. Given a parallel section $Y = \sum c^j X_j$ of E, the pullback of Y is the constant-coefficient linear combination $f^*Y = \sum c^j f^*X_j$ of the basis fields on N.

Thus each parallelism structure for a trivial bundle E over M induces a parallelism structure in the pullback of E by each C^∞ map f into M: $f^*Y \in \Gamma f^*E$ is defined to be parallel for each parallel $Y \in \Gamma E$.

Suppose that f: $N \to M$ misses a nonempty open set U in M. There exist nonparallel sections of E whose pullbacks are parallel sections of f^*E: if $X \in \Gamma E$ is parallel, change X inside U to get a new section X' of E which is not parallel. Since $X' = X$ on $f(N)$, $f^*X' = f^*X$, so f^*X' is parallel even though X' is not. This simple idea is the key to parallelism in nontrivial vector bundles.

To generalize the concept of parallelism it suffices to find a suitable class of C^∞ maps f into M such that f^*E is trivial for all vector bundles E over M, for then we can always find a global basis field for f^*E and apply definition 2.2 to sections of f^*E.

2.4 Let γ be a C^∞ curve in M; the restriction of a bundle E to the image of γ in M may be nontrivial; in fact it is not even a bundle if the image of γ is not a submanifold of M. The situation is entirely different for the pullback of E by γ.

Proposition Let γ be a C^∞ curve in M, and let E be a fiber bundle over M with structure group G. The pullback bundle γ^*E is trivial.

PROOF It suffices to prove that a bundle over an interval in the real line is trivial. Let Q be a bundle over an interval I, and assume that (π, φ) and (π, ψ) are bundle charts on Q over open subintervals (a, c) and (b, d), respectively; without loss of generality we may assume that $a < b < c < d$. The transition function $f_{\varphi, \psi}$ maps the interval (b, c) differentiably into G. Choose $e \in (b, c)$. Extend $f_{\varphi, \psi}|_{(b, e)}$ to a C^∞ curve $g: (b, d) \to G$. Define η from $\pi^{-1}(a, d)$ into the fiber F of E by $\eta|_{\pi^{-1}(p)} := \varphi|_{\pi^{-1}(p)}$ for $p \in (a, e)$, and $\eta|_{\pi^{-1}(p)} := g(p) \cdot \psi|_{\pi^{-1}(p)}$ for $p \in (b, d)$. The equality of g and $f_{\varphi, \psi}$ on (b, e) implies that η is well-defined and C^∞. The C^∞ map (π, η) is a bundle chart on Q over (a, d).

Since every compact interval in I can be covered by finitely many open intervals over which Q is trivial, iteration of the above procedure yields a trivialization of Q over every compact subinterval of I. By writing I as the union of a sequence of increasing compact subintervals, we then obtain triviality over I.

Exercise Prove that a fiber bundle over any contractible manifold is trivial.

2.5 **Definition** Let E be a vector bundle over M, and let γ be a C^∞ curve in M. By proposition 2.4, γ^*E admits a global basis field, and by 2.2, certain sections of γ^*E are then defined to be parallel with respect to this basis field. Call a section X of E along γ (1.59) *parallel in E along* γ if the pullback section γ^*X is parallel in γ^*E. The induced isomorphism $E_{\gamma(s)} \to E_{\gamma(t)}$ which sends $X(s)$ to $X(t)$ for each parallel-section X in E along γ is called *parallel transport* (or *parallel translation*) *in E along* γ.

2.6 Let E be a vector bundle over M. By 2.4 and 2.5, parallel transport can be defined along each C^∞ curve in M; this gives us the following axioms.

(1) Existence: For each $\xi \in E$ and each C^∞ curve $\gamma\colon [a, b] \to M$ with $\gamma(a) = \pi(\xi)$, there is a unique parallel field $\mathbb{P}_\gamma \xi$ in E along γ with initial value ξ: $\mathbb{P}_\gamma \xi \in \Gamma_\gamma E$ is parallel in E along γ, and $\mathbb{P}_\gamma \xi(a) = \xi$.

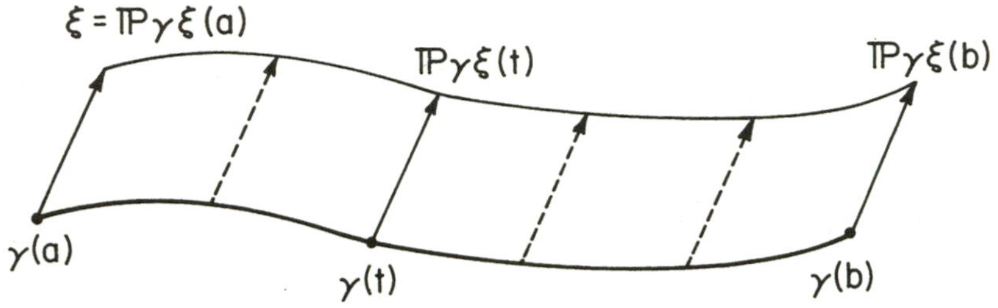

γ is a C^∞ curve in M, and $\mathbb{P}_\gamma \xi$ is a C^∞ curve in E

(2) Linearity and invertibility: Given a C^∞ curve $\gamma\colon [a, b] \to M$, the parallel transport map $E_{\gamma(a)} \to E_{\gamma(b)}$ along γ which sends $\xi = \mathbb{P}_\gamma \xi(a)$ to its parallel translate $\mathbb{P}_\gamma \xi(b)$ is a vector space isomorphism; its inverse is parallel transport along the so-called *reverse curve* γ^- defined by $\gamma^-(t) := \gamma(a - t + b)$, $a \le t \le b$.

For parallel transport to be a reasonable geometric concept, parallel transports along related curves must be meaningfully related; this is the content of the next three axioms.

(3) Parametrization independence: Let $\gamma\colon [a, b] \to M$ be a C^∞ curve, and let $\psi\colon [c, d] \to [a, b]$ be a C^∞ function such that $\psi(c) = a$ and $\psi(d) = b$; the images in M of the curves γ and $\alpha := \gamma \circ \psi$ are identical. Since $\mathbb{P}_\alpha \xi$ and $\mathbb{P}_\gamma \xi$ are lifts to E (1.59) of the curves α and γ, respectively, $\pi \circ \mathbb{P}_\alpha \xi(t) = \alpha(t) = \gamma \circ \psi(t) = \pi \circ (\mathbb{P}_\gamma \xi) \circ \psi(t)$ for all $\xi \in E_{\alpha(c)} = E_{\gamma(a)}$ and $t \in [c, d]$. We require that $\mathbb{P}_\alpha \xi(t) = (\mathbb{P}_\gamma \xi) \circ \psi(t)$ for all ξ and t.

(4) C^∞ dependence on initial conditions: For each open set U in M and each C^∞ map $f\colon TU \to M$ such that $f(0_p) = p$, $p \in U$, the following map is required to be C^∞:

$$\tilde{f}\colon TU \oplus \pi^{-1} U \to E$$

$$(v, \xi) \mapsto \mathbb{P}_\gamma \xi(1) \qquad \gamma(t) := f(tv)$$

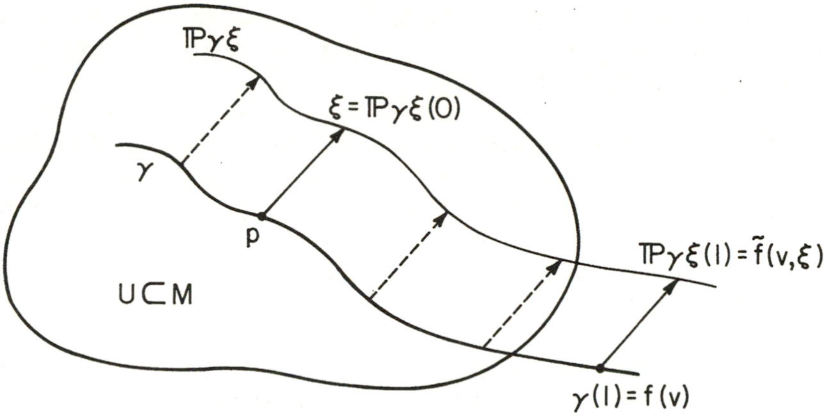

$\gamma(t) = f(tv)$ for some $v \in M_p$

(5) Initial uniqueness: Fix a point $p \in M$. Let β and γ be C^∞ curves in M emanating from $p = \beta(0) = \gamma(0)$ such that $\dot{\beta}(0) = \dot{\gamma}(0)$. For each $\xi \in E_p$, the parallel sections $\mathbb{P}_\beta \xi$ and $\mathbb{P}_\gamma \xi$ are C^∞ curves in E for which the initial tangent vectors are tangent to E at ξ. We require equality of the initial tangent vectors:

$$\dot{\widehat{\mathbb{P}_\beta \xi}}(0) = \dot{\widehat{\mathbb{P}_\gamma \xi}}(0)$$

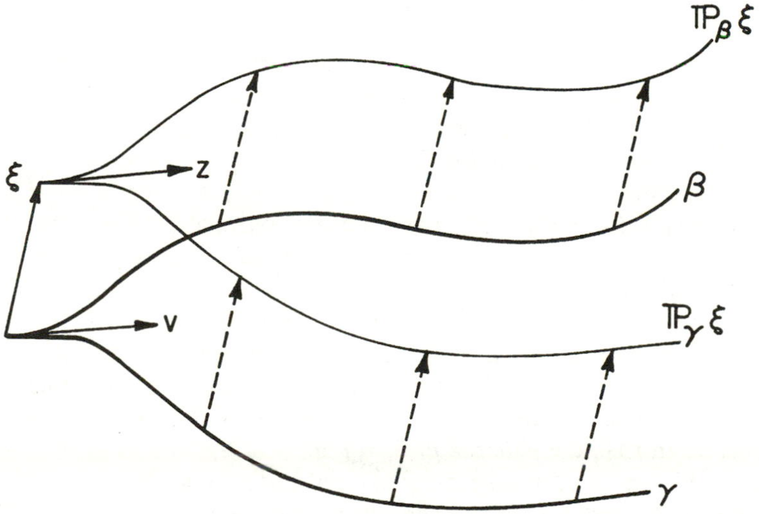

$\dot{\beta}(0) = v = \dot{\gamma}(0)$ and $\dot{\widehat{\mathbb{P}_\beta \xi}}(0) = z = \dot{\widehat{\mathbb{P}_\gamma \xi}}(0)$

An important example of the map in axiom (4) comes from the exponential map $\mathfrak{g} \to G$ of a Lie group G; let $f\colon TG \to G$ such that $f(X_g) := g \cdot \exp(X)$, $X \in \mathfrak{g}$, $g \in G$.

2.7 **Definition** Let E be a vector bundle over M. A system $\mathbb{P}$ of C^∞ lifts to E of the C^∞ curves in M will be called a *parallelism structure* (or a *system of parallel transport*) *in* E if axioms (1) to (5) are satisfied. A section X of E is *parallel* if $X \circ \gamma$ is parallel in E along γ for each C^∞ curve γ in M. The parallel section in E along γ with initial value ξ will be denoted by $\mathbb{P}_\gamma \xi$.

An axiomatic definition of parallel transport in TM was first given by W. Rinow in his lectures at Humboldt Universität in 1949; P. Dombrowski [Do2] defined parallel transport axiomatically for arbitrary vector bundles. These definitions differ formally from the definition given above in various ways; in general, Dombrowski's definition is equivalent to the one above, and both definitions are equivalent to Rinow's in the tangent bundle case.

2.8 **Examples**

(*a*) The *canonical trivial parallelism in* $M \times \mathbb{F}^m \to M$ is defined by letting each constant section of $C^\infty(M, \mathbb{F}^m) \cong \Gamma(M \times \mathbb{F}^m)$ be parallel. If E is trivial over M, then each bundle isomorphism $E \cong M \times \mathbb{F}^m$ induces a trivial parallelism structure in E. For instance, the *canonical left-invariant parallelism structure on a Lie group* G is the parallelism structure in TG for which the elements of $\mathfrak{g}$ are parallel, while the *canonical right-invariant parallelism structure on* G is the one in TG for which the elements of $\tilde{\mathfrak{g}}$ (2.1) are parallel. (The term *invariance* will be explained in 2.9.)

(*b*) If $\mathbb{P}_j$ is a parallelism structure in a vector bundle E_j over M_j, $j = 1$, 2, then $E_1 \times E_2 \to M_1 \times M_2$ has a natural parallelism structure: if X_j is parallel along γ_j in E_j, $j = 1, 2$, then (X_1, X_2) is parallel in $E_1 \times E_2$ along (γ_1, γ_2).

(*c*) Let $\mathbb{P}$ be a system of parallel transport in E, and let $f\colon N \to M$ be C^∞. For each $(p, \xi) \in f^*E$ and each C^∞ curve γ in N with $\gamma(a) = p$, the curve $(\gamma, \mathbb{P}_{f \circ \gamma}\xi)$ in f^*E is a section of f^*E along γ. The collection of all such lifts to f^*E of the C^∞ curves in N is a parallelism structure in f^*E, the *pullback of* $\mathbb{P}$ *to* f^*E.

(*d*) Let $\Delta\colon M \to M \times M$ be the diagonal embedding map $p \mapsto (p, p)$. By 1.29 and (*b*) and (*c*), parallelism structures in vector bundles E_j over M induce parallel transport in the Whitney sum $E_1 \oplus E_2 = \Delta^*(E_1 \times E_2)$: if X_j is parallel in E_j over γ in M, $j = 1, 2$, then $(X_1, X_2) \in \Gamma_\gamma(E_1 \oplus E_2)$ is parallel in $E_1 \oplus E_2$ along γ.

(*e*) ***Parallel translation on the standard sphere*** $S^n \subset \mathbb{R}^{n+1}$. Let $\langle u, v\rangle$ be the usual inner product of u and v in $\mathbb{R}^{n+1}$; write $u \perp v$ if $\langle u, v\rangle = 0$. The tangent bundle of the n-sphere $S^n = \{p \in \mathbb{R}^{n+1} \mid \langle p, p\rangle = 1\}$ is the submanifold $\{(p, v) \in \mathbb{R}^{n+1} \times \mathbb{R}^{n+1} \mid p \in S^n \text{ and } v \perp p\}$ of $\mathbb{R}^{2n+2}$, where $\langle p, v\rangle$ is defined by first identifying the two copies of $\mathbb{R}^{n+1}$.

Fix $\lambda \in \mathbb{R}$, and $p, q \in S^2$ such that $p \perp q$; set $\gamma(t) := \cos(\lambda t)p + \sin(\lambda t)q$. Denote by $R_t \in SO(3)$ the rotation of $\mathbb{R}^3$ such that $R_t p = \gamma(t)$. For each $v \in S^2_p$, define $\mathbb{P}_\gamma v$ to be the vector field along γ such that $\mathbb{P}_\gamma v(t) := R_{t*} v \in S^2_{\gamma(t)}$. The fields $\mathbb{P}_\gamma v$ are taken as the parallel fields along the constant-speed parametrizations γ of the great circles in S^2. For example, the tangent field $\dot\gamma$ is parallel along each such γ. In 2.42 we shall see how to extend this parallel transport uniquely to all C^∞ curves in S^2.

Now let $p \in S^n$, and let $u, v \in S^n_p$ be nonzero. Denote by C_u and C_v the great circles in S^n tangent to u and v, respectively. Let M be a great 2-sphere in S^n containing C_u and C_v; M is unique unless $C_u = C_v$. Define parallel translation of v along C_u in S^n by parallel translating v along C_u in the 2-sphere M. This now defines parallel translation along all the great circles in S^n; as with S^2, there is a unique extension to all the C^∞ curves in S^n. A parallel field (γ, X) has the properties that X has constant length, and if γ is a great circle, then the angle between X and γ' is constant, where the tangent field of γ is $\dot\gamma = (\gamma, \gamma')$ as a curve in $TS^n \subset \mathbb{R}^{n+1} \times \mathbb{R}^{n+1}$.

2.9 The parallelism structure on S^n is invariant under the natural action of $O(n+1)$ on S^n [W: 3.65(a)]: orthogonal mappings of $\mathbb{R}^{n+1}$ send great circles in S^n to great circles, and if X is parallel along a constant-speed parametrization γ of a great circle in S^n, then for all $g \in O(n+1)$, $g_* X$ is parallel along $g \circ \gamma$. Similarly, the parallelism structure on a Lie group G (2.8*a*) such that elements of $\mathfrak{g}$ are parallel is invariant under left translation; if each $X \in \tilde{\mathfrak{g}}$ is parallel, this structure is invariant under right translation.

Recall from 1.63*b* that if $\not{p}$ is the projection of a Lie group G onto the homogeneous space G/H, where H is a closed subgroup of G, then the natural action of G on G/H is $g \cdot \not{p}(h) = \not{p}(gh)$, $g, h \in G$.

Definition An *invariant parallelism structure on a homogeneous space* G/H is a parallelism structure in $T(G/H)$ such that for each C^∞ curve γ in G/H and for each $g \in G$, X parallel along γ implies that $g_* X$ is parallel along $g \circ \gamma$.

2.10 By 1.52*f*, a vector bundle chart $(\pi, \psi)\colon \pi^{-1}U \to U \times \mathscr{V}$ on a vector bundle E over M and a basis $\{\epsilon_j\}$ for $\mathscr{V}$ determine basis fields $\{\Psi_j\}$ for E over U by $\Psi_j|_p := (\pi, \psi)^{-1}(p, \epsilon_j)$, $p \in U$.

Proposition Let $\mathbb{P}$ be a parallelism structure in a vector bundle $\pi: E \to M$. Fix $p \in M$; choose a basis $\{\epsilon_j\}$ for E_p and a diffeomorphism h from a starlike neighborhood V of 0_p in M_p to an open set U in M such that $h(0_p) = p$. There exists a bundle chart $(\pi, \psi): \pi^{-1}U \to U \times E_p$ such that the basis fields Ψ_j determined by the basis $\{\epsilon_j\}$ for E_p are parallel along each curve of the form $h(tv)$, where $v \in V$ and $0 \leq t \leq 1$.

PROOF Let $\mathrm{pr}_2: TV \cong V \times M_p \to M_p$ be the second factor projection. Extend h to a C^∞ map $f: TU \to M$ such that $f(w) := h(v + \mathrm{pr}_2 \circ h_*|_v^{-1} w)$ if $w \in M_{h(v)}$, $v \in V$; what f does to w not of this form is irrelevant, so long as f is C^∞. For all $q \in U$, $f(0_q) = h(h^{-1}q + 0_p) = q$, so by axiom 2.6(4), the map $\tilde{f}: TU \oplus \pi^{-1}U \to E$ is C^∞. Define $\psi: \pi^{-1}U \to E_p$ by

$$\psi(\xi) := \tilde{f}\left(\frac{d}{dt}\bigg|_0 h(v - tv), \xi\right) = \tilde{f}(-h_* \mathscr{J}_v v, \xi)$$

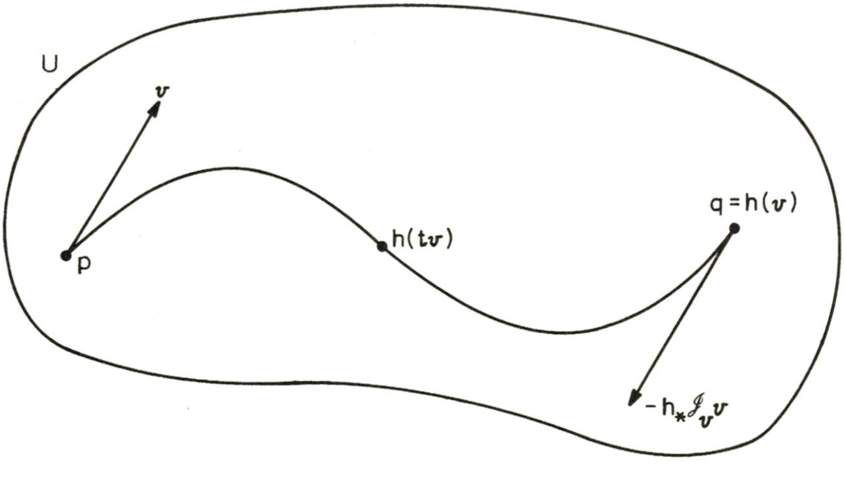

if $\pi(\xi) = h(v)$, $v \in V$, where $\mathscr{J}$ is the map from $TM \oplus TM$ to the vertical bundle $\mathscr{V}TM$ defined in 1.28 and 1.30. The map is C^∞ because $\tilde{f}$, h_*, and $\mathscr{J}$ are C^∞; it is fiberwise linear by axiom 2.6(2). The map (π, ψ) is a vector bundle chart on E over U because its inverse is C^∞: $(\pi, \psi)^{-1}(q, \zeta) = \tilde{f}(h^{-1}q, \zeta) \in E_q$, $(q, \zeta) \in U \times E_p$, by axiom 2.6(2).

Finally, $\Psi_j|_{h(v)} = (\pi, \psi)^{-1}(h(tv), \epsilon_j) = \tilde{f}(tv, \epsilon_j)$ for $v \in V$; in particular, $\Psi_j|_p = \epsilon_j$.

The bundle chart ψ and local basis field $\{\Psi_j\}$ determined by a basis for E_p and such a diffeomorphism h will be said to be *radially parallel with respect to h.*

2.11 **Definition** Let $\mathbb{P}$ be a parallelism structure in a vector bundle E over M. A vector subbundle E_1 of E is *invariant under* $\mathbb{P}$ if for every C^∞ curve $\gamma\colon [0, 1] \to M$ and every $\xi \in E_1|_{\gamma(0)}$, the parallel section $\mathbb{P}_\gamma \xi$ lies in E_1. In this case $\mathbb{P}$ is said to be *reducible to* E_1.

Exercise Show that if $\mathbb{P}$ is reducible from E to E_1, then the collection of all lifts of curves in M to parallel sections of E with initial values in E_1 is a parallelism structure in E_1.

2.12 **Proposition** Let $\mathbb{P}$ be a parallelism structure in a vector bundle E over a connected manifold M; if a subset $Q \subset E$ is invariant under parallel transport, and if for some $p \in M$, $Q \cap E_p$ is a linear subspace of E_p, then Q is a vector subbundle of E, and $\mathbb{P}$ is reducible to Q.

PROOF By the $\mathbb{P}$-invariance of Q, $Q \cap E_q$ is a linear subspace of E_q isomorphic to $Q \cap E_p$ for all $q \in M$. Vector subbundle charts on Q are obtained as the restrictions to Q of radially parallel vector bundle charts constructed as in proposition 2.10 for E.

2.13 **Definition** Let $\mathbb{P}$ be a system of parallel transport in a vector bundle $\pi\colon E \to M$. Define *parallel transport in the dual bundle* $\Pi\colon E^* \to M$ so that a lift ω to E^* of a curve γ in M is *parallel along* γ if $\omega(X)$ is constant for every parallel section X of E along γ.

We must check the axioms for parallel transport from 2.6:

(1) Let γ be a curve in M; given $\mu \in E^*_{\gamma(0)}$, define $\mathbb{P}_\gamma \mu \in \Gamma_\gamma E$ so that it is constant on each parallel section of E along γ:

$$(\mathbb{P}_\gamma \mu(t))(\mathbb{P}_\gamma \xi(t)) := \mu(\xi) \qquad \xi \in E_{\gamma(0)}$$

(2, 3) These follow directly from the corresponding axioms in E.

(4) Fix $f\colon TU \to M$ such that $f(0_p) = p$, $p \in U$; let $\tilde{f}$ be the corresponding map from $TU \oplus \pi^{-1}U$ to E, and let $\tilde{\tilde{f}}$ be the corresponding map from $TU \oplus \Pi^{-1}U$ to E^*; by assumption, $\tilde{f}$ is C^∞. Fix $p \in U$ and $v \in M_p$; let $\gamma(t) = f(tv)$. For all $\mu \in E^*_p$ and $\xi \in E_{\gamma(1)}$

$$\tilde{\tilde{f}}(v, \mu)(\xi) = (\mathbb{P}_\gamma \mu(1))(\xi) = \mu \circ \mathbb{P}_{\gamma^-} \xi(1) = \mu \circ \tilde{f}(-\dot{\gamma}(1), \xi)$$

where $\gamma^-(t) = \gamma(1-t)$ as in 2.6(2). Since f and $\tilde{f}$ are C^∞, so is $\tilde{\tilde{f}}$.

(5) Let β and γ be curves in M with $\dot{\beta}(0) = \dot{\gamma}(0)$; let ω and μ be parallel sections of E^* along β and γ respectively, with $\omega(0) = \mu(0)$.

Extend $\xi \in E_{\gamma(0)}$ to parallel sections X and Y of E along β and γ. By definition, $\omega_t(X_t) = \omega_0(\xi) = \mu_0(\xi) = \mu_t(Y_t)$. With respect to associated bundle charts (π, ψ) on E and (Π, φ) on E^* (1.38), $(\psi\omega_t)(\varphi X_t) = \omega_t(X_t)$ is constant, so

$$d\psi[\dot{\omega}(0)](\varphi\xi) + (\psi\omega_0)[d\varphi(\dot{X}(0))] = [(\psi\omega)(\varphi X)]'(0) = 0$$

where the terms on the left use the standard identification of each tangent space on a vector space with the vector space [W: 3.9]. Similarly, $d\psi[\dot{\mu}(0)](\varphi\xi) + (\psi\mu_0)[d\varphi(\dot{Y}(0))] = 0$; since $\omega_0 = \mu_0$ and $\dot{X}(0) = \dot{Y}(0)$ [by axiom (5) for E], $d\psi(\dot{\mu}(0))(\varphi\xi) = d\psi(\dot{\mu}(0))(\varphi\xi)$. This equality for all $\xi \in E_{\gamma(0)}$ implies $d\psi(\dot{\omega}(0)) = d\psi(\dot{\mu}(0))$; since $\Pi_*\dot{\omega}(0) = \dot{\beta}(0) = \dot{\gamma}(0) = \Pi_*\dot{\mu}(0)$, $\dot{\omega}(0) = \dot{\mu}(0)$.

2.14 **Definition** Let $E_1, \ldots, E_k$ be vector bundles over M, each endowed with a parallelism structure. Define *parallel transport in* $E_1 \otimes \cdots \otimes E_k$ so that if X_j is parallel along γ in E_j, $j = 1, \ldots, k$, then $X_1 \otimes \cdots \otimes X_k$ is parallel in $E_1 \otimes \cdots \otimes E_k$ along γ. Define *parallel transport* in $\mathrm{Hom}(E_1, E_2)$ by means of the isomorphism $\mathrm{Hom}(E_1, E_2) \cong E_1^* \otimes E_2$: if a section $\omega \otimes X$ is parallel in $E_1^* \otimes E_2$ along γ, then call it parallel along γ in $\mathrm{Hom}(E_1, E_2)$ also. If $X_1, \ldots, X_k$ are parallel sections of a vector bundle E along a curve γ in M, then define $X_1 \wedge \cdots \wedge X_k$ to be *parallel along* γ *in* $\Lambda(E)$.

Exercises Check the axioms for parallel transport for the systems defined above. Prove that a parallel section L of $\mathrm{Hom}(E_1, E_2)$ along γ has the property that if $X \in \Gamma_\gamma E_1$ is parallel along γ, then LX is parallel along γ in E_2. More generally, if any two of the lifts X, L, and LX of γ (into the respective bundles) are parallel, then so is the third; in this case it makes sense to refer to L as a *parallel bundle homomorphism* (see 1.52*d*). Similarly, given lifts X_j of γ to bundles E_j, $j = 1, \ldots, n$, if any n of $X_1, \ldots, X_n, X_1 \otimes \cdots \otimes X_n$ are parallel along γ, then so is the other.

Prove that the multiplication homomorphism (1.41*c*) in $\mathrm{End}\, E = \mathrm{Hom}(E, E)$ is parallel, and similarly for multiplication in $\mathfrak{gl}(E)$ (see 1.41*d* and the proof of 2.45).

Prove that parallel transport in E^* is the same as that in the canonically isomorphic bundle $\mathrm{Hom}(E, M \times \mathbb{F})$.

2.15 Parallel transport in a complex vector bundle E canonically induces parallel transport in the realification of E (1.20): if X is a parallel section of the complex vector bundle E along a curve γ in M, then X is parallel along γ in E thought of as a real vector bundle. The converse is false; let J be the complex structure of E represented as a field of endomorphisms of the real bundle E such that $J^2 = -\mathrm{id}$ (1.57, 1.58). Let there be given a system of parallel transport in the real vector bundle underlying E. Parallel transport between fibers in E is a complex linear map if and only if $J \in \Gamma$ End E is parallel [recall: End $E = \mathrm{Hom}(E, E)$].

Exercise Given parallel transport in a real vector bundle E over M, describe the induced parallel transport in the complexification $E^{\mathbb{C}} = E \otimes (M \times \mathbb{C})$ (1.22, 1.39*b*).

HOLONOMY IN VECTOR BUNDLES

Let E be a vector bundle over a manifold M. Parallel transport in E between two fibers was defined in terms of a parallel section of E along a C^∞ curve in M joining the basepoints. Different curves joining the basepoints may determine different parallel transport maps between the fibers; after some preparatory work a global measure for the path dependence of parallel transport in E will be defined and studied.

2.16 **Definition** Let β: $[a, b] \to M$ and γ: $[b, c] \to M$ be piecewise C^∞ curves such that $\beta(b) = \gamma(b)$. The *product curve* $\gamma * \beta$: $[a, c] \to M$ is the piecewise C^∞ curve defined by

$$\gamma * \beta(t) := \begin{cases} \beta(t) & a \le t \le b \\ \gamma(t) & b \le t \le c \end{cases}$$

The *reverse curve* γ^- is defined by $\gamma^-(t) := \gamma(b - t + c)$, $b \le t \le c$.

Suppose that a curve β is defined on $[a, b]$, while γ is defined on $[c, d]$; assume $\beta(b) = \gamma(c)$. If $b \neq c$, we cannot define $\gamma * \beta$ directly; however, the product of β and a suitable reparametrization of γ (or vice versa) does make sense. Axiom 2.6(3) guarantees that such reparametrizations have no effect on parallel transport.

2.17 **Definition** Assume that parallel transport has been defined in a vector bundle E over M. Let β: $[a, b] \to M$ and γ: $[b, c] \to M$ be C^∞ curves with $\beta(b) = \gamma(b)$. Define *parallel transport from* $E_{\beta(a)}$ *to* $E_{\gamma(c)}$ *along the piecewise* C^∞ *curve* $\gamma * \beta$ to be parallel transport along β from $E_{\beta(a)}$ to $E_{\beta(b)}$ followed by parallel transport along γ from $E_{\gamma(b)}$ to $E_{\gamma(c)}$. Similarly define parallel transport along the product of piecewise C^∞ curves.

It follows that parallel transport is associative: parallel transport along $\delta * (\gamma * \beta)$ equals parallel transport along $(\delta * \gamma) * \beta$. By axiom

2.6(3) for parallel transport, parallel transport along the reverse curve γ^- is the inverse of parallel transport along γ.

2.18 **Definition** Let $\mathbb{P}$ be a parallel transport system in a vector bundle E over M; fix $p \in M$. The *holonomy group of* $\mathbb{P}$ *at* p is the subgroup $G(p)$ of $GL(E_p)$ consisting of the parallel transport maps $E_p \to E_p$ around all piecewise C^∞ loops based at p; the group operation is composition along product paths.

2.19 **Proposition** Let $\mathbb{P}$ be parallel transport in a vector bundle E over a connected manifold M. Fix $p, q \in M$ and a piecewise C^∞ curve α in M joining p to q. The map from $G(p)$ to $G(q)$ which sends parallel transport around a loop γ at p to parallel transport around $\alpha * \gamma * \alpha^-$ is an isomorphism. Thus, up to isomorphism, the *holonomy group of* $\mathbb{P}$ is well-defined.

PROOF The given map is a homomorphism because parallel transport around $\alpha * (\gamma * \beta) * \alpha^-$ equals parallel transport around $(\alpha * \gamma * \alpha^-) * (\alpha * \beta * \alpha^-)$ for piecewise C^∞ loops β and γ at p. The inverse homomorphism sends parallel transport around a loop η at q to parallel transport around $\alpha^- * \eta * \alpha$.

2.20 **Proposition** Let $\mathbb{P}$ be a parallelism structure in a vector bundle E over a connected manifold M. The holonomy group G of $\mathbb{P}$ is trivial if and only if E is a trivial vector bundle and $\mathbb{P}$ is a trivial parallelism structure in E (2.8*a*).

PROOF $\Rightarrow$: Fix $p \in M$. Define $\psi: E \to E_p$ by $\xi \mapsto \mathbb{P}_\gamma \xi(1)$, where γ is a C^∞ curve in M such that $\gamma(0) = \pi(\xi)$ and $\gamma(1) = p$; ψ is well-defined because G is trivial, and is C^∞ by axiom 2.6(4) as applied in the proof of 2.10. The map (π, ψ) is then a parallel vector bundle isomorphism from E to $M \times E_p$ (1.52*d*, 2.14).

$\Leftarrow$: Every parallel field along a curve γ can be written in the form $X \circ \gamma$, where X is a globally parallel section of E; thus the parallel translate of $X \circ \gamma(0)$ around a loop γ is $X \circ \gamma(1) = X \circ \gamma(0)$.

2.21 **Examples**

(*a*) Both the canonical left- and right-invariant parallelisms on a connected Lie group G (2.8*a*) have trivial holonomy groups.

(*b*) It will be proved in 2.42 that the holonomy group of S^n is $SO(n)$.

2.22 **Definition** Let $\mathbb{P}$ be a system of parallel transport in a vector bundle E over M; fix $p \in M$. The *restricted holonomy group of* $\mathbb{P}$ *at* p is the subgroup $G_o(p)$ of the holonomy group $G(p)$ consisting of parallel transports around all null-homotopic piecewise C^∞ loops based at p.

2.23 **Lemma** Each homotopy class $c \in \pi_1(M, p)$ can be represented by a piecewise C^∞ loop (C^∞ is possible, but unnecessary for our purposes).

PROOF Let β: $[0, 1] \to M$ be a continuous loop representing the homotopy class $c \in \pi_1(M, p)$. By the compactness of the image of β in M, it can be covered by finitely many simply connected, connected open sets $U_1, \ldots, U_r$; let x_j: $U_j \to \mathbb{R}^n$ be a chart on U_j. Without loss of generality we may assume that the set $x_j(U_j)$ is a starlike neighborhood of the point $x_j \circ \beta(t_j)$ in $\mathbb{R}^n$, where $0 = t_o < t_1 < \cdots < t_r = 1$ are fixed numbers such that β maps each interval $[t_{j-1}, t_j]$ into U_j. Let l_j be any C^∞ parametrization of the line segment in $x_j(U_j) \subset \mathbb{R}^n$ joining $x_j \circ \beta(t_{j-1})$ to $x_j \circ \beta(t_j)$. The curve $(x_n^{-1} \circ l_n) * (x_{n-1}^{-1} \circ l_{n-1}) * \cdots * (x_1^{-1} \circ l_1)$: $[0, 1] \to M$ is then a piecewise C^∞ curve in M which is homotopic [with fixed endpoints (p, p)] to β, and thus represents $c \in \pi_1(M, p)$.

2.24 **Lemma** In the relative topology in $GL(E_p)$, the restricted holonomy group $G_o(p)$ is a path-connected, normal subgroup of $G(p)$, and there exists a homomorphism from $\pi_1(M, p)$ onto $G(p)/G_o(p)$.

PROOF If F: $[0, 1] \times [0, 1] \to M$ is a homotopy between γ and p such that each loop $f(s) := F(s, \cdot)$ is a piecewise C^∞ curve in M, then $s \mapsto \mathbb{P}_{f(s)}$ is a path in $G_o(p)$ from $\mathbb{P}_\gamma$ to $\{e\}$; thus $G_o(p)$ is path-connected. For all $h \in G(p)$, $h \circ \mathbb{P}_f \circ h^{-1}$ is a path from $h \circ \mathbb{P}_\gamma \circ h^{-1}$ to $\{e\}$, so $h \circ \mathbb{P}_\gamma \circ h^{-1} \in G_o(p)$, and $G_o(p)$ is normal in $G(p)$.

If $c \in \pi_1(M, p)$ is represented by a piecewise C^∞ loop γ, map c to the equivalence class of $\mathbb{P}_\gamma$. This map is clearly well-defined, and is a surjective homomorphism.

2.25 **Theorem** The holonomy group at $p \in M$ of a parallelism structure $\mathbb{P}$ in a vector bundle E over a connected manifold M is a Lie group, and its identity component is the restricted holonomy group at p.

PROOF A theorem of Kuranishi and Yamabe [Yam] says that a path-connected subgroup of a Lie group is a Lie group; thus the path-connected subgroup $G_o(p)$ of $GL(E_p)$ is a Lie group. Since M is connected (and by definition, second countable), its fundamental group is countable [W: 3, exercise 8]. Thus $G(p)/G_o(p)$ is countable by 2.24; by [W: 3.2(c)], $G(p)$ is a Lie group.

CONNECTIONS ON VECTOR BUNDLES

So far little use has been made of the differentiability assumption for parallel transport. In this section parallel transport will be reformulated so that calculus comes into play; in addition, the existence of a parallelism structure in every vector bundle will be clear in this setting.

2.26 **Definition** A *connection on a vector bundle* $E \xrightarrow{\pi} M$ is a vector subbundle $\mathscr{H}$ of $TE \to E$ such that

(i) $\mathscr{H}$ is complementary to the vertical bundle $\mathscr{V}E \to E$:

$$TE = \mathscr{H} \oplus \mathscr{V}E$$

(ii) $\mathscr{H}$ is homogeneous: $\mu_{c*}\mathscr{H}_\xi = \mathscr{H}_{c\xi}$, $c \in \mathbb{F}$, $\xi \in E$, where $\mu_c \xi := c\xi$.

A connection on the tangent bundle $TM \to M$ is usually referred to as a *connection on M*.

The bundles TE, $\mathscr{V}E$, and $\mathscr{H}$ are real vector bundles here, even if E is complex. Condition (i) just says that the restriction to $\mathscr{H}$ of π_*: $TE \to TM$ is an isomorphism along π.

Exercise Show that $(\pi_{TE}, \pi_*)|_{\mathscr{H}}$ is a fiber-preserving diffeomorphism from the composite bundle $\mathscr{H}$ over M to the bundle $E \oplus TM$.

2.27 **Definition** Let $\mathscr{H}$ be a connection on E over M. A section X of E along a map f: $N \to M$ is *horizontal with respect to* $\mathscr{H}$ (or *$\mathscr{H}$-horizontal*) if the subspace $X_* N_p$ is contained in $\mathscr{H}_{X(p)}$ for all $p \in N$; such an X is also called a *horizontal lift of f.*

Two special cases are worth singling out. A section Y of E is horizontal if and only if $Y_* M_p = \mathscr{H}_{Y(p)}$ for all $p \in M$, while a curve Z in E is horizontal if and only if its tangent field $\dot{Z}$ is a curve in $\mathscr{H}$.

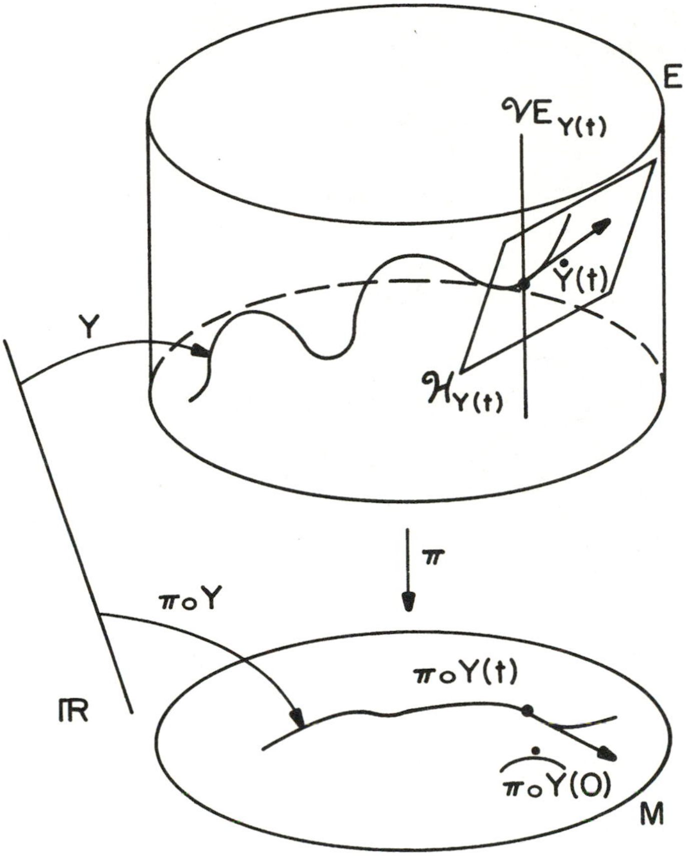

2.28 Theorem A parallelism structure $\mathbb{P}$ in a vector bundle E over M determines a connection $\mathscr{H}$ on E such that a curve X in E is a parallel section of E along the curve $\pi \circ X$ in M if and only if X is horizontal with respect to $\mathscr{H}$.

PROOF For each $\xi \in E$, define $\mathscr{H}_\xi$ to be the set of all tangent vectors at ξ of the form $\dot{\widehat{\mathbb{P}_\gamma \xi}}(0)$, where γ is a C^∞ curve in M emanating from $\pi(\xi)$. Define $\mathscr{H} := \bigcup_{\xi \in E} \mathscr{H}_\xi$.

Fix $p \in M$ and $\xi \in E_p$; the set $\mathscr{H}_\xi$ will be shown to be the image of a linear map from M_p to E_ξ, and therefore a vector space. Let (x, U) be a chart on M near p such that $x(p) = 0 \in \mathbb{R}^n$. Without loss of generality we may assume that $dx|_{M_p}^{-1} \circ x(U)$ is a starlike neighborhood V of 0 in M_p; set $h := x^{-1} \circ dx$ on V, and let $f\colon TU \to M$ be defined as in the proof of 2.10. It follows that for $v \in M_p$,

$$\frac{d}{dt}\bigg|_0 f(tv) = v$$

For each $v \in M_p$, let X_v be the parallel field in E with initial value ξ along the curve $f(tv)$, $0 \le t \le 1$; set $X(v, t) := X_v(t)$, $(v, t) \in M_p \times [0, 1]$. By axiom 2.6(4) for parallel transport, X is C^∞.

Axiom 2.6(3) implies that $X_v(t) = X_{tv}(1) = X(tv, 1)$ for $(v, t) \in M_p \times [0, 1]$; thus

$$\dot{X}_v(0) = \frac{d}{dt}\bigg|_0 X(tv, 1) = X_* \frac{d}{dt}\bigg|_0 (tv, 1) = X_*(\mathscr{I}_0 v, 0)$$

where $\mathscr{I}$ is the map defined for TM in 1.28. Hence the map k from M_p to E_ξ which sends v to $\dot{X}_v(0)$ is the composition of linear maps, and must be linear; its image $\mathscr{H}_\xi$ is therefore a linear subspace of E_ξ of dimension $\le \dim M$. But

$$\pi_* kv = \pi_* \dot{X}_v(0) = \dot{\widehat{\pi \circ X_v}}(0) = \frac{d}{dt}\bigg|_0 f(tv) = v$$

so $\pi_*|_{\mathscr{H}_\xi}$ is the inverse of k, and is thus an isomorphism. Since $\mathscr{V}E_\xi = \ker \pi_*|_\xi$, $E_\xi = \mathscr{H}_\xi \oplus \mathscr{V}E_\xi$ for all $\xi \in E$.

Now we need vector subbundle charts on $\mathscr{H}$. Let $\mathrm{pr}_2\colon \mathscr{V}E \to E$ be the projection $\mathscr{I}_\zeta \xi \mapsto \xi$ as in 1.28, and let $\pi_{TE}\colon TE \to E$ be the tangent bundle projection.

Fix a chart (x, U) on M, and let $(\pi, \psi)\colon \pi^{-1}U \to U \times \mathscr{W}$ be a vector bundle chart on E over U. Define a C^∞ map from $T(\pi^{-1}U)$ to $\mathscr{W}$ by

$$\tilde{\psi}(\eta) := \psi \circ \mathrm{pr}_2(\eta - \dot{\widehat{\mathbb{P}_\gamma \xi}}(0)), \qquad \eta \in E_\xi,\ \pi(\xi) \in U$$

where γ is any C^∞ curve in M with $\dot{\gamma}(0) = \pi_* \eta$; the initial tangent vector $\dot{\widehat{\mathbb{P}_\gamma \xi}}(0)$ is well-defined by axiom 2.6(5) for parallel transport. If $\eta \in \mathscr{H}_\xi$, then $\tilde{\psi}(\eta) = 0$. The map

$$(\pi_{TE}, dx \circ \pi_*, \tilde{\psi})\colon T(\pi^{-1}U) \to \pi^{-1}U \times \mathbb{R}^n \times \mathscr{W}$$

is a vector bundle chart on TE over $\pi^{-1}U$. Since $\tilde{\psi}(\mathscr{H}_\xi) = 0$, $(\pi_{TE}, dx \circ \pi_*, \tilde{\psi})(\eta) = (\xi, dx \circ \pi_*(\eta), 0)$ for $\eta \in \mathscr{H}_\xi$; thus the restriction $(\pi_{\mathscr{H}}, dx \circ \pi_*)$ of this vector bundle chart to $\mathscr{H}$ is a vector bundle chart on $\mathscr{H}$, so by 1.12*d* and 1.19, $\mathscr{H}$ will be a vector subbundle of TE once C^∞ compatibility of these charts is checked.

Let (y, W) be another chart on M with $U \cap W \neq \phi$. Fix $p \in U \cap W$ and $\xi \in E_p$. Let $\{e_j\}$ be the standard basis for $\mathbb{R}^n$; for each j,

$$(\pi_{\mathscr{H}}, dy \circ \pi_*) \circ (\pi_{\mathscr{H}}, dx \circ \pi_*)^{-1}(\xi, e_j) = (\pi_{\mathscr{H}}, dy \circ \pi_*) \circ \pi_*|_{\mathscr{H}}^{-1}\left(\frac{\partial}{\partial x^j}\bigg|_p\right)$$

$$= \left(\xi, dy\left(\frac{\partial}{\partial x^j}\bigg|_p\right)\right)$$

so the transition function is (id, $[\partial y^i/\partial x^j] \circ \pi$), which is C^∞.

The global splitting $TE = \mathscr{H} \oplus \mathscr{V}E$ follows from the pointwise splitting $E_\xi = \mathscr{H}_\xi \oplus \mathscr{V}E_\xi$ and the form of the charts constructed above.

Homogeneity: Fix $\xi \in E$, and let X be a parallel field in E with initial value ξ along a C^∞ curve γ in M with $\gamma(0) = \pi(\xi)$. By the linearity of parallel transport, the curve $\mu_c \circ X$ in E is also parallel along γ; the equality $\mu_{c*}\dot{X}(0) = \dot{\widehat{\mu_c \circ X}}(0)$ now implies that $\mu_{c*}\mathscr{H}_\xi \subset \mathscr{H}_{c\xi}$. But $\pi_* \mu_{c*}\dot{X}(0) = \pi_* \dot{X}(0)$ and π_* maps $\mathscr{H}_{c\xi}$ isomorphically onto $M_{\pi(\xi)}$, so $\mu_{c*}\mathscr{H}_\xi = \mathscr{H}_{c\xi}$.

If X is parallel in E along γ, then X is horizontal by the definition of $\mathscr{H}$.

Now let $Y \in \Gamma_\gamma E$ be a horizontal lift of a curve γ in M. If $\xi = Y(0)$, then $\mathbb{P}_\gamma \xi$ is the unique parallel section of E along γ with initial value ξ. Since both Y and $\mathbb{P}_\gamma \xi$ are horizontal and $\pi_* \dot{\widehat{\mathbb{P}_\gamma \xi}} = \dot{\widehat{\pi \circ \mathbb{P}_\gamma \xi}} = \dot{\gamma} = \pi_* \dot{Y}$, definition 2.26i for $\mathscr{H}$ implies $\dot{\widehat{\mathbb{P}_\gamma \xi}} = \dot{Y}$; but then $\mathbb{P}_\gamma \xi = Y$, so Y is parallel.

2.29 Now we prepare for the proof of the converse theorem.

Recall from 1.18 that given a vector bundle E over M and a C^∞ map $f: N \to M$, pr_2 is a bundle isomorphism from f^*E to E along f. The vertical bundle $\mathscr{V}f^*E$ is isomorphic to $\mathrm{pr}_2^*\mathscr{V}E$ by 1.26.

$$\begin{array}{ccc} f^*E & \xrightarrow{\mathrm{pr2}} & E \\ {\scriptstyle \mathrm{pr}_1}\downarrow & & \downarrow{\scriptstyle \pi} \\ N & \xrightarrow[f]{} & M \end{array}$$

Proposition Let $\mathscr{H}$ be a connection on a vector bundle E over M, and let $f: N \to M$ be C^∞. The subset $\tilde{\mathscr{H}} := \mathrm{pr}_{2*}^{-1}\mathscr{H}$ of Tf^*E is a connection on f^*E. A section Z of f^*E along a curve γ in N is horizontal with respect to $\tilde{\mathscr{H}}$ if and only if $\mathrm{pr}_2 Z \circ \gamma$ is horizontal in E along $f \circ \gamma$.

Proof Let $(\pi_{TE}, \varphi, \psi)$: $T(\pi^{-1}U) \to \pi^{-1}U \times \mathbb{R}^n \times \mathscr{W}$ be a vector bundle chart on TE over $\pi^{-1}U$, where U is an open set in M and $\mathscr{W}$ is the standard fiber of E (and f^*E); assume that ψ is zero on $\mathscr{H}|_{\pi^{-1}U}$, so that $(\pi_{\mathscr{H}}, \varphi)$: $\mathscr{H}|_{\pi^{-1}U} \to \pi^{-1}U \times \mathbb{R}^n$ is a vector bundle chart on $\mathscr{H}$ over $\pi^{-1}U$, where $\pi_{\mathscr{H}} = \pi_{TE}|_{\mathscr{H}}$. It follows that $\psi \circ \mathrm{pr}_{2*}$ is zero on $\tilde{\mathscr{H}}|_{(f\circ \mathrm{pr}_1)^{-1}U} \subset Tf^*E$. Given a chart (x, V) on N such that $f(V) \cap U$ is nonempty, $(\pi_{Tf^*E}, dx \circ \mathrm{pr}_{1*}, \psi \circ \mathrm{pr}_{2*})$ is a vector bundle chart on Tf^*E over $W := \mathrm{pr}_1^{-1}(V \cap f^{-1}U) \subset f^*E$; its restriction $(\pi_{\tilde{\mathscr{H}}}, dx \circ \mathrm{pr}_{1*})$ to $\tilde{\mathscr{H}}|_W$ is a vector bundle chart on $\tilde{\mathscr{H}}$ over W. Thus $\tilde{\mathscr{H}}$ is a vector subbundle of Tf^*E.

The form of the charts makes it clear that $Tf^*E = \tilde{\mathscr{H}} \oplus \mathscr{V}f^*E$. It is easy to check that if $v = \mathscr{H}v + \mathscr{V}v \in \mathscr{H}_\xi \oplus \mathscr{V}E_\xi$, $\xi \in E$, is the decomposition of tangent vectors on E into horizontal and vertical parts, then given a tangent vector (u, v) on f^*E at a point (p, ξ), $(u, v) = (u, \mathscr{H}v) + (0_p, \mathscr{V}v)$ is its decomposition into $\tilde{\mathscr{H}}$-horizontal and vertical components. This implies the statement about a section Z of f^*E along a curve in N.

By the definition of the vector bundle structure of f^*E (1.18), $\mathrm{pr}_2 \circ \mu_c = \mu_c \circ \mathrm{pr}_2 : f^*E \to E$ for all $c \in \mathbb{F}$; this being so,

$$\begin{aligned}\mathrm{pr}_{2*}\mu_{c*}\tilde{\mathscr{H}}_{(p,\xi)} &= \mu_{c*}\,\mathrm{pr}_{2*}\tilde{\mathscr{H}}_{(p,\xi)} = \mu_{c*}\mathscr{H}_\xi \\ &= \mathscr{H}_{c\xi} = \mathrm{pr}_{2*}\tilde{\mathscr{H}}_{(p,c\xi)}\end{aligned}$$

The identity $\pi_*\mu_{c*} = \pi_*$ and the injectivity of $\pi_*|_{\mathscr{H}}$ then imply that $\mu_{c*}\tilde{\mathscr{H}}_{(p,\xi)} = \tilde{\mathscr{H}}_{(p,c\xi)}$ as required.

Although $\mathscr{V}f^*E$ is isomorphic to $\mathrm{pr}_2^*\mathscr{V}E$ (functoriality), $\tilde{\mathscr{H}}$ is not in general isomorphic to $\mathrm{pr}_2^*\mathscr{H}$; in particular, $\tilde{\mathscr{H}}_{(p,\xi)} \cong N_p$, which is in general not isomorphic to $M_{f(p)} \cong \mathscr{H}_\xi$.

2.30 **Definition** Let $\mathscr{H}$ be a connection on a vector bundle E over M. The *horizontal lift of* $X \in \mathfrak{X}M$ is the unique vector field $\bar{X} \in \Gamma\mathscr{H} \subset \mathfrak{X}E$ such that $\bar{X}$ and X are π-related: $\pi_*\bar{X} = X \circ \pi$.

2.31 **Proposition** Let $\mathscr{H}$ be a connection on E over M. Given $X, Y \in \mathfrak{X}M$ and $f \in C^\infty M$,

(i) $\overline{X + Y} = \bar{X} + \bar{Y}$,
(ii) $\overline{fX} = f \circ \pi \cdot \bar{X}$, and
(iii) $\overline{[X, Y]}$ equals the horizontal component $\mathscr{H}[\bar{X}, \bar{Y}]$ of $[\bar{X}, \bar{Y}]$.

Proof Properties (i) and (ii) are obvious, while (iii) follows from the uniqueness of horizontal lifts and the equality $\pi_*[\bar{X}, \bar{Y}] = [X, Y] = \pi_*\overline{[X, Y]}$.

2.32 **Proposition** Let $\mathscr{H}$ be a connection on a vector bundle E over a manifold M. Let γ be a C^∞ curve in M, and let $\xi \in E_{\gamma(t_o)}$. There exists a unique horizontal lift X of γ to E such that $X(t_o) = \xi$.

Proof Let $\tilde{\mathscr{H}}$ be the connection induced on γ^*E from 2.29. Let $\bar{D}$ be the horizontal lift to γ^*E of $D = d/dt \in \mathfrak{X}([a, b])$, and let α be the integral curve of $\bar{D}$ in γ^*E such that $\alpha(t_o) = (t_o, \xi)$.

$$\begin{array}{ccc} \gamma^*E & \xrightarrow{\text{pr}2} & E \\ {\scriptstyle \text{pr}_1}\downarrow & & \downarrow{\scriptstyle \pi} \\ [a, b] & \xrightarrow[\gamma]{} & M \end{array}$$

The map $\text{pr}_1 \circ \alpha$ is an integral curve of D because of the equality

$$\dot{\widehat{\text{pr}_1 \circ \alpha}} = \text{pr}_{1*}\dot{\alpha} = \text{pr}_{1*}\bar{D} \circ \alpha = D \circ \text{pr}_1 \circ \alpha$$

Since $\text{pr}_1 \circ \alpha(t_o) = t_o$, $\text{pr}_1 \circ \alpha(t) = t$ for all $t \in [a, b]$. Thus α is a section of the bundle γ^*E over $[a, b]$. By the definition of α and $\bar{D}$, α is a horizontal section of γ^*E. By 2.29, $X := \text{pr}_2 \circ \alpha$ is a horizontal section of E along γ, and $X(t_o) = \xi$.

Uniqueness is a direct consequence of the uniqueness of the integral curve of a vector field through a point.

2.33 **Theorem** Let $\mathscr{H}$ be a connection on a vector bundle E over a manifold M. The system of $\mathscr{H}$-horizontal lifts to E of the C^∞ curves in M is a parallelism structure $\mathbb{P}$ in E. Furthermore, the connection on E determined by $\mathbb{P}$ as in theorem 2.28 is just $\mathscr{H}$.

Proof We must check the axioms for parallel transport.

(1): This is just proposition 2.32.

(2): Let $\gamma: [0, b] \to M$ be a C^∞ curve, with $p = \gamma(0)$ and $q = \gamma(b)$. If X is a horizontal lift of γ to E, then for all $c \in \mathbb{F}$, cX is also horizontal since $\dot{\widehat{cX}} = \mu_{c*}\dot{X}$ is a curve in the manifold $\mathscr{H}$ by 2.26ii. Since $(cX)(0) = c \cdot X(0) \in E_p$, the induced parallel transport map $L: E_p \to E_q$ along γ is homogeneous over $\mathbb{F}$; but by the differentiable dependence on the initial conditions of the integral curves of $\bar{D}$ in the proof of 2.32, L is also differentiable as a map from the manifold E_p to the manifold E_q. Calculate its differential dL applied to a tangent vector v at $0_p \in E_p$; since $L(0_p) = 0_q$,

$$dL(v) = \frac{d}{dt}\bigg|_0 L(tv) = \lim_{t \to 0} \frac{L(tv)}{t} = \lim_{t \to 0} \frac{tL(v)}{t} = L(v)$$

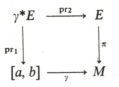

Thus L equals its differential at 0_p and is therefore linear over $\mathbb{R}$. In the case where $\mathbb{F} = \mathbb{C}$, real linearity plus complex homogeneity imply complex linearity. Thus L is $\mathbb{F}$-linear in either case.

Given $\xi \in E_p$, and the horizontal lift X of γ such that $X(0) = \xi$, let $X^-(t) := X(a - t + b)$; then X^- is the horizontal lift of γ^- with initial value $X(b)$. Hence L is invertible, and its inverse is parallel translation back along γ^-.

(3): Let $\gamma\colon [a, b] \to M$ be C^∞, and suppose that $\psi\colon [c, d] \to [a, b]$ is C^∞, with $\psi(c) = a$ and $\psi(d) = b$; set $\alpha := \gamma \circ \psi$. Fix $\xi \in E_{\gamma(a)}$; let X be the horizontal section of E along α such that $X(a) = \xi$, and let Y be the horizontal section of E along α such that $Y(c) = \xi$. For all $t \in [c, d]$, $\dot{Y}(t)$ and $\dot{X}(t)$ are horizontal. But

$$\begin{aligned}\pi_* \dot{Y} &= \dot{\alpha} = \psi' \cdot \dot{\gamma} \circ \psi = \psi' \cdot \pi_* \dot{X} \circ \psi = \pi_*(\psi' \cdot \dot{X} \circ \psi) \\ &= \pi_* X_* \dot{\psi} = \pi_* \dot{\widehat{X \circ \psi}}\end{aligned}$$

Thus $\dot{\widehat{(\mathrm{id}, Y)}}$ and $\dot{\widehat{(\mathrm{id}, X \circ \psi)}}$ are horizontal lifts to γ^*E of the same vector field $D = d/dt$ on $[c, d]$; they agree at a point, so they agree everywhere. Hence their integral curves (id, Y) and $(\mathrm{id}, X \circ \psi)$ also agree.

(4): Let U be an open set in M, and let $f\colon TU \to M$ be a C^∞ map such that $f(0_p) = p$, $p \in U$. Let $\pi_1(u, v) := u$ for $(u, v) \in TU \oplus TU$.

$$\begin{array}{ccc} (f \circ \pi_1)^*E & \xrightarrow{\mathrm{pr}_2} & E \\ \Big\downarrow \mathrm{pr}_1 & & \Big\downarrow \pi \\ TU \oplus TU & \xrightarrow[f \circ \pi_1]{} & M \end{array}$$

Define $Y \in \mathfrak{X}(TU \oplus TU)$ by

$$Y_{(u, v)} := \frac{d}{dt}\bigg|_0 (u + tv, v) = (\mathscr{I}_u v, 0)$$

The horizontal lift $\bar{Y}$ of Y to the bundle $(f \circ \pi_1)^*E$ has the property that for each $(u, v, \xi) \in (f \circ \pi_1)^*E$, $\bar{Y}_{(u, v, \xi)}$ is the initial tangent vector to the horizontal section of $(f \circ \pi_1)^*E$ with initial value (u, v, ξ) along the curve $s \mapsto (u + sv, v)$ in $TU \oplus TU$. In particular, for each $(v, \xi) \in f^*E$, the integral curve Z of $\bar{Y}$ through the point $(0, v, \xi)$ in $(f \circ \pi_1)^*E$ is the horizontal section of $(f \circ \pi_1)^*E$ along the curve $t \mapsto (tv, v)$ in $TU \oplus TU$ with initial value $(0, v, \xi)$; the dependence of Z on the initial condition $(0, v, \xi)$ is C^∞ by the usual theorems about the integral curves of vector fields [W: 1.48].

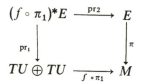

Finally, the curve $pr_2 \circ Z$ is the horizontal section of E along the curve $t \mapsto f(tv)$ in M with initial value ξ; since pr_2 from $(f \circ \pi_1)^*E$ to E is C^∞ and Z depends differentiably on the initial conditions, so does $pr_2 \circ Z$.

(5): Initial uniqueness is immediate, for if γ is a curve in M, $\xi \in E_{\gamma(0)}$, and X is the horizontal section of E along γ with initial value ξ, then $\dot{X}(0)$ is the unique element of $\mathscr{H}_\xi$ which is mapped by π_* to $\dot{\gamma}(0) \in M_{\gamma(0)}$.

Axiom (5) implies that the connection induced by the parallel transport system determined by $\mathscr{H}$ is just $\mathscr{H}$ itself.

2.34 Theorems 2.28 and 2.33 show that a system of parallel transport in a vector bundle is equivalent to a connection on the bundle; a connection is the infinitesimal version of parallel transport. This is analogous to the relationship between the family of solutions (that is, integral curves) of a first-order differential equation and the equation written as a vector field; in fact, the horizontal lift to E of a vector field X on M is the unique vector field $\bar{X}$ on E whose integral curves are the parallel sections of E along the integral curves of X.

Our first use of the equivalence between parallelism structures and connections will be an easy proof that every vector bundle admits a parallelism structure.

Lemma Let $\mathscr{H}^1$ and $\mathscr{H}^2$ be connections on E over M; let $f \in C^\infty M$ such that $0 \leq f \leq 1$. Given $p \in M$ and $\xi \in E_p$, define

$$\mathscr{H}_\xi := \{f(p)u_1 + (1 - f(p))u_2 \,|\, u_j \in \mathscr{H}^j_\xi, \text{ and } \pi_* u_1 = \pi_* u_2 \in M_p\}$$

and set $\mathscr{H} := \bigcup_{\xi \in E} \mathscr{H}_\xi$. The subset $\mathscr{H}$ of TE is a connection on E.

Proof The set $\mathscr{H}_\xi$ is obviously a linear subspace of E_ξ, and π_* maps $\mathscr{H}_\xi$ isomorphically to $M_{\pi(\xi)}$. Vector bundle charts are easy to construct which exhibit $\mathscr{H}$ as a subbundle of TE, and $\mathscr{H}$ is homogeneous because $\mathscr{H}^1$ and $\mathscr{H}^2$ are.

2.35 **Theorem** Let E be a vector bundle over a manifold M, which (as usual) is assumed to be paracompact. There exists a connection on E, and therefore there exists a parallelism structure on E.

Proof Let $\{U_\alpha\}_{\alpha \in \mathscr{A}}$ be a locally finite open covering of M by open sets U_α, over each of which E is trivial. Let $\{f_\alpha\}$ be a partition of unity subordinate to the open cover $\{U_\alpha\}$. Since $\pi^{-1}U_\alpha$ is a trivial open vector subbundle of E for each α, there exists a connection $\mathscr{H}^\alpha$

on $\pi^{-1}U_\alpha$. For each $\xi \in E$, set

$$\mathscr{H}_\xi := \bigcup_{u \in M_{\pi(\xi)}} \left\{ \sum_\alpha f_\alpha \circ \pi(\xi) v_\alpha \,\middle|\, v_\alpha \in \mathscr{H}^\alpha_\xi,\ \pi_* v_\alpha = u \right\}$$

$\mathscr{H} := \bigcup_{\xi \in E} \mathscr{H}_\xi$ is a connection on E by 2.33 applied to the $U_\alpha \cap U_\beta$.

2.36 **Definition** Let $\mathscr{H}$ be a connection on a vector bundle E over M, and let $\mathbb{P}$ be the system of parallel transport in E associated to $\mathscr{H}$ by theorem 2.33. For each $p \in M$, the *holonomy group of* $\mathscr{H}$ *at* p is defined to be the holonomy group $G(p)$ of $\mathbb{P}$ at p; if M is connected, the *holonomy group of* $\mathscr{H}$ is defined to be the holonomy group G of $\mathbb{P}$. The connection $\mathscr{H}$ is called a *trivial connection* if $\mathbb{P}$ is a trivial parallelism structure in E.

Exercise Calculate the canonical trivial connection on $M \times \mathbb{F}^m$, and the canonical left-invariant connection and canonical right-invariant connection on a Lie group G (2.8*a*).

2.37 **Proposition** The holonomy group G of a connection $\mathscr{H}$ on a vector bundle E over a connected manifold M is trivial if and only if E is a trivial vector bundle and $\mathscr{H}$ is a trivial connection.

PROOF Theorem 2.33 and proposition 2.20.

2.38 Let $\mathscr{H}$ be a connection on a vector bundle E over an n-dimensional manifold M. For each $\xi \in E$, the fiber $\mathscr{H}_\xi$ is an n-dimensional subspace of E_ξ, and these subspaces vary differentiably with ξ; thus $\mathscr{H}$ can be considered as a C^∞ distribution [W: 1.56] on the manifold E. If X is a parallel section of E, then by 2.28, for all $p \in M$ the linear subspace $X_* M_p$ of $E_{X(p)}$ equals the horizontal space $\mathscr{H}_{X(p)}$; this means that the embedded submanifold $X(M) \subset E$ is an integral manifold [W: 1.57] of the distribution $\mathscr{H}$.

Suppose that $\mathscr{H}$ is a trivial connection; then for each $\xi \in E$ there is a parallel section X of E such that $X_{\pi(\xi)} = \xi$. Thus through each point of E there passes an integral manifold of $\mathscr{H}$; in this case the distribution $\mathscr{H}$ is involutive [W: 1.59].

Definition A connection on a vector bundle is called a *flat connection* if it is involutive when considered as a C^∞ distribution. A *flat vector bundle* is a vector bundle endowed with a flat connection.

2.39 The discussion before definition 2.38 shows that a trivial connection is flat; the converse is true locally.

Proposition Let $\mathscr{H}$ be a flat connection on a vector bundle E over a connected manifold M. If N is a maximal connected integral manifold of $\mathscr{H}$ in E [W: 1.63], then the restriction of π to N is a C^∞ covering map, and $\mathscr{H}$ is locally a trivial connection.

PROOF Fix a maximal connected integral manifold N of $\mathscr{H}$ in E. The restriction of π to N is surjective, for given $\xi \in N$ and $q \in M$, let γ: $[0, 1] \to M$ be a C^∞ curve such that $\gamma(0) = \pi(\xi)$ and $\gamma(1) = q$; since $\mathbb{P}_\gamma \xi$ is horizontal, it lies in N, and $\pi \circ \mathbb{P}_\gamma \xi(1) = q$. The inverse function theorem implies that the C^∞ submersion $\pi|_N$ is a local diffeomorphism from N into M, and therefore a C^∞ covering map.

Let U be a connected, simply connected, open subset of M, and set $EU := \pi^{-1}U \subset E$; EU is an open vector subbundle of E over U. The intersection $N \cap EU$ is a disjoint union of connected, open submanifolds of N. Since π maps each component N_j of $N \cap EU$ diffeomorphically onto U, $\pi|_{N_j}^{-1}$ is a section of EU; this section is parallel since N_j is an integral manifold of $\mathscr{H}|_{EU}$.

There is an integral manifold of $\mathscr{H}|_{EU}$, that is, a parallel section of EU, through each point of EU; thus $\mathscr{H}|_{EU}$ is a trivial connection.

2.40 **Proposition** If $\mathscr{H}$ is a flat connection on a vector bundle E over a connected manifold M, then for each $p \in M$ the restricted holonomy group $G_o(p)$ of $\mathscr{H}$ at p (2.22) is trivial, and there is a homomorphism from $\pi_1(M, p)$ onto the holonomy group $G(p)$ of $\mathscr{H}$ at p.

PROOF Let γ: $[0, 1] \to M$ be a null-homotopic piecewise C^∞ loop at p. An argument similar to the proof of 2.23 shows that there is a homotopy F: $[0, 1] \times [0, 1] \to M$ with fixed endpoints between $F(0, \cdot) = p$ and $F(1, \cdot) = \gamma$ such that each intermediate loop $f(s) := F(s, \cdot)$, $0 \le s \le 1$, is piecewise C^∞. Given $\xi \in E_p$, let N be the maximal integral manifold of $\mathscr{H}$ through ξ; for each $s \in [0, 1]$, the horizontal lift $\mathbb{P}_{f(s)}\xi$ lies in N. Since N is a covering space of M by 2.39, the homotopy covering theorem [Gr: 5.3] implies that $(s, t) \mapsto \mathbb{P}_{f(s)}\xi(t)$ is a homotopy in N with fixed endpoints from the curve $\mathbb{P}_\gamma \xi$ to the constant curve ξ; in particular, $\mathbb{P}_\gamma\xi(1) = \xi$. Thus parallel transport around γ is trivial.

By 2.24, there is a homomorphism from $\pi_1(M, p)$ onto $G(p)/G_o(p) = G(p)/\{e\} = G(p)$.

2.41 **Theorem** Let $\mathscr{H}$ be a connection on a vector bundle E over a connected manifold M, and let f: $\tilde{M} \to M$ be the C^∞ universal covering of M. The connection $\mathscr{H}$ is flat if and only if f^*E is trivial and the pullback connection $\tilde{\mathscr{H}}$ on f^*E (2.29) is a trivial connection.

PROOF $\Rightarrow$: If $\mathscr{H}$ is flat, so is $\tilde{\mathscr{H}}$. By 2.40, the holonomy group $\tilde{G}(p)$ of $\tilde{\mathscr{H}}$ at $p \in \tilde{M}$ is the homomorphic image of $\pi_1(\tilde{M}, p) = \{e\}$, and hence is trivial. By 2.37, f^*E is trivial and $\tilde{\mathscr{H}}$ is a trivial connection.

$\Leftarrow$: Given $\xi \in E$, let N be an integral manifold of $\tilde{\mathscr{H}}$ through $(p, \xi) \in f^*E$; by 2.29, $\mathrm{pr}_2(N)$ is an integral manifold of $\mathscr{H}$ through ξ. Thus $\mathscr{H}$ is involutive.

2.42 ***The canonical connection on the standard sphere S^n*** Now we will calculate the connection determined by the canonical parallel transport on S^n (2.8*e*), and see that its holonomy group is $SO(n)$; the canonical connection on S^n is therefore not flat.

Let (γ, X) be a curve in $TS^n \subset \mathbb{R}^{n+1} \times \mathbb{R}^{n+1}$. By 2.8*e*, $\|\gamma\|^2 = \langle\gamma, \gamma\rangle = 1$, and $\langle\gamma, X\rangle = 0$, where $\langle\ ,\ \rangle$ is the usual inner product on $\mathbb{R}^{n+1}$. Thus if $\dot{\gamma}(t) = (\gamma(t), \gamma'(t))$ is the tangent field of γ, then $\langle\gamma, \gamma'\rangle = 0$, and $\langle\gamma', X\rangle + \langle\gamma, X'\rangle = \langle\gamma, X\rangle' = 0$; if the initial tangent vector is $\widehat{(\gamma, X)}(0) = ((\gamma, X), (\gamma', X'))(0) =: ((p, u), (v, w)) \in (TS^n)_{(p,u)} \subset TTS^n \subset \mathbb{R}^{4n+4}$, then $v \perp p$ and $\langle p, w\rangle + \langle u, v\rangle = 0$. It follows that

$$TTS^n = \{((p, u), (v, w)) \in TS^n \times \mathbb{R}^{2n+2} \mid v \perp p \text{ and } \langle p, w\rangle + \langle u, v\rangle = 0\} \tag{2-1}$$

Set $E := TS^n$, so that $TTS^n = TE$. The vertical space over $\xi := (p, u) \in E$ is $\mathscr{V}E_\xi = \{(\xi, (0, w)) \in E_\xi\}$ because $\pi_*(\xi, (v, w)) = (p, v) \in E_p$.

The horizontal space at $\xi = (p, u) \in E$ induced by the parallel transport from 2.8*e* is

$$\mathscr{H}_\xi = \{(\xi, (v, -\langle u, v\rangle p)) \in E_\xi\} \tag{2-2}$$

Proof: For a start, this is a vector space which is mapped isomorphically by $\pi_*|_\xi$ to $\{(p, v) \mid v \perp p\} = E_p = S^n_p$. Let $v \perp p$; we may assume $v \neq 0$, so set $\lambda := \|v\|$, $q := (1/\lambda)v \in S^n$, and $\gamma(t) := \cos(\lambda t)p + \sin(\lambda t)q$. Embed $\mathbb{R}^3$ into $\mathbb{R}^{n+1}$ so that $\mathbb{R}^3 \cap S^n$ is a great 2-sphere in S^n whose tangent space at p contains ξ and $(p, v) = \dot{\gamma}(0)$. Let $R_t \in SO(3)$ be the rotation of $\mathbb{R}^3$ such that $R_t p = \gamma(t)$; by definition, $\mathbb{P}_\gamma\xi(t) = R_{t*}\xi = (\gamma(t), R_t u)$ because R_t is linear. Since $u \perp p$ we can write $u = cq + r$, $c \in \mathbb{R}$, $r \perp p$, and $r \perp q$. Thus

$$R_t u = cR_t q + R_t r = cR_t\gamma\left(\frac{\pi}{2\lambda}\right) + r = c\gamma\left(\frac{\pi}{2\lambda} + t\right) + r$$

Hence $\dot{\widehat{\mathbb{P}_\gamma \xi}}(0) = (\xi, (v, c\gamma'(\pi/2\lambda)))$; but

$$c\gamma'\left(\frac{\pi}{2\lambda}\right) = -c\lambda p = -\langle cq, \lambda q\rangle p = -\langle u, v\rangle p$$

so $(\xi, (v, -\langle u, v\rangle p)) = \dot{\widehat{\mathbb{P}_\gamma \xi}}(0) \in \mathscr{H}_\xi$. *QED.*

In 2.8*e* parallel transport was defined on S^n, but only along the great circles; from this partial parallel transport we have now obtained a connection $\mathscr{H}$ on TS^n. By theorem 2.33 there is a parallelism structure on S^n: a section Z of $E = TS^n$ along an arbitrary C^∞ curve β in S^n is parallel in E along β if and only if Z is $\mathscr{H}$-horizontal; axiom (1) for parallel transport is now satisfied. If $Z = (\beta, X)$, then horizontality of Z means that $\dot{Z} = ((\beta, X), (\beta', X'))$ is a curve in $\mathscr{H}$, that is, X satisfies the differential equation

$$X' = -\langle X, \beta'\rangle \beta$$

which comes from the form $((p, u), (v, -\langle u, v\rangle p))$ for the points in $\mathscr{H}_{(p, u)}$. In particular, along great circles this is the same as before.

Given horizontal curves (β, X) and (β, Y) in $E = TS^n$,

$$\langle X, Y\rangle' = \langle X', Y\rangle + \langle X, Y'\rangle = -\langle X, \beta'\rangle\langle \beta, Y\rangle - \langle X, \beta\rangle\langle \beta', Y\rangle = 0$$

since $X \perp \beta$ and $Y \perp \beta$, so $\langle X, Y\rangle$ is constant for parallel (β, X) and (β, Y). Thus for every C^∞ curve β in S^n there is a C^∞ curve η in $O(n+1)$ such that $\mathbb{P}_\beta \xi(t) = (\beta(t), \eta(t)u)$, $\xi = (\beta(0), u) \in E_{\beta(0)}$. Since $\eta(0) = e \in O(n+1)$, η lies in the identity component $SO(n+1)$ of $O(n+1)$, that is, $\eta(t)$ is a rotation for all t.

Proof that the holonomy group of S^n is $SO(n)$: Let $p = e_0$, where $\{e_0,$

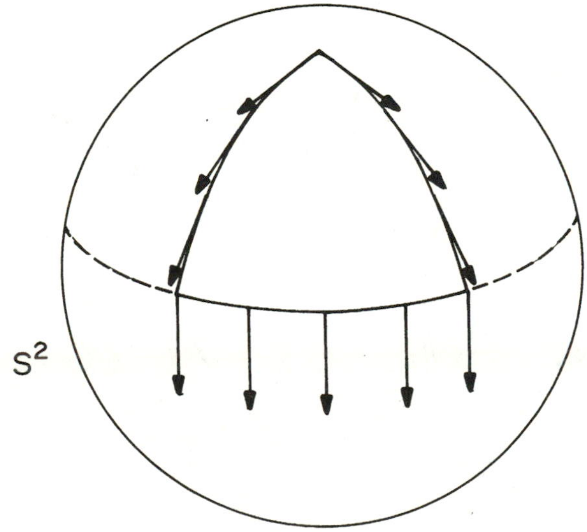

$\ldots, e_n\}$ is the usual orthonormal basis for $\mathbb{R}^{n+1}$. The isotropy group at p [W: 3.61] of $SO(n+1)$ acting on S^n is

$$H := \begin{bmatrix} 1 & 0 \\ 0 & SO(n) \end{bmatrix}$$

H is also the linear isotropy group [W: 3.61(4)] acting on the tangent space S^n_p at p. Since each g in the holonomy group $G(p)$ is a rotation of $\mathbb{R}^{n+1}$ which leaves S^n_p invariant, $G(p) \subset H$.

$G(p) \supset H$: Fix $g \in H$. Since the eigenvalues of g are of absolute value 1, there exists a matrix $h \in H$ such that $h^{-1}gh$ has the block diagonal form [Ha: 81]

$$h^{-1}gh = \begin{bmatrix} 1 & 0 & \cdots & 0 & 0 \\ 0 & R_{\theta 1} & \cdots & 0 & 0 \\ \vdots & \vdots & & \vdots & \vdots \\ 0 & 0 & \cdots & R_{\theta k} & 0 \\ 0 & 0 & \cdots & 0 & I \end{bmatrix} \qquad R_{\theta j} := \begin{bmatrix} \cos\theta_j & -\sin\theta_j \\ \sin\theta_j & \cos\theta_j \end{bmatrix}$$

where I is the $(n-2k) \times (n-2k)$ identity matrix.

Temporarily set $e_{-1} := e_0$. For $1 \le j \le k$, define

$$\alpha_j(t) := \cos(t)e_{2j-3} + \sin(t)e_{2j-1} \qquad 0 \le t \le \frac{\pi}{2}$$

$$\beta_j(t) := \cos(t)e_{2j-1} + \sin(t)e_{2j} \qquad 0 \le t \le \theta_j$$

$$\gamma_j(t) := \cos(t)e_{2j-3} + \sin(t)e_{2j} \qquad 0 \le t \le \frac{\pi}{2}$$

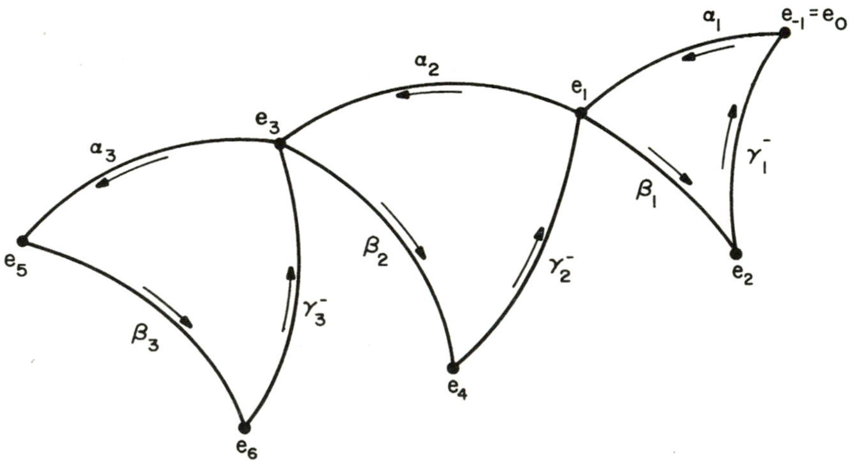

By the definition of parallel transport in S^n, parallel transport around the piecewise C^∞ loop

$$\eta := \gamma_1^- * \beta_1 * \gamma_2^- * \beta_2 * \cdots * \gamma_k^- * \beta_k * \alpha_k * \alpha_{k-1} * \cdots * \alpha_2 * \alpha_1$$

is just $h^{-1}gh$, so $h^{-1}gh \in G(p)$. But by the invariance of $\mathscr{H}$ under the action of $SO(n+1)$, $g = h \circ \mathbb{P}_\eta \circ h^{-1} \in G(hp)$, which equals $G(p)$ since $h \in H$. *QED*.

THE CURVATURE TENSOR

By 2.38 and 2.39, a connection $\mathscr{H}$ on E is locally a trivial connection if and only if it is involutive when considered as a C^∞ distribution on E; in this case it is called a flat connection (2.38). By the Frobenius theorem, failure of $\mathscr{H}$ to be flat implies that the vector space $\Gamma\mathscr{H} \subset \mathfrak{X}E$ of horizontal vector fields on E is not a Lie subalgebra of $\mathfrak{X}E$, for in this case there exist $X, Y \in \Gamma\mathscr{H}$ such that the vertical component $\mathscr{V}[X, Y] = [X, Y] - \mathscr{H}[X, Y] \in \Gamma\mathscr{V}E$ of the Lie bracket is nonzero.

2.43 **Definition** Let $\mathscr{H}$ be a connection on a vector bundle E over M. Given $U, V \in \mathfrak{X}M$, define

$$R(U, V)\xi := -\mathrm{pr}_2(\mathscr{V}[\bar{U}, \bar{V}]_\xi) \qquad \xi \in E$$

where $\bar{U}$ and $\bar{V}$ are the horizontal lifts to E of U and V (2.30), and $\mathrm{pr}_2\colon \mathscr{V}E \to E$ is the second factor projection from 1.28. The expression R is called the *curvature tensor field of* $\mathscr{H}$.

It follows that $R = 0$ if and only if $\mathscr{H}$ is flat since pr_2 is an isomorphism along π.

2.44 The proof of the next proposition provides an important interpretation of R in terms of parallel transport; in fact, this is essentially Cartan's definition of R.

Proposition Let $\mathscr{H}$ be a connection on E over M. The operator R defined in 2.43 is a section of the real homomorphism bundle $\mathrm{Hom}(\Lambda^2 TM, \mathrm{End}\, E) \cong \mathrm{Hom}(\Lambda^2 TM \otimes E, E) \cong \mathrm{Hom}(E, \mathrm{Hom}(\Lambda^2 TM, E)) \cong \Lambda^2 T^*M \otimes E \otimes E^*$; this justifies calling R a tensor field. Furthermore, if E is a complex vector bundle, then given $U, V \in \mathfrak{X}M$, $R(U, V)$ is a section of the complex algebra bundle End E, that is, for all $X \in \Gamma E$, $R(U, V)(iX) = iR(U, V)X$; if the complex vector bundle E is thought of as a real vector bundle with a complex structure $J \in \Gamma$ End E such that $J^2 = -\mathrm{id}$ (1.58), then this says that $R(U, V) \circ J = J \circ R(U, V)$.

Proof Given $U, V \in \mathfrak{X}M$ and $X \in \Gamma E$, the map $p \mapsto (R(U, V)X)_p :=$ $R(U, V)(X_p) = -\text{pr}_2 \mathscr{V}[\bar{U}, \bar{V}]_{X(p)}$ is a section of E because $\text{pr}_2 \mathscr{V}[\bar{U}, \bar{V}]_{X(p)}$ is a point of E_p which depends differentiably on $p \in M$.

By 2.31, given U, V, and $W \in \mathfrak{X}M$, and $X \in \Gamma E$,

$$\mathscr{V}[\overline{U + V}, \bar{W}]_X = \mathscr{V}[\bar{U} + \bar{V}, \bar{W}]_X = \mathscr{V}[\bar{U}, \bar{W}]_X + \mathscr{V}[\bar{V}, \bar{W}]_X$$

so $R(U + V, W)X = R(U, W)X + R(V, W)X$ since pr_2 is a bundle homomorphism from $\mathscr{V}E$ to E along $\pi_E\colon E \to M$.

Directly from the definition of R, $R(U, V)X + R(V, U)X = 0$.

Given $f \in C^\infty M$, $U, V \in \mathfrak{X}M$, and $X \in \Gamma E$,

$$\begin{aligned}\mathscr{V}[\overline{fU}, \bar{V}]_X &= \mathscr{V}[f \circ \pi \cdot \bar{U}, \bar{V}]_X \\ &= (f \circ \pi \cdot \mathscr{V}[\bar{U}, \bar{V}])_X - \mathscr{V}(\bar{V}(f \circ \pi) \cdot \bar{U})_X \\ &= f \cdot \mathscr{V}[\bar{U}, \bar{V}]_X\end{aligned}$$

since $\mathscr{V}\bar{U} = 0$, so $R(fU, V)X = fR(U, V)X$.

Thus, given $X \in \Gamma E$, the map $R(\cdot, \cdot)X$ is alternating and bilinear over $C^\infty M$, and is therefore a section of $\text{Hom}(\Lambda^2 TM, E)$.

Linearity of R over $C^\infty(M, \mathbb{F})$ in the third argument: Since for all $U, V \in \mathfrak{X}M$, $R(U, V)$ maps ΓE to ΓE, it suffices to show that for each $p \in M$ and $u, v \in M_p$, the map $R(u, v)$ from E_p to E_p is linear Choose $U, V \in \mathfrak{X}M$ such that $U_p = u$, $V_p = v$, and $[U, V] = 0$ near p. Denote by $\{\varphi_t\}$ and $\{\psi_t\}$ the local 1-parameter groups of U and V, respectively [W: 1.49]. For each small $t > 0$, the product γ_t of the

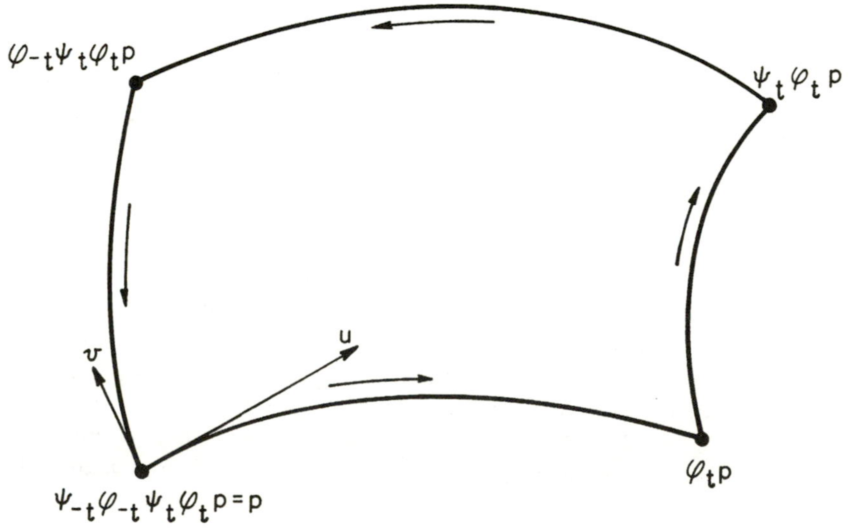

curves $\varphi_\tau p$, $\psi_\tau \varphi_t p$, $\varphi_{-\tau}\psi_t\varphi_t p$, and $\psi_{-\tau}\varphi_{-t}\psi_t\varphi_t p$, $0 \le \tau \le t$, in the order listed is a piecewise C^∞ loop at p because $[U, V] = 0$ near p; the initial tangent vector is u, while the terminal tangent vector is $-v$. If $\{\bar{\varphi}_t\}$ is the local 1-parameter group of $\bar{U}$, then $\pi \circ \bar{\varphi}_t = \varphi_t \circ \pi$, and for all ξ in E near $\pi^{-1}p$, $\bar{\varphi}_t(\xi) = \mathbb{P}_\beta\, \xi(t)$, where $\beta(t) := \varphi_t \circ \pi(\xi)$, and similarly for the local 1-parameter group $\{\bar{\psi}_t\}$ for $\bar{V}$. Hence for each $\xi \in E_p$ and each small $t > 0$,

$$\alpha_\xi(t) := \bar{\psi}_{-\sqrt{t}} \circ \bar{\varphi}_{-\sqrt{t}} \circ \bar{\psi}_{\sqrt{t}} \circ \bar{\varphi}_{\sqrt{t}}(\xi) \in E_p \tag{2-3}$$

is the parallel translate of ξ around the loop $\gamma_{\sqrt{t}}$. By [W: 2, exercise 6] the one-sided derivative $\dot{\alpha}_\xi(0)$ exists and equals $[\bar{U}, \bar{V}]_\xi$. Since α_ξ lies in E_p, $\dot{\alpha}_\xi(0)$ is vertical; hence $R(u, v)\xi = -\mathrm{pr}_2\,\dot{\alpha}_\xi(0) \cong -\dot{\alpha}_\xi(0) \in E_p$.

Define a curve g in $GL(E_p)$ by $g(t)(\xi) := \alpha_\xi(t)$ for small $t > 0$. The one-sided derivative $\dot{g}(0) \in \mathfrak{gl}(E_p)$ exists because $\dot{\alpha}_\xi(0)$ exists for all $\xi \in E_p$, and $R(u, v)\xi = -\dot{\alpha}_\xi(0) = -\dot{g}(0)(\xi)$. Thus the transformation $R(u, v)$ of E_p belongs to $\mathfrak{gl}(E_p)$, and is therefore linear over $\mathbb{F}$.

2.45 In 2.25 it was proved that the holonomy group $G(p)$ of a connection (that is, a parallelism structure) on a vector bundle E over M is a Lie subgroup of $GL(E_p)$ for each $p \in M$. It follows that the Lie algebra $\mathfrak{g}(p)$ of $G(p)$ is a Lie subalgebra of $\mathfrak{gl}(E_p)$. Set $\mathfrak{g}(E) := \bigcup_{p \in M} \mathfrak{g}(p)$.

Proposition The subset $\mathfrak{g}(E)$ of $\mathfrak{gl}(E)$ is a Lie algebra subbundle of $\mathfrak{gl}(E)$. The parallel transport induced on $\mathfrak{gl}(E)$ by the parallel transport in E is reducible to $\mathfrak{g}(E)$ (2.11), and the associated connection $\mathscr{H}$ on $\mathfrak{gl}(E)$ is reducible to $\mathfrak{g}(E)$ in the sense that for each $L \in \mathfrak{g}(E) \subset \mathfrak{gl}(E)$, the horizontal space $\mathscr{H}_L$ in the tangent space $\mathfrak{gl}(E)_L$ is contained in the subspace $\mathfrak{g}(E)_L$. In fact, the restriction $\mathscr{H}|_{\mathfrak{g}(E)}$ is the connection of the reduced parallel transport in $\mathfrak{g}(E)$.

PROOF Parallel transport in E induces parallel transport in the vector bundle $\mathfrak{gl}(E) = \mathrm{End}\, E = \mathrm{Hom}(E, E)$ by 2.14. By 2.11, if the subset $\mathfrak{g}(E)$ of $\mathfrak{gl}(E)$ is invariant under parallel transport in $\mathfrak{gl}(E)$, then it is a vector subbundle of $\mathfrak{gl}(E)$, and the parallel transport in $\mathfrak{gl}(E)$ is reducible to $\mathfrak{g}(E)$.

Let γ be a C^∞ curve in M, $p := \gamma(0)$. For each t, let μ_t be the isomorphism from $G(p)$ to $G(\gamma(t))$ defined in 2.19: $\mu_t h = \mathbb{P}_t \circ h \circ \mathbb{P}_t^{-1}$, $h \in G(p)$, where $\mathbb{P}_t$ here denotes parallel translation along γ from E_p to $E_{\gamma(t)}$. Given a parallel section $X \in \Gamma_\gamma E$ and $h \in G(p)$, $\mu_t h(X_t) = \mathbb{P}_t \circ h(X_0)$.

Now let $L \in \mathfrak{g}(p)$; $L = \dot{g}(0)$ for some C^∞ curve g in $G(p)$, so if $X \in \Gamma_\gamma E$ is parallel, then

$$(\mu_{t*} L)(X_t) = \frac{\partial}{\partial s}\bigg|_0 \mu_t g(s)(X_t) = \frac{\partial}{\partial s}\bigg|_0 \mathbb{P}_t \circ g(s)(X_0)$$
$$= d\mathbb{P}_t \circ L(X_0) = \mathbb{P}_t \circ L(X_0)$$

since $\mathbb{P}_t\colon E_p \to E_{\gamma(t)}$ is linear. Thus $t \mapsto (\mu_{t*} L)(X_t)$ is parallel in E along γ for each parallel X along γ. By definition 2.14, $\mu_{t*} L$ is parallel along γ in $\mathfrak{gl}(E) = \mathrm{Hom}(E, E)$. Thus $\mathfrak{g}(E)$ is a $\mathbb{P}$-invariant subset of $\mathfrak{gl}(E)$; by 2.11, $\mathfrak{g}(E)$ is a vector subbundle of $\mathfrak{gl}(E)$, and parallel transport in $\mathfrak{gl}(E)$ is reducible to $\mathfrak{g}(E)$.

Now we must check that $\mathfrak{g}(E)$ is a Lie algebra subbundle of $\mathfrak{gl}(E)$. If K and L are parallel sections of the vector bundle $\mathfrak{gl}(E) =$ End E along a curve γ, then so are their composites $K \circ L$ and $L \circ K$, which implies that the Lie bracket $[K, L] = K \circ L - L \circ K$ is also parallel along γ; thus the Lie bracket operator $[\ ,\ ]$ is a parallel section of the bundle $\mathrm{Hom}(\Lambda^2 \mathfrak{gl}(E), \mathfrak{gl}(E))$.

Fix $p \in M$ and a diffeomorphism h from a starlike neighborhood V of 0_p in M_p to an open set U in M such that $h(0_p) = p$. Let $(\pi, \psi)\colon \pi^{-1}U \to U \times \mathfrak{gl}(E)_p$ be the radially parallel vector bundle chart on $\mathfrak{gl}(E)$ determined by h as in 2.10; by definition, ψ is constant when applied to a parallel section of $\mathfrak{gl}(E)$ along any curve γ of the form $\gamma(t) = h(tv)$, $v \in V$, $0 \le t \le 1$. Since $[\ ,\ ]$ is parallel, $\psi[K, L] = [K_0, L_0] = [\psi K, \psi L]$ for every pair of parallel sections K and L of $\mathfrak{gl}(E)$ along each such curve γ; thus ψ preserves the Lie bracket, and therefore (π, ψ) is a Lie algebra bundle chart on $\mathfrak{gl}(E)$. Since the vector subbundle $\mathfrak{g}(E)$ of $\mathfrak{gl}(E)$ is invariant under parallel transport, (π, ψ) restricts to a Lie algebra bundle chart on $\mathfrak{g}(E)$.

2.46 **Definition** The Lie algebra $\mathfrak{g}(p)$ of the holonomy group $G(p)$ of a connection $\mathscr{H}$ on a vector bundle E over M, $p \in M$, is called the *holonomy algebra of* $\mathscr{H}$ *at* p. The Lie algebra bundle $\mathfrak{g}(E)$ is called the *holonomy algebra bundle of* $\mathscr{H}$ (here M is assumed to be connected).

2.47 Given a connection $\mathscr{H}$ on a vector bundle E over M, $p \in M$, and $u, v \in M_p$, the proof of 2.44 shows that $R(u, v)$ belongs to the holonomy algebra of $\mathscr{H}$ at p. In general, not every element of $\mathfrak{g}(p)$ must be of the form $R(u, v)$, $u, v \in M_p$. For example, it is possible to construct a connection $\mathscr{H}$ on E which is not flat globally, but which is a trivial connection (and therefore flat) over a neighborhood of p. Although $R(u, v) = 0$ for all $u, v \in M_p$, the algebra $\mathfrak{g}(p)$ is nonzero if M is connected, because since $\mathscr{H}$ is not flat globally, there is a nonzero curvature transformation $R(w, x)$ at some $q \in M$. Since $\mathfrak{g}(q)$ contains $R(w, x)$, it is nonzero, and therefore the isomorphic Lie algebra $\mathfrak{g}(p)$ is also nonzero; in fact, the parallel translate of $R(w, x)$ along any curve joining q to p is a nonzero element of $\mathfrak{g}(p)$.

Theorem (Ambrose-Singer) Let $\mathscr{H}$ be a connection on a vector bundle E over a connected manifold M. Fix $p \in M$. The holonomy algebra $\mathfrak{g}(p)$ is the union over all $q \in M$ of the set of all $\mathbb{P}_\gamma R(u, v)$, $u, v \in M_q$, where γ is a C^∞ curve in M joining q to p.

PROOF [AS] The proof will appear in 9.21.

2.48 **Definition** Let E be a vector bundle over a connected manifold M, and let G be a Lie subgroup of the structure group $GL(\mathscr{V})$ of E. A *G-connection on E* is a connection on E whose holonomy group is contained in G.

Exercise Let G be a Lie subgroup of the structure group $GL(\mathscr{V})$ of a vector bundle E over a connected manifold M. Prove that a connection $\mathscr{H}$ on E is a G-connection if and only if there is a Lie algebra subbundle $\mathfrak{g}(E)$ of $\mathfrak{gl}(E)$ with fiber $\mathfrak{g}$ isomorphic to the Lie algebra of G such that for all $U, V \in \mathfrak{X}M$, $R(U, V)$ is a section of the subbundle $\mathfrak{g}(E)$; in this case, $\mathfrak{g}(E)$ contains the holonomy algebra bundle of $\mathscr{H}$, and $\mathscr{H}$ is reducible to $\mathfrak{g}(E)$.

COVARIANT DERIVATIVE OPERATORS

2.49 **Definition** Let $\mathscr{H}$ be a connection on a vector bundle E over M; given $v \in TE = \mathscr{H} \oplus \mathscr{V}E$, write $v = \mathscr{H}v + \mathscr{V}v$. Define the *connection map κ of $\mathscr{H}$* by $\kappa\colon TE \to E$, $\kappa(v) := \mathrm{pr}_2(\mathscr{V}v) = \mathrm{pr}_2(v - \mathscr{H}v)$, where $\mathrm{pr}_2\colon \mathscr{V}E \to E$ is the second factor projection from 1.28: for all $p \in M$ and $\zeta, \xi \in E_p$, $\mathrm{pr}_2(\mathscr{I}_\zeta \xi) := \xi$.

2.50 It follows from 2.43 that the curvature tensor of $\mathscr{H}$ satisfies the identity $R(U, V)\xi = -\kappa[U, V]_\xi$, $U, V \in \mathfrak{X}M$, $\xi \in E$. The reader should verify (2.44) that the tensoriality of R in the first two arguments essentially depended on the trivial fact that κ is a vector bundle homomorphism along π.

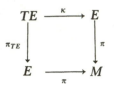

$$\begin{array}{ccc} TE & \xrightarrow{\kappa} & E \\ {\scriptstyle \pi_{TE}}\downarrow & & \downarrow{\scriptstyle \pi} \\ E & \xrightarrow[\pi]{} & M \end{array}$$

Of equal importance is the fact that κ also respects the vector bundle structure of TE over TM defined in 1.32.

Proposition Let $\mathscr{H}$ be a connection on a vector bundle E over M. The connection map κ of $\mathscr{H}$ is a vector bundle homomorphism from TE to E along the map π_{TM}.

$$\begin{array}{ccc} TE & \xrightarrow{\kappa} & E \\ {\scriptstyle \pi_*}\downarrow & & \downarrow{\scriptstyle \pi} \\ TM & \xrightarrow[\pi_{TM}]{} & M \end{array}$$

Proof Fix $u = \dot{\gamma}(0) \in M_p$, γ *a* C^∞ curve in M. If $v \in \pi_*^{-1}u \subset TE$ is tangent to E at $\zeta \in E_p$, then

$$v = \mathscr{H}v + \mathscr{V}v = \dot{\widehat{\mathbb{P}_\gamma \zeta}}(0) + \mathscr{I}_\zeta \kappa(v)$$

because π_* maps $\mathscr{H}_\xi$ isomorphically to M_p (2.26), and pr_2 maps $\mathscr{V}E_\xi$ isomorphically to E_p with inverse $\mathscr{I}_\xi$ (1.28). Let (π, ψ) be a vector bundle chart on E near p such that ψ is constant on each parallel section of E along γ (2.10); since $\pi \circ \mathbb{P}_\gamma \eta = \gamma$ for all $\eta \in E_p$ and $\dot{\gamma}(0) = u$,

$$\begin{aligned}(\pi, \psi)_* v = (u, \psi_* \mathscr{I}_\zeta \kappa v) &= \left(u, \frac{d}{dt}\bigg|_0 \psi(\zeta + t\kappa v)\right) \\ &= \frac{d}{dt}\bigg|_0 (\pi, \psi)(\mathbb{P}_\gamma \zeta(t) + t(\mathbb{P}_\gamma \kappa v)(t)) \\ &= (\pi, \psi)_* \frac{d}{dt}\bigg|_0 \mathbb{P}_\gamma(\zeta + t\kappa v)(t)\end{aligned}$$

so

$$v = \frac{d}{dt}\bigg|_0 \mathbb{P}_\gamma(\zeta + t\kappa v)(t)$$

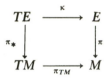

Recall from 1.32 that if $A(\zeta, \xi) := \zeta + \xi$ for $(\zeta, \xi) \in E \oplus E$, then addition of vectors in the bundle $\pi_*\colon TE \to TM$ is defined by $x + y := A_*(x, y)$. Thus, if

$$w = \frac{d}{dt}\bigg|_0 \mathbb{P}_\gamma(\xi + t\kappa w)(t)$$

is another point in TE over $u = \dot{\gamma}(0) \in TM$, then

$$\begin{aligned} v + w = A_*(v, w) &= \frac{d}{dt}\bigg|_0 (\mathbb{P}_\gamma(\zeta + t\kappa v)(t) + \mathbb{P}_\gamma(\xi + t\kappa w)(t)) \\ &= \frac{d}{dt}\bigg|_0 (\mathbb{P}_\gamma(\zeta + \xi)(t) + t\mathbb{P}_\gamma(\kappa v + \kappa w)(t)) \\ &= \overset{\bullet}{\widehat{\mathbb{P}_\gamma(\zeta + \xi)}}(0) + \mathscr{J}_{(\zeta+\xi)}(\kappa v + \kappa w) \end{aligned}$$

Thus $\kappa(v + w) = \kappa v + \kappa w$. Similarly, $cv = \overset{\bullet}{\widehat{\mathbb{P}_\gamma c\zeta}}(0) + \mathscr{J}_{c\zeta} c\kappa v$, $c \in \mathbb{F}$, so $\kappa(cv) = c\kappa v$.

This proposition implies that (π_{TE}, κ) is an isomorphism from the vector bundle $\pi_*\colon TE \to TM$ to the vector bundle $E \oplus E$ over M along the map $\pi_{TM}\colon TM \to M$; this has an interesting application. Recall from 1.57 that the vector bundle $E \oplus E$ has a natural complex structure: $J(\zeta, \xi) := (-\xi, \zeta)$. Since each fiber of TE over TM is isomorphic to a fiber of $E \oplus E$ by the map (π_{TE}, κ), we can pull this complex structure back to each fiber of TE over TM. Thus each connection on E determines a complex bundle structure on TE over TM. Observe that if E is a complex vector bundle, then the natural complex vector bundle structure from 1.32 (with respect to which (π_{TE}, κ) is an isomorphism along π_{TM}) has nothing to do with the connection-induced complex structure, just as the natural complex structure $i(\zeta, \xi) := (i\zeta, i\xi)$ on $E \oplus E$ has nothing to do with the complex structure on $E \oplus E$ given by $J(\zeta, \xi) := (-\xi, \zeta)$.

As a special case, a connection on TM induces a complex structure on the vector bundle $\pi_*\colon TTM \to TM$. Now consider the other vector bundle structure on TTM over TM; $\pi_{TTM}\colon TTM \to TM$ is the tangent bundle of the manifold TM. In this case, the map (π_*, κ) is an isomorphism from TTM to $TM \oplus TM$ along π.

$$\begin{array}{ccc} TTM & \xrightarrow{(\pi_*,\,\kappa)} & TM \oplus TM \\ {\scriptstyle \pi_{TTM}}\big\downarrow & & \big\downarrow \\ TM & \xrightarrow[\pi]{} & M \end{array}$$

This map too induces a complex structure on TTM over TM [Do1] because of the complex structure $(u, v) \mapsto (-v, u)$ on $TM \oplus TM$. This time the complex structure is in the tangent bundle of the manifold TM, so it is an example of an almost complex structure on a manifold (1.58).

Exercises

1. Prove that $\pi_*|_{\mathscr{H}}: \mathscr{H} \to TM$ is a real vector subbundle of $\pi_*: TE \to TM$.
2. Give a second proof that the curvature tensor R (2.43) of a connection on E is linear over $C^\infty(M, \mathbb{F})$ in the third argument.
3. Given a connection on TM, prove that $(\pi_{TTM}, \pi_*, \kappa)$ is a fiber-preserving diffeomorphism from the composite bundle TTM over M to the bundle $TM \oplus TM \oplus TM$.

2.51 **Definition** Let E be an $\mathbb{F}$-vector bundle over a manifold M, and let $f: N \to M$ be C^∞. A map $\nabla: \mathfrak{X}N \times \Gamma_f E \to \Gamma_f E$ is called a *covariant derivative operator in E along f* if given $U, V \in \mathfrak{X}N$ and $X, Y \in \Gamma_f E$,

(i) $\nabla_{(U+V)}X = \nabla_U X + \nabla_V X$

(ii) $\nabla_{gU} X = g\nabla_U X, \qquad g \in C^\infty N$

(iii) $\nabla_U(X + Y) = \nabla_U X + \nabla_U Y$

(iv) $\nabla_U hX = U(h)X + h\nabla_U X, \qquad h \in C^\infty(N, \mathbb{F})$

The expression $\nabla_U X$ is called the *covariant derivative of X with respect to U*. For each $X \in \Gamma_f E$, the map $\nabla X \in \mathrm{Hom}(\mathfrak{X}N, \Gamma_f E) \cong \Gamma\, \mathrm{Hom}(TN, f^*E)$ defined by $(\nabla X)(U) := \nabla_U X$ is called the *covariant differential of X*; for $u \in N_p$, $p \in N$, set $\nabla_u X := (\nabla_U X)_p$, where $U \in \mathfrak{X}N$ satisfies $U_p = u$ (this is well-defined by 1.53).

If $N = M$ and $f = \mathrm{id}$, then ∇ is called a *covariant derivative operator in E*; if $E = TM$, then ∇ is called a *covariant derivative operator on M*. By abuse of language a covariant derivative operator in E is called a *connection* in E because of the next theorem.

2.52 **Theorem** Let $\mathscr{H}$ be a connection on a vector bundle E over a manifold M, with connection map $\kappa: TE \to E$. For each C^∞ map $f: N \to M$, $\nabla_U X := \kappa(X_* U)$, $U \in \mathfrak{X}N$, $X \in \Gamma_f E$, defines a covariant derivative operator in E along f. A section $X \in \Gamma_f E$ is parallel if and only if $\nabla_U X = 0$ for all $U \in \mathfrak{X}N$. Finally (the chain rule), if $f: N \to M$ and $g: L \to N$ are C^∞, then (1.61) $\nabla(X \circ g) = (\nabla X) \circ g_*$ for $X \in \Gamma_f E$, where the first ∇ is along the map $f \circ g$, and the second ∇ is along f; in terms of covariant derivatives, $\nabla_U(X \circ g) = \nabla_{g_* U} X$, $U \in \mathfrak{X}L$.

PROOF The map $\nabla_U X = \kappa \circ X_* \circ U$ clearly belongs to $\Gamma_f E$. The rest of the proof comprises the next four lemmas.

2.53 **Lemma** With the hypotheses from 2.52, $X \in \Gamma_f E$ is parallel along f if and only if $\nabla_U X = 0$ for all $U \in \mathfrak{X}N$.

PROOF By 2.49, $\mathscr{H}$ is the kernel of the bundle homomorphism κ; but by 2.27 and 2.28, X is parallel along f if and only if $X_* TN \subset \mathscr{H}$.

2.54 **Lemma** With the hypotheses from 2.52, for each $X \in \Gamma_f E$ and $g: L \to N$, $\nabla(X \circ g) = (\nabla X) \circ g_*$.

PROOF Since $X \in \Gamma_f E$, $X \circ g \in \Gamma_{f \circ g} E$; thus for all $u \in TL$,

$$\nabla_u(X \circ g) = \kappa \circ (X \circ g)_* u = \kappa \circ X_* \circ g_* u = \kappa X_*(g_* u) = \nabla_{g_* u} X$$

2.55 **Lemma** With the hypotheses from 2.52, for each $u \in TN$, $X, Y \in \Gamma_f E$, and $h \in C^\infty(N, \mathbb{F})$,

$$\nabla_u(X + hY) = \nabla_u X + u(h)Y_p + h(p)\nabla_u Y \qquad u \in N_p$$

PROOF Addition and scalar multiplication are in the $\mathbb{F}$-vector bundle $\pi_*: TE \to TM$. If $u = \dot{\gamma}(0)$, γ a C^∞ curve in N, then by 1.32,

$$X_* u + Y_* u = A_*(X_* u, Y_* u) := \frac{d}{dt}\bigg|_0 (X \circ \gamma(t) + Y \circ \gamma(t))$$

$$= \frac{d}{dt}\bigg|_0 (X + Y) \circ \gamma(t) = (X + Y)_* u$$

Since κ is a bundle homomorphism along the map π_{TM} by 2.50,

$$\nabla_u(X + Y) = \kappa(X + Y)_* u = \kappa(X_* u + Y_* u) = \nabla_u X + \nabla_u Y$$

Since $h \circ \gamma$ is C^∞, $h \circ \gamma(t) = h(p) + tu(h) + o(t)$, $p := \gamma(0)$. Assume that X is parallel along γ; by the proof of 2.50,

$$(hX)_* u = \frac{d}{dt}\bigg|_0 ((h(p) + tu(h) + o(t))X \circ \gamma(t))$$

$$= h(p)X_* u + \mathscr{I}_{(hX)_p} u(h) X_p$$

Therefore $\nabla_u hX = u(h)X_p$ because $X_* u \in \mathscr{H}$.

Now suppose that $\{X_j\}$ is a local basis field for E near p such that each $X_j \circ \gamma$ is parallel. Given $Y \in \Gamma_f E$, let $Y = \sum Y^j X_j$ be the local expression for Y near p. For $h \in C^\infty(N, \mathbb{F})$,

$$\nabla_u hY = \sum \nabla_u hY^j X_j = \sum (u(h)Y^j(p) + h(p)u(Y^j))X_j|_p$$

$$= u(h) \sum (Y^j X_j)_p + h(p) \sum u(Y^j) X_j|_p$$

$$= u(h)Y_p + h(p)\nabla_u Y$$

2.56 **Definition** Let ∇ be the covariant derivative operator in E over M associated with a connection $\mathscr{H}$ on E. For each $f\colon N \to M$, the covariant derivative operator along f defined in theorem 2.52 is called the *pullback of* ∇ *along* f, and is also denoted by ∇.

Exercise Prove that $\tilde{\nabla}_U X = f^*(\nabla_U \operatorname{pr}_2 X)$ for $U \in \mathfrak{X}N$ and $X \in \Gamma f^*E$, where $\tilde{\nabla}$ is the covariant derivative operator in f^*E determined by the pullback connection $\tilde{\mathscr{H}}$ on f^*E.

2.57 If ∇ is the covariant derivative operator in E determined by a connection $\mathscr{H}$ on E over M, then the equation $\nabla_v X = \kappa X_* v$, $X \in \Gamma E$, $v \in TM$, implies that $\nabla_v X$ depends only on the behavior of X near $\pi(v)$. If γ is a C^∞ curve in M, then by 2.54, $\nabla_{\dot{\gamma}} X = \nabla_{\gamma_* D} X = \nabla_D(X \circ \gamma)$, $D = d/dt \in \mathfrak{X}\mathbb{R}$, so in fact $\nabla_v X$ can be calculated once X is known along any C^∞ curve in M with initial tangent vector v; this can be done directly in terms of parallel transport.

Proposition Let ∇ be the covariant derivative operator in E along $f\colon N \to M$ determined by a connection $\mathscr{H}$ on E. Let γ be a C^∞ curve in N; denote by $\mathbb{P}_t$ the parallel transport from $E_{f \circ \gamma(0)}$ to $E_{f \circ \gamma(t)}$ along $f \circ \gamma$. If $X \in \Gamma_f E$, then with $D = d/dt \in \mathfrak{X}\mathbb{R}$,

$$\nabla_{\dot{\gamma}(0)} X = \nabla_{D_0}(X \circ \gamma) = \frac{d}{dt}\bigg|_0 \mathbb{P}_t^{-1} X \circ \gamma(t)$$

PROOF Fix a parallel basis $\{Z_j\}$ for E along $f \circ \gamma$, and let $X \circ \gamma = \sum X^j Z_j$. For all t,

$$\mathbb{P}_t^{-1} X \circ \gamma(t) = \mathbb{P}_t^{-1} \sum X^j(t) Z_j|_t = \sum X^j(t) Z_j|_0$$

Thus

$$\nabla_{\dot{\gamma}(0)} X = \nabla_{\gamma_* D_0} X = \nabla_{D_0}(X \circ \gamma) = \sum \nabla_{D_0} X^j Z_j = \sum X^{j\prime}(0) Z_j|_0$$

$$= \frac{d}{dt}\bigg|_0 \mathbb{P}_t^{-1} X \circ \gamma(t)$$

2.58 Now for the converse to 2.52.

Theorem Let ∇ be a covariant derivative operator in a vector bundle $\pi: E \to M$. There exists a connection $\mathscr{H}$ on E such that given $X \in \Gamma E$ and $v \in TM$, $X_* v \in \mathscr{H}$ if and only if $\nabla_v X = 0$.

PROOF For each $\xi \in E$, define

$$\mathscr{H}_\xi := \{X_* v - \mathscr{I}_\xi \nabla_v X \mid X \in \Gamma E \text{ such that } X_{\pi(\xi)} = \xi,\ v \in M_{\pi(\xi)}\}$$

and set $\mathscr{H} := \bigcup_{\xi \in E} \mathscr{H}_\xi$. Let $\pi_{\mathscr{H}}(\mathscr{H}_\xi) := \xi$.

For each ξ, $\mathscr{H}_\xi$ is a linear subspace of E_ξ which is mapped isomorphically to $M_{\pi(\xi)}$ by $\pi_*|_\xi$; thus $E_\xi = \mathscr{H}_\xi \oplus \mathscr{V} E_\xi$.

If (x, U) is a chart on M, then $(\pi_{\mathscr{H}}, dx \circ \pi_*)$ is a vector bundle chart on $\mathscr{H}$ over $\pi^{-1}U$; it can be extended to a vector bundle chart on TE over $\pi^{-1}U$ as in the proof of 2.28. The bundle $\mathscr{H}$ is homogeneous because $(cX)_* v - \mathscr{I}_{c\xi} \nabla_v cX = \mu_{c*}(X_* v - \mathscr{I}_\xi \nabla_v X)$ for $c \in \mathbb{F}$. Thus $\mathscr{H}$ is a connection on E. The last part of the theorem follows directly from the definition of $\mathscr{H}$.

2.59 **Corollary** A covariant derivative operator ∇ in a vector bundle E over M determines a parallelism structure $\mathbb{P}$ in E such that given a curve γ in M and a section $X \in \Gamma_\gamma E$, X is parallel along γ if and only if $\nabla_{\dot{\gamma}} X = 0$.

Exercises

1. Fix a chart (x, U) on M, a vector bundle chart (π, ψ) on E over U, and the local basis $\{Y_j\}$ for E over U such that $\psi^i(Y_j) = \delta_{ij}$; a section $Z \in \Gamma E$ is parallel along a curve γ in U if and only if

$$(Z^k \circ \gamma)' + \sum_{i,j} dx^i(\dot{\gamma}) \cdot Z^j \circ \gamma \cdot \Gamma^k_{ij} \circ \gamma = 0 \qquad k = 1, \ldots, m$$

where $Z = \sum Z^j Y_j$ on U, and $\Gamma^k_{ij} := \psi^k \circ \nabla Y_j(\partial/\partial x^i) \in C^\infty(U, \mathbb{F})$.

2. Prove the corollary directly without invoking 2.58.
3. By analogy with 2.36, prove directly that every vector bundle admits a covariant derivative operator.
4. Theorems 2.28, 2.33, 2.52, 2.58, and 2.59 show that $\mathbb{P}$ in E, $\mathscr{H}$ on

E, and ∇ in E are equivalent geometric concepts; prove that the connection map κ is equivalent to $\mathbb{P}$, $\mathscr{H}$, and ∇.

For more details on the relationships among $\mathbb{P}$, $\mathscr{H}$, ∇, and κ, see [BrC], [Be], [Di], [Do*1*], [GKM], or [Pa].

2.60 Examples

(*a*) The coordinate basis fields $D_j := \partial/\partial u^j \in \mathfrak{X}\mathbb{R}^n$ are parallel, so if $Y = \sum Y^j D_j \in \mathfrak{X}\mathbb{R}^n$, then $\nabla_V Y = \sum V(Y^j)D_j$, $V \in \mathfrak{X}\mathbb{R}^n$; this is the ordinary vector derivative on $\mathbb{R}^n$.

(*b*) For each $c \in \mathbb{R}$, set $\nabla_X Y := c[X, Y]$, X, Y in the Lie algebra $\mathfrak{g}$ of a Lie group G; extend ∇ to a connection on G by 2.51: given $Z \in \mathfrak{X}G$, express Z as a linear combination over $C^\infty G$ of left-invariant vector fields, and then use the product rule to define $\nabla_X Z$, $X \in \mathfrak{g}$. This covariant derivative operator is left-invariant in the sense that $L_{g*}\nabla_X Y = \nabla_X L_{g*} Y$ for all $g \in G$, X, $Y \in \mathfrak{X}G$; equivalently, the parallel transport on G determined by ∇ is left-invariant (2.9). The choice $c = 0$ yields the canonical left-invariant connection on G from 2.8*a*.

Similarly, if ∇ is a connection on G for which there exists $c \in \mathbb{R}$ such that $\nabla_X Y = c[X, Y]$ for all right-invariant fields X, $Y \in \tilde{\mathfrak{g}}$ on G, then ∇ is a *right-invariant connection:* $R_{g*}\nabla_X Y = \nabla_X R_{g*} Y$ for all $g \in G$ and X, $Y \in \mathfrak{X}G$. The choice $c = 0$ yields the canonical right-invariant connection on G from 2.8*a*. *Exercise:* In this case, $\nabla_U V = [U, V]$ for all U, $V \in \mathfrak{g}$.

A connection on G is *bi-invariant* if it is both left- and right-invariant.

(*c*) Let $X \in \mathfrak{X}S^n$; extend X to a vector field on $\mathbb{R}^{n+1}$, which will also be denoted by X. Set $\tilde{X} := \mathrm{pr}_2 X$, so that for each $p \in \mathbb{R}^{n+1}$, $X_p = (p, \tilde{X}_p) \in \mathbb{R}^{n+1} \times \mathbb{R}^{n+1}$. For $v \in \mathbb{R}^{n+1}_p$, $X_* v = ((p, \tilde{X}_p), (v, d\tilde{X}(v)))$, and the Euclidean covariant derivative of X with respect to v from (*a*) is $\bar{\nabla}_v X = (p, d\tilde{X}(v))$.

Let κ be the connection map for the connection $\mathscr{H}$ on TS^n defined in 2.42. If $p \in S^n$ and $v \in S^n_p$, then the horizontal component

$$\mathscr{H}X_* v = ((p, \tilde{X}_p), (v, -\langle \tilde{X}_p, v\rangle p)) = ((p, \tilde{X}_p), (v, \langle d\tilde{X}(v), p\rangle p))$$

by eqs. (2-1) and (2-2) from 2.42, so

$$\begin{aligned} \nabla_v X = \kappa X_* v &= (p, \mathrm{pr}_2(X_* v - \mathscr{H}X_* v)) \\ &= (p, \mathrm{pr}_2((p, \tilde{X}_p), (0, d\tilde{X}(v) - \langle d\tilde{X}(v), p\rangle p))) \\ &= (p, d\tilde{X}(v) - \langle d\tilde{X}(v), p\rangle p) = (\bar{\nabla}_v X)^\perp \in S^n_p, \end{aligned}$$

the component of the Euclidean covariant derivative $\bar{\nabla}_v X$ tangent to

S^n at p. This close relation with the canonical connection on $\mathbb{R}^{n+1}$ is the reason for calling ∇ the canonical connection on S^n.

The connection ∇ on S^n is invariant under the action of $O(n+1)$: $g_* \nabla_v X = \nabla_v g_* X$, $g \in O(n+1)$, $X \in \mathfrak{X}S^n$, where the right-hand ∇ is along the map $g: S^n \to S^n$; this is another way of expressing the fact (2.8*e*) that parallel transport on S^n is invariant under $O(n+1)$.

(*d*) More generally, a covariant derivative operator ∇ on a homogeneous space G/H is *invariant*, $g_* \nabla_v X = \nabla_v g_* X$, $g \in G$, $v \in T(G/H)$, $X \in \mathfrak{X}(G/H)$, if and only if the associated parallel transport $\mathbb{P}$ on G/H is invariant as in 2.9; this happens if and only if $\mathscr{H}$ is invariant: $g_{**}\mathscr{H} = \mathscr{H}$, $g \in G$.

2.61 In 2.13 we saw that parallel transport in a vector bundle E determines parallel transport in the dual bundle E^*; now we see how the corresponding covariant derivatives are related.

Proposition Let ∇ be a covariant derivative operator in a vector bundle E over M. The corresponding covariant derivative operator in E^* is given by $(\overset{*}{\nabla}_v \omega)(X) = v\omega(X) - \omega(\nabla_v X)$ for $v \in TM$, $\omega \in \Gamma E^*$, and $X \in \Gamma E$; expressed another way, this is just the product rule:

$$v(\omega X) = (\overset{*}{\nabla}_v \omega)(X) + \omega(\nabla_v X)$$

where the first term is really the covariant derivative in $M \times \mathbb{F}$.

PROOF Let $v = \dot{\gamma}(0)$, γ a C^∞ curve in M. By 2.13, if $\mu \in \Gamma_\gamma E^*$ and $Y \in \Gamma_\gamma E$ are parallel, then $(\overset{*}{\nabla}_D \mu)(Y) = 0 = D\mu(Y) - \mu(\nabla_D Y)$, $D = d/dt \in \mathfrak{X}\mathbb{R}$; the desired equality follows since both $(\nabla_v \omega)(X)$ and $v\omega(X) - \omega(\nabla_v X)$ are tensorial in v and X, and a derivative in ω.

From now on, write ∇ instead of $\overset{*}{\nabla}$.

2.62 Proposition Let $\overset{j}{\nabla}$ be a covariant derivative operator in a vector bundle E_j over M, $j = 1, 2$. The corresponding covariant derivative operators in $E_1 \otimes E_2$ and $\mathrm{Hom}(E_1, E_2)$ are given by

$$\overset{\otimes}{\nabla}_V(X_1 \otimes X_2) = (\overset{1}{\nabla}_V X_1) \otimes X_2 + X_1 \otimes \overset{2}{\nabla}_V X_2 \qquad V \in \mathfrak{X}M,\ X_j \in \Gamma E_j$$

and $$(\overset{H}{\nabla}_V L)(X) = \overset{2}{\nabla}_V(LX) - L(\overset{1}{\nabla}_V X) \qquad L \in \Gamma\, \mathrm{Hom}(E_1, E_2),\ X \in \Gamma E_1$$

that is (the product rule),

$$\overset{2}{\nabla}_V(LX) = (\overset{H}{\nabla}_V L)(X) + L(\overset{1}{\nabla}_V X)$$

Let $E = E_1$. The covariant derivative operator in $\Lambda(E)$ is

$$\nabla_V(X_1 \wedge \cdots \wedge X_r) = \sum_{j=1}^{r} X_1 \wedge \cdots \wedge (\nabla_V X_j) \wedge \cdots \wedge X_r \qquad X_j \in \Gamma E$$

PROOF An exercise similar to the proof of 2.61.

From now on, write ∇ instead of $\overset{1}{\nabla}$, $\overset{2}{\nabla}$, $\overset{\otimes}{\nabla}$, and $\overset{H}{\nabla}$.

Exercise Compare ∇ in $\text{Hom}(E, M \times \mathbb{F})$ with ∇ in E^*.

THE STRUCTURE EQUATION

Given a connection ∇ in a vector bundle E over M, and $U, V \in \mathfrak{X}M$, the expressions ∇_U and ∇_V are differential operators on ΓE. In general these operators do not commute.

2.63 **Definition** Let ∇ be a covariant derivative operator in a vector bundle E along a map f: $N \to M$. For $U, V \in \mathfrak{X}N$ and $X \in \Gamma_f E$, define

$$C(U, V)X := [\nabla_U, \nabla_V]X - \nabla_{[U, V]}X = \nabla_U \nabla_V X - \nabla_V \nabla_U X - \nabla_{[U, V]}X$$

The first two terms of $C(U, V)X$ measure the failure of the operators ∇_U and ∇_V to commute, while the third term is what is needed for C to be tensorial (see 2.64). Another interpretation is that C measures the failure of ∇ to be a Lie algebra homomorphism from $\mathfrak{X}N$ to the Lie algebra of differential operators on $\Gamma_f E$. Most important, in 2.66 we shall see that if $N = M$ and $f = \text{id}$, then C is the curvature tensor R of the connection $\mathscr{H}$ associated with ∇.

2.64 **Proposition** The operator C is tensorial in all three arguments, and is alternating in the first two arguments.

Proof Given U, V, $W \in \mathfrak{X}N$, X, $Y \in \Gamma_f E$, $g \in C^\infty N$, and $h \in C^\infty(N, \mathbb{F})$,

$$\begin{aligned} C(gU, V)X &= \nabla_{gU}\nabla_V X - \nabla_V \nabla_{gU} X - \nabla_{[gU, V]} X \\ &= g\nabla_U \nabla_V X - \nabla_V g\nabla_U X - \nabla_{(g[U, V] - V(g)U)} X \\ &= g\nabla_U\nabla_V X - V(g)\nabla_U X - g\nabla_V\nabla_U X - g\nabla_{[U, V]}X \\ &\quad + V(g)\nabla_U X \\ &= gC(U, V)X \end{aligned}$$

Similarly,

$$C(U + V, W)X = C(U, W)X + C(V, W)X$$

Clearly $C(U, V)X = -C(V, U)X$, so C is also tensorial in the second argument. Analogous calculations establish that C is tensorial in the third slot:

$$C(U, V)hX = hC(U, V)X$$
$$C(U, V)(X + Y) = C(U, V)X + C(U, V)Y$$

2.65 Now we shall see that C is natural with respect to pullbacks.

Proposition Let ∇ be a connection in E over M, and let $\overset{f}{\nabla}$ be the pullback of ∇ along $f\colon N \to M$ (2.56). If C and $\overset{f}{C}$ are the tensor fields defined in 2.63 for ∇ and $\overset{f}{\nabla}$, respectively, then

$$C(f_* U, f_* V)Y = \overset{f}{C}(U, V)Y \in \Gamma_f E \qquad U, V \in \mathfrak{X}N, \ Y \in \Gamma_f E$$

Proof Fix U, $V \in \mathfrak{X}N$ and $Y \in \Gamma_f E$. Suppose that U and V are f-related to vector fields W and X on M, respectively; similarly, assume that there is a section $Z \in \Gamma E$ such that $Y = Z \circ f$. In this special case, by the chain rule for covariant derivatives (2.54)

$$\begin{aligned} \nabla_U \nabla_V Y &= \nabla_U \nabla_V Z \circ f = \nabla_U \nabla_{f_* V} Z = \nabla_U \nabla_{X \circ f} Z = \nabla_U (\nabla_X Z) \circ f \\ &= \nabla_{f_* U} \nabla_X Z = \nabla_{W \circ f} \nabla_X Z = (\nabla_W \nabla_X Z) \circ f \end{aligned}$$

Similarly, $\nabla_V \nabla_U Y = (\nabla_X \nabla_W Z) \circ f$, and $\nabla_{[U, V]} Y = (\nabla_{[W, X]} Z) \circ f$.

Thus,

$$\begin{aligned}\overset{f}{C}(U, V)Y &= \overset{f}{\nabla}_U \overset{f}{\nabla}_V Y - \overset{f}{\nabla}_V \overset{f}{\nabla}_U Y - \overset{f}{\nabla}_{[U, V]} Y \\ &= (\nabla_W \nabla_X Z - \nabla_X \nabla_W Z - \nabla_{[W, X]} Z) \circ f = (C(W, X)Z) \circ f \\ &= C(W \circ f, X \circ f)Z \circ f = C(f_* U, f_* V)Y\end{aligned}$$

Although not all vector fields on N are f-related to vector fields on M, the set of all $W \circ f$, $W \in \mathfrak{X}M$, does contain local basis fields for $\Gamma_f TM$ over $C^\infty N$; for example, given $U \in \mathfrak{X}N$ and a chart x on M, on f^{-1} (domain of x) in N, $f_* U = \sum U^j \cdot (\partial/\partial x^j) \circ f$, where the functions U^j are C^∞ on an open set in N. Similarly, the set of all $Z \circ f$, $Z \in \Gamma E$, contains local basis fields for $\Gamma_f E$ over $C^\infty(N, \mathbb{F})$. Since C and $\overset{f}{C}$ are both tensorial, this proves the required identity.

2.66 Proposition If R is the curvature tensor field of the connection $\mathscr{H}$ on E corresponding to ∇ in E, then $R = C$, that is,

$$R(U, V)X = \nabla_U \nabla_V X - \nabla_V \nabla_U X - \nabla_{[U, V]} X \qquad U, V \in \mathfrak{X}M, \ X \in \Gamma E$$

PROOF By tensoriality, at a point $p \in M$, $R(U, V)X$ and $C(U, V)X$ depend just on $u := U_p$, $v := V_p$, and $\xi := X_p$.

Let D_j be the usual jth coordinate field on $\mathbb{R}^2$, and let f be a C^∞ map from a neighborhood of 0 in $\mathbb{R}^2$ into M such that $f_* D_1|_0 = u$, $f_* D_2|_0 = v$. Define a section Y of E along f such that $Y_0 = \xi$ and

(i) Y is parallel along the curve $t \mapsto f(0, t)$,

(ii) for all t, Y is parallel along the curve $s \mapsto f(s, t)$.

By (ii), 2.65, and the equality $[D_1, D_2] = 0$ on $\mathbb{R}^2$,

$$\begin{aligned}C(u, v)\xi &= C(f_* D_1|_0, f_* D_2|_0)\xi \\ &= (\nabla_{D_1} \nabla_{D_2} Y - \nabla_{D_2} \nabla_{D_1} Y - \nabla_{[D_1, D_2]} Y)_0 = \nabla_{D_1|_0} \nabla_{D_2} Y\end{aligned}$$

By 2.57,

$$\nabla_{D_1|_0} \nabla_{D_2} Y = \left.\frac{d}{ds}\right|_0 \mathbb{P}_s^{-1}(\nabla_{D_2} Y)_{(s, 0)}$$

where $\mathbb{P}_s\colon E_p \to E_{f(s, 0)}$ denotes parallel translation in E along the curve $f(\sigma, 0)$, $0 \leq \sigma \leq s$; similarly, if $\mathbb{P}_{s,t}\colon E_{f(s, 0)} \to E_{f(s, t)}$ denotes parallel translation in E along the curve $f(s, \tau)$, $0 \leq \tau \leq t$,

$$(\nabla_{D_2} Y)_{(s, 0)} = \left.\frac{\partial}{\partial t}\right|_0 \mathbb{P}_{s,t}^{-1}(Y_{(s, t)})$$

Thus

$$C(u, v)\xi = \frac{\partial^2}{\partial s\,\partial t}\bigg|_0 \mathbb{P}_s^{-1}\mathbb{P}_{s,t}^{-1} Y_{(s,t)}$$

$$= \lim_{s,t\to 0} \frac{\mathbb{P}_s^{-1}\mathbb{P}_{s,t}^{-1} Y_{s,t} - \mathbb{P}_s^{-1}\mathbb{P}_{s,0}^{-1} Y_{s,0} - \mathbb{P}_0^{-1}\mathbb{P}_{0,t}^{-1} Y_{0,t} + \mathbb{P}_0^{-1}\mathbb{P}_{0,0}^{-1} Y_{0,0}}{st}$$

$$= \lim_{s,t\to 0} \frac{\mathbb{P}_s^{-1}\mathbb{P}_{s,t}^{-1} Y_{s,t} - \xi}{st}$$

since $Y_{s,0} = \mathbb{P}_s\xi$, $Y_{0,t} = \mathbb{P}_{0,t}\xi$, and $\mathbb{P}_{s,0} = \mathrm{id}$. Therefore

$$C(u, v)\xi = \lim_{s\to 0} \frac{\mathbb{P}_s^{-1}\mathbb{P}_{s,s}^{-1} Y_{s,s} - \xi}{s^2}$$

Substitute $\sqrt{t}$ for s, let t go to zero, and compare with eq. (2-3) from 2.44; in the terminology from 2.44, $C(u, v)\xi = -\dot{\alpha}_\xi(0) = R(u, v)\xi$.

2.67 Theorem The structure equation: Let ∇ be a connection in a vector bundle E over M, with curvature tensor R. For each C^∞ map $f: N \to M$, and all $U, V \in \mathfrak{X}N$, $X \in \Gamma_f E$,

$$R(f_* U, f_* V)X = \nabla_U \nabla_V X - \nabla_V \nabla_U X - \nabla_{[U,V]} X$$

where here ∇ is the covariant derivative operator along f.

PROOF Apply 2.65 and 2.66.

2.68 ***The curvature tensor of the canonical connection on S^n*** Since the canonical connection on S^n is invariant under the transitive group action of $SO(n+1)$ on S^n, it suffices to calculate R at a single point, say $p = e_0$. Let $u := (p, e_1)$ and $v := (p, e_2) \in S^n_p$; it follows from the holonomy discussion in 2.42 that if $w \in S^n_p$ is orthogonal to u and v, then $R(u, v)w = 0$. Thus it is enough to calculate $R(u, v)u$ and $R(u, v)v$, and for this we may as well work in S^2.

For all vectors (p, x) and $(p, y) \in S^2_p$, define $X, Y \in \mathfrak{X}S^2$ by

$$X_q := (q, x - \langle x, q\rangle q) \cong x - \langle x, q\rangle q \in \mathbb{R}^3$$
$$Y_q := (q, y - \langle y, q\rangle q) \cong y - \langle y, q\rangle q \in \mathbb{R}^3$$

From now on suppress the basepoint q. By 2.60c, $(\nabla_X Y)_q = (\bar{\nabla}_X Y)_q^\perp$, where $\bar{\nabla}$ is the canonical connection on $\mathbb{R}^3$. But for all $q \in S^2$,

$$(\bar{\nabla}_X Y)_q = \frac{d}{dt}\bigg|_0 Y(q + t(x - \langle x, q\rangle q))$$

$$= \frac{d}{dt}\bigg|_0 (y - \langle y, q + t(x - \langle x, q\rangle q)\rangle(q + t(x - \langle x, q\rangle q)))$$

$$= 2\langle x, q\rangle\langle y, q\rangle q - \langle x, y\rangle q - \langle y, q\rangle x \in \mathbb{R}^3_q$$

Thus

$$(\nabla_X Y)_q = (\bar{\nabla}_X Y)_q - \langle \bar{\nabla}_X Y)_q, q\rangle q$$

$$= \langle x, q\rangle\langle y, q\rangle q - \langle y, q\rangle x \in S^2_q$$

In particular, the covariant differential $(\nabla Y)_p = 0$.

Define $U_q := e_1 - \langle e_1, q\rangle q$ and $V_q := e_2 - \langle e_2, q\rangle q$; since $u = (p, e_1)$, the formula above implies

$$\bar{\nabla}_u \nabla_V V = \frac{d}{dt}\bigg|_0 ((\nabla_V V)(p + te_1))$$

$$= \frac{d}{dt}\bigg|_0 (\langle e_2, p + te_1\rangle^2(p + te_1) - \langle e_2, p + te_1\rangle e_2) = 0$$

hence $\nabla_u \nabla_V V = 0$. Similarly,

$$\bar{\nabla}_v \nabla_U V = \frac{d}{dt}\bigg|_0 ((\nabla_U V)(p + te_2))$$

$$= \frac{d}{dt}\bigg|_0 (\langle e_1, p + te_2\rangle\langle e_2, p + te_2\rangle(p + te_2)$$

$$- \langle e_2, p + te_2\rangle e_1)$$

$$= -e_1 = -u$$

Since $\langle e_1, p\rangle = 0$, $\nabla_v \nabla_U V = \bar{\nabla}_v \nabla_U V = -u$. Finally, $(\nabla V)_p = 0$, so

$$R(u, v)v = \nabla_u \nabla_V V - \nabla_v \nabla_U V - \nabla_{[U, V]_p} V = u$$

By symmetry, $R(v, u)u = v$, so $R(u, v)u = -R(v, u)u = -v$.

The formulas

$$R(u, v)u = -v \qquad R(u, v)v = u \qquad R(u, v)w = 0 \qquad \text{for } w \perp u, w \perp v$$

express $R(u, v)$ for all $p \in S^n$ and all orthonormal $u, v \in S^n_p$ because $O(n + 1)$ acts transitively on S^n and preserves the canonical inner product and connection on both $\mathbb{R}^{n+1}$ and S^n.

Exercise Verify theorem 2.47 for the canonical connection on S^n.

2.69 Proposition Let ∇ be a connection in E, with its natural extension to $\Lambda(E)$. Given $U, V \in \mathfrak{X}M$, the operator $R(U, V) \in \Gamma \operatorname{End} \Lambda(E)$ acts as a derivation: for all $X_j \in \Gamma E$,

$$R(U, V)(X_1 \wedge \cdots \wedge X_r) = \sum_{j=1}^{r} X_1 \wedge \cdots \wedge R(U, V)X_j \wedge \cdots \wedge X_r$$

Proof Assume $r = 2$; the general proof is similar. By tensoriality, we may assume that $[U, V] = 0$. By 2.62,

$$\begin{aligned} R(U, V)(X \wedge Y) &= \nabla_U \nabla_V (X \wedge Y) - \nabla_V \nabla_U (X \wedge Y) \\ &= \nabla_U((\nabla_V X) \wedge Y + X \wedge \nabla_V Y) \\ &\quad - \nabla_V((\nabla_U X) \wedge Y + X \wedge \nabla_U Y) \\ &= (\nabla_U \nabla_V X) \wedge Y + (\nabla_V X) \wedge \nabla_U Y + (\nabla_U X) \wedge \nabla_V Y \\ &\quad + X \wedge \nabla_U \nabla_V Y - (\nabla_V \nabla_U X) \wedge Y - (\nabla_U X) \wedge \nabla_V Y \\ &\quad - (\nabla_V X) \wedge \nabla_U Y - X \wedge \nabla_V \nabla_U Y \\ &= (R(U, V)X) \wedge Y + X \wedge R(U, V)Y \end{aligned}$$

2.70 Proposition Let ∇ be a connection in E. The curvature tensors in $\Lambda(E)$ and $\Lambda(E^*)$ are related by the identity $R(U, V)\mu = -\mu \circ R(U, V) = -{}^tR(U, V)(\mu)$ for $\mu \in \Gamma\Lambda(E^*)$ and $U, V \in \mathfrak{X}M$, where ${}^tR(U, V) \in \Gamma \operatorname{End} \Lambda(E^*)$ is the transpose of $R(U, V) \in \Gamma \operatorname{End} \Lambda(E)$.

Proof Let $\mu \in \Gamma\Lambda^r E^*$ and $X \in \Gamma\Lambda^r E$. Fix $U, V \in \mathfrak{X}M$; by tensoriality we may assume that $[U, V] = 0$. By 2.61,

$$\begin{aligned} (\nabla_U \nabla_V \mu)(X) &= U((\nabla_V \mu)(X)) - (\nabla_V \mu)(\nabla_U X) \\ &= UV\mu(X) - (\nabla_V \mu)(\nabla_U X) - (\nabla_U \mu)(\nabla_V X) - \mu(\nabla_U \nabla_V X) \end{aligned}$$

Combining this with the similar expression for $-(\nabla_V \nabla_U \mu)(X)$ we obtain

$$(R(U, V)\mu)(X) = -\mu(\nabla_U \nabla_V X - \nabla_V \nabla_U X) = -\mu \circ R(U, V)(X)$$

THE SPACE OF CONNECTIONS ON A VECTOR BUNDLE

Fix a vector bundle E over a manifold M. If $\overset{1}{\nabla}$ and $\overset{2}{\nabla}$ are connections on E, then for all $t \in \mathbb{R}$, the operator $\nabla := t\overset{1}{\nabla} + (1 - t)\overset{2}{\nabla}$ is also a connection on

E: $\nabla_U X = t\overset{1}{\nabla}_U X + (1-t)\overset{2}{\nabla}_U X$, $U \in \mathfrak{X}M$, $X \in \Gamma E$. Thus [Gr: 9] the space of connections on E is a real affine space. We will now see that this affine space is modeled on the vector space $A^1(M, \operatorname{End} E)$ of endomorphism-valued 1-forms on M (1.55).

2.71 **Proposition** Let ∇ be a connection on E, and let $\mathscr{D} \in A^1(M, \operatorname{End} E) \cong A^1(M) \otimes \operatorname{End} \Gamma E \cong \Gamma \operatorname{Hom}(TM, \operatorname{End} E)$. The operator $\bar{\nabla} := \nabla + \mathscr{D}$ on $\mathfrak{X}M \times \Gamma E$ is a connection on E. Conversely, if $\bar{\nabla}$ and ∇ are connections on M, then $\mathscr{D} := \bar{\nabla} - \nabla$ belongs to $A^1(M, \operatorname{End} E)$.

PROOF Given $U \in \mathfrak{X}M$, $X \in \Gamma E$, and $g \in C^\infty(M, \mathbb{F})$,

$$\begin{aligned}\bar{\nabla}_U gX &= \nabla_U gX + \mathscr{D}(U)gX \\ &= U(g)X + g\nabla_U X + g\mathscr{D}(U)X \\ &= U(g)X + g\bar{\nabla}_U X\end{aligned}$$

Similarly, $\bar{\nabla}_U(X+Y) = \bar{\nabla}_U X + \bar{\nabla}_U Y$. Linearity over $C^\infty M$ in the first argument is checked in the same way, as is the converse.

2.72 **Definition** If $\bar{\nabla}$ and ∇ are connections on a vector bundle E over a manifold M, $\mathscr{D} := \bar{\nabla} - \nabla \in A^1(M, \operatorname{End} E)$ is called the *connection difference* 1*-form* (or *connection difference tensor field*).

2.73 **Proposition** Let $\mathscr{H}$ and $\bar{\mathscr{H}}$ be connections on a vector bundle E over M, with corresponding covariant derivative operators ∇ and $\bar{\nabla}$ in E; let $\mathscr{D} := \bar{\nabla} - \nabla$ be the connection difference form. For all $\xi \in E$,

$$\bar{\mathscr{H}}_\xi = \{z - \mathscr{J}_\xi \mathscr{D}(\pi_* z)\xi \mid z \in \mathscr{H}_\xi\}$$

PROOF Fix $p \in M$, $\xi \in E_p$, and $z \in \mathscr{H}_\xi$. Let $X \in \Gamma E$ such that $(\nabla X)_p = 0$ and $X_p = \xi$. If $u = \pi_* z \in M_p$, then $z = X_* u \in \mathscr{H}_\xi$.

Denote by $\bar{\kappa}$: $TE \to E$ the connection map of $\bar{\nabla}$; then

$$\bar{\kappa} X_* u = \bar{\nabla}_u X = \nabla_u X + \mathscr{D}(u)\xi = \mathscr{D}(u)\xi = \bar{\kappa}\mathscr{J}_\xi \mathscr{D}(u)\xi$$

so $\bar{\kappa}(z - \mathscr{J}_\xi \mathscr{D}(\pi_* z)\xi) = 0$, and $z - \mathscr{J}_\xi \mathscr{D}(\pi_* z)\xi \in \bar{\mathscr{H}}_\xi$.

Since π_* maps $\bar{\mathscr{H}}_\xi$ and $\mathscr{H}_\xi$ isomorphically to M_p and $\pi_* z$ equals $\pi_*(z - \mathscr{J}_\xi \mathscr{D}(\pi_* z)\xi)$, this yields all of $\bar{\mathscr{H}}_\xi$.

So far in our discussion we have used only the vector bundle structure of End E; to bring holonomy into the picture we need a Lie algebra structure on the fibers, so from now on we will think of the connection difference form $\mathscr{D}$ as a 1-form on M with values in the Lie algebra bundle $\mathfrak{gl}(E)$ (1.41*e*), $\mathscr{D} \in A^1(M, \mathfrak{gl}(E))$.

2.74 Let E be a vector bundle over a connected manifold M with standard fiber a vector space $\mathscr{V}$; let G be a Lie subgroup of $GL(\mathscr{V})$. Recall from 2.48 that a G-connection on E is a connection $\mathscr{H}$ on E whose holonomy group (2.19) is contained in G. Equivalently, given a Lie algebra $\mathfrak{g}$, a connection $\mathscr{H}$ on E is a G-connection for some Lie group G with Lie algebra $\mathfrak{g}$ if and only if its holonomy algebra bundle (2.46) is a Lie algebra subbundle $\mathfrak{g}(E)$ of $\mathfrak{gl}(E)$ with fiber isomorphic to the Lie algebra $\mathfrak{g}$. In this case the curvature tensor of $\mathscr{H}$ is a 2-form on M with values in $\mathfrak{g}(E)$, $R \in A^2(M, \mathfrak{g}(E))$, and the extension of $\mathscr{H}$ to a connection on $\mathfrak{gl}(E)$ (2.14, 2.62) is reducible to $\mathfrak{g}(E)$ (2.45). Two G-connections on E might have different holonomy algebra bundles.

Proposition Let $\mathscr{H}$ be a connection on a vector bundle E over a connected manifold M, with covariant derivative operator ∇ in E. Assume $\mathscr{H}$ is a G-connection on E, with holonomy algebra bundle $\mathfrak{g}(E)$ in $\mathfrak{gl}(E)$. If $\mathscr{D} \in A^1(M, \mathfrak{g}(E))$ is a 1-form with values in $\mathfrak{g}(E)$, then $\bar{\nabla} := \nabla + \mathscr{D}$ is also a G-connection on E with holonomy algebra bundle $\mathfrak{g}(E)$.

PROOF Let $\bar{\mathscr{H}}$ and $\mathscr{H}$ be the extensions to $\mathfrak{gl}(E)$ of the respective connections on E; by assumption, if $\zeta \in \mathfrak{g}(E) \subset \mathfrak{gl}(E)$, then the horizontal space $\mathscr{H}_\zeta$ is tangent to the submanifold $\mathfrak{g}(E)$ at ζ (2.45). By 2.43, for each $v \in \mathscr{H}_\zeta$, the vector $v - \mathscr{I}_\zeta \hat{\mathscr{D}}(\pi_* v)\zeta$ belongs to $\bar{\mathscr{H}}_\zeta$, where $\hat{\mathscr{D}}$ is the difference of the connections $\bar{\nabla}$ and ∇ applied to sections of the bundle $\mathfrak{gl}(E)$. Given $U \in \mathfrak{X}M$, $L \in \Gamma\mathfrak{gl}(E)$, and $X \in \Gamma E$,

$$\begin{aligned}(\hat{\mathscr{D}}(U)L)(X) &= (\bar{\nabla}_U L - \nabla_U L)(X)\\ &= \bar{\nabla}_U LX - L\bar{\nabla}_U X - \nabla_U LX + L\nabla_U X\\ &= \mathscr{D}(U)LX - L\mathscr{D}(U)X = [\mathscr{D}(U), L](X)\end{aligned}$$

Therefore $\hat{\mathscr{D}}(U)L$ is the Lie bracket $[\mathscr{D}(U), L]$ in the bundle $\mathfrak{gl}(E)$. Since $\mathscr{D}$ is $\mathfrak{g}(E)$-valued and $\mathfrak{g}(E)$ is a Lie algebra subbundle of $\mathfrak{gl}(E)$, If $L \in \Gamma\mathfrak{g}(E)$, then $\hat{\mathscr{D}}(U)L = [\mathscr{D}(U), L]$ is a section of $\mathfrak{g}(E)$. Hence, given $\zeta \in \mathfrak{g}(E)$ and $v \in \mathscr{H}_\zeta$, the vector $v - \mathscr{I}_\zeta \hat{\mathscr{D}}(\pi_* v)\zeta = v - \mathscr{I}_\zeta[\mathscr{D}(\pi_* v), \zeta] \in \bar{\mathscr{H}}_\zeta \subset \mathfrak{gl}(E)_\zeta$ lies in the subspace $\mathfrak{g}(E)_\zeta$. Thus $\bar{\mathscr{H}}_\zeta$ is tangent to $\mathfrak{g}(E)$ at ζ, and the connection $\bar{\mathscr{H}}$ on $\mathfrak{gl}(E)$ is reducible to $\mathfrak{g}(E)$.

This proposition gives the following information about the space of connections on a vector bundle E over a connected manifold M. Fix a Lie subgroup G of the general linear group of the fiber of E, and suppose there exists a Lie algebra subbundle $\mathfrak{g}(E)$ of $\mathfrak{gl}(E)$ with standard fiber isomorphic to the Lie algebra of G. The space of connections on E

with holonomy algebra bundle $\mathfrak{g}(E)$ (such connections exist by [KN: II.8.2]) is an affine subspace of the space of connections on E which is modeled on the vector space $A^1(M, \mathfrak{g}(E))$.

2.75 At this point it is useful to study the exterior covariant derivative associated with a connection on a vector bundle.

Definition Given a connection ∇ in a vector bundle E over M, define the *exterior covariant derivative operator with respect to* ∇ to be the collection of maps $d^\nabla\colon A^r(M, E) \to A^{r+1}(M, E)$ such that

$$[d^\nabla\psi](U) := \nabla_U \psi \qquad \psi \in A^0(M, E),\ U \in \mathfrak{X}M$$

$$[d^\nabla\psi](U_0, \ldots, U_r) := \sum_j (-1)^j \nabla_{U_j}\psi(U_0, \ldots, \hat{U}_j, \ldots, U_r)$$

$$+ \sum_{j<k} (-1)^{j+k}\psi([U_j, U_k], U_0, \ldots, \hat{U}_j, \ldots, \hat{U}_k, \ldots, U_r)$$

for $\psi \in A^r(M, E)$ and $U_j \in \mathfrak{X}M$.

2.76 **Proposition** If ∇ is the product connection in the bundle $E = M \times \mathbb{F}$, then d^∇ is the usual exterior derivative operator d. Exterior differentiation with respect to ∇ is natural with respect to maps, that is, if $\bar{\nabla}$ is the pullback of ∇ along a map $f\colon N \to M$, then $d^{\bar{\nabla}} f^*\psi = f^* d^\nabla \psi$ for $\psi \in A(M, E)$. Furthermore, if E is an algebra bundle and ∇ is a connection such that multiplication is parallel (that is, if X and Y are parallel sections along a curve γ, then $X \cdot Y$ is required to be parallel along γ), then d^∇ satisfies the product rule

$$d^\nabla(\varphi \cdot \psi) = (d^\nabla\varphi) \cdot \psi + (-1)^r \varphi \cdot d^\nabla\psi \qquad \varphi \in A^r(M, E),\ \psi \in A(M, E)$$

Proof The proofs are analogous to the standard proofs in the special case where $E = M \times \mathbb{R}$ [W: 2.20, 2.23].

2.77 The operator $d^{\nabla 2} = d^\nabla \circ d^\nabla$ is not always zero, unlike d^2.

Proposition For all $X \in \Gamma E$, $d^{\nabla 2}X(U, V) = R(U, V)X$. Thus $d^{\nabla 2} = 0$ if and only if the bundle (E, ∇) is flat.

Proof For $U, V \in \mathfrak{X}M$ and $X \in \Gamma E$,

$$(d^{\nabla 2}X)(U, V) = \nabla_U d^\nabla X(V) - \nabla_V d^\nabla X(U) - d^\nabla X([U, V])$$

$$= \nabla_U \nabla_V X - \nabla_V \nabla_U X - \nabla_{[U, V]} X = R(U, V)X$$

2.78 Proposition Let $\bar{\nabla}$ and ∇ be connections on a vector bundle E over a manifold M, and let $\mathscr{D} := \bar{\nabla} - \nabla \in A^1(M, \mathrm{gl}(E))$ be the connection difference form. If $\bar{R}$ and R are the curvature tensors of $\bar{\nabla}$ and ∇, respectively, then $\bar{R} - R = d^\nabla \mathscr{D} + \frac{1}{2}[\mathscr{D}, \mathscr{D}]$ (1.56*b*).

Proof Given $U, V \in \mathfrak{X}M$ and $X \in \Gamma E$,

$$\begin{aligned}
\bar{R}(U, V)X &= \bar{\nabla}_U(\nabla_V X + \mathscr{D}(V)X) - \bar{\nabla}_V(\nabla_U X + \mathscr{D}(U)X) - \bar{\nabla}_{[U, V]}X \\
&= R(U, V)X + \mathscr{D}(U)\nabla_V X + \nabla_U \mathscr{D}(V)X + \mathscr{D}(U)\mathscr{D}(V)X \\
&\quad - \mathscr{D}(V)\nabla_U X - \nabla_V \mathscr{D}(U)X - \mathscr{D}(V)D(U)X - \mathscr{D}([U, V])X \\
&= R(U, V)X + (\nabla_U \mathscr{D}(V))(X) - (\nabla_V \mathscr{D}(U))(X) \\
&\quad - \mathscr{D}([U, V])X + [\mathscr{D}(U), \mathscr{D}(V)]X \\
&= R(U, V)X + (d^\nabla \mathscr{D}(U, V))(X) + (\tfrac{1}{2}[\mathscr{D}, \mathscr{D}](U, V))(X)
\end{aligned}$$

Exercises Give a second proof of 2.74.

2.79 Proposition The Bianchi identity: If R is the curvature tensor field of a connection ∇ in a vector bundle E over M, then $d^\nabla R = 0$.

Proof Given $U, V, W \in \mathfrak{X}M$, we must show that

$$\begin{aligned}
&\nabla_U R(V, W) + \nabla_V R(W, U) + \nabla_W R(U, V) \\
&\qquad - R([U, V], W) - R([V, W], U) - R([W, U], V) = 0
\end{aligned}$$

Since the expression on the left is tensorial in U, V, and W, we may assume that $[U, V] = [U, W] = [V, W] = 0$. Thus for $X \in \Gamma E$,

$$\begin{aligned}
&(d^\nabla R(U, V, W))(X) \\
&\quad = \nabla_U R(V, W)X - R(V, W)\nabla_U X + \nabla_V R(W, U)X \\
&\qquad - R(W, U)\nabla_V X + \nabla_W R(U, V)X - R(U, V)\nabla_W X \\
&\quad = \nabla_U \nabla_V \nabla_W X - \nabla_U \nabla_W \nabla_V X - \nabla_V \nabla_W \nabla_U X \\
&\qquad + \nabla_W \nabla_V \nabla_U X + \nabla_V \nabla_W \nabla_U X - \nabla_V \nabla_U \nabla_W X \\
&\qquad - \nabla_W \nabla_U \nabla_V X + \nabla_U \nabla_W \nabla_V X + \nabla_W \nabla_U \nabla_V X \\
&\qquad - \nabla_W \nabla_V \nabla_U X - \nabla_U \nabla_V \nabla_W X + \nabla_V \nabla_U \nabla_W X \\
&\quad = 0
\end{aligned}$$

Exercise Rephrase the proof as follows: for all $X \in \Gamma E$, define

$RX \in A^2(M, E)$ by $(RX)(U, V) := R(U, V)X$; then

$$(d^\nabla R)(U, V, W)(X) = (d^\nabla(RX) - R \wedge d^\nabla X)(U, V, W)$$
$$= (d^\nabla \circ d^{\nabla 2}(X) - d^{\nabla 2} \circ d^\nabla(X))(U, V, W) = 0$$

CHARACTERISTIC CLASSES

The Bianchi identity has a very important application to the study of the de Rham cohomology of a manifold M. Given a connection on a vector bundle E over M, we will construct certain elements of $H^*(M, \mathbb{R})$; although the differential forms which represent these cohomology classes depend on the connection used in their construction, the classes themselves do not. Instead they depend only on the vector bundle E; for this reason they are called *characteristic classes of E*. It is to be emphasized that characteristic classes of E are cohomology classes on M. Characteristic classes will be dealt with again briefly in chaps. 3, 8, and 9. For more details, see [KN: XII] or [MiS].

2.80 To illustrate the idea, let us consider the simplest possible case. Fix a connection ∇ on a vector bundle E over M. For each $p \in M$ and u, $v \in M_p$, the curvature operator $R(u, v)$ is an endomorphism of the vector space E_p; its trace is well-defined, and can be calculated using any basis for E_p. The map $(u, v) \mapsto \operatorname{tr} R(u, v)$ is an alternating bilinear form on M_p. Define a 2-form on M by $\mu(U, V) := \operatorname{tr} R(U, V)$ for all $U, V \in \mathfrak{X}M$. By the Bianchi identity (2.79), if the brackets of U, V, W are zero, then

$$d\mu(U, V, W) = U \operatorname{tr} R(V, W) - V \operatorname{tr} R(U, W) + W \operatorname{tr} R(U, V)$$
$$= \operatorname{tr}(\nabla_U R(V, W) - \nabla_V R(U, W) + \nabla_W R(U, V)) = 0$$

The step in which ∇ is moved past tr will be justified later in a more general setting. Thus μ is a closed 2-form on M, and therefore represents a de Rham cohomology class [W: 4.17]. Although it is not obvious, this class depends only on E and the trace function on End E. Now we look at more general functions similar to the trace.

2.81 **Definition** Let $\mathbb{F} = \mathbb{R}$ or $\mathbb{C}$. A *polynomial of degree k* on a vector space $\mathscr{V}$ over $\mathbb{F}$ is a map $f: \mathscr{V} \to \mathbb{F}$ such that with respect to a basis $\{\mu^j\}$ for $\mathscr{V}^*$, there exist constants $c_{j_1 \cdots j_k}$ for which

$$f(v) = \sum c_{j_1 \cdots j_k} \mu^{j_1}(v) \cdots \mu^{j_k}(v) \qquad v \in \mathscr{V}$$

We may assume that the coefficients $c_{j_1 \cdots j_k}$ are symmetric in the indices $j_1, \ldots, j_k$. The *polarization of* $f = \sum c_{j_1 \cdots j_k} \mu^{j_1} \cdots \mu^{j_k}$ is the polynomial $\bar{f} := \sum c_{j_1 \cdots j_k} \mu^{j_1} \otimes \cdots \otimes \mu^{j_k}$ on $\mathscr{V} \otimes \cdots \otimes \mathscr{V}$ (where the coefficients are assumed to be symmetric in the indices). If $\mathfrak{g}$ is the Lie algebra of a Lie group G, then an *invariant polynomial on* $\mathfrak{g}$ is a polynomial which is invariant under the adjoint action of G on $\mathfrak{g}$: $f \circ \mathrm{Ad}_g = f$ for all $g \in G$ (W: 3.46).

The polarization of a polynomial is independent of the choice of basis $\{\mu^j\}$ for $\mathscr{V}^*$; a polynomial on $\mathfrak{g}$ is invariant if and only if its polarization is invariant.

For example, both the trace and the determinant functions are invariant polynomials on $\mathfrak{gl}(n, \mathbb{F})$. More generally, the equation

$$\det(\lambda I + A) =: \sum_{k=0}^{n} f_k(A)\lambda^{n-k} \qquad A \in \mathfrak{gl}(n, \mathbb{F}),\ \lambda \in \mathbb{F}$$

defines invariant polynomials $f_0, \ldots, f_n$. It is easily seen that

$$f_0 = 1, \qquad f_1(A) = \operatorname{tr} A, \quad f_2(A) = \sum_{j<k} \det \begin{bmatrix} A_{jj} & A_{jk} \\ A_{kj} & A_{kk} \end{bmatrix},$$

$$\ldots, \qquad f_{n-1}(A) = \sum_j \text{cofactor of } A_{jj}, \qquad f_n(A) = \det A$$

2.82 **Proposition** Let E be a vector bundle over M with standard fiber equal to a vector space $\mathscr{V}$. If f is an invariant polynomial on $\mathfrak{gl}(\mathscr{V})$, then f determines a function (also denoted by f) on the Lie algebra bundle $\mathfrak{gl}(E)$. As a section of $\otimes(\mathfrak{gl}(E))^*$, f is parallel with respect to the natural extension of any connection in E.

PROOF For each $p \in M$, fix a basis b for E_p and its dual basis b^* for E_p^*; b and b^* determine an isomorphism $\hat{b}$ from $\mathfrak{gl}(E_p)$ to $\mathfrak{gl}(\mathscr{V})$. For each $L \in \mathfrak{gl}(E_p)$, define $f(L)$ to be $f \circ \hat{b}(L)$; this is well-defined by the invariance of f.

If β is a parallel basis field for E along a curve γ in M, then each parallel section L of $\mathfrak{gl}(E)$ along γ is a constant-coefficient linear combination of the basis fields in β and the dual basis fields in β^* by 2.14. Hence $\hat{\beta}(L)$ is a constant curve in $\mathfrak{gl}(\mathscr{V})$, which means that $f(L) := f \circ \hat{\beta}(L)$ is constant along γ; by 2.13 and 2.14, f is parallel.

2.83 Consider the general linear algebra bundle $\mathfrak{gl}(E)$ of a vector bundle E over a manifold M. Since $\mathfrak{gl}(E)$ is one of the summands of the tensor bundle $\otimes(\mathfrak{gl}(E))$ (1.41*a*), each $\mathfrak{gl}(E)$-valued differential form $\psi \in A(M, \mathfrak{gl}(E))$ (1.54) also belongs to $A(M, \otimes(\mathfrak{gl}(E)))$. Thus given $\psi_1, \ldots, \psi_k \in A(M, \mathfrak{gl}(E))$, the product $\psi_1 \otimes \cdots \otimes \psi_k$ is defined (1.55, 1.56*c*); the point here is that although the ψ_j takes values in $\mathfrak{gl}(E)$, the multiplication to be used here is $\otimes$, not [,].

Now let f be an invariant polynomial of degree k on $\mathfrak{gl}(\mathscr{V})$, where $\mathscr{V}$ is the standard fiber of E. The expression $\tilde{f}(\psi_1 \otimes \cdots \otimes \psi_k)$ is a well-defined $\mathbb{F}$-valued differential form on M by 2.81 and 2.82, where $\tilde{f}$ is the polarization of f; from now on we will delete the tilde and simply write $f(\psi_1 \otimes \cdots \otimes \psi_k)$. By 1.56$c$ and the symmetry of the polarization of f, $f(\psi_1 \otimes \cdots \otimes \psi_k)$ satisfies the usual commutation relations for differential forms in the ψ_j.

Lemma Suppose that ∇ is a connection in a vector bundle E over M, that $\psi_j \in A^{r_j}(M, \mathfrak{gl}(E))$, and f is an invariant polynomial on $\mathfrak{gl}(\mathscr{V})$, where $\mathscr{V}$ is the standard fiber of E; then

$$\begin{aligned} d(f(\psi_1 \otimes \cdots \otimes \psi_k)) &= f(d^\nabla(\psi_1 \otimes \cdots \otimes \psi_k)) \\ &= f\left(\sum_j (-1)^{r_1 + \cdots + r_j} \psi_1 \otimes \cdots \otimes d^\nabla \psi_j \otimes \cdots \otimes \psi_k\right) \end{aligned}$$

PROOF Apply 2.76 and 2.82.

2.84 The curvature tensor R of a connection ∇ in a vector bundle E over M can be considered as a section of $\mathrm{Hom}(\Lambda^2 TM, \mathfrak{gl}(E))$, that is, $R \in A^2(M, \mathfrak{gl}(E))$. Given a real invariant polynomial f on $\mathfrak{gl}(\mathscr{V})$, the expression $f(R) := f(R \otimes \cdots \otimes R)$ is therefore a well-defined differential form on M. By 2.83 and the Bianchi identity (2.79), $f(R)$ is a closed differential form on M; by the de Rham theorem, it represents a real cohomology class of M.

Theorem (Chern-Weil) Let E be a vector bundle over M, with standard fiber equal to a vector space $\mathscr{V}$ over $\mathbb{F}$. If f is a real invariant polynomial of degree k on $\mathfrak{gl}(\mathscr{V})$ and if R is the curvature tensor of a connection in E, then the differential form $f(R)$ on M is closed; the de Rham cohomology class of $f(R)$ is independent of the choice of connection in E, that is, if R_1 and R_2 are the curvature tensors of two connections in E, then the forms $f(R_1)$ and $f(R_2)$ are cohomologous.

PROOF See [Ch4] or [KN: XII.1.1].

Exercise Prove that the characteristic classes of vector bundles are natural with respect to pullback: if $h\colon N \to M$ is C^∞, then the characteristic class of h^*E determined by an invariant polynomial f on $\mathfrak{gl}(\mathscr{V})$ equals the pullback of the characteristic class of E determined by f.

2.85 **Examples**

(*a*) The *Chern polynomials* on $\mathfrak{gl}(m, \mathbb{C})$ are defined by (see sec. 2.81)

$$\sum_{k=0}^{m} c_k(A)\lambda^{m-k} := \det\left(\lambda I - \frac{1}{2\pi i} A\right) \qquad A \in \mathfrak{gl}(m, \mathbb{C})$$

The *Chern classes* of a complex vector bundle E over a manifold M are the classes $c_j(E) \in H^{2j}(M; \mathbb{R})$ determined by the Chern polynomials c_j and the curvature tensor of any connection in E. The fact that the $c_j(E)$ are real cohomology classes follows from the independence of $c_j(E)$ on the connection and the fact that a complex vector bundle admits a $U(m)$-connection (8.18, 9.7*h*), for the polynomials c_j are real when restricted to the unitary algebra $\mathfrak{u}(m)$.

(*b*) Let E be a real vector bundle of rank m over M; it can be shown [KN: XII.4] that the Chern classes $c_j(E^{\mathbb{C}})$, j odd, of the complexification $E^{\mathbb{C}}$ of E (1.23) are zero. The *jth Pontrjagin class of E* is defined to be $(-1)^j c_{2j}(E^{\mathbb{C}}) \in H^{4j}(M; \mathbb{R})$, where $c_j(E^{\mathbb{C}})$ is the jth Chern class of $E^{\mathbb{C}}$.

It can be shown [KN: XII] that the Chern classes of a complex vector bundle and the Pontrjagin classes of a real vector bundle over M are integral cohomology classes of M, that is, $c_j(E)$ belongs to $H^{2j}(M, \mathbb{Z})$ for each complex vector bundle E over M.

THE TANGENT BUNDLE: LINEAR CONNECTIONS

The remainder of the chapter will be devoted to connections on the tangent bundle of a manifold.

2.86 **Definition** A connection $\mathscr{H}$ on TM is called a *linear connection on M*; by abuse of notation, the associated covariant derivative operator ∇ in TM is also called a *linear connection on M*.

2.87 If γ is a curve in M, then $\dot{\gamma}$ is a section of TM along γ; curves γ for which $\dot{\gamma}$ is parallel with respect to a linear connection play a central role in the geometry of TM.

Definition Let ∇ be a linear connection on M. A curve γ in M is called a *geodesic* of ∇ if the tangent curve $\dot{\gamma}$ in TM is parallel; in this case γ is also said to be *autoparallel.*

2.88 **Proposition** Let ∇ be a linear connection on M, and let $v \in TM$. There exists a geodesic γ of ∇ in M with initial tangent vector $\dot{\gamma}(0) = v$. Furthermore, γ is unique up to the domain of definition.

Proof By definition, γ is a geodesic if and only if $\nabla_D \dot{\gamma} = 0$, where $D = d/dt \in \mathfrak{X}\mathbb{R}$, and ∇ is the pullback connection along the map γ. With respect to a chart x on M (2.59, exercise 1),

$$\begin{aligned}\nabla_D \dot{\gamma} &= \sum \nabla_D(D(x^j \circ \gamma) \cdot X_j \circ \gamma)\\ &= \sum (D^2(x^j \circ \gamma) \cdot X_j \circ \gamma + D(x^j \circ \gamma)\nabla_D(X_j \circ \gamma))\end{aligned}$$

where $X_j := \partial/\partial x^j$. But

$$\begin{aligned}\nabla_D(X_j \circ \gamma) = \nabla_{\dot{\gamma}} X_j &= \sum_i D(x^i \circ \gamma)\nabla_{X_i \circ \gamma} X_j\\ &= \sum_{i,k} D(x^i \circ \gamma) \cdot \Gamma^k_{ij} \circ \gamma \cdot X_k \circ \gamma\end{aligned}$$

where $\Gamma^k_{ij} := dx^k(\nabla_{X_i} X_j)$; Γ^k_{ij} is called a *Christoffel symbol.*

Thus the equation for a geodesic is the second-order system of differential equations

$$D^2(x^k \circ \gamma) + \sum_{i,j} D(x^i \circ \gamma) \cdot D(x^j \circ \gamma) \cdot \Gamma^k_{ij} \circ \gamma = 0 \qquad k = 1, \ldots, n$$

For such a system existence, uniqueness, and C^∞ dependence on initial conditions are all guaranteed by standard theorems on differential equations; existence is guaranteed only locally unless all the Γ^k_{ij} are zero, making the system linear.

2.89 **Proposition** Let γ: $(a, b) \to M$ be a nonconstant geodesic of a linear connection on M, and let f: $(c, d) \to (a, b)$ be C^∞. The curve $\gamma \circ f$ is a geodesic if and only if f is affine.

Proof By 2.55, $\nabla_D \dot{\overline{\gamma \circ f}} = \nabla_D(f' \cdot \dot{\gamma} \circ f) = f'' \cdot \dot{\gamma} \circ f + f' \cdot \nabla_D(\dot{\gamma} \circ f)$. But $\nabla_D(\dot{\gamma} \circ f) = \nabla_{f_* D}\dot{\gamma} = f' \cdot \nabla_{D \circ f}\dot{\gamma} = 0$, so $\gamma \circ f$ is a geodesic if and only if $f'' = 0$.

2.90 It may be impossible to extend a geodesic γ: $(a, b) \to M$ to a geodesic defined on all of $\mathbb{R}$.

Definition A linear connection on M is called *complete* if all its geodesics can be extended to geodesics with domain $\mathbb{R}$.

Examples

2.91 (*a*) The canonical geodesics in $\mathbb{R}^n$ are the constant-speed parametrizations of the straight lines; the canonical linear connection on $\mathbb{R}^n$ is complete.

(*b*) Let ∇ be the linear connection on a Lie group G such that $\nabla_X Y = c[X, Y]$, $X, Y \in \mathfrak{g}$, where $c \in \mathbb{R}$ is fixed. The geodesics of ∇

are the integral curves of the left-invariant vector fields, for if $\gamma(t) = g \exp(tX)$, $g \in G$, $X \in \mathfrak{g}$, then $\dot\gamma = X \circ \gamma$, and thus

$$\nabla_D \dot\gamma = \nabla_D(X \circ \gamma) = \nabla_{\dot\gamma} X = (\nabla_X X)_\gamma = 0$$

This linear connection is complete because each 1-parameter subgroup of G is the image of a homomorphism from $\mathbb{R}$ to G.

(*c*) The canonical geodesics in S^n are the constant-speed parametrizations of the great circles of S^n; the canonical linear connection on S^n is complete.

(*d*) The restriction to a proper open submanifold of a connected manifold M of a linear connection on M is incomplete.

2.92 The equation $\nabla_D \dot\gamma = 0$ for a geodesic is a second-order differential equation in γ; in the terminology of elementary differential equations [BD: 3.2, prob. 2], the independent variable is missing (see also the local coordinate version in the proof of 2.88), so we can consider the equation as a first-order equation in $\dot\gamma$. In fact, $\dot\gamma$ is an integral curve of a vector field on TM; this should also be compared with the reduction of order process for converting an nth-order equation to a system of n first-order equations [Ar: 9.1] [BD: 7.1].

Definition A vector field S on TM is called a *spray* if $\pi_* \circ S = \text{id}$ and S is homogeneous: $S \circ \mu_c = c \cdot \mu_{c*} S$, $c \in \mathbb{R}$, where $\mu_c v := cv$.

The identity $\pi_* S = \text{id}$ means that S is a section of the bundle $\pi_*: TTM \to TM$; the homogeneity condition should be compared with the homogeneity condition for a connection $\mathscr{H}$ in 2.26, and with the definition of scalar multiplication in the bundle $\pi_*: TTM \to TM$ in 1.32.

2.93 Theorem Let ∇ be a linear connection on M. There is a unique ∇-horizontal spray S on TM. A curve γ in M is a geodesic of ∇ if and only if $\dot\gamma$ is an integral curve of S.

PROOF For all $v \in TM$, $S_v := \pi_*|_{\mathscr{H}_v}^{-1}(v)$ is the only horizontal vector at v which is mapped by π_* to v. The map S defined this way is C^∞ because $(\pi_{\mathscr{H}}, \pi_*)$ is a diffeomorphism from $\mathscr{H}$ to $TM \oplus TM$ (see exercise 2 in 2.50), and $(\pi_{\mathscr{H}}, \pi_*)(S_v) = (v, v)$.

By 1.32, for all $v \in TM$ and $c \in \mathbb{R}$, $\pi_* \mu_{c*} S_v = v$, so $\pi_*(c\mu_{c*} S_v) = cv = \pi_* S_{cv}$; but $c\mu_{c*} S_v \in \mu_{c*}\mathscr{H}_v = \mathscr{H}_{cv}$, so $c\mu_{c*} S_v = S_{cv}$.

Now let α be an integral curve of S, and let $\gamma := \pi \circ \alpha$. Since $\dot\alpha$ is a curve in $\mathscr{H}$, α is parallel in TM along γ. But

$$\dot{\gamma} = \dot{\widehat{\pi \circ \alpha}} = \pi_* \dot{\alpha} = \pi_* S \circ \alpha = \alpha$$

so $\dot{\gamma}$ is parallel along γ, and therefore γ is a geodesic.

Conversely, if γ is a geodesic in M, then the integral curve α of S through $\dot{\gamma}(0)$ projects down to the geodesic in M with initial value $\dot{\gamma}(0)$ (by the argument above). By 2.88, the geodesics γ and $\pi \circ \alpha$ agree since they have the same initial tangent vector.

2.94 **Definition** The spray S given in 2.93 is called the *geodesic spray of* ∇.

2.95 Now let us study sprays in general.

Definition Let S be a spray on TM. For each $v \in TM$ let $\alpha_v: I_v \to TM$ be the maximal integral curve of S [W: 1.48] with initial value v, where I_v is an open interval about 0. Set $\widetilde{TM} := \{v \in TM \mid 1 \in I_v\}$ and $\tilde{M}_p := M_p \cap \widetilde{TM}$, $p \in M$. Define the *exponential map of* S from $\widetilde{TM}$ to M by $\exp v := \pi \circ \alpha_v(1)$; denote $\exp|_{\tilde{M}_p}$ by $\exp_p$.

For example, let S be the geodesic spray of the linear connection $\nabla_X Y = c[X, Y]$, $X, Y \in \mathfrak{g}$ ($c \in \mathbb{R}$ fixed) on a Lie group G; $\exp_e$ is the usual exponential map from $G_e \cong \mathfrak{g}$ to G, and $\exp_g = L_g \circ \exp_e \circ L_{g*}^{-1}$, $g \in G$. Calculate exp for the multiplicative group of positive real numbers.

The exponential map is the model for the map f in axiom 2.6(4) for parallel transport (see the comment at the end of 2.6).

2.96 By [W: 1.48(e)], $\widetilde{TM}$ is an open subset of TM. It can be shown (see [AM: 2.1.15] and [GKM: 8.6]) that $\mathscr{D} := \bigcup_{v \in TM} I_v \times \{v\} \subset \mathbb{R} \times TM$ is open, and that the *flow of* S, that is, the map $\psi: \mathscr{D} \to TM$ such that $\psi(t, v) := \alpha_v(t)$, is C^∞; this implies that $\exp = \pi \circ \psi(1, \cdot)$ is C^∞.

Proposition Let $\exp: \widetilde{TM} \to M$ be the exponential map of a spray S on TM. If α_v is the integral curve of S with initial value $v \in TM$, then $\exp tv = \pi \circ \alpha_v(t)$ for all $t \in I_v$. For each $p \in M$, $\tilde{M}_p$ is a starlike neighborhood of 0 in M_p. Finally, $\exp_p$ has maximal rank at 0 in M_p, and therefore maps a neighborhood of 0 in M_p diffeomorphically to a neighborhood of p in M.

Proof Fix $v \in TM$, and $t \in I_v$. The curve $\mu_t \circ \alpha_v \circ \mu_t$ in TM is an integral curve of S, where $\mu_t w := tw$, for

$$\dot{\widehat{\mu_t \circ \alpha_v \circ \mu_t}}(s) = \mu_{t*} \dot{\widehat{\alpha_v \circ \mu_t}}(s) = t\mu_{t*} \dot{\alpha}_v(ts)$$
$$= t\mu_{t*} S \circ \alpha_v \circ \mu_t(s) = S \circ \mu_t \circ \alpha_v \circ \mu_t(s)$$

by the homogeneity of S. But $\alpha_{tv}(0) = tv = t\alpha_v(0) = \mu_t \circ \alpha_v \circ \mu_t(0)$, so $\alpha_{tv}(s) = \mu_t \circ \alpha_v \circ \mu_t(s) = t\alpha_v(ts)$. Thus $t \in I_v$ implies $1 \in I_{tv}$ because $\alpha_{tv}(1) = t\alpha_v(t)$ is defined, and then

$$\exp tv = \pi \circ \alpha_{tv}(1) = \pi(t\alpha_v(t)) = \pi \circ \alpha_v(t)$$

In particular, if $v \in \widetilde{TM}$, then $\exp tv$ is defined for $0 \le t \le 1$.

Finally, for all $v \in M_p$,

$$\exp_{p*} \mathscr{I}_0 v = \left.\frac{d}{dt}\right|_0 \exp_p tv = v$$

so $\exp_{p*}|_0$ is an isomorphism; now apply the inverse function theorem.

2.97 Corollary Let $\exp: \widetilde{TM} \to M$ be the exponential map of the geodesic spray S of a linear connection ∇ on M. The maximal geodesic γ of ∇ with $\dot{\gamma}(0) = v \in TM$ is given by $\gamma(t) = \exp tv$, $t \in I_v$.

The following statements are equivalent:

(i) The spray S is a complete vector field on TM [W: 1.49].
(ii) The linear connection ∇ is complete (2.90).
(iii) $\widetilde{TM} = TM$.

PROOF If γ is a geodesic and $\dot{\gamma}(0) = v$, then by 2.93, $\alpha_v = \dot{\gamma}$, so $\gamma(t) = \pi \circ \alpha_v(t) = \exp tv$.

(i)$\Leftrightarrow$(ii): The domain of α_v equals that of $\gamma = \pi \circ \alpha_v$; in particular, α_v is defined on $\mathbb{R}$ if and only if γ is defined on $\mathbb{R}$.

(ii)$\Leftrightarrow$(iii): Let $v \in TM$. If $\exp tv$ is defined for all $t \in \mathbb{R}$, then $\exp v = \exp(1 \cdot v)$ is defined, so $v \in \widetilde{TM}$. Conversely, if $\widetilde{TM} = TM$, then $\exp tv$ is defined for all $(t, v) \in \mathbb{R} \times TM$.

2.98 Theorem (Ambrose-Palais-Singer) Let $S \in \mathfrak{X}TM$ be a spray. There exists a linear connection ∇ on M for which S is the geodesic spray.

PROOF It suffices to construct an appropriate connection on TM. Given $p \in M$ and $v, w \in M_p$, we need to find an initial direction in TM for the extension of w to a parallel section along each curve in M with initial tangent vector v. Start off with the parallel section $s \mapsto \mathscr{I}_{sv} w$ in $T(M_p)$ along the ray $s \mapsto sv$ in M_p. Follow this section by the tangent map $\exp_{p*}: T(\tilde{M}_p) \to TM$ induced by the exponential map of S; the result is a section of TM along the curve $s \mapsto \exp sv$ in M. The initial direction for parallel translation of w along v is then defined to be the initial tangent vector to this curve in TM.

In other words, for each point $p \in M$ and each $w \in M_p$, set

$$\mathscr{H}_w := \left\{ \left.\frac{d}{ds}\right|_0 \exp_{p*} \mathscr{I}_{sv} w \,\middle|\, v \in M_p \right\}$$

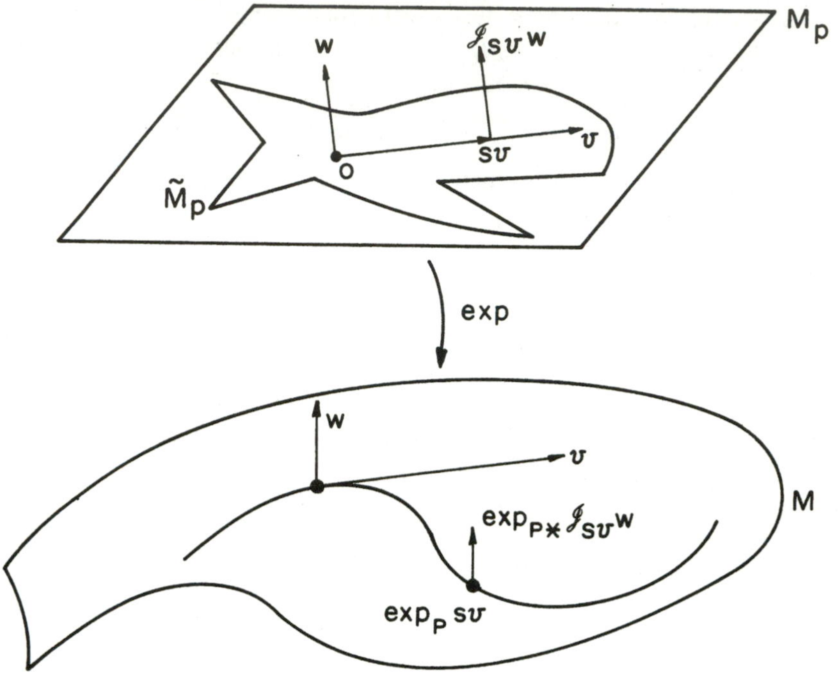

We must now show that $\mathscr{H} := \bigcup_{w \in TM} \mathscr{H}_w$ is a connection on TM, and that its geodesic spray is S.

Given $w \in M_p$, the proof that $\mathscr{H}_w$ is a vector space is virtually the same as the corresponding proof in 2.28. For $(v, s) \in \tilde{M}_p \times [0, 1]$, define $W(v, s) = W_v(s) := \exp_{p*} \mathscr{J}_{sv} w$; since $W(v, s) = W(sv, 1)$,

$$\frac{d}{ds}\bigg|_0 \exp_{p*} \mathscr{J}_{sv} w = \dot{W}_v(0) = \frac{d}{ds}\bigg|_0 W(sv, 1) = W_*(\mathscr{J}_0 v, 0)$$

so $\mathscr{H}_w$ is the image of a linear map, hence is a vector space of dimension less than or equal to the dimension n of M. But

$$\pi_* \frac{d}{ds}\bigg|_0 \exp_{p*} \mathscr{J}_{sv} w = \frac{d}{ds}\bigg|_0 \pi \circ \exp_{p*} \mathscr{J}_{sv} w = \frac{d}{ds}\bigg|_0 \exp_p sv = v$$

so π_* maps $\mathscr{H}_w$ isomorphically onto M_p.

The maps $\exp: \widetilde{TM} \to M$ and $\mathscr{J}: TM \oplus TM \to \mathscr{V}TM$ are C^∞, $\exp$ has maximal rank near the zero section of TM, and $\mathscr{J}$ has maximal rank everywhere, so $\mathscr{H}$ is a submanifold of TTM. Vector subbundle charts are easy to construct. Thus $\mathscr{H}$ is a vector subbundle of TTM which is complementary to the vertical subbundle $\mathscr{V}TM$.

If $c \in \mathbb{R}$, then $\mu_{c*}\mathscr{H}_w \subset \mathscr{H}_{cw}$ by the calculation

$$\mu_{c*} \left.\frac{d}{ds}\right|_0 \exp_{p*} \mathscr{I}_{sv} w = \left.\frac{d}{ds}\right|_0 \mu_c \circ \exp_{p*} \mathscr{I}_{sv} w = \left.\frac{d}{ds}\right|_0 \exp_{p*} \mathscr{I}_{sv} cw$$

Equality of $\mu_{c*}\mathscr{H}_w$ and $\mathscr{H}_{cw}$ follows because π_* maps both spaces isomorphically onto M_p.

Thus $\mathscr{H}$ is a connection on TM. Now we must show that the geodesic spray of $\mathscr{H}$ is S. Since

$$\pi_* \left.\frac{d}{ds}\right|_0 \exp_{p*} \mathscr{I}_{sv} w = w = \pi_* S_w$$

for all w in TM, it suffices to show that the vector S_w belongs to the space $\mathscr{H}_w$. If α is the integral curve of S through w, then $\pi \circ \alpha$ is the geodesic of S with tangent field α, and so

$$\alpha(s) = \dot{\widehat{\pi \circ \alpha}}(s) = \left.\frac{d}{dt}\right|_0 \pi \circ \alpha(s + t)$$

$$= \left.\frac{d}{dt}\right|_0 \exp_p(sw + tw) = \exp_{p*} \mathscr{I}_{sw} w$$

Hence

$$S_w = \dot{\alpha}(0) = \left.\frac{d}{ds}\right|_0 \exp_{p*} \mathscr{I}_{sw} w \in \mathscr{H}_w$$

Exercise [Do*1*] Let ∇ be the covariant derivative operator of $\mathscr{H}$. Show that $2\nabla_X Y = \mathrm{pr}_2([S, \mathscr{I}Y]_X + X_* Y) + [X, Y]$ for all X, $Y \in \mathfrak{X}M$, where $\mathscr{I}Y$ is the vertical field on TM (1.64) such that $(\mathscr{I}Y)_v = \mathscr{I}_v(Y_p)$ for $v \in TM$, $p = \pi(v) \in M$.

2.99 If S is the geodesic spray of a linear connection ∇ on M, then it does not follow that ∇ is the linear connection constructed in 2.98. This brings us to the structure of the space of linear connections on M, a topic already looked at briefly (2.71) for arbitrary vector bundles. Given linear connections $\bar{\nabla}$ and ∇ on M, write the connection difference form $\mathscr{D} := \bar{\nabla} - \nabla \in A^1(M, \mathrm{End}\, TM) \cong \mathrm{Hom}(TM \otimes TM, TM)$ as a (1, 2)-tensor field: $\mathscr{D}(U, V) := \bar{\nabla}_U V - \nabla_U V$.

Proposition Linear connections $\bar{\nabla}$ and ∇ on M have the same geodesic spray if and only if $\mathscr{D} = \bar{\nabla} - \nabla$ is skew-symmetric.

PROOF For each $w \in TM$, $S_w = \pi_* |_{\mathscr{H}_w}^{-1} w$, while by 2.73, $\bar{S}_w = \pi_* |_{\bar{\mathscr{H}}_w}^{-1} w = \pi_* |_{\mathscr{H}_w}^{-1} w - \mathscr{I}_w \mathscr{D}(w, w)$. Thus $\bar{S} = S$ if and only if $\mathscr{D}(w, w) = 0$ for all $w \in TM$.

Exercise Prove directly without using the connections or the sprays that $\bar{\nabla}$ and ∇ have the same geodesics if and only if $\bar{\nabla} - \nabla$ is skew-symmetric.

2.100 **Definition** Let ∇ be a linear connection on M. The *torsion tensor field of* ∇ is defined by

$$T(U, V) := \nabla_U V - \nabla_V U - [U, V] \qquad U, V \in \mathfrak{X}M$$

If $T = 0$, then ∇ is called *torsion-free*, or *symmetric*.

The word "symmetric" refers to the Christoffel symbols from 2.88: $T = 0$ if and only if $\Gamma_{ij}^k = \Gamma_{ji}^k$, where $\Gamma_{ij}^k := dx^k(\nabla_{X_i} X_j)$, X_i being the ith-coordinate basis field $\partial/\partial x^i$ for a chart x on M.

Exercise Let $I \in \Gamma$ End TM be the identity tensor field, $Iv = v$, and let d^∇ be the exterior covariant derivative operator for ∇ from 2.75; prove that $d^\nabla I$ is the torsion tensor T for ∇.

The canonical linear connections on $\mathbb{R}^n$ and S^n are torsion-free. If ∇ is a linear connection on M, then $\nabla - \frac{1}{2}T$ is torsion-free.

2.101 **Theorem** Linear connections ∇ and $\bar{\nabla}$ on M are equal if and only if they have the same geodesics and torsion fields.

PROOF Let $\mathscr{D} = \mathscr{D}_s + \mathscr{D}_a$ be the decomposition of the connection difference field into symmetric and alternating parts. By 2.99, $\mathscr{D}_s$ is zero if and only if ∇ and $\bar{\nabla}$ have the same geodesics. But

$$\begin{aligned} 2\mathscr{D}_a(U, V) &= \mathscr{D}(U, V) - \mathscr{D}(V, U) = \bar{\nabla}_U V - \nabla_U V - \bar{\nabla}_V U + \nabla_V U \\ &= \bar{T}(U, V) - T(U, V) \end{aligned}$$

2.102 ***The torsion-free bi-invariant linear connection on a Lie group*** Let $\hat{\nabla}$ be the canonical left-invariant connection on a Lie group G (2.60*b*), $\hat{\nabla} Y = 0$ for $Y \in \mathfrak{g}$. Since the torsion tensor of $\hat{\nabla}$ is $\hat{T}(X, Y) = -[X, Y]$, the connection $\nabla_X Y = \hat{\nabla}_X Y - \frac{1}{2}\hat{T}(X, Y) = \frac{1}{2}[X, Y]$ for $X, Y \in \mathfrak{g}$ is torsion-free; it has the same geodesics as $\hat{\nabla}$, and is left-invariant.

Let us calculate $\hat{\nabla}_U V$ for right-invariant fields $U, V \in \tilde{\mathfrak{g}}$ on G. Let Y be the left-invariant field on G such that $Y_e = V_e$; for all $g \in G$,

$V_g = R_{g*} V_e = R_{g*} L_{g^{-1}*} Y_g = (\mathrm{Ad}_{g^{-1}} Y)_g =: \sum Y^j(g) E_j|_g$, where $\{E_j\}$ is a fixed basis for $\mathfrak{g}$. Thus for all $g \in G$ (cf. [W: 3.31]),

$$\begin{aligned}
\hat{\nabla}_{U_g} V &= \sum U_g(Y^j) E_j|_g \\
&= \frac{d}{dt}\bigg|_0 Y^j(\exp(tU)g) \cdot E_j|_g \\
&= \frac{d}{dt}\bigg|_0 (\mathrm{Ad}_{g^{-1}\exp(-tU)} Y)_g \\
&= \frac{d}{dt}\bigg|_0 L_{\exp(-tU)*} (\mathrm{Ad}_{g^{-1}\exp(-tU)} Y)_{\exp(tU)g} \\
&= \frac{d}{dt}\bigg|_0 L_{\exp(-tU)*} V_{\exp(tU)g} = (\mathscr{L}_U V)_g = [U, V]_g
\end{aligned}$$

where $\mathscr{L}$ is the Lie derivative operator on G [W: 2.24, 2.25]. Since $\hat{\nabla}_U V = [U, V]$ for $U, V \in \tilde{\mathfrak{g}}$, $\hat{T}(U, V) = [U, V]$ for $U, V \in \tilde{\mathfrak{g}}$. Hence $\nabla_U V = \hat{\nabla}_U V - \frac{1}{2}\hat{T}(U, V) = \frac{1}{2}[U, V]$ for $U, V \in \tilde{\mathfrak{g}}$, which means that ∇ is also right-invariant. Since ∇ is both left- and right-invariant, it is bi-invariant.

2.103 Theorem The structure equations for a linear connection: Let ∇ be a linear connection on M, and let $f: N \to M$ be a C^∞ map; denote the covariant derivative operator along f by ∇.

(i) $T(f_* U, f_* V) = \nabla_U f_* V - \nabla_V f_* U - f_*[U, V] \qquad U, V \in \mathfrak{X}N$

(ii) $R(f_* U, f_* V)X = \nabla_U \nabla_V X - \nabla_V \nabla_U X - \nabla_{[U, V]} X \qquad X \in \Gamma_f TM$

PROOF Equation (ii) was proved for arbitrary vector bundles in 2.67; the proof of (i) is similar (but simpler), and is left as an exercise.

2.104 Corollary 1 Let S be a spray on TM, and let ∇ be the linear connection on M constructed using the exponential map of S in 2.98; then ∇ is torsion-free.

PROOF The proof uses the pullback covariant derivative operator along the map $\exp: \widetilde{TM} \to M$. Fix u and v in M_p, and let $\mathscr{J}u$ and $\mathscr{J}v$ be the usual parallel extensions of u and v to vector fields on the vector space M_p, $(\mathscr{J}u)_w = \mathscr{J}_w u$ for $w \in M_p$ (1.28, 1.64). The maps

$\exp_* \mathscr{J}u$ and $\exp_* \mathscr{J}v$ are vector fields along exp. By 2.103,

$$\begin{aligned} T(u, v) &= \nabla_u \exp_* \mathscr{J}v - \nabla_v \exp_* \mathscr{J}u - \exp_*[\mathscr{J}u, \mathscr{J}v] \\ &= \kappa(\exp_* \mathscr{J}v)_* u - \kappa(\exp_* \mathscr{J}u)_* v - \exp_* 0 \\ &= \kappa \frac{d}{ds}\bigg|_0 \exp_* \mathscr{J}_{su} v - \kappa \frac{d}{ds}\bigg|_0 \exp_* \mathscr{J}_{sv} u \end{aligned}$$

which is zero by the definition of $\mathscr{H}$ and κ.

2.105 **Corollary 2** (Ambrose, Palais, and Singer) Let S be a spray on TM, and let T be an alternating tensor field of type (1, 2) on M. There exists a unique linear connection ∇ on M with geodesic spray S and torsion tensor field T.

PROOF Let $\bar{\nabla}$ be the torsion-free linear connection on M with geodesic spray S (2.104). The linear connection $\nabla := \bar{\nabla} + \frac{1}{2}T$ has geodesic spray S and torsion field T.

Exercise Show that ∇ can be expressed by the formula (2.98)

$$2\nabla_X Y = \mathrm{pr}_2([S, \mathscr{J}Y]_X + X_* Y) + [X, Y] + T(X, Y) \qquad X, Y \in \mathfrak{X}M$$

2.106 **Theorem** Let ∇ be a torsion-free linear connection on M. Fix $p \in M$ and a basis $\{\epsilon_j\}$ for M_p. There exists a chart x on M near p such that, with $X_j := \partial/\partial x^j$,

(i) $x(p) = 0$

(ii) $X_j|_p = \epsilon_j$

(iii) $(\nabla X_j)_p = 0$

(iv) $(\nabla\, dx^j)_p = 0$

where in (iv), ∇ is the associated connection in T^*M.

PROOF Set $x := \exp_p|_U^{-1}$, where U is an open neighborhood of 0 in M_p which is mapped diffeomorphically to an open neighborhood of p in M (2.96). Property (i) is immediate from the definition (if the chart is required to be $\mathbb{R}^n$-valued, rather than M_p-valued, identify M_p with $\mathbb{R}^n$ by means of the basis $\{\epsilon_j\}$ for M_p).

(ii): By the definition of $X_j = \partial/\partial x^j$ [W: 1.19], on U $\exp_{p*} \mathscr{J}\epsilon_j = X_j \circ \exp_p$, so $X_j|_p = \exp_{p*} \mathscr{J}_0 \epsilon_j = \epsilon_j$.

(iii): If $l(t) := tv \in M_p$ for $t \in \mathbb{R}$, $v \in M_p$, then

$$X_{j*}v = \widehat{X_j \circ l}(0) = \left.\frac{d}{ds}\right|_0 \exp_{p*} \mathscr{I}_{sv}\, \epsilon_j \in \mathscr{H}_{X_j|p}$$

so $\nabla_v X_j = \kappa X_{j*} v = 0$.

(iv): For all $v \in M_p$,

$$(\nabla_v\, dx^i)(\epsilon_j) = v \cdot dx^i(X_j) - dx^i(\nabla_v X_j) = v\delta_{ij} - dx^i(0) = 0$$

Exercise Compare the induced vector bundle chart x_* on TM with the chart from 2.10 constructed using $h := \exp_p|_U$.

2.107 **Theorem** The Bianchi identities for a linear connection: if R is the curvature tensor field of a torsion-free linear connection ∇ on M, then for all $X, Y, Z \in \mathfrak{X}M$,

(i) $R(X, Y)Z + R(Y, Z)X + R(Z, X)Y = 0$

(ii) $(\nabla_X R)(Y, Z) + (\nabla_Y R)(Z, X) + (\nabla_Z R)(X, Y) = 0$

Proof By tensoriality we may assume that $[X, Y] = [Y, Z] = [Z, X] = 0$, so that $\nabla_X Y = \nabla_Y X + T(X, Y) = \nabla_Y X$, and so forth; thus

$$\begin{aligned} &R(X, Y)Z + R(Y, Z)X + R(Z, X)Y \\ &\quad = \nabla_X\nabla_Y Z - \nabla_Y\nabla_X Z + \nabla_Y\nabla_Z X - \nabla_Z\nabla_Y X + \nabla_Z\nabla_X Y - \nabla_X\nabla_Z Y \\ &\quad = \nabla_X(\nabla_Y Z - \nabla_Z Y) + \nabla_Y(\nabla_Z X - \nabla_X Z) + \nabla_Z(\nabla_X Y - \nabla_Y X) = 0 \end{aligned}$$

Similarly, by 2.62 and the general Bianchi identity (2.79),

$$\begin{aligned} &(\nabla_X R)(Y, Z) + (\nabla_Y R)(Z, X) + (\nabla_Z R)(X, Y) \\ &\qquad = \nabla_X R(Y, Z) - R(\nabla_X Y, Z) - R(Y, \nabla_X Z) \\ &\qquad\quad + \nabla_Y R(Z, X) - R(\nabla_Y Z, X) - R(Z, \nabla_Y X) \\ &\qquad\quad + \nabla_Z R(X, Y) - R(\nabla_Z X, Y) - R(X, \nabla_Z Y) \\ &\qquad = (d^\nabla R)(X, Y, Z) = 0 \end{aligned}$$

THE TANGENT BUNDLE: AFFINE CONNECTIONS

Associated with a linear connection on M is a different sort of parallel transport which ignores the linear structure of the tangent spaces on M.

2.108 **Definition** Let $\mathscr{H}$ be a connection on TM. The *associated affine* *connection on TM* is the vector subbundle $\mathscr{AH}$ of TTM for which

$$\mathscr{AH}_w := \{z - \mathscr{I}_w \pi_* z \mid z \in \mathscr{H}_w\} \qquad w \in TM$$

A section W of TM along a C^∞ curve γ in M is *affine parallel* if $\dot{W}$ is a curve in $\mathscr{AH}$.

2.109 The bundle $\mathscr{AH}$ is not a connection on TM in the sense of 2.26 because it is not homogeneous:

$$\mu_{c*}(z - \mathscr{I}_w \pi_* z) = \mu_{c*} z - \mathscr{I}_{cw} \pi_* cz \qquad c \in \mathbb{R}$$

which does not belong to $\mathscr{AH}_{cw}$ if $c \neq 1$. Thus the differential equation for affine parallel transport along a curve in M must be different from that for linear parallel transport.

Proposition A section W of TM along a curve γ is affine parallel if and only if $\nabla_D W = -\dot{\gamma}$, where ∇ is the pullback along γ of the linear connection associated with $\mathscr{H}$, and $D = d/dt \in \mathfrak{X}\mathbb{R}$.

PROOF Since $\pi \circ W = \gamma$, $\pi_* \dot{W} = \dot{\gamma}$; thus for each t there exists a vector $Z_t \in M_{\gamma(t)}$ such that

$$\dot{W}(t) = \pi_* |^{-1}_{\mathscr{H}W(t)}(\dot{\gamma}(t)) + \mathscr{I}_{W(t)} Z_t$$

Therefore $\dot{W}(t)$ belongs to $\mathscr{AH}_{W(t)}$ if and only if $Z_t = -\dot{\gamma}(t)$; this holds if and only if $\nabla_D W = -\dot{\gamma}$.

For example, if γ is a geodesic, a necessary and sufficient condition for a section W along γ to be affine parallel is that $W(t) = V(t) - (t - c)\dot{\gamma}(t)$, $c \in \mathbb{R}$, where V is $\mathscr{H}$-parallel along γ. In particular, the zero section of TM is not affine parallel.

From this example we see that given a geodesic γ in M, the map $M_{\gamma(a)} \to M_{\gamma(b)}$ which sends w to its affine parallel translate along γ is actually an affine map: $w \mapsto \mathbb{P}_\gamma w - (b - a)\dot{\gamma}(b)$. For arbitrary γ the formula is more complicated, but the map is still affine. If we consider TM as a bundle over M of affine spaces rather than linear spaces, then $\mathscr{AH}$-parallel transport preserves the structure of the fibers. This theme will occur again in chap. 9.

2.110 **Comments** Hermann Weyl generalized the Levi-Civita connection of a Riemannian manifold (which will appear in chap. 3) to obtain linear parallel transport; since the Christoffel symbols (2.88, 2.100) for the Levi-Civita connection are symmetric, $\Gamma^k_{ij} = \Gamma^k_{ji}$, Weyl kept this requirement in his definition of parallel transport. Shortly afterward, Élie Cartan dropped the symmetry requirement to obtain arbitrary linear parallel transport in TM, and also introduced affine parallel transport, which he referred to as an "affine connection." Unfortunately, Weyl had already used the phrase affine connection to refer to his linear parallel transport.

Today the term affine connection is usually used in Weyl's sense; that is, it refers to linear parallel transport rather than to Cartan's genuine affine parallel transport. To complicate the situation further, more general affine parallel transport can be defined [KN: III.3] using an arbitrary $(1, 1)$-tensor field L on M to "twist" the affine horizontal bundle $\mathscr{A}\mathscr{H}$:

$$L\mathscr{A}\mathscr{H}_w := \{z - \mathscr{J}_w L\pi_* z \mid z \in \mathscr{H}_w\} \qquad w \in TM$$

Just to add a little more confusion, the term affine connection is now occasionally used to refer to a flat $(R = 0)$, torsion-free $(T = 0)$ linear connection [GKM: 2.7(vi)].

2.111   Recall that the *affine group* [W: 3.3(i)] of a real vector space $\mathscr{V}$ is the semidirect $A(\mathscr{V}) := \mathscr{V} \times GL(\mathscr{V})$; as a manifold, $A(\mathscr{V})$ is just the Cartesian product $\mathscr{V} \times GL(\mathscr{V})$, and $A(\mathscr{V})$ acts on $\mathscr{V}$ by the rule $(v, g) \cdot w := v + gw$, $(v, g) \in A(\mathscr{V})$, $w \in \mathscr{V}$. It follows that multiplication in $A(\mathscr{V})$ is given by $(v, g) \cdot (w, h) := (v + gw, gh)$.

> **Definition** Let $\mathscr{H}$ be a linear connection on TM, with affine connection $\mathscr{A}\mathscr{H}$. The *affine holonomy group of $\mathscr{H}$* at $p \in M$ is the subgroup of the affine group $A(M_p)$ consisting of all affine automorphisms of M_p induced by affine parallel transport around the piecewise C^∞ loops at p. If $\kappa: TTM \to TM$ is the connection map of $\mathscr{H}$, then the *affine connection map of $\mathscr{H}$* is $\mathscr{A}\kappa := \kappa + \pi_*$. The *affine curvature field of $\mathscr{H}$* is defined by $\mathscr{A}R(U, V)W := -\mathscr{A}\kappa[\hat{U}, \hat{V}]_w$ for U, V, $W \in \mathfrak{X}M$, where $\hat{U}$ and $\hat{V}$ are the $\mathscr{A}\mathscr{H}$-horizontal lifts of U and V.

The name "affine connection map" is justified by the fact that $\mathscr{A}\kappa$ is zero on $\mathscr{A}\mathscr{H}$, and is $\mathrm{pr}_2 = \kappa$ on $\mathscr{V}TM$.

2.112   Now we would like to calculate $\mathscr{A}R$ in terms of the linear connection ∇.

> **Lemma** Let U, $V \in \mathfrak{X}M$, with vertical lifts (1.64) $\mathscr{J}U$ and $\mathscr{J}V$, $\mathscr{H}$-horizontal lifts $\bar{U}$ and $\bar{V}$, and $\mathscr{A}\mathscr{H}$-horizontal lifts $\hat{U}$ and $\hat{V}$ on TM.
>
> (i) $[\bar{U}, \mathscr{J}V] = \mathscr{J}\nabla_U V$
>
> (ii) $[\mathscr{J}U, \mathscr{J}V] = 0$
>
> (iii) $\hat{U} = \bar{U} - \mathscr{J}U$

PROOF (i): Let $\{\psi_t\}$ be the 1-parameter group for U, with lift $\{\bar{\psi}_t\}$ for $\bar{U}$; denote by $\mathscr{L}$ the Lie derivative operator [W: 2.24]. Given $p \in M$ and $w \in M_p$, $[\bar{U}, \mathscr{J}V]_w$ equals

$$(\mathscr{L}_{\bar{U}}\mathscr{J}V)_w = \frac{d}{dt}\bigg|_0 \bar{\psi}_{-t*}\mathscr{J}_{\bar{\psi}_t w}V_{\psi_t p} = \frac{\partial^2}{\partial t\,\partial s}\bigg|_0 \bar{\psi}_{-t}(\bar{\psi}_t w + sV_{\psi_t p})$$

$$= \frac{\partial^2}{\partial t\,\partial s}\bigg|_0 (w + s\bar{\psi}_{-t}V_{\psi_t p}) = \frac{d}{dt}\bigg|_0 \mathscr{J}_w\bar{\psi}_{-t}V_{\psi_t p} = \mathscr{J}_w\nabla_{U_p}V$$

by 2.57.

(ii): Fix $p \in M$, and let $\iota\colon M_p \to TM$ be the inclusion; set $u := U_p$ and $v := V_p$. The vector fields $\mathscr{J}u$ and $\mathscr{J}v$ on M_p are constant, so $[\mathscr{J}u, \mathscr{J}v] = 0$. But $\mathscr{J}u$ and $\mathscr{J}v$ are ι-related to $\mathscr{J}U$ and $\mathscr{J}V$, respectively, so $[\mathscr{J}U, \mathscr{J}V]$ is zero on $\iota M_p \subset TM$.

(iii): This follows directly from 2.108.

2.113 **Theorem** If ∇ is a linear connecton on M with linear curvature tensor field R, affine curvature field $\mathscr{A}R$, and torsion tensor field T, then

$$\mathscr{A}R(U, V)W = T(U, V) + R(U, V)W \qquad U, V, W \in \mathfrak{X}M$$

PROOF Let $\bar{U}$ and $\bar{V}$ be the $\mathscr{H}$-horizontal lifts of U and V, and let $\hat{U}$, $\hat{V}$ be the $\mathscr{A}\mathscr{H}$-horizontal lifts of U and V.

$$\begin{aligned}\mathscr{A}R(U, V)W &= -\mathscr{A}\kappa[\hat{U}, \hat{V}]_W \\ &= -(\kappa + \pi_*)([\bar{U}, \bar{V}] - [\bar{U}, \mathscr{J}V] - [\mathscr{J}U, \bar{V}] + [\mathscr{J}U, \mathscr{J}V])_W \\ &= R(U, V)W - [U, V] + \nabla_U V - \nabla_V U \\ &= T(U, V) + R(U, V)W\end{aligned}$$

Just as $R \in \Gamma\,\mathrm{Hom}(\Lambda^2 TM, \mathfrak{gl}(TM))$ measures the infinitesimal linear holonomy of $\mathscr{H}$, $\mathscr{A}R\colon \Gamma\Lambda^2 TM \to \Gamma\mathfrak{a}(TM)$ measures the infinitesimal affine holonomy of $\mathscr{H}$; here $\mathfrak{a}(TM) := \bigcup_{p \in M} \mathfrak{a}(M_p)$ is the affine algebra bundle of TM, $\mathfrak{a}(M_p)$ being the Lie algebra of the affine group $A(M_p)$. The semidirect product structure of $A(M_p)$ is reflected in $\mathfrak{a}(M_p)$: the underlying vector space for $\mathfrak{a}(M_p) = M_p \rtimes \mathfrak{gl}(M_p)$ is the direct sum $M_p \oplus \mathfrak{gl}(M_p)$, and $(v, L) \cdot w = v + Lw$, $(v, L) \in \mathfrak{a}(M_p)$, $w \in M_p$. This is just the decomposition given by $\mathscr{A}R(u, v)w = T(u, v) + R(u, v)w$: the translation part of $\mathscr{A}R(u, v) \in \mathfrak{a}(M_p)$ is $T(u, v)$, and the linear part is $R(u, v)$.

The affine curvature field $\mathscr{A}R$ is tensorial in all three arguments if and only if T equals zero; in this case, $\mathscr{A}R = R$.

Exercise Let $\mathscr{V}$ be a real vector space; prove that the Lie algebra structure of $\mathfrak{a}(\mathscr{V})$ is given by

$$[(v, X), (w, Y)] = (Xw - Yv, [X, Y]) \qquad v, w \in \mathscr{V},\ X, Y \in \mathfrak{gl}(\mathscr{V})$$

Furthermore, $\exp\colon \mathfrak{a}(\mathscr{V}) \to A(\mathscr{V})$ is given by

$$\exp(v, X) = \left(\sum_{n=0}^{\infty} \frac{X^n v}{(n+1)!}, \exp(X) \right)$$
$$= \left(\exp(X) \int_0^1 \exp(-Xs)\, ds \cdot v, \exp(X) \right)$$

Hint: Exponentiate directly, or consider the "nonhomogeneous linear" equation $\dot{x} = Xx + v$ on $\mathscr{V}$ [BD: 7.10].

AFFINE TRANSFORMATIONS

Recall from 2.110 that the terms linear connection and affine connection are often used interchangeably to refer to a connection in TM; this is not true for many objects associated with a connection in TM, so in keeping with tradition, such objects will be referred to as affine even though the connection itself will be called a linear connection for the sake of precision.

2.114 **Definition** Let $\bar{\nabla}$ and ∇ be linear connections on manifolds $\bar{M}$ and M, respectively. A map $f\colon \bar{M} \to M$ is called *affine* if

$$f_* \bar{\nabla}_X Y = \nabla_X f_* Y \qquad X, Y \in \mathfrak{X}M$$

An *affine transformation of* (M, ∇) is an affine diffeomorphism of M.

For example, if $f\colon \bar{M} \to M$ is a C^∞ covering of manifolds, then for each linear connection ∇ on M there is a unique $\bar{\nabla}$ on $\bar{M}$ for which f is an affine covering map.

Exercise Show that an affine map preserves the associated affine connections.

2.115 **Proposition** A map $f\colon (\bar{M}, \bar{\nabla}) \to (M, \nabla)$ is affine if and only if f sends geodesics to geodesics and $f_* \bar{T}(U, V) = T(f_* U, f_* V)$, $U, V \in \mathfrak{X}M$; in this case, $f \circ \overline{\exp} = \exp \circ f_*$, and $f_* \bar{R}(U, V)W = R(f_* U, f_* V) f_* W$.

PROOF Apply 2.114 and 2.101 for the first part; for the second part, use the second structure equation 2.103ii.

2.116 **Proposition** Let f and g be affine maps from $(\bar{M}, \bar{\nabla})$ to (M, ∇), where $\bar{M}$ is connected. If $f_*|_p = g_*|_p$ for some $p \in \bar{M}$, then $f = g$.

PROOF Let $A = \{q \in \bar{M} \mid f_*|_q = g_*|_q\}$. Since f_* and g_* are continuous and TM is Hausdorff, A is closed [Ke: 3A]. Now let $q \in A$, and let $U \subset M_q$ be a neighborhood of 0 which is mapped diffeomorphically to a neighborhood of q in M by $\exp_q$ (2.96). For all $u \in U$, $f \circ \overline{\exp}(u) = \exp(f_* u) = \exp(g_* u) = g \circ \overline{\exp}(u)$, so $\exp(U)$ is an open neighborhood of q in A; thus A is open. The connectivity of $\bar{M}$ now implies that $A = \bar{M}$ if A is nonempty.

2.117 The set of all affine transformations of (M, ∇) is a group under composition.

Theorem The group $A(M, \nabla)$ of affine transformations of (M, ∇) is a Lie group.

PROOF See [N*1*], [HM], or [KN: VI.1.5].

2.118 **Definition** An *affine vector field on* (M, ∇) is an element $X \in \mathfrak{X}M$ such that the local 1-parameter group of X [W: 1.49] consists of local affine maps of (M, ∇).

2.119 **Proposition** The set of complete affine vector fields on (M, ∇) is a Lie subalgebra of $\mathfrak{X}M$ isomorphic to the Lie algebra $\mathfrak{a}(M, \nabla)$ of the Lie group $A(M, \nabla)$ of affine transformations of (M, ∇).

PROOF If $X \in \mathfrak{X}M$ is a complete affine field, then its 1-parameter group $\{\psi(t)\}$ is a 1-parameter subgroup of $A(M, \nabla)$, and $\hat{X} := \dot{\psi}(0)$ belongs to $A(M, \nabla)_e \cong \mathfrak{a}(M, \nabla)$; conversely, $\hat{X} \in \mathfrak{a}(M, \nabla)$ determines a complete affine field X on M by

$$X_p := \frac{d}{dt}\bigg|_0 \exp(t\hat{X})p$$

Given $\hat{X}, \hat{Y} \in \mathfrak{a}(M, \nabla)$,

$$\begin{aligned}\exp(t[\hat{X}, \hat{Y}]) &= \exp \circ \operatorname{ad}_{t\hat{X}} \hat{Y} = \operatorname{Ad}_{\exp(t\hat{X})}\hat{Y} \\ &= \frac{\partial}{\partial s}\bigg|_0 \exp(t\hat{X}) \exp(s\hat{Y}) \exp(-t\hat{X})\end{aligned}$$

by [W: 3.46, 3.47], so

$$\begin{aligned}\frac{\partial}{\partial t}\bigg|_0 \exp(t[\hat{X}, \hat{Y}])p &= \frac{\partial^2}{\partial t\, \partial s}\bigg|_0 \exp(t\hat{X}) \exp(s\hat{Y}) \exp(-t\hat{X})p \\ &= \frac{\partial}{\partial t}\bigg|_0 \exp(t\hat{X})_* Y_{\exp(-t\hat{X})p} \\ &= -(\mathscr{L}_X Y)_p = -[X, Y]_p\end{aligned}$$

Therefore the complete affine fields on M are mapped isomorphically to $\mathfrak{a}(M, \nabla)$ by $X \mapsto -\hat{X}$.

2.120 **Lemma** Let $X \in \mathfrak{X}M$, with local 1-parameter group $\{\psi_t\}$. Suppose there exists $p \in M$ such that $X_p = 0$ and $(\nabla X)_p = 0$; then for all t, $\psi_{t*}|_p = \mathrm{id}|_{M_p}$.

PROOF If $X_p = 0$, then $\psi_t p = p$ for all t. Let $V \in \mathfrak{X}M$, and set $\eta(t) := \psi_{t*} V_p \in M_p$. By hypothesis $(T(X, V) - \nabla_X V + \nabla_V X)_p = 0$, so

$$\dot{\eta}(0) = \frac{d}{dt}\bigg|_0 \psi_{t*} V_p = -\frac{d}{dt}\bigg|_0 \psi_{-t*} V_{\psi_t p} = -[X, V]_p = 0$$

by 2.100. But then $\dot{\eta}(t) = \psi_{t*}\dot{\eta}(0) = 0$ for all t, so η is constant, and $\psi_{t*} V_p = V_p$ for all t.

2.121 **Theorem** The affine transformation group $A(M, \nabla)$ of a connected manifold M of dimension n with a given linear connection ∇ has dimension less than or equal to $n^2 + n$.

PROOF Fix $p \in M$. The vector space homomorphism from $\mathfrak{a}(M, \nabla)$ to $M_p \oplus \mathfrak{gl}(M_p)$ which sends X to $(X, \nabla X)_p$ is injective, for if X goes to zero, then by 2.120, $\psi_{t*}|_p = \mathrm{id}|_{M_p} = I_*|_p$, where I is the identity map of M, so by 2.116, $\psi_t = I$ for all t, and $X = 0$.

2.122 There is a useful tensor criterion for affinity of vector fields.

Definition Given a linear connection ∇ on M and $X \in \mathfrak{X}M$, define

$$\begin{aligned}(\mathscr{L}_X\nabla)(U, V) &:= \mathscr{L}_X(\nabla_U V) - \nabla_{\mathscr{L}_X U} V - \nabla_U \mathscr{L}_X V \qquad U, V \in \mathfrak{X}M \\ &= [X, \nabla_U V] - \nabla_{[X, U]} V - \nabla_U [X, V]\end{aligned}$$

2.123 **Proposition** For all $X \in \mathfrak{X}M$, the expression $\mathscr{L}_X\nabla$ from 2.122 is a tensor field of type (1, 2). Let T be the torsion of ∇.

(i) The field X is affine $\Leftrightarrow \mathscr{L}_X\nabla = 0$
(ii) Assume $T = 0$; X is affine $\Leftrightarrow \nabla_U \nabla X = R(U, X) \qquad U \in \mathfrak{X}M$

PROOF The proof that $\mathscr{L}_X\nabla$ is a tensor field is standard.

(i): Let $\{\psi_t\}$ be the local 1-parameter group of X. For $U \in \mathfrak{X}M$, define $U_t := \psi_{-t*} U \circ \psi_t \in \mathfrak{X}M$. If X is affine, then for all $U, V \in \mathfrak{X}M$,

$$(\nabla_U V)_t = \nabla_{U \circ \psi_t}(\psi_{-t*} V) = \nabla_{\psi_{t*} U_t}(V_t \circ \psi_{-t}) = \nabla_{U_t} V_t$$

by the chain rule (2.54); and therefore, since $[X, U] = d/dt|_0 \, U_t$,

$$\begin{aligned}[X, \nabla_U V] &= \lim_{t\to 0} \frac{1}{t}((\nabla_U V)_t - \nabla_U V) \\ &= \lim_{t\to 0} \frac{1}{t}(\nabla_{U_t} V_t - \nabla_{U_t} V + \nabla_{U_t} V - \nabla_U V) \\ &= \nabla_U[X, V] + \nabla_{[X, U]} V\end{aligned}$$

Conversely, for U, $V \in \mathfrak{X}M$ and $t \in \mathbb{R}$, define $W(t) := \psi_{t*}\nabla_U(\psi_{-t*}V) \in \mathfrak{X}M$. We want $W(t) = \nabla_U V$ for all t. With the notation above,

$$\begin{aligned}W(t) &= \psi_{t*}\nabla_U \psi_{-t*}V = \psi_{t*}\nabla_{\psi_{t*}U_t \circ \psi_{-t}}(V_t \circ \psi_{-t}) \\ &= \psi_{t*}\nabla_{U_t \circ \psi_{-t}} V_t = \psi_{t*}(\nabla_{U_t} V_t) \circ \psi_{-t} = (\nabla_{U_t} V_t)_{-t}\end{aligned}$$

Therefore,

$$\begin{aligned}\dot{W}(0) &= \lim \frac{1}{t}((\nabla_{U_t} V_t)_{-t} - \nabla_U V) \\ &= \lim \frac{1}{t}((\nabla_{U_t} V_t)_{-t} - \nabla_{U_t} V_t + \nabla_{U_t} V_t - \nabla_U V_t + \nabla_U V_t - \nabla_U V) \\ &= -[X, \nabla_U V] + \nabla_{[X, U]} V + \nabla_U[X, V] = -(\mathscr{L}_X \nabla)(U, V) = 0\end{aligned}$$

by hypothesis. But

$$\begin{aligned}\dot{W}(t) = \frac{d}{ds}\bigg|_0 W(t+s) &= \frac{d}{ds}\bigg|_0 \psi_{t*}\psi_{s*}\nabla_U \psi_{-s*}\psi_{-t*}V \\ &= \psi_{t*}\left(\frac{d}{ds}\bigg|_0 \psi_{s*}\nabla_{U_t}\psi_{-s*}V_t\right) \circ \psi_{-t}\end{aligned}$$

which vanishes by the work above with U_t and V_t substituted for U and V. Thus W is constant over each $p \in M$, and so $W(t) = \nabla_U V$.

(ii): For U, $V \in \mathfrak{X}M$,

$$\begin{aligned}(R(X, U) + \nabla_U \nabla X)(V) &= \nabla_X \nabla_U V - \nabla_U \nabla_X V - \nabla_{[X, U]} V \\ &\quad + \nabla_U \nabla_V X - \nabla_{\nabla_U V} X \\ &= [X, \nabla_U V] - \nabla_{[X, U]} V - \nabla_U[X, V]\end{aligned}$$

by the assumption that $T = 0$. Now use (i).

Exercises

1. Fix a connection $\mathscr{H}$ on TM; let $\mathscr{H}$ also denote the projection of TTM onto the bundle $\mathscr{H}$: $X = \mathscr{H}X + \mathscr{V}X \mapsto \mathscr{H}X$, $X \in \mathfrak{X}TM$. Define $\mathscr{L}_X \mathscr{H}$ for all $X \in \mathfrak{X}TM$. Given $U \in \mathfrak{X}M$, let $\bar{U} \in \mathfrak{X}TM$ be the $\mathscr{H}$-horizontal lift of U. Compare $\mathscr{L}_U \nabla$ and $\mathscr{L}_{\bar{U}} \mathscr{H}$.
2. Prove that the set of affine vector fields on (M, ∇) is a Lie subalgebra of $\mathfrak{X}M$.
3. If ∇ is torsion-free and $X \in \mathfrak{X}M$ is affine, prove that $\mathscr{L}_X R = \mathscr{L}_X \nabla R = \cdots = \mathscr{L}_X \nabla^k R = 0$ for all k.
4. A linear connection on a homogeneous space G/H is invariant (2.60*d*) if and only if each $g \in G$ is an affine transformation of G/H; in this case $(G/H, \nabla)$ is called an *affine homogeneous space*. Prove that for each $X \in \mathfrak{g}$, the field $X^* \in \mathfrak{X}(G/H)$ from 1.63*c* is an affine field on $(G/H, \nabla)$ if ∇ is invariant.

CHAPTER

THREE

RIEMANNIAN VECTOR BUNDLES

Fix a nonvanishing section L of the tensor bundle $\otimes(E)$ of a vector bundle E over a manifold M. Not all parallel transport systems in E leave L invariant. For example, if $J \in \Gamma$ End E is a complex structure on a real vector bundle E, parallel transport in the real bundle E leaves J invariant if and only if it is actually parallel transport in the complex vector bundle (E, J) (2.15).

In particular, if L is parallel, then parallel transport around a loop in M based at $p \in M$ must leave the tensor $L_p \in \otimes(E_p)$ invariant. This imposes a restriction on the holonomy group of the connection.

In this chapter we study the classical example of such a tensor field—a Riemannian metric; if a Riemannian metric is parallel with respect to a connection, then the holonomy group of the connection is a subgroup of the orthogonal group.

All vector bundles in this chapter are assumed to be real.

RIEMANNIAN METRICS

3.1 **Definitions** A *Riemannian metric on a vector bundle E* over a manifold M is a section g of $(E \otimes E)^*$ such that for all $p \in M$, g_p is an inner product on the vector space E_p. A *Riemannian vector bundle* (E, g) is a vector bundle E endowed with a Riemannian metric g. A *Riemannian manifold* (M, g) is a manifold M with a Riemannian metric g on TM; g is also called a *Riemannian metric on M*.

Notation Given ζ, $\xi \in E_p$, $p \in M$, write $\langle\zeta, \xi\rangle$ for $g(\zeta, \xi) := g_p(\zeta \otimes \xi) \in \mathbb{R}$, and set $\|\zeta\|^2 := \langle\zeta, \zeta\rangle$; by the definition of an inner product, $\|\zeta\|^2 > 0$ if $\zeta \neq 0$.

Given $X, Y \in \Gamma E$, it follows that $\langle X, Y\rangle \in C^\infty M$.

3.2 **Definition** Let (E_j, g_j) be a Riemannian vector bundle over a manifold M_j, $j = 1, 2$. A vector bundle homomorphism $h: E_1 \to E_2$ along a map $f: M_1 \to M_2$ is *isometric* if $\langle h\zeta, h\xi\rangle_2 = \langle\zeta, \xi\rangle_1$ for all $\zeta \otimes \xi \in E_1 \otimes E_1$. Let (M_1, g_1) and (M_2, g_2) be Riemannian manifolds; a map f from M_1 to M_2 is *isometric* if $f_*: TM_1 \to TM_2$ is isometric, that is, if $f^*g_2 = g_1$. A *Riemannian covering* is an isometric C^∞ covering map of Riemannian manifolds. An *isometry* is an isometric diffeomorphism of Riemannian manifolds.

It follows immediately that an isometric map of Riemannian manifolds is an immersion [W: 1.27].

3.3 **Proposition** Every vector bundle admits a Riemannian metric.

Proof (cf. 2.35) Let $\{U_\alpha\}_{\alpha \in \mathscr{A}}$ be a locally finite open covering of M such that E is trivial over each U_α; let (π, ψ_α) be a vector bundle chart on E over U_α. Define a Riemannian metric on $\pi^{-1}U_\alpha$ by $g_\alpha(\zeta, \xi) := \langle\psi_\alpha\zeta, \psi_\alpha\xi\rangle$, where $\langle\ ,\ \rangle$ denotes the usual inner product on $\mathbb{R}^m$. Let $\{f_\alpha\}$ be a partition of unity subordinate to $\{U_\alpha\}$, and extend the local section $f_\alpha \cdot g_\alpha$ of $(E \otimes E)^*$ so that it is zero off U_α. The section $g := \sum_\alpha f_\alpha \cdot g_\alpha$ of $(E \otimes E)^*$ is a Riemannian metric on E.

3.4 **Examples**

(*a*) The *canonical Riemannian metric on* $M \times \mathbb{R}^m$ is $\langle(p, \zeta), (p, \xi)\rangle := \langle\zeta, \xi\rangle$, the inner product of ζ and ξ in $\mathbb{R}^m$. The *canonical Riemannian metric on the manifold* $\mathbb{R}^n$ is $\langle(p, u), (p, v)\rangle := \langle u, v\rangle$; the usual vector bundle isomorphism $T\mathbb{R}^n = \mathbb{R}^n \times \mathbb{R}^n$ is isometric.

(*b*) Given an inner product $\langle\ ,\ \rangle_{\mathfrak{g}}$ on the Lie algebra $\mathfrak{g}$ of a Lie group G, define a Riemannian metric on G by $\langle X_g, Y_g\rangle := \langle X, Y\rangle_{\mathfrak{g}}$ for all $X, Y \in \mathfrak{g}$ and $g \in G$. This metric is *left-invariant*, that is, each left translation L_g of G, $g \in G$, is an isometry: $L_g^*\langle\ ,\ \rangle = \langle\ ,\ \rangle$. Similarly, a *right-invariant metric on G* is one for which all right translations R_g, $g \in G$, are isometries: $R_g^*\langle\ ,\ \rangle = \langle\ ,\ \rangle$; in this case, $\langle U_g, V_g\rangle = \langle U, V\rangle_{\tilde{\mathfrak{g}}}$, U, V in the Lie algebra $\tilde{\mathfrak{g}}$ of right-invariant fields (1.63*b*), where $\langle\ ,\ \rangle_{\tilde{\mathfrak{g}}}$ is an inner product on $\tilde{\mathfrak{g}}$. A metric on G is *bi-invariant* if it is left-invariant and right-invariant; for example, the canonical metric on $\mathbb{R}^n$ is bi-invariant.

A left-invariant Riemannian metric $\langle\ ,\ \rangle$ on G is also right-invariant, hence bi-invariant, if and only if the associated inner product on $\mathfrak{g}$ is invariant under Ad_g for all $g \in G$, where Ad is the *adjoint representation of G on* $\mathfrak{g}$ [W: 3.46(2)]: $\mathrm{Ad}_g = (L_g \circ R_{g^{-1}})_*|_e$. *Proof:* By the left invariance of $\langle\ ,\ \rangle$,

$$\begin{aligned}\langle R_{g*}X, R_{g*}Y\rangle &= \langle L_{g^{-1}*}R_{g*}X, L_{g^{-1}*}R_{g*}Y\rangle \\ &= \langle \mathrm{Ad}_{g^{-1}}\, X, \mathrm{Ad}_{g^{-1}}\, Y\rangle_{\mathfrak{g}}\end{aligned}$$

for all $g \in G$, $X, Y \in \mathfrak{g}$. *QED.*

Infinitesimally, this says that $\mathrm{ad}_X := \mathrm{Ad}_*|_e X \in \mathfrak{gl}(\mathfrak{g})$, $X \in \mathfrak{g}$, is skew-symmetric with respect to $\langle\ ,\ \rangle$, for

$$\begin{aligned}\langle \mathrm{ad}_X Y, Z\rangle &= \frac{d}{dt}\bigg|_0 \langle \mathrm{Ad}_{\exp(tX)} Y, Z\rangle \\ &= \frac{d}{dt}\bigg|_0 \langle Y, \mathrm{Ad}_{\exp(-tX)} Z\rangle = -\langle Y, \mathrm{ad}_X Z\rangle\end{aligned}$$

for all $X, Y, Z \in \mathfrak{g}$. Conversely, since $\mathrm{Ad}_{\exp(X)} = \exp(\mathrm{ad}_X) \in GL(\mathfrak{g})$ for $X \in \mathfrak{g}$ [W: 3.46(5)] (that is, the following diagram commutes),

$$\begin{array}{ccc} \mathfrak{g} & \xrightarrow{\ \mathrm{ad}\ } & \mathfrak{gl}(\mathfrak{g}) \\ {\scriptstyle \exp}\downarrow & & \downarrow{\scriptstyle \exp} \\ G & \xrightarrow{\ \mathrm{Ad}\ } & GL(\mathfrak{g}) \end{array}$$

$$\begin{aligned}\langle \mathrm{Ad}_{\exp(X)} Y, Z\rangle &= \sum_{k=0}^{\infty} \left\langle \frac{(\mathrm{ad}_X)^k}{k!} Y, Z \right\rangle = \sum_{k=0}^{\infty} \left\langle Y, \frac{(-\mathrm{ad}_X)^k}{k!} Z \right\rangle \\ &= \langle Y, \mathrm{Ad}_{\exp(-X)} Z\rangle\end{aligned}$$

Thus if G is connected and each ad_X is skew-symmetric with respect to $\langle\ ,\ \rangle$, then $\langle\ ,\ \rangle$ is bi-invariant by 2.95, 2.96, and the fact [W: 3.18] that G is generated by a neighborhood of e.

(*c*) The *canonical metric on* S^n is defined by letting the embedding of S^n into $\mathbb{R}^{n+1}$ be isometric; that is,

$$\langle (p, u), (p, v)\rangle_{S^n} := \langle (p, u), (p, v)\rangle_{\mathbb{R}^{n+1}} = \langle u, v\rangle$$

for $(p, u), (p, v) \in TS^n$. The group $O(n + 1)$ acts on S^n by isometries:

$$\langle g_*(p, u), g_*(p, v)\rangle_{S^n} = \langle gu, gv\rangle = \langle u, v\rangle = \langle (p, u), (p, v)\rangle_{S^n}$$

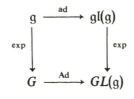

(*d*) More generally, a Riemannian metric on a homogeneous space $M = G/H$ is *invariant* if each $g \in G$ acts on G/H as an isometry; in this case $(G/H, \langle\ ,\ \rangle)$ is called a *Riemannian homogeneous space.* Suppose that H is the isotropy group of $p \in M$. Let $\wp\colon G \to M$ be the projection map; $\wp(g) = gp$ for all $g \in G$; in particular, $\wp(H) = \wp(e) = p$. Invariance of $\langle\ ,\ \rangle$ means that $\langle g_* u, g_* v\rangle_{gp} = \langle u, v\rangle_p$ for $g \in G$, u, $v \in M_p$; for instance, $\langle h_* u, h_* v\rangle_p = \langle u, v\rangle_p$, $h \in H$, so $\langle\ ,\ \rangle_p$ is invariant under the *linear isotropy representation* $\lambda\colon H \to GL(M_p)$, $h \mapsto h_*|_p$ [W: 3.61]. Conversely, an inner product $\langle\ ,\ \rangle_p$ on M_p which is invariant under the subgroup $\lambda H \subset GL(M_p)$ can be extended to an invariant Riemannian metric on M by $\langle g_* u, g_* v\rangle_{gp} := \langle u, v\rangle_p$ for all $g \in G$, u, $v \in M_p$. This is well-defined because if $h \in H$, and u, $v \in M_p$, then

$$\langle g_* u, g_* v\rangle_{gp} := \langle u, v\rangle_p = \langle h_* u, h_* v\rangle_p =: \langle (gh)_* u, (gh)_* v\rangle_{gp}$$

Thus an invariant metric on $M = G/H$ is equivalent to a λH-invariant inner product on the tangent space M_p, $p = \wp(H) \in M$. This will be dealt with further in chap. 6.

(*e*) Real projective space $\mathbb{R}P^n$ is diffeomorphic to the quotient space $S^n/\mathbb{Z}_2$; the projection map $f\colon S^n \to \mathbb{R}P^n$, $f(p) := \{p, -p\}$, is a C^∞ covering. Define a metric on $\mathbb{R}P^n$ by $\langle f_*(p, u), f_*(p, v)\rangle_{\mathbb{R}P^n} := \langle (p, u), (p, v)\rangle_{S^n}$; this is well-defined because $\mathbb{Z}_2$ acts on S^n by isometries: $\langle (p, u), (p, v)\rangle_{S^n} = \langle (-p, -u), (-p, -v)\rangle_{S^n}$. The projection f is now a Riemannian covering map.

(*f*) Fix a Riemannian metric $\langle\ ,\ \rangle$ and a linear connection ∇ on a manifold M, with connection map $\kappa\colon TTM \to TM$ (2.49). The *Sasaki* (or *connection*) *metric on the manifold TM* is defined by $\widetilde{\langle U, V\rangle} := \langle \pi_* U, \pi_* V\rangle + \langle \kappa U, \kappa V\rangle$, U, $V \in \mathfrak{X}TM$ [Sa] [Do*l*]. The reader should note carefully that $\widetilde{\langle\ ,\ \rangle}$ is a metric on the bundle TTM over TM, not on the bundle TM over M. *Exercises:* Prove the following: The embedding of each tangent space M_p into TM is isometric with respect to the canonical extension to the manifold M_p (as in 3.4*a*) of the inner product $\langle\ ,\ \rangle_p$ on the vector space M_p. The zero section (1.52*c*) of TM embeds M isometrically into TM. The complex structure $J \in \Gamma\,\mathrm{End}\,TTM$ defined before the exercises in 2.50 acts isometrically.

3.5 **Definition** Given Riemannian vector bundles (E_j, g_j) over manifolds M_j, $j = 1, 2$, the *Riemannian product metric on* $E_1 \times E_2$ is defined by

$$\langle (\zeta_1, \zeta_2), (\xi_1, \xi_2)\rangle := \langle \zeta_1, \xi_1\rangle_1 + \langle \zeta_2, \xi_2\rangle_2$$

If $M = M_1 \times M_2$ and g is the product metric on TM determined by metrics g_1 and g_2 on TM_1 and TM_2, then (M, g) is called the *Riemannian product of* (M_1, g_1) *and* (M_2, g_2).

For example, $\mathbb{R}^2$ is canonically the Riemannian product of $\mathbb{R}$ and $\mathbb{R}$; by iteration, $\mathbb{R}^n$ is canonically the Riemannian product of n copies of $\mathbb{R}$, where $\mathbb{R}$ and $\mathbb{R}^n$ are given the usual inner products.

3.6 **Definition** Given Riemannian vector bundles (E_j, g_j), $j = 1, 2$, over a manifold M, the *Riemannian metric on the Whitney sum* $E_1 \oplus E_2 = \Delta^*(E_1 \times E_2)$ is the unique metric on $E_1 \oplus E_2$ such that pr_2 is an isometric vector bundle homomorphism along the diagonal embedding map Δ of M into $M \times M$, $\Delta(p) = (p, p)$.

$$\begin{array}{ccc} E_1 \oplus E_2 & \xrightarrow{\mathrm{pr}_2} & E_1 \times E_2 \\ \downarrow & & \downarrow \\ M & \xrightarrow[\Delta]{} & M \times M \end{array}$$

The *Riemannian metric on the tensor product* $E_1 \otimes E_2$ is defined by

$$\langle \zeta_1 \otimes \zeta_2, \xi_1 \otimes \xi_2 \rangle := \langle \zeta_1, \xi_1 \rangle_1 \cdot \langle \zeta_2, \xi_2 \rangle_2$$

3.7 **Definition** The *Riemannian metric on the exterior algebra bundle* $\Lambda(E)$ of a Riemannian vector bundle (E, g) over M (1.41b) is defined by

$$\langle \zeta_1 \wedge \cdots \wedge \zeta_r, \xi_1 \wedge \cdots \wedge \xi_s \rangle := \begin{cases} \det[\langle \zeta_i, \xi_j \rangle] & \text{if } r = s \\ 0 & \text{otherwise} \end{cases}$$

3.8 Let (E, g) be a Riemannian vector bundle over M. For each $p \in M$, g_p is *nonsingular*, that is, $\langle \zeta, \xi \rangle = 0$ for all $\zeta \in E_p$ implies $\xi = 0$; this implies that the map $E \to E^*$ which sends each vector ζ to the covector $\langle \zeta, \cdot \rangle$ is a vector bundle isomorphism.

Definition Define the *musical isomorphisms with respect to* $\langle\ ,\ \rangle$ by $\flat: E \to E^*$, $\zeta^\flat := \langle \zeta, \cdot \rangle$, and $\#: E^* \to E$ such that $\#$ is the inverse of $\flat$. The expression $\zeta^\flat$ is read "ζ flat," and for $\mu \in E^*$, $\mu^\#$ is read "μ sharp."

In the classical tensor calculus, the isomorphism $\flat$ "lowered the indices," for given a basis $\{\epsilon_j\}$ for E_p and the dual basis $\{\omega^i\}$ for E_p^*, if $\zeta = \sum \zeta^i \epsilon_i \in E_p$, then $\zeta^\flat = \sum \zeta_j \omega^j$, where $\zeta_j = \sum_i \langle \epsilon_i, \epsilon_j \rangle \zeta^i$. The inverse isomorphism $\#$ raised the indices.

3.9 **Definition** The *Riemannian metric on the dual E^* of a Riemannian vector bundle* (E, g) is defined by

$$\langle \eta, \mu \rangle_{E^*} := \langle \eta^\#, \mu^\# \rangle_E \qquad \eta \otimes \mu \in E^* \otimes E^*$$

Exercises

1. Given $\eta, \mu \in E_p^*$ and an orthonormal basis $\{\epsilon_j\}$ for E_p, show that $\langle \eta, \mu \rangle = \sum \eta(\epsilon_j)\mu(\epsilon_j)$; for $\xi \in E_p$, show that $\langle \mu, \xi^\flat \rangle = \mu(\xi) = \langle \mu^\#, \xi \rangle$.
2. Prove that $\flat$ and $\#$ are isometric vector bundle isomorphisms.

3.10 **Definition** Given Riemannian vector bundles (E_j, g_j) over M, the *Riemannian metric on the homomorphism bundle* $\mathrm{Hom}(E_1, E_2)$ over M is defined by letting the isomorphism $\mathrm{Hom}(E_1, E_2) \cong E_1^* \otimes E_2$ be isometric.

Exercises

1. Given $K, L \in \Gamma\, \mathrm{Hom}(E_1, E_2)$ and an orthonormal basis $\{\epsilon_j\}$ for $E_1|_p$, $\langle K, L \rangle(p) := \sum \langle K\epsilon_j, L\epsilon_j \rangle_2$: Compare E^* and $\mathrm{Hom}(E, M \times \mathbb{R})$.
2. Let K be a symmetric (self-adjoint) transformation of an inner product space $(\mathscr{V}, \langle\ ,\ \rangle)$, and let L be skew-symmetric. Prove that $2\langle K, L \rangle = \mathrm{tr}[K, L] = 0$, so $K \perp L$.

3.11 **Definition** Let (M, g) be a Riemannian manifold; for each $f \in C^\infty M$, the *gradient of f* is the vector field $\nabla f := (df)^\#$ on M:

$$\langle \nabla f, v \rangle := \langle df^\#, v \rangle = df(v) = vf \qquad v \in TM$$

3.12 **Definition** Given an isometric immersion $f\colon \bar{M} \to M$ of Riemannian manifolds, and given a vector field $X \in \Gamma_f TM$ along f, let $\mu \in A^1\bar{M}$ be the 1-form such that $\mu(V) := \langle f_* V, X \rangle$, $V \in \mathfrak{X}\bar{M}$. The *tangential component of X* is $X^\top := f_*\mu^\# \in \Gamma_f TM$. The *orthogonal component of X* is $X^\perp := X - X^\top \in \Gamma_f TM$.

Exercise For all $p \in \bar{M}$, $X_p^\top$ belongs to $f_*\bar{M}_p$, and $X_p^\perp$ is orthogonal to $f_*\bar{M}_p$.

3.13 **Definition** Given an inner product space $(\mathscr{V}, \langle\ ,\ \rangle)$, define a *linear map* $\Lambda^2\mathscr{V} \to \mathrm{gl}(\mathscr{V})$ by $u \wedge v(x) := \langle v, x \rangle u - \langle u, x \rangle v$, $x \in \mathscr{V}$. If (E, g) is a Riemannian vector bundle, define a *vector bundle homomorphism* $\Lambda^2 E \to \mathrm{gl}(E)$ to be the homomorphism above on each fiber.

Given u, v, w, and $x \in \mathscr{V}$,

$$\langle u \wedge v(w), x\rangle = \langle\langle v, w\rangle u - \langle u, w\rangle v, x\rangle$$

$$= \det\begin{bmatrix}\langle u, x\rangle & \langle u, w\rangle \\ \langle v, x\rangle & \langle v, w\rangle\end{bmatrix} = \langle u \wedge v, x \wedge w\rangle$$

In particular, $\langle u \wedge v(w), x\rangle = -\langle u \wedge v, w \wedge x\rangle = -\langle w, u \wedge v(x)\rangle$, so $u \wedge v$ is skew-symmetric with respect to $\langle\ ,\ \rangle$; in other words, $u \wedge v$ belongs to the orthogonal algebra $\mathfrak{o}(\mathscr{V}) \subset \mathfrak{gl}(\mathscr{V})$ with respect to $\langle\ ,\ \rangle$. It follows immediately that the linear map $\Lambda^2\mathscr{V} \to \mathfrak{o}(\mathscr{V})$ is a vector space isomorphism. *Exercise:* Calculate its inverse.

On the vector bundle level, this implies that $\Lambda^2 E$ is isomorphic to the *orthogonal algebra bundle* $\mathfrak{o}(E) \subset \mathfrak{gl}(E)$ of all skew-symmetric transformations of the fibers of (E, g).

Note: This isomorphism is not isometric with respect to the Riemannian metrics on $\Lambda^2 E$ and $\mathfrak{o}(E) \subset \mathfrak{gl}(E)$ (which is isomorphic to $\mathrm{Hom}(E, E)$ as a Riemannian vector bundle), for as an element of $\mathfrak{o}(E)$, $u \wedge v = u^\flat \otimes v - v^\flat \otimes u$, and

$$\|u^\flat \otimes v - v^\flat \otimes u\|^2 = 2(\|u\|^2\|v\|^2 - \langle u, v\rangle^2) = 2\|u \wedge v\|^2$$

3.14 **Definition** Let (M, g) be an n-dimensional, oriented Riemannian manifold. The *Riemannian volume element on* (M, g) is the unique volume element (1.66) ω in the orientation of M such that $\|\omega\|^2 \in C^\infty M$ equals 1.

Uniqueness of ω follows from the fact that $\Lambda^n T^*M$ is a line bundle. If $\{\epsilon_j\}$ is a positively oriented orthonormal basis for M_p, then $\omega_p(\epsilon_1, \ldots, \epsilon_n) = 1$.

3.15 **Lemma** If (E, g) is a Riemannian vector bundle over M, then locally there exists an orthonormal basis field for E (1.52f).

PROOF Apply the Gram-Schmidt orthonormalization process to the elements of an arbitrary local basis field for E.

3.16 **Proposition** Let $\omega \in A^n M$ be the Riemannian volume element on an oriented Riemannian manifold (M, g). If $\{\omega^i\}$ is a positively oriented, orthonormal local basis field for T^*M, then $\omega = \omega^1 \wedge \cdots \wedge \omega^n$.

PROOF $\|\omega^1 \wedge \cdots \wedge \omega^n\|^2 = \det[\langle \omega^i, \omega^j\rangle] = \det(I) = 1$.

3.17 **Definition** The *Hodge star operator* on an n-dimensional, oriented

Riemannian manifold (M, g) is the section $\bigstar \in \Gamma\,\mathrm{End}(\Lambda T^*M)$ such that

$$(\bigstar\mu)(X) := \langle \mu \wedge X^\flat, \omega\rangle \qquad \mu \in A^rM,\ X \in \Gamma\Lambda^{n-r}TM$$

where $\omega \in A^nM$ is the Riemannian volume element, and $X^\flat$ is as defined in 3.8.

3.18 **Proposition** The star operator $\bigstar$ on an oriented, n-dimensional Riemannian manifold (M, g) is an isometric bundle isomorphism from ΛT^*M to itself, and maps $\Lambda^r T^*M$ isomorphically to $\Lambda^{n-r}T^*M$. In particular, $\bigstar\omega = 1 \in A^0M$, $\bigstar 1 = \omega$, and $\bigstar(\omega^1 \wedge \cdots \wedge \omega^r) = \omega^{r+1} \wedge \cdots \wedge \omega^n$ if $\{\omega^i\}$ is a positively oriented orthonormal basis for M_p^*. On A^rM, $\bigstar^2 = (-1)^{r(n-r)}$. Next, $\langle \eta, \mu\rangle = \bigstar(\eta \wedge \bigstar\mu) = \bigstar(\mu \wedge \bigstar\eta)$ for η, $\mu \in A(M)$. Finally, if $\{\epsilon_j\}$ and $\{\omega^i\}$ are dual orthonormal bases for M_p and M_p^*, then $\bigstar(\omega^j \wedge \mu) = (-1)^r \iota_{\epsilon_j}\bigstar\mu$ for $\mu \in \Lambda^r M_p^*$, where ι is interior multiplication [W: 2.21], $\iota_X\mu := \mu(X, \cdot, \ldots, \cdot)$ for $X \in \mathfrak{X}M$, $\mu \in A(M)$.

PROOF Clearly, $\bigstar$ maps $\Lambda^r T^*M$ into $\Lambda^{n-r}T^*M$. By 3.9, given $\mu \in A^rM$ and $\eta \in A^{n-r}M$, $\langle \bigstar\mu, \eta\rangle = (\bigstar\mu)(\eta^\#) = \langle \mu \wedge \eta, \omega\rangle$. Thus, if $\{\omega^i\}$ is as in the hypothesis, then

$$\langle \bigstar(\omega^{-1} \wedge \cdots \wedge \omega^r), \omega^{j_1} \wedge \cdots \wedge \omega^{j_{n-r}}\rangle = 0$$

if

$$\{j_1, \ldots, j_{n-r}\} \neq \{r+1, \ldots, n\};$$

hence $\bigstar(\omega^1 \wedge \cdots \wedge \omega^r) = \omega^{r+1} \wedge \cdots \wedge \omega^n$, and this implies the remaining parts of the proposition.

Every manifold is locally orientable, with exactly two choices of local orientation. Thus locally a Riemannian manifold has two star operators, one for each choice of local orientation. Such a local star operator can be extended to a global star operator if and only if M is orientable.

RIEMANNIAN CONNECTIONS

3.19 **Definition** A *Riemannian connection* in a Riemannian vector bundle (E, g) over M is a connection ∇ in E such that g is a parallel section of the vector bundle $E^* \otimes E^*$.

Exercise Let ∇ be a Riemannian connection in a Riemannian vector bundle (E, g), and fix $p \in M$. Prove that the holonomy group of ∇ at p is a subgroup of the orthogonal group $O(E_p, g_p)$.

3.20 **Proposition** A connection ∇ in a Riemannian vector bundle (E, g) is Riemannian if and only if the *Ricci identity* holds:

$$U\langle X, Y\rangle = \langle \nabla_U X, Y\rangle + \langle X, \nabla_U Y\rangle \qquad U \in \mathfrak{X}M,\ X, Y \in \Gamma E$$

PROOF By 2.61 and 2.62,

$$(\nabla_U g)(X, Y) = U\langle X, Y\rangle - \langle \nabla_U X, Y\rangle - \langle X, \nabla_U Y\rangle$$

3.21 **Proposition** A Riemannian vector bundle admits a (nonunique) Riemannian connection.

PROOF (cf. 3.3 and 2.35) Let $\{U_\alpha\}_{\alpha \in \mathscr{A}}$ be a locally finite open covering of M such that E is trivial over each U_α. Fix a local orthonormal basis field $\{\overset{\alpha}{Z}_j\}$ for E over U_α, and define $\overset{\alpha}{\nabla}_U Y := \sum_j U\langle \overset{\alpha}{Z}_j, Y\rangle \overset{\alpha}{Z}_j$ over U_α; extend $\overset{\alpha}{\nabla}$ to be zero off U_α. Given a partition of unity $\{f_\alpha\}$ subordinate to $\{U_\alpha\}$, define $\nabla := \sum f_\alpha \cdot \overset{\alpha}{\nabla}$; as in 2.36 (cf. 2.59, exercise 3), ∇ is a connection in E. Furthermore,

$$\begin{aligned} U\langle X, Y\rangle &= \sum_{\alpha, j} f_\alpha \cdot U(\langle X, \overset{\alpha}{Z}_j\rangle\langle \overset{\alpha}{Z}_j, Y\rangle) \\ &= \sum_\alpha f_\alpha \cdot (\langle \overset{\alpha}{\nabla}_U X, Y\rangle + \langle X, \overset{\alpha}{\nabla}_U Y\rangle) \\ &= \langle \nabla_U X, Y\rangle + \langle X, \nabla_U Y\rangle \end{aligned}$$

3.22 **Examples**

(*a*) The canonical connection on $M \times \mathbb{R}^m$ is a Riemannian connection for the canonical Riemannian metric; in particular, this is true in $T\mathbb{R}^n$.

(*b*) If $\langle\ ,\ \rangle$ is a left-invariant Riemannian metric on a Lie group G, then $\nabla Y := 0$, $Y \in \mathfrak{g}$, defines a Riemannian connection.

Suppose that $\langle\ ,\ \rangle$ is bi-invariant (3.4*b*). For all $c \in \mathbb{R}$, $\nabla_X Y := c[X, Y]$, $X, Y \in \mathfrak{g}$, yields a Riemannian connection, for [W: 3.47] $\mathrm{ad}_X Z = [X, Z]$, and therefore by 3.4*b*,

$$X\langle Y, Z\rangle = 0 = \langle c[X, Y], Z\rangle + \langle Y, c[X, Z]\rangle \qquad X, Y, Z \in \mathfrak{g}$$

(*c*) The canonical connection on S^n is Riemannian with respect to the canonical Riemannian metric from 3.4*c*, for if X and Y are parallel along a curve γ in S^n, then by 2.42, $\langle X, Y\rangle$ is constant, that is, $\langle X, Y\rangle$ is a parallel section of $S^n \times \mathbb{R}$ along γ. But then $\langle\ ,\ \rangle$ is parallel along γ by 2.13.

Exercise Verify the Ricci identity directly on S^n.

3.23 **Proposition** Let ∇ be a Riemannian connection in (E, g) over M. The associated connections in E^*, $\otimes(E)$, and $\Lambda(E)$ are Riemannian with respect to the associated Riemannian metrics in these bundles.

PROOF The proofs are purely computational, and are left as an exercise (see 2.13, 2.14, 3.6, 3.7, and 3.9).

3.24 **Proposition** Let ∇ be a Riemannian connection in (E, g) over M. The curvature tensor field R of ∇ satisfies the identity

$$\langle R(U, V)X, Y\rangle + \langle X, R(U, V)Y\rangle = 0 \qquad U, V \in \mathfrak{X}M,\ X, Y \in \Gamma E$$

PROOF Fix $U, V \in \mathfrak{X}M$; we may assume that $[U, V] = 0$ by the tensoriality of R. By 3.20 and 2.66, since $[U, V] = 0$,

$$\begin{aligned}\langle R(U, V)X, X\rangle &= \langle \nabla_U \nabla_V X, X\rangle - \langle \nabla_V \nabla_U X, X\rangle \\ &= U\langle \nabla_V X, X\rangle - \langle \nabla_V X, \nabla_U X\rangle \\ &\quad - V\langle \nabla_U X, X\rangle + \langle U_U X, \nabla_V X\rangle \\ &= \tfrac{1}{2}(UV\langle X, X\rangle - VU\langle X, X\rangle) = 0\end{aligned}$$

3.25 **Corollary** Let ∇ be a Riemannian connection in (E, g) over M. For $U, V \in \mathfrak{X}M$, the curvature operator $R(U, V)$ is a section of the orthogonal algebra bundle $\mathfrak{o}(E)$ of (E, g). If M is connected, the holonomy group of ∇ is a subgroup of $O(m)$, m being the rank of E.

PROOF For each $p \in M$, $\mathfrak{o}(E_p)$ consists of the skew-symmetric linear transformations of the inner product space (E_p, g_p); hence $R(u, v) \in \mathfrak{o}(E_p)$ for $u, v \in M_p$ by 3.24. If M is connected, the holonomy algebra of ∇ lies in $\mathfrak{o}(m) \subset \mathfrak{gl}(m, \mathbb{R})$, so the holonomy group $\subset GL(m, \mathbb{R})$ must be contained in $O(m)$.

3.26 **Proposition** The space of Riemannian connections on a Riemannian vector bundle (E, g) is parametrized by the space $A^1(M, \mathfrak{o}(E))$, where $\mathfrak{o}(E)$ is the orthogonal algebra bundle of (E, g).

PROOF By 2.74, if ∇ is a Riemannian connection in E, and if $\mathscr{D} \in A^1(M, \mathfrak{o}(E))$, then the connection $\bar{\nabla} := \nabla + \mathscr{D}$ is also Riemannian. Conversely, if $\bar{\nabla}$ and ∇ are Riemannian connections in (E, g), then

$$\begin{aligned}\langle(\bar{\nabla} - \nabla)_U X, X\rangle &= \langle \bar{\nabla}_U X, X\rangle - \langle \nabla_U X, X\rangle \\ &= \tfrac{1}{2}U\|X\|^2 - \tfrac{1}{2}U\|X\|^2 = 0\end{aligned}$$

3.27 **Proposition** The Riemannian volume element and star operator on an oriented Riemannian manifold (M, g) are parallel with respect to each Riemannian linear connection. The musical isomorphisms (3.8) $\flat \in \Gamma \operatorname{Hom}(E, E^*)$ and $\# \in \Gamma \operatorname{Hom}(E^*, E)$ for dual Riemannian vector bundles E and E^* (3.9) are parallel with respect to associated Riemannian connections in E and E^*.

PROOF Let ω^i be a parallel 1-form along a curve γ in M, $i = 1, \ldots, n$, such that $\{\omega^i|_0\}$ is a positively oriented orthonormal basis for $M^*_{\gamma(0)}$. By 3.16 the Riemannian volume form ω equals $\omega^1 \wedge \cdots \wedge \omega^n$ along γ; hence it is parallel along γ by 2.14. Next, $\bigstar(\omega^1 \wedge \cdots \wedge \omega^r) = \omega^{r+1} \wedge \cdots \wedge \omega^n$, and both $\omega^1 \wedge \cdots \wedge \omega^r$ and $\omega^{r+1} \wedge \cdots \wedge \omega^n$ are parallel along γ, so $\bigstar$ is parallel along γ by 2.14.

Similar proofs work for $\flat$ and $\#$.

THE LEVI-CIVITA CONNECTION

3.28 **The fundamental theorem of Riemannian geometry** Let (M, g) be a Riemannian manifold; there is a unique torsion-free Riemannian connection on (M, g).

PROOF Uniqueness: Alternately using the Ricci identity for ∇ and the fact that ∇ is torsion-free, we see that for $X, Y, Z \in \mathfrak{X}M$,

$$\begin{aligned}
\langle \nabla_X Y, Z \rangle &= X\langle Y, Z\rangle - \langle Y, \nabla_X Z\rangle \\
&= X\langle Y, Z\rangle - \langle Y, \nabla_Z X + [X, Z]\rangle \\
&= X\langle Y, Z\rangle - Z\langle Y, X\rangle + \langle \nabla_Z Y, X\rangle + \langle Y, [Z, X]\rangle \\
&= X\langle Y, Z\rangle - Z\langle X, Y\rangle \\
&\quad + \langle \nabla_Y Z + [Z, Y], X\rangle + \langle Y, [Z, X]\rangle \\
&= X\langle Y, Z\rangle - Z\langle X, Y\rangle + Y\langle Z, X\rangle - \langle Z, \nabla_Y X\rangle \\
&\quad - \langle X, [Y, Z]\rangle + \langle Y, [Z, X]\rangle \\
&= X\langle Y, Z\rangle + Y\langle Z, X\rangle - Z\langle X, Y\rangle + \langle Z, [X, Y]\rangle \\
&\quad + \langle Y, [Z, X]\rangle - \langle X, [Y, Z]\rangle - \langle \nabla_X Y, Z\rangle
\end{aligned}$$

Thus

$$\begin{aligned}
\langle \nabla_X Y, Z\rangle = \tfrac{1}{2}(X\langle Y, Z\rangle &+ Y\langle Z, X\rangle - Z\langle X, Y\rangle \\
&+ \langle Z, [X, Y]\rangle + \langle Y, [Z, X]\rangle - \langle X, [Y, Z]\rangle) \qquad (3\text{-}1)
\end{aligned}$$

Since this identity holds for every torsion-free Riemannian connection and g_p is nonsingular for all $p \in M$, ∇ is unique.

Existence: Fix $X, Y \in \mathfrak{X}M$, and define $\mu: \mathfrak{X}M \to C^\infty M$ by setting $\mu(Z)$ equal to the right side of eq. (3-1) for $Z \in \mathfrak{X}M$; the map μ is obviously additive. Given $f \in C^\infty M$,

$$X\langle Y, fZ\rangle + \langle Y, [fZ, X]\rangle = f \cdot (X\langle Y, Z\rangle + \langle Y, [Z, X]\rangle)$$

and similarly for the other two pairs of terms on the right side of eq. (3-1), so $\mu(fZ) = f\mu(Z)$, and μ is linear over $C^\infty M$. Thus $\mu \in A^1 M$. Define $\nabla_X Y := \mu^\# \in \mathfrak{X}M$, that is (3.8 and 3.11), $\nabla_X Y$ is the unique vector field on M such that $\langle \nabla_X Y, Z\rangle = \mu(Z)$ for all $Z \in \mathfrak{X}M$. It is easily checked that ∇ is a torsion-free Riemannian connection.

3.29 **Definition** The unique torsion-free Riemannian connection on a Riemannian manifold (M, g) is called the *Levi-Civita connection of g*.

For a treatment of the Levi-Civita connection from the point of view of the calculus of variations, see [Be].

Exercise Let ∇ be the Levi-Civita connection of a Riemannian metric $\langle\ ,\ \rangle$ on a manifold M; let $\widetilde{\langle\ ,\ \rangle}$ be the associated Sasaki metric (3.4*f*) on the manifold TM, with Levi-Civita connection $\tilde{\nabla}$. Verify that the geodesic spray $S \in \mathfrak{X}TM$ of ∇ is $\tilde{\nabla}$-autoparallel: $\tilde{\nabla}_S S = 0$. Thus if γ is a geodesic of ∇, then $\dot{\gamma}$ is a geodesic of $\tilde{\nabla}$.

Unless otherwise stated, a linear connection on a Riemannian manifold will always be assumed to be the Levi-Civita connection.

3.30 **Proposition** The Levi-Civita connection is natural with respect to isometries, that is (2.114), an isometry is an affine map: if $f: \bar{M} \to M$ is an isometry of Riemannian manifolds, then

$$f_* \bar{\nabla}_U V = \nabla_U f_* V \qquad U, V \in \mathfrak{X}\bar{M}$$

In particular, if $\bar{M}$ is connected and if f and g are isometries from $\bar{M}$ to M such that f_* and g_* agree at some $p \in M$, then $f = g$.

Proof Let $U, V, W \in \mathfrak{X}\bar{M}$ be f-related to $X, Y, Z \in \mathfrak{X}M$, respectively. Since f is an isometry.

$$\begin{aligned} U\overline{\langle V, W\rangle} &= U\langle f_* V, f_* W\rangle = U(\langle Y, Z\rangle \circ f) \\ &= f_* U\langle Y, Z\rangle = (X\langle Y, Z\rangle) \circ f \end{aligned}$$

and similarly for the other terms in eq. (3-1) for $\overline{\langle \bar{\nabla}_U V, W \rangle}$. Therefore,

$$\langle f_* \bar{\nabla}_U V, f_* W \rangle = \overline{\langle \bar{\nabla}_U V, W \rangle} = \langle \nabla_X Y, Z \rangle \circ f$$
$$= \langle \nabla_{f_* U} Y, f_* W \rangle = \langle \nabla_U f_* V, f_* W \rangle$$

for all $W \in \mathfrak{X}\bar{M}$, so $f_* \bar{\nabla}_U V = \nabla_{f_* U} Y = \nabla_U f_* V$. Equality of the isometrics f and g follows from 2.116.

With some reasonable extra conditions, the Levi-Civita connection is the unique connection on Riemannian manifolds which is natural with respect to isometries (see [Ep] and [St]).

3.31 **Corollary 1** If $f: \bar{M} \to M$ is a Riemannian covering of Riemannian manifolds, then f is affine.

PROOF It suffices to prove the identity locally. But $(\bar{M}, \bar{g})$ is locally isometric to (M, g), so we can apply 3.30.

3.32 **Corollary 2** If $f: \bar{M} \to M$ is an isometric immersion of Riemannian manifolds, then (3.12)

$$f_* \bar{\nabla}_U V = (\nabla_U f_* V)^{\top} \qquad U, V \in \mathfrak{X}\bar{M}$$

PROOF By [W: 1.35], for each $p \in \bar{M}$ there is a neighborhood A of p in $\bar{M}$ such that $f: A \to f(A)$ is an isometry with respect to the metrics induced by the inclusions $A \subset \bar{M}$ and $f(A) \subset M$. Now apply 3.30 to verify the local identity on A.

3.33 **Proposition** Let p be a point in a Riemannian manifold (M, g). There exist charts x on M about p such that $x(p) = 0, \{X_j|_p := \partial/\partial x^j|_p\}$ is an orthonormal basis for M_p, $(\nabla X_j)_p = 0$, and $(\nabla dx^i)_p = 0$.

PROOF In the proof of 2.106 it suffices to assume that the basis $\{\epsilon_j\}$ for M_p is orthonormal.

Such a chart on M is called a *normal chart*.

3.34 **Examples**

(*a*) The canonical connection on $\mathbb{R}^n$ is the Levi-Civita connection for the canonical Riemannian metric; its curvature tensor is zero.

(*b*) For this example, let W, X, Y, Z be left-invariant vector fields on a fixed Lie group G. The Levi-Civita connection of a left-invariant metric on G is

$$\langle \nabla_X Y, Z\rangle = \tfrac{1}{2}(\langle [X, Y], Z\rangle - \langle [Y, Z], X\rangle + \langle [Z, X], Y\rangle)$$

Restrict ∇ to a real bilinear map from $\mathfrak{g} \times \mathfrak{g}$ to $\mathfrak{g}$ with symmetric and alternating parts S and A, respectively. Clearly $A(X, Y) = \tfrac{1}{2}[X, Y]$ and $\langle S(X, Y), Z\rangle = \tfrac{1}{2}(\langle [Z, X], Y\rangle + \langle X, [Z, Y]\rangle)$. This implies that

$$\begin{aligned}\langle R(W, X)Y, Z\rangle = {} & \tfrac{1}{4}(\langle Y, [[W, X], Z]\rangle - \langle [[W, X], Y], Z\rangle \\ & + \langle W, [[Y, Z], X]\rangle - \langle [[Y, Z], W], X\rangle) \\ & + \tfrac{1}{4}(\langle [W, Y], [X, Z]\rangle - \langle [W, Z], [X, Y]\rangle) \\ & + \tfrac{1}{2}\langle [W, X], [Y, Z]\rangle + \langle S(W, Y), S(X, Z)\rangle \\ & - \langle S(W, Z), S(X, Y)\rangle\end{aligned}$$

If the metric is bi-invariant, then by 3.4*b*, $S = 0$ and $\nabla_X Y = \tfrac{1}{2}[X, Y]$; the converse holds if G is connected. In this case, $R(W, X)Y = -\tfrac{1}{4}[[W, X], Y]$, and

$$\langle R(W, X)Y, Z\rangle = -\tfrac{1}{4}\langle [W, X], [Y, Z]\rangle.$$

By 2.102, the geodesics are just the integral curves of the left-invariant vector fields on G.

(*c*) By 2.60*c*, the canonical connection ∇ on S^n is the tangent component of the canonical connection $\bar{\nabla}$ on $\mathbb{R}^{n+1}$; by 3.32, this is the Levi-Civita connection of the canonical Riemannian metric on S^n. *Exercise:* Check directly that $T = 0$.

Fix $p \in S^n$ and orthonormal tangent vectors u and v at p; under the isomorphism $\mathfrak{o}(S^n_p) \cong \Lambda^2 S^n_p$ from 3.13, $R(u, v) = u \wedge v$, for by 2.68, $R(u, v)w = \langle v, w\rangle u - \langle u, w\rangle v = u \wedge v(w)$ for all $w \in S^n_p$. Thus as a section of $\mathrm{End}(\Lambda^2 TS^n)$, R is just the identity map, $R(U \wedge V) = U \wedge V$ for $U, V \in \mathfrak{X}S^n$.

(*d*) Let (M_j, g_j) be Riemannian manifolds, $j = 1, 2$, and let (M, g) be the Riemannian product manifold (3.5). The curvature tensor of g satisfies the identity

$$\langle R(W, X)Y, Z\rangle = \langle R_1(W_1, X_1)Y_1, Z_1\rangle_1 + \langle R_2(W_2, X_2)Y_2, Z_2\rangle_2$$

where $W = (W_1, W_2) \in \mathfrak{X}M$, $W_j \in \mathfrak{X}M_j$, and so forth.

3.35 Proposition The curvature tensor field for the Levi-Civita connection on a Riemannian manifold (M, g) satisfies the following identities for W, X,

$Y, Z \in \mathfrak{X}M$:

$$R(X, Y)Z + R(Y, X)Z = 0$$
$$R(X, Y)Z + R(Y, Z)X + R(Z, X)Y = 0$$
$$\langle R(W, X)Y, Z\rangle + \langle R(W, X)Z, Y\rangle = 0$$
$$\langle R(W, X)Y, Z\rangle = \langle R(Y, Z)W, X\rangle$$

PROOF The first identity holds for all connections by 2.44, the second holds for all torsion-free connections by 2.107, and the third holds for all Riemannian connections by 3.24. The fourth identity is a formal algebraic consequence of the others:

$$\begin{aligned}\langle R(W, X)Y, Z\rangle &= -\langle R(W, X)Z, Y\rangle \\ &= \langle R(X, Z)W + R(Z, W)X, Y\rangle \\ &= -\langle R(X, Z)Y, W\rangle - \langle R(Z, W)Y, X\rangle \\ &= \langle R(Z, Y)X + R(Y, X)Z, W\rangle \\ &\quad + \langle R(W, Y)Z + R(Y, Z)W, X\rangle \\ &= 2\langle R(Y, Z)W, X\rangle + \langle R(X, Y)W, Z\rangle \\ &\quad + \langle R(Y, W)X, Z\rangle \\ &= 2\langle R(Y, Z)W, X\rangle - \langle R(W, X)Y, Z\rangle\end{aligned}$$

3.36 **Definition** The *curvature operator of* (M, g) is the curvature tensor field R of g interpreted as a section of $\mathrm{End}(\Lambda^2 TM)$ by the third identity from 3.35 and the isomorphism $\mathfrak{o}(TM) \cong \Lambda^2 TM$ (3.13):

$$\langle R(W \wedge X), Y \wedge Z\rangle_{\Lambda^2 TM} := \langle R(W, X)Z, Y\rangle_{TM}$$

for $W, X, Y, Z \in \mathfrak{X}M$, that is, given $W, X \in \mathfrak{X}M$, $R(W \wedge X) = \mu^\#$, where μ is the element of $\Gamma\Lambda^2 T^*M$ such that $\mu(Y \wedge Z) := \langle R(W, X)Z, Y\rangle_{TM}$ (cf. the proof of 3.28).

For example, by 3.34*c* the curvature operator of S^n is $I \in \Gamma\ \mathrm{End}(\Lambda^2 TS^n)$.

The curvature operator is symmetric (that is, self-adjoint) by 3.35:

$$\begin{aligned}\langle R(W \wedge X), Y \wedge Z\rangle &= \langle R(W, X)Z, Y\rangle = \langle R(Y, Z)X, W\rangle \\ &= \langle R(Y \wedge Z), W \wedge X\rangle\end{aligned}$$

3.37 Assume $\dim M \geq 2$; fix $p \in M$. Given $u \wedge v \in \Lambda^2 M_p - \{0\}$, set

$$K_{u \wedge v} := \frac{\langle R(u \wedge v), u \wedge v \rangle}{\|u \wedge v\|^2}$$

Since $u \wedge v \neq 0$, u and v span a 2-plane $\sigma \subset M_p$. If $h \in GL(\sigma)$, then $hu = au + bv$, $hv = cu + dv$, where $ad - bc \neq 0$; thus $hu \wedge hv = (ad - bc)u \wedge v = (\det h)u \wedge v$, so

$$K_{hu \wedge hv} = \frac{\langle R(hu \wedge hv), hu \wedge hv \rangle}{\|hu \wedge hv\|^2} = \frac{(\det h)^2 \langle R(u \wedge v), u \wedge v \rangle}{(\det h)^2 \|u \wedge v\|^2} = K_{u \wedge v}$$

Definition The *sectional curvature of* (M, g) *with respect to a 2-plane* $\sigma \subset M_p$, $p \in M$, is

$$K_\sigma := K_{u \wedge v} = \frac{\langle R(u \wedge v), u \wedge v \rangle}{\|u \wedge v\|^2} = \frac{\langle R(u, v)v, u \rangle}{\|u\|^2 \|v\|^2 - \langle u, v \rangle^2}$$

where $u, v \in M_p$ span σ.

If (M, g) is flat, that is, if $R = 0$, then $K = 0$. Conversely [GKM: 3.6(17)], if $K = 0$, then $R = 0$.

3.38 **Examples**

(*a*) The sectional curvature of $\mathbb{R}^n$ is zero for all 2-planes σ.

(*b*) By 3.34*b* the sectional curvature of a left-invariant metric on a Lie group G equals

$$K_{U \wedge V} = \tfrac{1}{2}(\langle [[U, V], U], V \rangle + \langle U, [V, [U, V]] \rangle) - \tfrac{3}{4}\|[U, V]\|^2 + \|S(U, V)\|^2 - \langle S(U, U), S(V, V) \rangle$$

where $U, V \in \mathfrak{g}$ are orthonormal and S is the symmetric part of ∇.

If the metric is bi-invariant, then $S = 0$ and ad_U is skew-symmetric with respect to the metric (3.4*b*), so $K_{U \wedge V} = \tfrac{1}{4}\|[U, V]\|^2 \geq 0$ for orthonormal $U, V \in \mathfrak{g}$. Thus the sectional curvature of a bi-invariant metric on a Lie group G is nonnegative.

(*c*) The unit sphere S^n has constant sectional curvature 1, for if $0 \neq u \wedge v \in \Lambda^2 S^n_p$, then by 3.34*c*,

$$K_{u \wedge v} = \frac{\langle R(u, v)v, u \rangle}{\|u \wedge v\|^2} = \frac{\|u \wedge v\|^2}{\|u \wedge v\|^2} = 1$$

Exercises

1. Define a Riemannian metric on S^n by embedding it into $\mathbb{R}^{n+1}$ as a sphere of radius $\rho > 0$. The curvature tensor of this metric is $R = (1/\rho^2)I \in \Gamma\, \mathrm{End}(\Lambda^2 TS^n)$, so $K_\sigma = 1/\rho^2$ for all σ.

2. What is the sectional curvature of $\mathbb{R}P^n$ with the metric from 3.4*e*?

(*d*) If (M, g) is the Riemannian product of (M_j, g_j), $j = 1, 2$, then by 3.34*d*, given $u = (u_1, u_2)$, $v = (v_1, v_2) \in M_{(p_1, p_2)}$,

$$K_{u \wedge v} = \frac{\langle R_1(u_1, v_1)v_1, u_1\rangle + \langle R_2(u_2, v_2)v_2, u_2\rangle}{(\|u_1\|^2 + \|u_2\|^2)(\|v_1\|^2 + \|v_2\|^2) - (\langle u_1, v_1\rangle + \langle u_2, v_2\rangle)^2}$$

It follows that if (M_1, g_1) and (M_2, g_2) have nonnegative sectional curvature, then so does (M, g). The product metric on M fails to have positive sectional curvature, even if both g_1 and g_2 have positive sectional curvature; for example, if u is tangent to M_1 and v is tangent to M_2, then $K_{u \wedge v} = 0$. A famous question posed by H. Hopf is whether there exists any metric of positive sectional curvature on $S^2 \times S^2$. More generally, if M_1 and M_2 admit metrics of positive sectional curvature, is there a metric of positive sectional curvature on $M_1 \times M_2$?

3.39 **Definition** A *Lorentz inner product* on a real $(n + 1)$-dimensional vector space $\mathscr{V}$ is a symmetric bilinear form β on $\mathscr{V}$ such that there exists a basis $\{\epsilon_0, \ldots, \epsilon_n\}$ for $\mathscr{V}$ for which

$$\beta(\epsilon_0, \epsilon_0) = -1 \qquad \beta(\epsilon_0, \epsilon_j) = 0 \qquad \beta(\epsilon_i, \epsilon_j) = \delta_{ij} \qquad i, j > 0$$

A *Lorentz metric* on an $(n + 1)$-dimensional manifold M is a section $\beta \in \Gamma(TM \otimes TM)^*$ such that β_p is a Lorentz inner product on M_p for each $p \in M$. A *Lorentz connection on* (M, β) is a connection for which β is parallel.

More generally, we could define Lorentz metrics and connections on arbitrary real vector bundles over M, but this will not be needed.

A Lorentz inner product is not definite; in fact there are nonzero vectors v for which $\beta(v, v) = 0$. Nevertheless, a Lorentz inner product is nonsingular (3.8): if $\beta(u, v) = 0$ for all $v \in \mathscr{V}$, then $u = 0$.

Suppose that dim $M > 1$ and that there is a line subbundle E of TM, that is, E is a vector subbundle of TM with 1-dimensional fibers. Let $\langle\ ,\ \rangle$ be a Riemannian metric on M and let $E^\perp$ be the subbundle of TM orthogonal to E: $E^\perp = \{v \in TM \mid \langle v, w\rangle = 0,\ w \in E\}$. Define a Lorentz metric on M such that $\beta = \langle\ ,\ \rangle$ on $E^\perp$, $\beta = -\langle\ ,\ \rangle$ on E, and $\beta(u, v) = 0$ for $u \in E$, $v \in E^\perp$. Conversely, using a partition of unity one can show that if M admits a Lorentz metric, then there exists a line subbundle of TM. Thus not every manifold admits a Lorentz metric.

3.40 **Theorem** Let β be a Lorentz metric on M; there exists a unique torsion-free Lorentz connection on M, called the Levi-Civita connection of β.

PROOF The proof of 3.28 never used the positive definiteness of $\langle\ ,\ \rangle$, just its nonsingularity. Since β is nonsingular, the same proof works here; in particular, eq. (3-1) from 3.28 holds for ∇ (with β substituted for $\langle\ ,\ \rangle$).

3.41 Given (r, u), $(s, v) \in \mathbb{R} \times \mathbb{R}^n$, define $\beta((r, u), (s, v)) := -rs + \langle u, v\rangle$, where $\langle\ ,\ \rangle$ is the usual inner product on $\mathbb{R}^n$; β is a Lorentz inner product on $\mathbb{R}^{n+1}$. For each nonzero vector $v \in \mathbb{R}^{n+1}$, $\beta(v, v)$ is negative, zero, or positive, depending on whether the usual Euclidean angle between v and the first coordinate axis is less than, equal to, or greater than $\pi/4$. Extend β to a Lorentz metric on the manifold $\mathbb{R}^{n+1}$ (see 3.4*a*) by $\beta((p, u), (p, v)) := \beta(u, v)$ for (p, u), $(p, v) \in T\mathbb{R}^{n+1}$. By eq. (3-1) the Levi-Civita connection of β is just the canonical connection on $\mathbb{R}^{n+1}$.

Let H^n be the component $\{(r, u) \in \mathbb{R} \times \mathbb{R}^n \mid r^2 = \|u\|^2 + 1, r > 0\}$ of the hyperboloid $\beta^{-1}(-1)$, and denote the inclusion map of H^n into $\mathbb{R}^{n+1}$ by $\imath$. If $\xi \in T\mathbb{R}^{n+1}$ is tangent to H^n and nonzero, then $\beta(\xi, \xi) > 0$, which implies that $g := \imath^*\beta$ is a Riemannian metric on H^n.

Definition The Riemannian manifold (H^n, g) is called *n-dimensional hyperbolic space.*

Exercise Show that $TH^n = \{(p, v) \in T\mathbb{R}^{n+1} \mid p \in H^n \text{ and } \beta(p, v) = 0\}$.

Checking the proofs of 3.30 and 3.32 we see that the Levi-Civita connection ∇ of (H^n, g) is the tangent component of the Levi-Civita connection $\bar{\nabla}$ of $(\mathbb{R}^{n+1}, \beta)$. Specifically, for all $y \in \mathbb{R}^{n+1}$ define $Y \in \mathfrak{X}H^n$ by $Y_p := (p, y + \beta(y, p)p) \cong y + \beta(y, p)p$. For all $y, z \in \mathbb{R}^{n+1}$,

$$(\nabla_Y Z)_p = \left(\frac{d}{dt}\bigg|_0 Z(p + t(y + \beta(y, p)p))\right)^{\mathrm{T}} \qquad p \in H^n$$

$$= (2\beta(y, p)\beta(z, p)p + \beta(z, y)p + \beta(z, p)y)^{\mathrm{T}}$$

$$= \beta(z, p)\beta(y, p)p + \beta(z, p)y$$

Further computation as in 2.68 and 3.34*c* shows that the curvature operator $R \in \Gamma\,\mathrm{End}(\Lambda^2 TH^n)$ of ∇ equals minus the identity: for U, $V \in \mathfrak{X}H^n$, $R(U \wedge V) = -U \wedge V$. The sectional curvature K of H^n is then

$$K_{u \wedge v} = \frac{\langle R(u, v)v, u\rangle}{\|u \wedge v\|^2} = -\frac{\|u \wedge v\|^2}{\|u \wedge v\|^2} = -1$$

for each nonzero $u \wedge v \in \Lambda^2 TH^n$.

Furthermore, for each $p \in H^n$ the holonomy algebra $\mathfrak{g}(p)$ of g at $p \in H^n$ is $\mathfrak{g}(p) = \{R(u, v) \in \mathfrak{o}(H^n_p) \mid u, v \in H^n_p\} = \mathfrak{o}(H^n_p)$. Hence the holonomy group G of (H^n, g) is a subgroup of $O(n)$ with Lie algebra $\mathfrak{o}(n)$; since H^n is simply connected, 2.24 implies $G = SO(n)$.

3.42 By 3.38*d*, if (M_1, g_1) and (M_2, g_2) are manifolds of nonpositive curvature, then so is their Riemannian product. By analogy with Hopf's question in the positive curvature case (3.38*d*), one asks whether the existence of metrics of negative curvature on M_1 and M_2 implies the existence of some metric of negative curvature on $M_1 \times M_2$. If M_1 and M_2 are compact, the answer is known: there cannot exist any metric of negative curvature on $M_1 \times M_2$ [BO] [Pn] regardless of what sorts of metrics M_1 and M_2 admit.

Exercises

1. Let $p, q \in H^n$ such that $\beta(p, q) = 0$, and let $\lambda \in \mathbb{R}$. Verify that $\gamma(t) := \cosh(\lambda t)p + \sinh(\lambda t)q$ is the geodesic in (H^n, g) with initial tangent vector $(p, \lambda q) \in TH^n$. Prove that a nonconstant curve α in H^n is a geodesic if and only if $g(\dot{\alpha}, \dot{\alpha})$ is constant and the image of α is the intersection of H^n and a 2-dimensional linear subspace of $\mathbb{R}^{n+1}$. Is ∇ complete?
2. (Schur) Let R be the curvature tensor on a connected Riemannian manifold (M, g) of dimension ≥ 3. Assume there exists $f \in C^\infty M$ such that $R(U, V) = f \cdot U \wedge V$ for all $U, V \in \mathfrak{X}M$. Using the Bianchi identities from 2.107, prove that f is constant; thus M is a space of constant sectional curvature.

3.43 Let R be the curvature tensor field on a manifold M with connection ∇. For fixed $u, v \in M_p$, $R(\cdot, u)v = [w \mapsto R(w, u)v]$ is an endomorphism of M_p, for which the trace is well-defined. Given $U, V \in \mathfrak{X}M$, the function tr $R(\cdot, U)V$ is C^∞ on M, and the map $(U, V) \mapsto \operatorname{tr} R(\cdot, U)V$ is a (0, 2)-tensor field on M.

Definition The (0, 2)-tensor field $\mathscr{R}$ defined on a manifold M with connection ∇ by $\mathscr{R}(U, V) := \operatorname{tr} R(\cdot, U)V$ for $U, V \in \mathfrak{X}M$ is called the *Ricci tensor field of* ∇.

3.44 **Proposition** The Ricci tensor field $\mathscr{R}$ on a Riemannian manifold (M, g) is symmetric: $\mathscr{R}(U, V) = \mathscr{R}(V, U)$ for all $U, V \in \mathfrak{X}M$.

PROOF Given an orthonormal basis $\{\epsilon_j\}$ for M_p and $u, v \in M_p$,

$$\begin{aligned}\mathscr{R}(u, v) &= \sum \langle R(\epsilon_j, u)v, \epsilon_j\rangle = \sum \langle R(v, \epsilon_j)\epsilon_j, u\rangle \\ &= \sum \langle R(\epsilon_j, v)u, \epsilon_j\rangle = \mathscr{R}(v, u)\end{aligned}$$

Exercise Fix an orthonormal basis $\{\epsilon_j\}$ for M_p; prove that the matrix of the endomorphism $v \mapsto \sum R(v, \epsilon_j)\epsilon_j$ of M_p is $[\mathscr{R}(\epsilon_i, \epsilon_j)]$.

3.45 **Definition** The *Ricci curvature of* (M, g) with respect to a nonzero vector $v \in TM$ is $r(v) := \mathscr{R}(v, v)/\|v\|^2$, and the *scalar curvature of* (M, g) is the trace of $\mathscr{R}$ with respect to g:

$$s(p) := \sum \mathscr{R}(\epsilon_j, \epsilon_j) = \sum_{i,j} \langle R(\epsilon_i, \epsilon_j)\epsilon_j, \epsilon_i \rangle = \sum_{i \neq j} K_{\epsilon_i \wedge \epsilon_j}$$

where $\{\epsilon_j\}$ is an orthonormal basis for M_p. A manifold of constant Ricci curvature is called an *Einstein manifold*.

Exercise The Ricci curvature is a C^∞ function on the unit sphere bundle $T_1 M := \{v \in TM \mid \|v\|^2 = 1\}$, and the scalar curvature is a C^∞ function on M. What can be said about r and s on a Riemannian product manifold, for example, on $S^n \times S^m$?

THE METRIC STRUCTURE OF A RIEMANNIAN MANIFOLD

3.46 **Definition** Let $\gamma: [a, b] \to M$ be a piecewise C^∞ curve in a Riemannian manifold (M, g). The *arc length of* γ is

$$l(\gamma) := \int_a^b \|\dot{\gamma}(t)\| \, dt$$

It follows that if $\varphi: [c, d] \to [a, b]$ is a surjective, monotone, piecewise C^∞ function, then $l(\gamma \circ \varphi) = l(\gamma)$; for example, $l(\gamma^-) = l(\gamma)$.

3.47 **Theorem** Let (M, g) be a connected Riemannian manifold. Let ρ: $M \times M \to \mathbb{R}$ such that $\rho(p, q) := \inf[l(\gamma)]$, where the inf is taken over all piecewise C^∞ curves γ joining p to q. For each $p \in M$ there exists $\delta > 0$ such that if $v \in M_p$ with $\|v\| < \delta$, then $\rho(p, \exp_p v) = \|v\|$.

PROOF Fix $p \in M$. By 2.96, $\exp_p$ maps a neighborhood V of 0 in M_p diffeomorphically onto a neighborhood U of p in M; we may assume that $V = \{v \in M_p \mid \|v\| < \delta\}$ for some $\delta > 0$. Let $v \in V - \{0\}$, and set $\gamma(t) := \exp tv$, $0 \leq t \leq 1$. The proof will consist in showing that for each piecewise C^∞ curve β joining p to $\exp_p v$, $l(\beta) \geq \|v\| = l(\gamma)$, for then the inequalities $l(\gamma) \leq \inf l(\beta) \leq l(\gamma)$ imply that $l(\gamma) = \inf l(\beta) = \rho(p, \exp_p v)$. *Exercise:* Why is it only necessary to consider curves β from p to $\exp_p v$ which remain inside U?

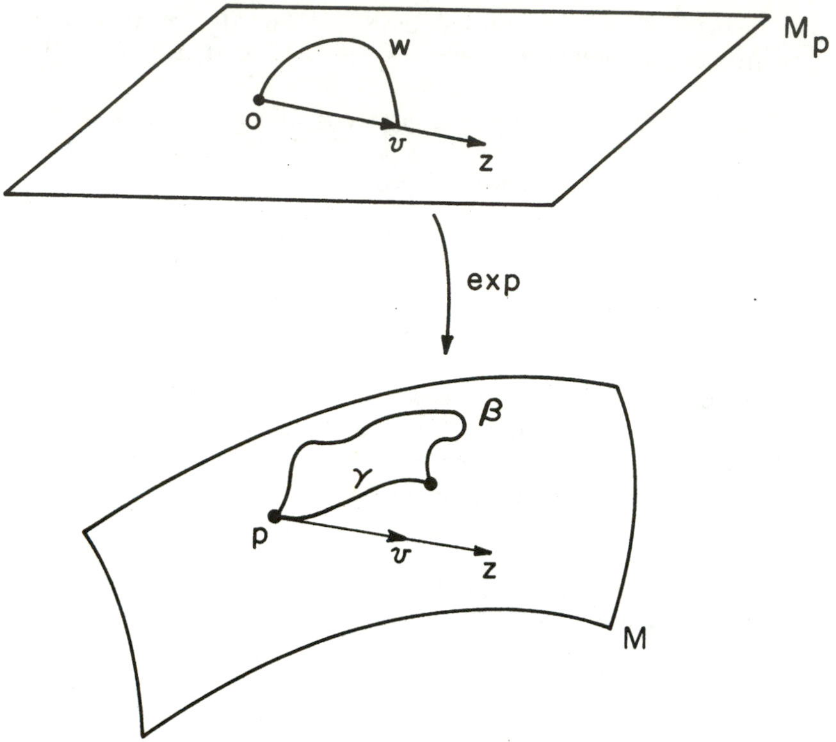

Since $\dot{\gamma}$ is parallel, $\|\dot{\gamma}(t)\| = \|\dot{\gamma}(0)\| = \|v\|$, so $l(\gamma) = \int_0^1 \|\dot{\gamma}(t)\|\, dt = \|v\|$. Next, let w: $[0, a] \to V$ be the piecewise C^∞ curve such that $\exp \circ w(t) = \beta(t)$. It follows that where $\|w\|$ is differentiable, the absolute value of $\|w\|'$ equals $|\langle \dot{w}, w\rangle/\|w\|| = \|\dot{w}^r\|$, where $\dot{w}^r = (\langle \dot{w}, w\rangle/\|w\|^2)w$ is the radial component of the tangent field $\dot{w}$. But by the Gauss lemma [GKM: 4.4], if $z \in (M_p)_{w(t)}$ is orthogonal to $\dot{w}^r(t)$, then $\exp_* z$ is orthogonal to the radial component (with respect to $\exp_p$) of $\dot{\beta}(t)$; this implies that $\|\dot{\beta}\| \geq \|\dot{\beta}^r\| = \|\dot{w}^r\|$, so

$$l(\beta) = \int_0^a \|\dot{\beta}(t)\|\, dt \geq \int_0^a |\|w\|'(t)|\, dt \geq \|w(a)\| = \|v\|$$

The theorem says that geodesics realize distance locally in a Riemannian manifold; the great circles in S^n show why geodesics generally realize distance only locally.

3.48 Corollary The function ρ on $M \times M$ from 3.47 is a metric on the set M, and the induced metric topology on M coincides with the manifold topology of M.

PROOF It is obvious that ρ is nonnegative and symmetric, that ρ satisfies the triangle inequality, and that $\rho(p, p) = 0$ for all $p \in M$. Thus ρ is a metric if $\rho(p, q) > 0$ for distinct $p, q \in M$. Since M is Hausdorff there are disjoint open neighborhoods U and W of p and q, respectively. Choose $\delta > 0$ such that if $v \in M_p$ and $\|v\| < \delta$, then $\exp_p v \in U$ and $\rho(p, \exp_p v) = \|v\|$. Given a piecewise C^∞ curve β joining p to q, if b is the smallest number t such that $\beta(t) \notin U$, then $l(\beta) \geq l(\beta|_{[0,b]}) \geq \delta$. Thus $\rho(p, q) \geq \delta > 0$.

This also implies that every neighborhood of p in M contains a metric ball of positive radius about p. Now let $B_r p := \{q \in M \mid \rho(p, q) < r\}$ be a metric ball in M, $r > 0$, and let $V \subset M_p$ be an open set about 0 mapped diffeomorphically by $\exp_p$ to an open subset of M about p; $\exp_p$ maps $\{v \in V \mid \|v\| < r\}$ diffeomorphically to an open subset of $B_r(p)$ because given $v \in M_p$,

$$\rho(p, \exp_p v) \leq \int_0^1 \|v\| \, dt = \|v\|$$

3.49 Theorem (Myers-Steenrod [MS]) Let $f: \bar{M} \to M$ be a map between Riemannian manifolds; f is an isometry of Riemannian manifolds (3.2) if and only if f is an isometry of metric spaces, that is [Ke] f is a surjective, distance-preserving map.

PROOF $\Rightarrow$: Since f is isometric, $\rho(f(p), f(q)) \leq \bar{\rho}(p, q)$; because f is an isometry (3.2), f^{-1} exists and is also isometric, so the reverse inequality holds. Thus f is onto and respects $\bar{\rho}$ and ρ.

$\Leftarrow$: A metric space isometry is automatically a homeomorphism. To show f is C^∞, we construct f_*. Given a unit tangent vector v on $\bar{M}$, set $\bar{\gamma}(t) := \overline{\exp}\, tv$ for small t, and $\gamma := f \circ \bar{\gamma}$. Since $\bar{\gamma}$ is a geodesic, 3.47 implies there exists $\delta > 0$ such that if $|s|, |t| < \delta$, then $|s - t| = \bar{\rho}(\bar{\gamma}(s), \bar{\gamma}(t)) = \rho(\gamma(s), \gamma(t))$. Thus γ locally realizes distance in M, and is therefore a geodesic (3.47). Set $k(v) := \dot{\gamma}(0)$, and for each $c \in \mathbb{R}$, set $k(cv) := ck(v)$. The fiberwise homogeneous map $k: T\bar{M} \to TM$ induced by f has the property that $f \circ \overline{\exp} = \exp \circ k$ on a neighborhood of the zero section of $T\bar{M}$ ($\overline{\exp}$ and $\exp$ might not be complete). For each $q \in M$ there is a neighborhood V_q of 0 in M_q which is mapped diffeomorphically by $\exp_q$ to a neighborhood of q in M; set $\widetilde{\exp}_q := \exp|_{V_q}$. Then on a neighborhood of the zero section of $T\bar{M}$, $k(v) = \widetilde{\exp}_{f \circ \pi(v)}{}^{-1} \circ f \circ \overline{\exp}\, v$, so k is continuous near the zero section of $T\bar{M}$. By homogeneity, it is continuous on $T\bar{M}$.

Fix $p \in \bar{M}$. Given $u, v \in \bar{M}_p$,

$$\overline{\exp}_* \mathscr{I}_u v = \left.\frac{d}{dt}\right|_0 \overline{\exp}(u + tv)$$

is tangent to $\bar{M}$ at $\overline{\exp}_p\, u$. Let $\bar{\gamma}_{u,v}$ be the geodesic in $\bar{M}$ with initial tangent vector $\overline{\exp}_* \mathscr{I}_u v$. Thus if $h_{u,v}(t) := \overline{\exp}_p^{-1} \circ \bar{\gamma}_{u,v}(t)$, then

$$\left.\frac{d}{dt}\right|_0 k \circ h_{u,v}(t) = \left.\frac{d}{dt}\right|_0 \widetilde{\exp}_{f(p)}{}^{-1} \circ f \circ \bar{\gamma}_{u,v}(t) = \widetilde{\exp}_{f(p)*}{}^{-1}\, k\, \overline{\exp}_* \mathscr{I}_u v$$

exists and is continuous in u and v near 0 in $\bar{M}_p$. But for all u, v near

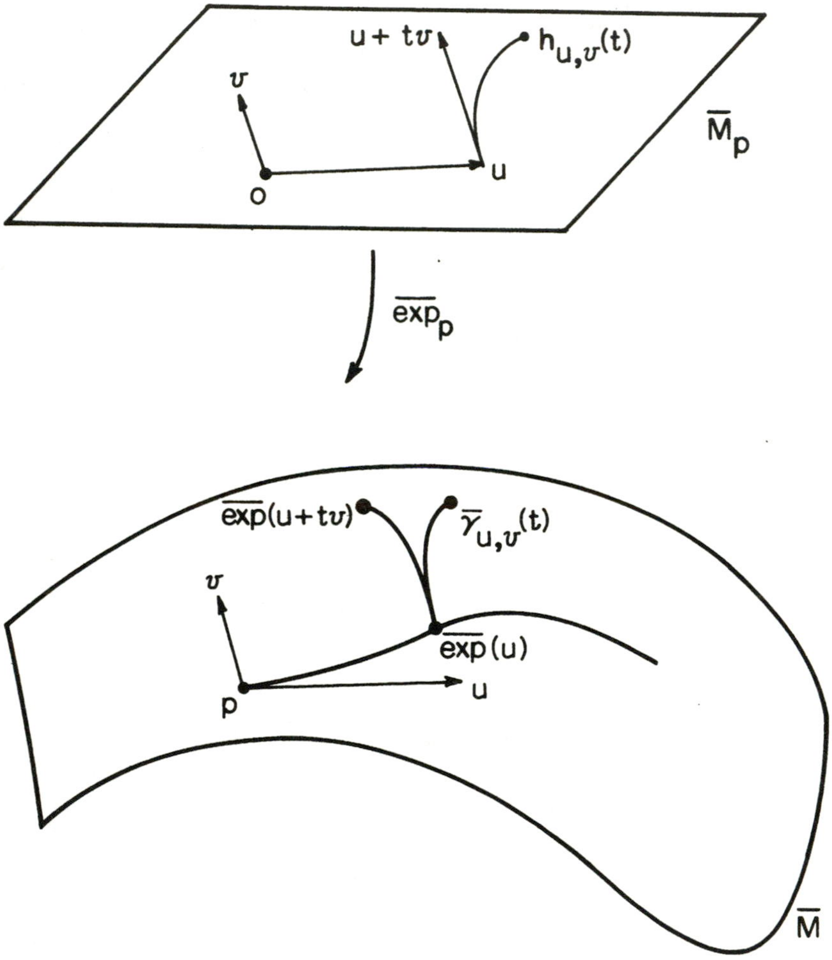

0 in $\bar{M}_p$, $h_{u,v}(t) - (u + tv)$ is $o(t)$, so the directional derivative

$$\left.\frac{d}{dt}\right|_0 k(u + tv) = \widetilde{\exp}_{f(p)*}{}^{-1}\, k\, \overline{\exp}_* \mathscr{I}_u v$$

exists and is continuous in u and v. Hence $k|_{\bar{M}_p}$ is C^1 on a neighborhood of 0 in $\bar{M}_p$; its differential at 0 is just $k|_{\bar{M}_p}$, which must then be a linear map, and therefore C^∞. The local identity $f = \exp \circ k \circ \exp_p^{-1}$ then shows that f is C^∞ near p, with induced tangent map k.

Finally, since $f_* = k$, for all u and v in $\bar{M}_p$

$$2\langle f_*u, f_*v\rangle = \|f_*u + f_*v\|^2 - \|f_*u\|^2 - \|f_*v\|^2$$
$$= \overline{\|u + v\|}^2 - \overline{\|u\|}^2 - \overline{\|v\|}^2 = 2\overline{\langle u, v\rangle}$$

3.50 **Theorem** (Myers-Steenrod) The isometry group $I(M)$ of a Riemannian manifold with finitely many components is a Lie group with respect to the compact open topology. If M is compact, so is $I(M)$.

PROOF See [MS].

3.51 By 3.30, every isometry of a Riemannian manifold (M, g) is affine with respect to the Levi-Civita connection. By 2.121 it follows that the isometry group of a connected, n-dimensional Riemannian manifold has dimension less than or equal to $n^2 + n$. We shall now improve this estimate.

Definition A vector field X on a Riemannian manifold (M, g) is a *Killing field* if and only if its local 1-parameter group consists of local isometries of g.

For example, the rotation vector fields on S^n are Killing fields. Every constant vector field on $\mathbb{R}^n$ is a Killing field, and these project down to Killing fields on the *flat torus* T^n, which is defined to be the Riemannian homogeneous space (3.4c) $\mathbb{R}^n/H$, where H is the subgroup of $\mathbb{R}^n$ with integer coordinates, with the metric $\langle\ ,\ \rangle$ such that the projection map is a Riemannian covering map. For the usual 2-torus in $\mathbb{R}^3$, only the rotation fields are Killing.

Exercises Show that T^n is a compact Lie group, and that $\langle\ ,\ \rangle$ is a bi-invariant metric. Find some Killing fields on hyperbolic space H^n.

3.52 **Proposition** A vector field X on (M, g) is a Killing field if and only if $\mathscr{L}_X g = 0$; this is equivalent to skew-symmetry of $\nabla X \in \operatorname{End} \mathfrak{X}M$ with respect to g.

PROOF If $\{\psi_t\}$ is the local 1-parameter group of X, then

$$\left.\frac{d}{ds}\right|_t \psi_s^* g = \left.\frac{d}{ds}\right|_0 \psi_{s+t}^* g = \psi_t^* \left.\frac{d}{ds}\right|_0 \psi_s^* g = \psi_t^* \mathscr{L}_X g$$

so $\psi_t^* g = g$ for all t if and only if $\mathscr{L}_X g = 0$. But for $U, V \in \mathfrak{X}M$,

$$(\mathscr{L}_X g)(U, V) = X\langle U, V\rangle - \langle [X, U], V\rangle - \langle U, [X, V]\rangle$$
$$= \langle \nabla_U X, V\rangle + \langle U, \nabla_V X\rangle$$

since g is parallel and ∇ is torsion-free.

3.53 As in 2.119, the Lie algebra $\mathfrak{i}(M)$ of the isometry group $I(M)$ of a connected Riemannian manifold (M, g) is isomorphic to the Lie subalgebra of $\mathfrak{X}M$ of complete Killing fields on M.

Proposition The isometry group of a connected n-dimensional Riemannian manifold has dimension less than or equal to $\binom{n+1}{2}$.

PROOF Fix $p \in M$. By the proof of 2.121 the map from $\mathfrak{i}(M)$ to $M_p \oplus \mathfrak{gl}(M_p)$ which sends X to $(X, \nabla X)_p$ is injective. Since ∇X is skew-symmetric with respect to g, $(\nabla X)_p$ belongs to the orthogonal algebra $\mathfrak{o}(M_p)$, which has dimension $\binom{n}{2}$.

3.54 **Theorem** (Hopf-Rinow) The following statements are equivalent for a connected Riemannian manifold (M, g) with distance function ρ:

(i) The metric space (M, ρ) is complete.
(ii) The Levi-Civita connection of g is complete, that is (2.90, 2.97), for all $p \in M$, $\exp_p$ is defined on the tangent space M_p.
(iii) For some $p \in M$, $\exp_p$ is defined on M_p.
(iv) Every bounded subset of M has compact closure in M.

Furthermore, if (i) to (iv) hold, then any two points $p, q \in M$ can be joined by a geodesic in M of length $\rho(p, q)$.

PROOF See [GKM: 5.3] or [KN: IV.4.1].

3.55 **Examples**

(*a*) Given $(p, u), (p, v) \in \mathbb{R}^n_p$, $\rho(\exp_p tu, \exp_p tv) = \|t(u - v)\|$; $\mathbb{R}^n$ is complete, and its isometry group is the semidirect product $\mathbb{R}^n \rtimes O(n)$, the so-called group of *Euclidean motions*.
(*b*) Let $(p, u), (p, v) \in S^n_p$ be unit tangent vectors. By the spherical law of cosines [Fi: 13.8]

$$\rho(\exp_p tu, \exp_p tv) = \cos^{-1}(\cos^2 t + \langle u, v\rangle \sin^2 t)$$

which is less than $\|t(u - v)\|$ for $t \neq 0$, $u \neq \pm v$. This is an example of a general phenomenon: given a 2-plane $\sigma \subset M_p$ on a Riemannian manifold (M, g), let γ_1 and γ_2 be geodesics in M such that $\dot\gamma_1(0)$ and $\pm\dot\gamma_2(0)$ are distinct unit vectors in σ; by [BC: 9.5], [GKM: 6.3(4)], or

[Hi: 10.3], the sectional curvature $K_\sigma > 0$ if and only if $\rho(\gamma_1(t), \gamma_2(t)) < \|t\dot{\gamma}_1(0) - t\dot{\gamma}_2(0)\|$ for small nonzero t.

The geodesics are defined for all time, so S^n is complete; this is also obvious (3.54) because S^n is a compact metric space.

The isometry group of S^n is $O(n+1)$. *Proof*: $O(n+1)$ acts on S^n by isometries (3.4*c*). Now let $h \in I(S^n)$; fix $p = e_0 \in S^n$ and the orthonormal basis $\{(p, e_j)\}$ for S^n_p. Set $(q, \epsilon_j) := h_*(p, e_j) \in S^n_q$, $q := h(p)$. Let $k \in O(n+1)$ take p to q and e_j to ϵ_j, $j \geq 1$. Since $k \in I(S^n)$ and $k_*|_p = h_*|_p$, $h = k \in O(n+1)$ by 3.30. *QED*.

(*c*) Let $(p, u), (p, v) \in H^n_p$ be unit tangent vectors. By the hyperbolic law of cosines [Fi: 12.9]

$$\rho(\exp_p tu, \exp_p tv) = \cosh^{-1}(\cosh^2 t - \langle u, v\rangle \sinh^2 t)$$

which is greater than $\|t(u - v)\|$ for $t \neq 0$ and $u \neq \pm v$. More generally, as in (*b*), given geodesics γ_1 and γ_2 in a Riemannian manifold (M, g), where $\dot{\gamma}_1(0)$ and $\pm\dot{\gamma}_2(0)$ are distinct unit vectors in a 2-plane $\sigma \subset M_p$, the sectional curvature $K_\sigma < 0$ if and only if $\rho(\gamma_1(t), \gamma_2(t)) > \|t\dot{\gamma}_1(0) - t\dot{\gamma}_2(0)\|$ for small nonzero t.

The geodesics are defined for all time, so H^n is complete. *Exercise*: Prove directly that H^n is a complete metric space.

The isometry group of H^n is the *proper Lorentz group* $O_+(1, n)$ of all linear transformations of $\mathbb{R}^{n+1}$ which leave invariant the Lorentz inner product β on $\mathbb{R}^{n+1}$ from 3.41, and which also leave each component of the hyperboloid $\beta^{-1}(-1)$ invariant. The proof is similar to the case of S^n in (*b*).

(*d*) Every Riemannian homogeneous space is complete. *Proof:* Fix $p \in M$. There exists $d > 0$ such that every *normal geodesic* (that is, a geodesic γ with $\|\dot{\gamma}\| = 1$) in $M = G/H$ through p is defined at least

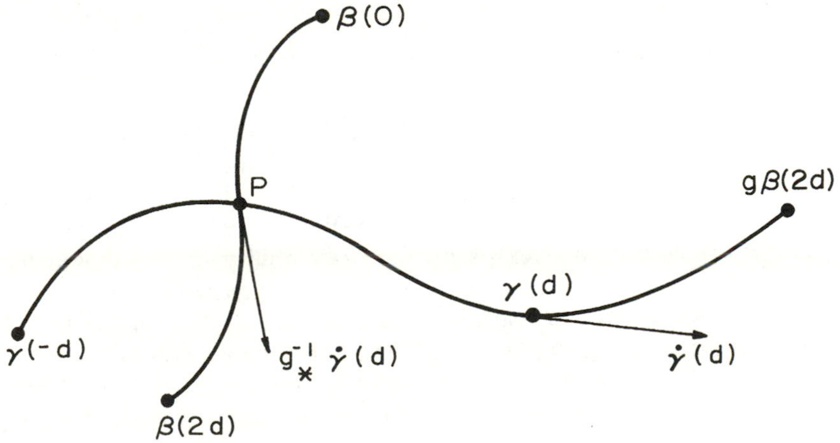

on $[-d, d]$. It follows that if γ is a normal geodesic emanating from p, and if $g \in G$ maps p to $\gamma(d)$, then the curve $\alpha(t) = \exp_p(g_*^{-1}(t-d)\dot{\gamma}(d))$ is a geodesic defined on $[0, 2d]$. But for $0 \le t \le d$, $\alpha(t) = g^{-1}\gamma(t)$, so the curve γ_1 which equals γ on $[-d, d]$ and $g\alpha$ on $[0, 2d]$ is an extension of γ to a geodesic on the interval $[-d, 2d]$. Similarly, γ can be extended to a geodesic on $[-2d, 2d]$, and by iteration, to a geodesic on $\mathbb{R}$. Thus $\exp_p$ is defined on M_p, so by the Hopf-Rinow theorem, M is complete. *QED*. In particular, every left-invariant metric on a Lie group G is complete. *Exercise:* If the metric on G is bi-invariant, then $\exp\colon \mathfrak{g} \to G$ is surjective. Is this true in general?

THE GAUSS-BONNET THEOREM

In 2.84 we saw that if E is a real vector bundle over M with standard fiber $\mathscr{V}$, then each invariant polynomial f on $\mathfrak{gl}(\mathscr{V})$ determines a real cohomology class on M which is in some sense characteristic of the bundle E. If $R \in A^2(M, \mathfrak{gl}(E))$ is the curvature tensor of a connection in E and if f is of degree k, then $f(R \otimes \cdots \otimes R) \in A^{2k}M$ is a closed $2k$-form on M which represents the characteristic class of E determined by f; in particular, the class is independent of the choice of connection in E.

Now we shall see that a Riemannian metric on E determines an additional characteristic class of E if E is oriented and has even rank. As with the Chern classes from 2.85*a*, an appropriate universal normalization makes this class integral.

3.56 **Definition** Let ∇ be a Riemannian connection in a Riemannian vector bundle (E, g) over M. Define the *curvature form* $\Omega \in A^2(M, \Lambda^2 E^*)$ by

$$\Omega(U, V)(X, Y) := \langle R(U, V)X, Y\rangle \qquad U, V \in \mathfrak{X}M,\ X, Y \in \Gamma E$$

The Bianchi identity $d^\nabla R = 0$ from 2.79 implies that $d^\nabla \Omega = 0$.

3.57 Suppose (E, g) is an oriented Riemannian vector bundle of rank m over M. Let ω be the unique section of $\Lambda^m E^*$ such that for all $p \in M$, $\omega_p = \omega^1 \wedge \cdots \wedge \omega^m$, where $\{\omega^i\}$ is a positively oriented orthonormal basis for E_p^*; such a section exists by the orientability of E^*, and in fact represents the orientation of E^* (1.65). By analogy with 3.14 and 3.16, call ω the *Riemannian volume form of the oriented bundle* (E, g). As in 3.27, ω is parallel with respect to every Riemannian connection in (E, g).

If $\mu \in \Gamma\Lambda^m E^*$ and if $\{\epsilon_j\}$ is a positively oriented orthonormal basis for E_p, $p \in M$, then $\mu_p = \mu(\epsilon_1, \ldots, \epsilon_m) \cdot \omega_p$; thus μ_p is a multiple of ω_p.

More generally, μ equals a C^∞ function on M times ω, $\mu = f_\mu \cdot \omega$, where $f_\mu(p) = \mu(\epsilon_1, \ldots, \epsilon_m)$ for any positively oriented orthonormal basis $\{\epsilon_j\}$ for E_p. Thus ω defines a vector bundle isomorphism between the trivial line bundles $\Lambda^m E^*$ and $M \times \mathbb{R}$; this should be compared with the star operator in the tangent bundle case.

Now suppose $\eta \in A^r(M, \Lambda^m E^*)$ is an r-form on M with values in $\Lambda^m E^*$. Given $U_1, \ldots, U_r \in \mathfrak{X}M$, $\eta(U_1, \ldots, U_r)$ is a section of $\Lambda^m E^*$, and can therefore be identified with a C^∞ function on M. In this way the bundle isomorphism $\Lambda^m E^* \cong M \times \mathbb{R}$ induced by ω allows us to consider $\eta \in A^r(M, \Lambda^m E^*)$ as an ordinary real-valued r-form on M:

$$\eta \in A^r(M, \Lambda^m E^*) \cong A^r(M, M \times \mathbb{R}) = A^r M$$

For example, if the rank of E is $m = 2k$, and if $\Omega \in A^2(M, \Lambda^2 E^*)$ is the curvature form of a Riemannian connection in (E, g), then $\Omega^k \in A^m(M, \Lambda^m E^*) \cong A^m M$ can be thought of as a real-valued m-form on M.

Each element of $A^r(M, \Lambda^m E^*)$ can be decomposed as a product $\eta \otimes \omega$, where $\eta \in A^r M$. If ∇ is a Riemannian connection in E, then since ω is parallel, $d^\nabla(\eta \otimes \omega) = (d^\nabla \eta) \otimes \omega = (d\eta) \otimes \omega$. In particular, $d^\nabla(\eta \otimes \omega) = 0$ if and only if η is closed, in which case we say $\eta \otimes \omega$ is *closed*; similarly, $\eta \otimes \omega = d^\nabla(\mu \otimes \omega)$ if and only if $\eta = d\mu$, in which case we say that $\eta \otimes \omega$ is *exact*.

Exercise The Bianchi identity $d^\nabla \Omega = 0$ from 3.56 implies that $\Omega^k \in A^m M$ is closed.

3.58 **Definition** Let (E, g) be an oriented Riemannian vector bundle over M of rank $m = 2k$. Identify $\Lambda^m E^*$ with $M \times \mathbb{R}$, and $A^m(M, \Lambda^m E^*)$ with $A^m M$ by means of the Riemannian volume form of (E, g) as in 3.57. Given a Riemannian connection ∇ in (E, g), the *Chern-Euler form of* ∇ is

$$\chi := \frac{(-\Omega)^k}{k!(2\pi)^k} \in A^m M$$

where $\Omega \in A^2(M, \Lambda^2 E^*)$ is the curvature form of ∇ from 3.56.

It can be shown that the cohomology class of the Chern-Euler form of ∇ is independent of the Riemannian metric g on E and the Riemannian connection ∇ in (E, g); it does however depend on the choice of orientation of E because the orientation of E determines the isomorphism $\Lambda^m E^* \cong M \times \mathbb{R}$. The cohomology class of χ is called the *Euler class of the oriented vector bundle E.*

The Euler class is natural (see the exercise in 2.84): if f maps N to M, then the Euler class of f^*E is the pullback by f of the Euler class of E. In addition, the Euler class of the Whitney sum $E_1 \oplus E_2$ of oriented vector bundles over M is the product of the Euler classes of E_1 and E_2 (analogous results hold for the Chern and Pontrjagin classes from 2.85). Finally, the Euler class of E is integral: if the rank of E is $2k$, then the Euler class of E is in $H^{2k}(M; \mathbb{Z})$.

3.59 **Theorem** (Gauss-Bonnet-Allendoerfer-Weil-Chern) Let M be a $2k$-dimensional, oriented, compact Riemannian manifold. Let χ be the Chern-Euler form of a Riemannian connection on M. The integral over M of χ is the Euler characteristic of M.

The proof of the theorem will consist of secs. 3.61 to 3.71.

3.60 First let us look at the special case of a surface. In this case, $\chi = -\Omega/2\pi \in A^2(M, \Lambda^2 T^*M)$. Let $\{\epsilon_1, \epsilon_2\}$ be a positively oriented orthonormal basis for M_p. For all $u, v \in M_p$,

$$\begin{aligned}\Omega(u, v)(\epsilon_1, \epsilon_2) &= \langle R(u, v)\epsilon_1, \epsilon_2\rangle \\ &= -\langle R(\epsilon_1, \epsilon_2)\epsilon_2, \epsilon_1\rangle\omega(u, v) = -K(p)\omega(u, v)\end{aligned}$$

where K is the sectional curvature function on M and ω is the Riemannian area form on M. Therefore

$$\int_M \chi = \frac{1}{2\pi}\int_M K\omega$$

This gives us the classical Gauss-Bonnet theorem: the integral over a compact, oriented Riemannian surface of its curvature function is 2π times the Euler characteristic of the surface.

3.61 A large part of Chern's proof of the Gauss-Bonnet theorem makes sense in a more general setting. From now until the end of the chapter, let E be an oriented Riemannian vector bundle over M, and let E_1 be the unit sphere bundle of vectors of length 1 in E; denote by ι the inclusion map of E_1 into E, and by I the identity map of E.

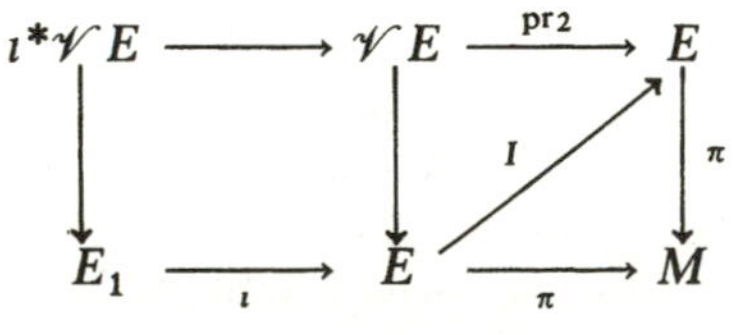

Recall from 1.27 that the pullback bundle π^*E over E is isomorphic to the vertical bundle $\mathscr{V}E$ by the map $\mathscr{I}$:

$$\mathscr{I}_\zeta \xi = \frac{d}{dt}\bigg|_0 (\zeta + t\xi) \in \mathscr{V}E \qquad (\zeta, \xi) \in \pi^*E$$

By 1.60, the $C^\infty E$ modules $\Gamma\pi^*E$ and $\Gamma_\pi E$ are naturally isomorphic, where $X \in \Gamma\pi^*E$ corresponds to $\text{pr}_2 X \in \Gamma_\pi E$, and $s \in \Gamma_\pi E$ corresponds to the section π^*s of π^*E defined by $(\pi^*s)_\xi := (\xi, s_\xi)$ for $\xi \in E$.

3.62 Pull the Riemannian metric on E back to a Riemannian metric on the vector bundle π^*E over E: $\langle X, Y\rangle := \langle \text{pr}_2 X, \text{pr}_2 Y\rangle$, $X, Y \in \Gamma\pi^*E$. Fix a Riemannian connection $\bar\nabla$ in E; let ∇ be the pullback of $\bar\nabla$ to a Riemannian connection in π^*E. By the exercise in 2.56, if $U \in \Gamma E$ and $X \in \Gamma\pi^*E$, then $\nabla_U X = \pi^*\bar\nabla_U(\text{pr}_2 X) \in \Gamma\pi^*E$, where the connection on the right side is the original connection in E along the map π.

It follows that $\nabla_U \pi^*I = \mathscr{V}U \in \Gamma\mathscr{V}E$ for all $U \in \mathfrak{X}E$, since

$$\pi^*\nabla_U I = \pi^*\kappa I_* U = \kappa U = \mathscr{V}U \in \Gamma\mathscr{V}E$$

by 2.49 and 2.52. Similarly, the structure equation from 2.67 implies that the curvature tensors R and $\bar R$ of ∇ and $\bar\nabla$ are related by the identity $\text{pr}_2 R(U, V)X = \bar R(\pi_* U, \pi_* V)\text{pr}_2 X$ for $U, V \in \mathfrak{X}E$ and $X \in \Gamma\pi^*E$.

Given $\mu \in A^r(M, \Lambda^s E^*)$, define $\pi^*\mu \in A^r(E, \Lambda^s\pi^*E^*)$ by

$$\pi^*\mu(U_1, \ldots, U_r)(X_1, \ldots, X_s) := \mu(\pi_* U_1, \ldots, \pi_* U_r)(\text{pr}_2 X_1, \ldots, \text{pr}_2 X_s),$$

$U_j \in \mathfrak{X}E$, $X_j \in \Gamma\pi^*E$; if any U_j is vertical, that is, if $\pi_* U_j = 0$, then $\pi^*\mu(U_1, \ldots, U_r) = 0$. If Ω is the curvature form of $\bar\nabla$ (3.56), then it follows from the paragraph above that for $U, V \in \mathfrak{X}E$, $X, Y \in \Gamma\pi^*E$,

$$\pi^*\Omega(U, V)(X, Y) = \langle \bar R(\pi_* U, \pi_* V)\text{pr}_2 X, \text{pr}_2 Y\rangle = \langle R(U, V)X, Y\rangle$$

3.63 Define the *canonical 1-form* $\eta \in A^0(E, \pi^*E)$ by $\eta(X) := \langle \pi^*I, X\rangle$ for $X \in \Gamma\mathscr{V}E \cong \Gamma\pi^*E$. If $X = \mathscr{I}_Y Z$ for $Y, Z \in \Gamma E$, then

$$\eta(X) = \langle \pi^*I, X\rangle_Y = \langle (\pi^*I)_Y, X\rangle = \langle \mathscr{I}_Y Y, \mathscr{I}_Y Z\rangle = \langle Y, Z\rangle$$

by 3.62. Next define the *connection 1-form* $\theta := d^\nabla\eta \in A^1(E, \pi^*E^*)$. For $U \in \mathfrak{X}E$ and $X \in \Gamma\mathscr{V}E$,

$$\begin{aligned}\theta(U)(X) &= (\nabla_U\eta)(X) = \nabla_U(\eta X) - \eta(\nabla_U X) = U\langle \pi^*I, X\rangle - \langle \pi^*I, \nabla_U X\rangle \\ &= \langle \nabla_U \pi^*I, X\rangle + \langle \pi^*I, \nabla_U X\rangle - \langle \pi^*I, \nabla_U X\rangle \\ &= \langle \nabla_U \pi^*I, X\rangle = \langle \mathscr{V}U, X\rangle\end{aligned}$$

For example, if $U \in \Gamma\mathscr{V}E \subset \mathfrak{X}E$, then $\theta(U)(X) = \langle U, X\rangle$.

3.64 Now suppose that v is tangent to the unit sphere bundle $E_1 \subset E$. Since $\langle \pi^*I, \pi^*I\rangle = 1$ on E_1, $\theta(v)(\pi^*I) = \frac{1}{2}v\langle \pi^*I, \pi^*I\rangle = 0$.

3.65 Finally, for $U, V \in \mathfrak{X}E$ and $X \in \Gamma\mathscr{V}E$, by 2.75, 3.62, and 3.64,

$$\begin{aligned}(d^{\nabla}\theta(U, V))(X) &= (\nabla_U(\theta V) - \nabla_V(\theta U) - \theta([U, V]))(X) \\ &= U\langle \mathscr{V}V, X\rangle - \langle \mathscr{V}V, \nabla_U X\rangle - V\langle \mathscr{V}U, X\rangle \\ &\quad + \langle \mathscr{V}U, \nabla_V X\rangle - \langle \mathscr{V}[U, V], X\rangle \\ &= \langle \nabla_U \mathscr{V}V - \nabla_V \mathscr{V}U - \mathscr{V}[U, V], X\rangle \\ &= \langle \nabla_U \nabla_V \pi^*I - \nabla_V \nabla_U \pi^*I - \nabla_{[U,V]}\pi^*I, X\rangle \\ &= \langle R(U, V)\pi^*I, X\rangle = \pi^*\Omega(U, V)(\pi^*I, X)\end{aligned}$$

3.66 Now we will see that the restriction to E_1 of the pullback $\pi^*\chi$ of the Chern-Euler form

$$\chi = \frac{(-\Omega)^k}{k!\,(2\pi)^k} \in A^m(M, \Lambda^m E^*)$$

of a Riemannian vector bundle E of rank $m = 2k$ is exact.

For $1 \le j \le k$, define

$$\Pi_j := \eta \wedge \theta^{2j-1} \wedge \pi^*\Omega^{k-j} \in A^{m-1}(E, \Lambda^m \pi^* E^*)$$

$$c_j := \frac{(-1)^{k+j+1}(j-1)!}{\pi^k 2^{k-j+1}(k-j)!\,(2j-1)!}$$

and set

$$\Pi := \sum_{j=1}^{k} c_j \Pi_j \in A^{m-1}(E, \Lambda^m \pi^* E^*)$$

The wedge product is that defined in 1.55, and satisfies the commutation rule from 1.56*d*: if μ is an (r, j)-form and ν is an (s, k)-form, then $\mu \wedge \nu = (-1)^{rs+jk}\nu \wedge \mu$. Thus the (2, 2)-form $\pi^*\Omega$ on E commutes with all forms. Similarly, by 2.76; $d^{\nabla}(\mu \wedge \nu) = (d^{\nabla}\mu) \wedge \nu + (-1)^r \mu \wedge d^{\nabla}\nu$ for μ and ν as above.

With these rules, the first step in the proof that $d^{\nabla}\Pi = \pi^*\chi$ on E_1 is completely formal: by 3.63 and 3.56,

$$\begin{aligned}d^{\nabla}\Pi &= \sum c_j d^{\nabla}\Pi_j \\ &= \sum c_j(d^{\nabla}\eta \wedge \theta^{2j-1} \wedge \pi^*\Omega^{k-j} + (2j-1)\eta \wedge d^{\nabla}\theta \wedge \theta^{2j-2} \wedge \pi^*\Omega^{k-j}) \\ &= \sum c_j(\theta^{2j} \wedge \pi^*\Omega^{k-j} + (2j-1)\theta^{2j-2} \wedge \eta \wedge d^{\nabla}\theta \wedge \pi^*\Omega^{k-j})\end{aligned}$$

3.67 Given $X, Y \in \Gamma\iota^*\pi^*E$, set $\theta(X) := \theta(\cdot)(X) \in A^1 E_1$, and $\pi^*\Omega(X, Y) := \pi^*\Omega(\cdot, \cdot)(X, Y) \in A^2 E_1$. Choose a local orthonormal basis field $\{X_i\}$ for $\iota^*\pi^*E$ over E_1 such that $X_{2j-1} = \pi^*I$. With this choice,

$\eta(X_i) = \langle X_i, \pi^*I\rangle = 1$ for $i = 2j - 1$, and 0 otherwise (3.63). In the following calculation write Ω instead of $\pi^*\Omega$, and if σ is a permutation of $\{1, \ldots, m\}$, write σ_i in place of X_{σ_i}. By 1.55 and 3.65,

$$\begin{aligned}
&2^{k-j}(\theta^{2j-2} \wedge \eta \wedge d^\nabla\theta \wedge \Omega^{k-j})(X_1, \ldots, X_m) \\
&\quad = \sum_\sigma \operatorname{sgn}(\sigma)\theta(\sigma_1) \wedge \cdots \wedge \theta(\sigma_{2j-2}) \wedge \eta(\sigma_{2j-1}) \wedge d^\nabla\theta(\sigma_{2j}) \\
&\qquad \wedge \Omega(\sigma_{2j+1}, \sigma_{2j+2}) \wedge \cdots \wedge \Omega(\sigma_{m-1}, \sigma_m) \\
&\quad = \sum_{\sigma_{2j-1} = 2j-1} \operatorname{sgn}(\sigma)\theta(\sigma_1) \wedge \cdots \wedge \theta(\sigma_{2j-2}) \wedge \Omega(\pi^*I, \sigma_{2j}) \\
&\qquad \wedge \Omega(\sigma_{2j+1}, \sigma_{2j+2}) \wedge \cdots \wedge \Omega(\sigma_{m-1}, \sigma_m) \\
&\quad = \frac{1}{2(k-j+1)} \sum_\sigma \operatorname{sgn}(\sigma)\theta(\sigma_1) \wedge \cdots \wedge \theta(\sigma_{2j-2}) \\
&\qquad \wedge \Omega(\sigma_{2j-1}, \sigma_{2j}) \wedge \cdots \wedge \Omega(\sigma_{m-1}, \sigma_m) \\
&\quad = \frac{2^{k-j+1}}{2(k-j+1)} \theta^{2j-2} \wedge \Omega^{k-j+1}(X_1, \ldots, X_m)
\end{aligned}$$

Therefore

$$\theta^{2j-2} \wedge \eta \wedge d^\nabla\theta \wedge \pi^*\Omega^{k-j} = \frac{1}{k-j+1}\theta^{2j-2} \wedge \pi^*\Omega^{k-j+1}$$

on E_1. By 3.66, on E_1

$$\begin{aligned}
d^\nabla\Pi &= \sum_1^k c_j\left(\theta^{2j} \wedge \pi^*\Omega^{k-j} + \frac{2j-1}{k-j+1}\theta^{2j-2} \wedge \pi^*\Omega^{k-j+1}\right) \\
&= \frac{(-\pi^*\Omega)^k}{k!\,(2\pi)^k} + \sum_1^{k-1}\left(c_j + \frac{2j+1}{k-j}c_{j+1}\right)\theta^{2j} \wedge \pi^*\Omega^{k-j} + c_k\theta^m \\
&= \pi^*\chi
\end{aligned}$$

since $\theta(\pi^*I) = 0$ on E_1 by 3.64.

3.68 Before proceeding to the special case where $E = TM$, we can establish one more result in the general case. Recall from 3.57 that if the form $\Pi \in A^{m-1}(E_1, \Lambda^m\iota^*\pi^*E^*)$ is evaluated in the second argument on a local orthonormal positively oriented basis field for $\iota^*\pi^*E$, the result is a local $(m-1)$-form on E_1, which can then be integrated over a fiber of E_1 (which is isometrically the unit $(m-1)$-sphere S^{m-1}). The claim now is that for each $p \in M$, the integral of $\Pi \in A^{m-1}E_1$ over the $(m-1)$-sphere $E_1|_p$ equals -1. *Proof:* Fix $u \in E_1|_p$, and let $\{X_j\}$ be a positively oriented local orthonormal basis field for $\iota^*\pi^*E$ over E_1 near u such that $X_1 = \pi^*I$. Since tangent vectors to the fiber $E_1|_p$ are vertical and

$\pi^*\Omega = 0$ on vertical vectors, we need to consider only the kth term of Π; thus by 3.63,

$$\begin{aligned}\Pi(X_1, \ldots, X_m) &= c_k \Pi_k(X_1, \ldots, X_m)\\ &= c_k \sum_{\sigma} \operatorname{sgn}(\sigma)\eta(X_{\sigma_1}) \wedge \theta(X_{\sigma_2}) \wedge \cdots \wedge \theta(X_{\sigma_m})\\ &= c_k \sum_{\sigma_1 = 1} \operatorname{sgn}(\sigma)\theta(X_{\sigma_2}) \wedge \cdots \wedge \theta(X_{\sigma_m})\\ &= c_k(m-1)!\,\theta(X_2) \wedge \cdots \wedge \theta(X_m)\end{aligned}$$

If $v_1, \ldots, v_{m-1}$ are tangent to $E_1|_p$, then since they are vertical, 3.63 implies $\theta(v_i)(X_j) = \langle v_i, X_j\rangle$. Therefore

$$\begin{aligned}&\Pi(v_1, \ldots, v_{m-1})(X_1, \ldots, X_m)\\ &\quad= c_k(m-1)! \sum_{\tau} \operatorname{sgn}(\tau)\langle v_{\tau_1}, X_2\rangle \cdots \langle v_{\tau_{m-1}}, X_m\rangle\\ &\quad= c_k(m-1)!\det[\langle v_i, X_j\rangle]\\ &\quad= c_k(m-1)!\,\omega(v_1, \ldots, v_{m-1})\end{aligned}$$

where ω is the volume form on the sphere $E_1|_p$. Since the volume of the unit $(2k-1)$-sphere is $2\pi^k/(k-1)!$, the expression for c_k in 3.66 implies that the integral of Π over $E_1|_p$ is -1. *QED.*

The constant c_k was chosen so this result would hold; the other c_j were then chosen iteratively to obtain the cancellation in 3.67.

3.69 The reference for this section is Milnor's book [Mi2] on differential topology, except where otherwise stated.

Let N be an oriented, compact, n-dimensional manifold, and let $f: N \to N$ be C^∞. A point $p \in N$ is called a *regular value of f* if $f_*|_q$ is nonsingular for each $q \in f^{-1}p$; by Sard's theorem, the set of regular $p \in N$ is dense in N. For each regular $p \in N$, set

$$\deg(f; p) := \sum_{q \in f^{-1}p} \operatorname{sign}(f_*|_q) \in \mathbb{Z}$$

where $\operatorname{sign}(f_*|_q)$ is 1 or -1 depending on whether $f_*|_q$ preserves or reverses the orientation. The integer $\deg(f; p)$ is independent of the choice of regular value p, and is called the *degree of f*, written $\deg(f)$.

3.70 It can be shown [GHV: I.6.1, 6.3] that for each n-form μ on N,

$$\int_N f^*\mu = \deg(f) \int_N \mu$$

3.71 Now let χ be the Chern-Euler form on a compact, oriented, n-dimensional Riemannian manifold M, and let

$$\Pi \in A^{n-1}(T_1 M, \Lambda^n \pi^* T^* M)$$

be the form from 3.66 such that $d^\nabla \Pi = \pi^* \chi$ on $T_1 M$. Fix a vector field U on M with finitely many zeros, and let $Z \subset M$ be the zero set of U. Let $Y := U/\|U\| \in \Gamma T_1(M - Z)$ be the field of length 1 on $M - Z$ determined by U. We may choose U so that $\lim_{t \to 0+} Y \circ \gamma(t)$ exists for every geodesic γ emanating from each zero of U.

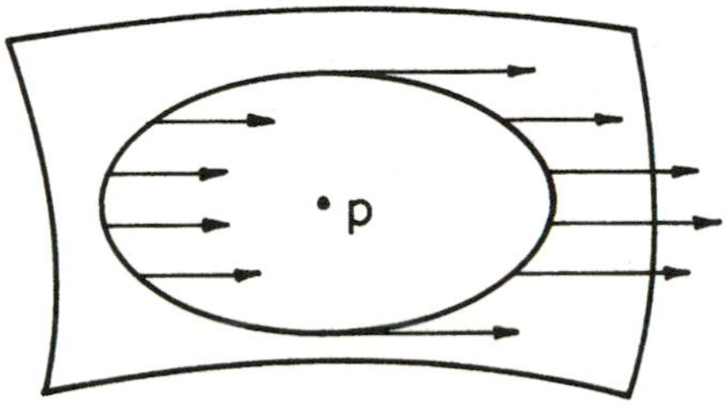

A unit vector field Y on $\partial B_\epsilon p$

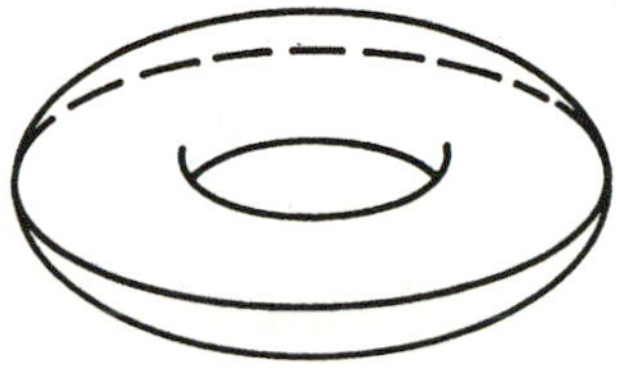

Y as a section of $(T_1 M)|_{\partial B_\epsilon p}$

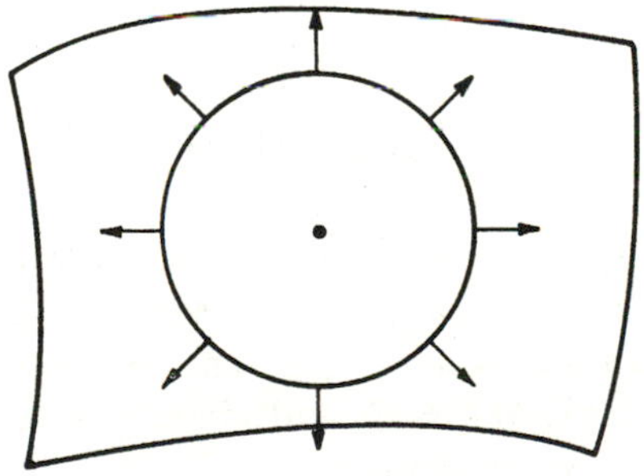

Another unit vector field on $\partial B_\epsilon p$

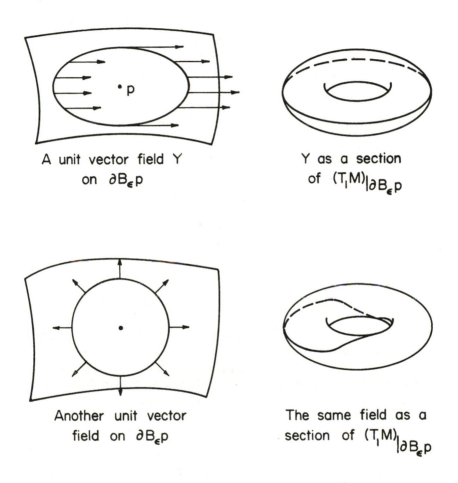

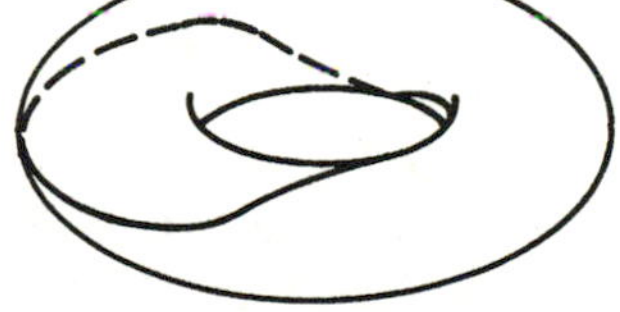

The same field as a section of $(T_1 M)|_{\partial B_\epsilon p}$

There exists $\epsilon > 0$ such that for each zero $p \in Z$ of U, the set $\{v \in M_p \mid \|v\| \leq \epsilon\}$ is mapped diffeomorphically by $\exp_p$ to the closure $\overline{B_\epsilon p}$ in M of the metric ball $B_\epsilon p$ about p of radius ϵ, and such that the distance in M between any two zeros of U is bigger than 2ϵ. Fix $p \in Z$, and let F be the unit sphere $(T_1 M)_p$ in M_p. Define $\varphi: F \times [0, \epsilon] \to M$ by $\varphi(v, t) := \exp_p tv$. The image of φ is $\overline{B_\epsilon p}$, and φ maps $F \times (0, \epsilon]$ diffeomorphically to $\overline{B_\epsilon p} - \{p\}$.

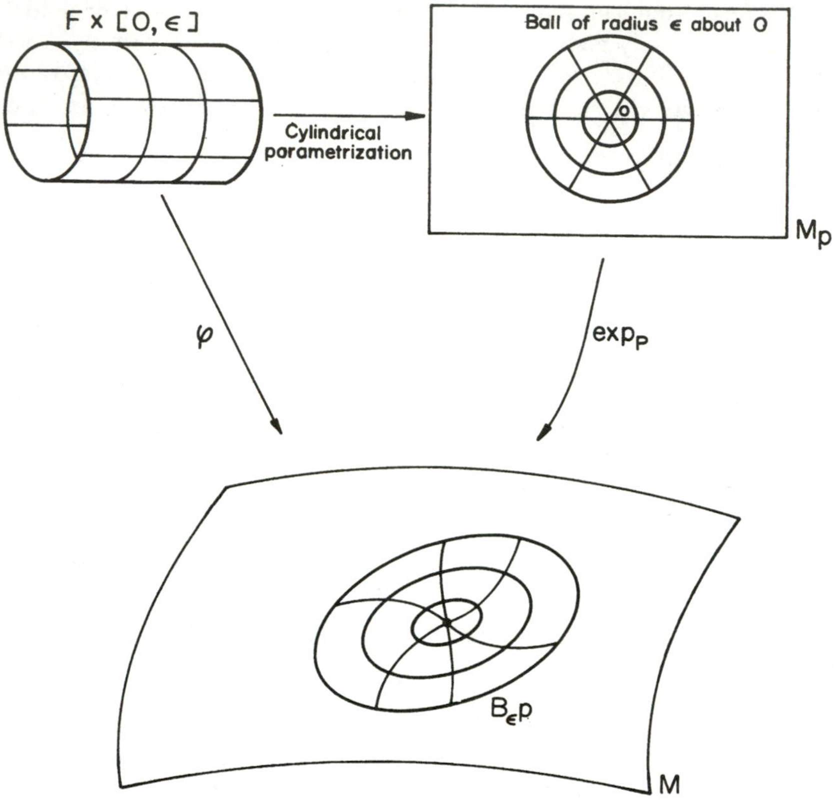

The C^∞ map $Y \circ \varphi \colon F \times (0, \epsilon] \to T_1 M$ is a section of $T_1 M$ along φ; it has a unique extension to a section $W \colon F \times [0, \epsilon] \to T_1 M$ along φ, even though Y cannot be extended continuously to $B_\epsilon p$. The extension W maps $F \cong F \times \underline{\{0\}}$ differentiably to $(T_1 M)_p = F$, and since p is the only zero of U in $B_\epsilon p$, the degree of the map W from F to F equals the degree of the map $d(\exp_p^{-1}) \circ Y \circ \exp_p$ from F to F; this degree is called

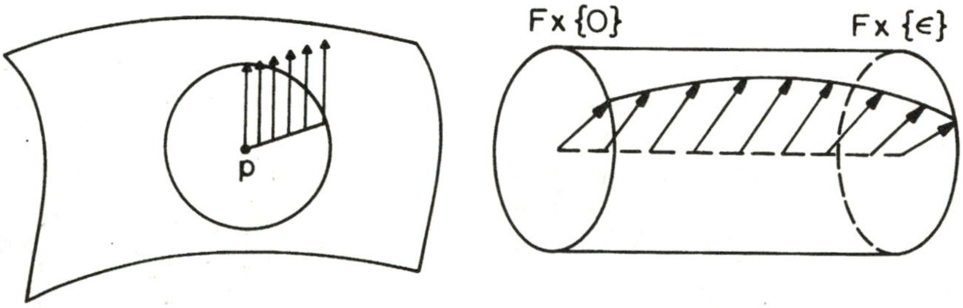

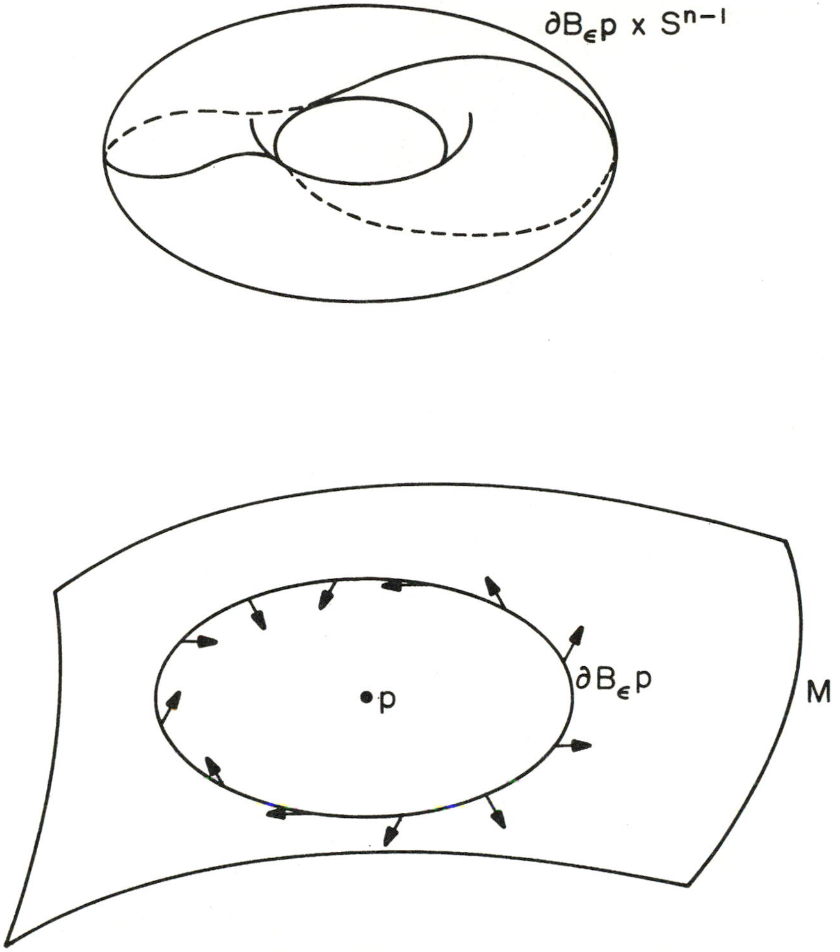

Figure 3-1

the *index of the vector field U at* $p \in M$ [Mi2]. *Exercise:* The vector field in Fig. 3-1 has index 2 at p.

By 3.67, the pullback $\pi^*\chi$ to $T_1 M$ of the Chern-Euler form on M equals the exterior covariant derivative $d^\nabla \Pi$ of the form Π defined in 3.66. By the remark at the end of 3.57, $\pi^*\chi$ then equals the ordinary exterior derivative $d\Pi$ when $\pi^*\chi$ and Π are both thought of as ordinary differential forms on $T_1 M$ by means of the Riemannian volume form in the bundle $\iota^*\pi^*TM$ over $T_1 M$. Since Y is a section of $T_1(M - Z)$, on $M - Z$ the Chern-Euler form χ equals $(\pi \circ Y)^*\chi = Y^*d\Pi = dY^*\Pi$.

Hence by 3.70,

$$\int_{\overline{B_\epsilon p}} \chi = \int_{\overline{B_\epsilon p}-\{p\}} \chi = \int_{\varphi(F\times(0,\epsilon])} dY^*\Pi = \int_{F\times(0,\epsilon]} d(Y\circ\varphi)^*\Pi$$

$$= \int_{F\times[0,\epsilon]} dW^*\Pi = \int_{\partial(F\times[0,\epsilon])} W^*\Pi = \int_{F\times\{\epsilon\}} W^*\Pi - \int_{F\times\{0\}} W^*\Pi$$

$$= \int_{\partial B_\epsilon p} Y^*\Pi - \int_F W^*\Pi = \int_{\partial B_\epsilon p} Y^*\Pi - \deg(W)\int_F \Pi$$

But by 3.68, $\int_F \Pi = -1$, so since $\deg(W)$ is the index of the vector field U at p, which will be denoted by $\mathrm{ind}_p U$,

$$\int_{\overline{B_\epsilon p}} = \int_{\partial B_\epsilon p} Y^*\Pi + \mathrm{ind}_p U$$

To complete the proof of the Gauss-Bonnet theorem, set $N := \bigcup_{p\in Z} B_\epsilon p$; then

$$\int_M \chi = \int_{M-N} \chi + \sum_{p\in Z} \int_{\overline{B_\epsilon p}} \chi$$

$$= \int_{\partial(M-N)} Y^*\Pi + \sum_{p\in Z}\left(\int_{\partial B_\epsilon p} Y^*\Pi + \mathrm{ind}_p U\right)$$

$$= -\int_{\partial N} Y^*\Pi + \int_{\partial N} Y^*\Pi + \sum_{p\in Z} \mathrm{ind}_p U = \sum_{p\in Z} \mathrm{ind}_p U$$

By the Poincaré-Hopf theorem [Mi2] the index sum over all the zeros of the vector field U on M equals the Euler characteristic of M. *QED*

This version of Chern's proof [Ch1] of the theorem is due to Gromoll.

3.72 Cohn-Vossen [CV] generalized the original 2-dimensional Gauss-Bonnet theorem to the inequality $0 \le \int_M \chi \le$ Euler characteristic of M for noncompact, orientable surfaces M. This was in turn generalized in several directions by Walter [Wa], and Greene and Wu [GW].

For the arbitrary Riemannian vector bundle case, see [Mi3] and [GHV].

3.73 **Exercise** The *Pfaffian* of a skew-symmetric matrix A is defined by setting $\mathrm{Pf}(A) := 0$ if $A \in \mathfrak{o}(2m-1)$, and

$$\mathrm{Pf}(A) := \frac{1}{2^m m!} \sum_\sigma \mathrm{sgn}(\sigma) A_{\sigma_1 \sigma_2} \cdots A_{\sigma_{2m-1} \sigma_{2m}} \qquad A \in \mathfrak{o}(2m)$$

For example, if A has 2×2 blocks

$$\begin{bmatrix} 0 & \lambda_j \\ -\lambda_j & 0 \end{bmatrix}$$

on the diagonal, and zeros elsewhere, then $\mathrm{Pf}(A) = \lambda_1 \cdots \lambda_m$. Show that (cf. 2.81) Pf is an invariant polynomial on the Lie algebra $\mathfrak{o}(m)$ of the rotation group $SO(m)$. By analogy with 2.84 and 2.85, express the Chern-Euler form χ of an oriented Riemannian vector bundle (E, g) over M in terms of the Pfaffian.

CHAPTER

FOUR

HARMONIC THEORY

In chap. 3 we saw how a Riemannian metric on a manifold determines a natural differential operator, the Levi-Civita connection; this operator was of fundamental importance in the study of the geometry of the metric. In this chapter we will see other operators determined by a Riemannian metric, and study some of their basic geometric properties.

THE BASIC DIFFERENTIAL OPERATORS

We begin with another look at the exterior derivative operator on $A(M)$; this should be compared with 2.76.

4.1 **Proposition** Let ∇ be a torsion-free linear connection on M. The exterior derivative operator $d\colon A(M) \to A(M)$ satisfies the identity

$$d\mu(X_0, \ldots, X_r) = \sum_{j=0}^{r} (-1)^j (\nabla_{X_j}\mu)(X_0, \ldots, \hat{X}_j, \ldots, X_r)$$

for $\mu \in A^r M$ and $X_j \in \mathfrak{X}M$, where the entry $\hat{X}_j$ is to be deleted. In other words, if $\{\epsilon_j\}$ is a basis for M_p and $\{\omega^i\}$ is the dual basis for M_p^*, then

$$d\mu = \sum_i \omega^i \wedge \nabla_{\epsilon_i}\mu$$

PROOF Fix $p \in M$. Because both sides are tensorial we may assume $[X_i, X_j]_p = (\nabla_{X_i} X_j)_p = 0$ (2.100); thus by 2.61 and [W: 2.25],

$$\begin{aligned} d\mu(X_0, \ldots, X_r)(p) &= \sum_{j=0}^{r} (-1)^j X_j \mu(X_0, \ldots, \hat{X}_j, \ldots, X_r)(p) \\ &= \sum_{j=0}^{r} (-1)^j (\nabla_{X_j} \mu)(X_0, \ldots, \hat{X}_j, \ldots, X_r)(p) \\ &= \sum_{j=0}^{r} (-1)^j \sum_{i=1}^{n} \omega^i(X_j) \cdot (\nabla_{\epsilon_i} \mu)(X_0, \ldots, \hat{X}_j, \ldots, X_r)(p) \\ &= \sum_i (\omega^i \wedge \nabla_{\epsilon_i} \mu)(X_0, \ldots, X_r)(p) \end{aligned}$$

4.2 **Definition** The *divergence of a vector field* $X \in \mathfrak{X}M$ with respect to a linear connection ∇ on M is $\operatorname{div} X := \operatorname{tr} \nabla X \in C^\infty M$.

If $\{\epsilon_j\}$ and $\{\omega^i\}$ are dual bases for M_p and M_p^*, then

$$\operatorname{div} X(p) = \operatorname{tr}[(\nabla X)_p] = \sum \omega^i(\nabla_{\epsilon_i} X)$$

4.3 **Definition** Let ∇ be the Levi-Civita connection on a Riemannian manifold (M, g). The *divergence of a C^∞ function* on M is defined to be zero. The *divergence of a $(0, s)$-tensor field* Z on M, $s > 0$, is the $(0, s-1)$-tensor field $\operatorname{div} Z$ on M such that at $p \in M$,

$$(\operatorname{div} Z)_p := \sum_i \iota_{\epsilon_i} \nabla_{\epsilon_i} Z$$

where $\{\epsilon_i\}$ is an orthonormal basis for M_p and ι denotes interior multiplication [W: 2.21]. Given $v_1, \ldots, v_{s-1} \in M_p$,

$$(\operatorname{div} Z)_p(v_1, \ldots, v_{s-1}) = \sum (\nabla_{\epsilon_i} Z)(\epsilon_i, v_1, \ldots, v_{s-1})$$

It is easy to check that $\operatorname{div} Z$ is well-defined.

Given $X \in \mathfrak{X}M$, recall from 3.8 that $X^\flat \in A^1 M$ is defined by $X^\flat(Y) := \langle X, Y \rangle$, $Y \in \mathfrak{X}M$; with respect to an orthonormal basis $\{\epsilon_j\}$ for M_p, since the musical isomorphisms are parallel (3.27),

$$\begin{aligned} (\operatorname{div} X)(p) &= \sum \langle \nabla_{\epsilon_j} X, \epsilon_j \rangle = \sum \iota_{\epsilon_j} (\nabla_{\epsilon_j} X)^\flat \\ &= \sum \iota_{\epsilon_j} \nabla_{\epsilon_j} X^\flat = (\operatorname{div} X^\flat)(p) \end{aligned}$$

Exercise Calculate $\operatorname{div} X$ if X is a Killing field on (M, g) (3.51).

4.4 **Definition** Let p be a point in a Riemannian manifold (M, g). A moving frame $\{E_j\}$ on M near p will be called an *adapted moving frame near* p if it is radially parallel with respect to $\exp_p$ (2.10) and if $\{E_j|_p\}$ is orthonormal.

An adapted moving frame is thus obtained by radial parallel translation of an orthonormal basis for M_p along the geodesics emanating from p; it follows that $(\nabla E_j)_p = 0$. The dual local basis for T^*M near p satisfies the same conditions as $\{E_j\}$. Since ∇ is the Levi-Civita connection, $[E_i, E_j]_p = (\nabla_{E_i} E_j - \nabla_{E_j} E_i)_p = 0$.

4.5 **Proposition** Let (M, g) be a Riemannian manifold. With respect to the local Riemannian volume element ω and local star operator $\bigstar$ determined by a local orientation of M,

$$\bigstar \operatorname{div} X = (\operatorname{div} X)\omega = d\iota_X \omega = \mathscr{L}_X \omega \qquad X \in \mathfrak{X}M$$

where $\mathscr{L}$ is the Lie derivative operator on M [W: 2.25(d)].

Proof Let $\{E_j\}$ be a positively oriented, adapted moving frame near $p \in M$ (4.4) with dual frame field $\{\omega^i\}$. By 4.1, 3.16, and 3.18,

$$\begin{aligned}\mathscr{L}_X \omega = d\iota_X \omega + \iota_X d\omega = d\iota_X \omega &= \sum_i \omega^i \wedge \nabla_{\epsilon_i} \iota_X \omega \qquad (\epsilon_i := E_i|_p) \\ &= \sum_{i,j} (-1)^{j-1} \omega^i \wedge \nabla_{\epsilon_i}(\omega^j(X)\omega^1 \wedge \cdots \wedge \widehat{\omega^j} \wedge \cdots \wedge \omega^n) \\ &= \sum (-1)^{j-1} \epsilon_i(\omega^j X)\omega^i \wedge \omega^1 \wedge \cdots \wedge \widehat{\omega^j} \wedge \cdots \wedge \omega^n \\ &= \sum \omega^i(\nabla_{\epsilon_i} X)\omega = (\operatorname{div} X)\omega = \bigstar \operatorname{div} X\end{aligned}$$

4.6 The operator div (4.2) on the space of $(0, \cdot)$-tensor fields on a Riemannian manifold (M, g) maps the subspace $A^r M$ of differential r-forms to the subspace $A^{r-1}M$; the restriction of div to $A(M)$ is important and is usually denoted by a special symbol.

Definition Define $\delta\mu := -\operatorname{div} \mu$ for $\mu \in A(M)$.

The minus sign in the definition is explained by the next proposition.

4.7 If (M, g) is a compact Riemannian manifold, then [W: 4.10(8)] the integral over M of a continuous function is well-defined. In particular, if η and μ are differential forms on M, then the function $\langle \eta, \mu \rangle$ (3.7, 3.9) can be integrated over M; this defines an inner product $\int_M \langle \eta, \mu \rangle$ on the real vector space $A(M)$.

Proposition The operator $\delta = -\operatorname{div}|_{A(M)}$ on a Riemannian manifold

(M, g) of dimension n is given by $\delta\mu = (-1)^{n(r+1)+1}\bigstar d\bigstar\mu$, $\mu \in A^r M$, where $\bigstar$ is the local star operator of either local orientation of M; as a result, $\delta^2 = 0$. If M is compact, then $\delta \in \text{End}\, A(M)$ is the adjoint of $d \in \text{End}\, A(M)$ with respect to the integration inner product on $A(M)$:

$$\int_M \langle \delta\eta, \mu\rangle = \int_M \langle \eta, d\mu\rangle \qquad \eta, \mu \in A(M)$$

PROOF Fix $p \in M$ and dual orthonormal bases $\{\epsilon_j\}$ and $\{\omega^i\}$ for M_p and M_p^*. By 3.18, 3.27, and 4.1, $(\text{div}\, \mu)_p$ equals

$$\sum_j (-1)^{r(n-r)} \iota_{\epsilon_j} \bigstar\bigstar \nabla_{\epsilon_j} \mu = (-1)^{n(r+1)} \sum \bigstar(\omega^j \wedge \nabla_{\epsilon_j} \bigstar\mu)$$
$$= (-1)^{n(r+1)}(\bigstar d\bigstar\mu)_p$$

Suppose M is orientable; choose an orientation, with Riemannian volume element ω and star operator $\bigstar$. Given $\eta \in A^{r-1}M$ and $\mu \in A^r M$, [W: 4.10(7, 8)], Stokes' theorem, and 3.18 imply

$$\int_M \langle d\eta, \mu\rangle = \int_M \langle d\eta, \mu\rangle\omega = \int_M d\eta \wedge \bigstar\mu$$
$$= \int_M [d(\eta \wedge \bigstar\mu) + (-1)^r \eta \wedge d\bigstar\mu]$$
$$= \int_M (-1)^{r+(n-r+1)(r-1)} \eta \wedge \bigstar\bigstar d\bigstar\mu$$
$$= \int_M \langle \eta, (-1)^{n(r+1)+1} \bigstar d\bigstar\mu\rangle\omega$$
$$= \int_M \langle \eta, (-1)^{n(r+1)+1} \bigstar d\bigstar\mu\rangle$$
$$= \int_M \langle \eta, \delta\mu\rangle$$

Now suppose M is nonorientable. Set $\tilde{M} := \{(p, \sigma) \mid p \in M$ and σ is an orientation for $M_p\}$; define $\psi(p, \sigma) := p$. The set $\tilde{M}$ has a natural topology as a 2-fold covering space of M. Make $\tilde{M}$ into a C^∞ covering space of M by defining $f \circ \psi$ to be C^∞ on $\tilde{M}$ for all $f \in C^\infty M$. With $\langle\ ,\ \rangle := \psi^*\langle\ ,\ \rangle$, $\tilde{M}$ is an orientable Riemannian covering of M, called the *Riemannian orientation covering of M*. For each $f \in C^\infty M$, $\int_{\tilde{M}} f \circ \psi = 2 \int_M f$; in addition, $\psi^* \text{div}\, \mu =$

div $\psi^*\mu$ since ψ is affine (3.31). Thus

$$2\int_M \langle d\eta, \mu\rangle = \int_{\tilde{M}} \langle \psi^* d\eta, \psi^*\mu\rangle = \int_{\tilde{M}} \langle d\psi^*\eta, \psi^*\mu\rangle$$

$$= \int_{\tilde{M}} \langle \psi^*\eta, \delta\psi^*\mu\rangle = \int_{\tilde{M}} \langle \psi^*\eta, -\operatorname{div} \psi^*\mu\rangle$$

$$= 2\int_M \langle \eta, -\operatorname{div} \mu\rangle = 2\int_M \langle \eta, \delta\mu\rangle$$

Exercises

1. Show that $\psi\colon \tilde{M} \to M$ is a principal $\mathbb{Z}_2$-bundle over M.
2. Show that for all $X \in \mathfrak{X}M$, div $X = -\delta(X^\flat)$.
3. Show that δ commutes with isometries.
4. Extend the proposition to the differential forms with compact support on a Riemannian manifold.

4.8 **Definition** Let $\mu \in A^r M$; μ is *closed* if $d\mu = 0$, and *coclosed* if $\delta\mu = 0$. If there exists $\eta \in A^{r-1}M$ such that $\mu = d\eta$, then μ is *exact*, while if $\mu = \delta\nu$ for some $\nu \in A^{r+1}M$, then μ is *coexact*.

By 4.1 and 4.7, a parallel form is both closed and coclosed. An exact form is closed; the converse is true locally by the Poincaré lemma [W: 4.18].

4.9 **Definition** The *Laplace-Beltrami operator* (or *Laplacian*) on a Riemannian manifold M is the second-order differential operator $\Delta := d\delta + \delta d$ on ΛT^*M. A form $\mu \in A(M)$ is *harmonic* if $\Delta\mu = 0$.

The following properties of Δ are immediate [W: 6.1–6.3]:

(i) $\Delta = (d + \delta)^2$, and Δ preserves the degree of a form.

(ii) Δ commutes with d, δ, $\bigstar$, and isometries.

(iii) If M is compact, then Δ is self-adjoint with respect to the integration inner product:

$$\int_M \langle \Delta\eta, \mu\rangle = \int_M \langle \eta, \Delta\mu\rangle \qquad \eta, \mu \in A(M)$$

(iv) $d\mu = \delta\mu = 0 \Leftrightarrow (d + \delta)\mu = 0 \Rightarrow \Delta\mu = 0$. In particular, a closed, coclosed form is harmonic; the converse is false because δ is in general not the adjoint of d. However, if M is compact, then δ is the adjoint of d (4.7), so $(d + \delta)^2\mu = \Delta\mu = 0$ implies $(d + \delta)\mu = 0$; hence in this case a harmonic form is closed and coclosed. Observe that parallel implies harmonic; the converse holds in some special cases (4.27).

(v) For $f \in C^\infty M$, $\Delta f = \delta df = -\text{div}\, df = -\text{div}\, \nabla f$ (4.3), where $\nabla f = df^\#$ is the gradient vector field of f (3.11).

More generally, the "adjoint" δ^∇ of the exterior covariant derivative operator d^∇ (2.75) of a Riemannian connection can be defined on vector-valued differential forms on M, and then the Laplacian $\Delta^\nabla = d^\nabla \delta^\nabla + \delta^\nabla d^\nabla$ also makes sense [Bn].

4.10 Let M be compact. The fact (4.9iv) that the Laplacian Δ and its "square root" $d + \delta$ have the same kernel, namely the harmonic forms, has an interesting consequence. Suppose that $\mathscr{V}$ and $\mathscr{W}$ are inner product spaces, and let $L\colon \mathscr{V} \to \mathscr{W}$ be a linear map with adjoint L^*, $\langle v, L^*w\rangle = \langle Lv, w\rangle$ for $v \in \mathscr{V}$ and $w \in \mathscr{W}$; the *index of* L is defined to be $\text{ind}(L) := \dim(\ker L) - \dim(\ker L^*)$ if this is finite. Let us calculate the index of the operator

$$d + \delta\colon A^{\text{even}}M \to A^{\text{odd}}M$$

where $A^{\text{even}}M$ is the space $\bigoplus_{k \geq 0} A^{2k}M$ of even-degree forms on M, and $A^{\text{odd}}M$ is the space of odd-degree forms on M; the adjoint of $d + \delta$ is the map $\delta + d$ from $A^{\text{odd}}M$ to $A^{\text{even}}M$. Since the harmonic forms on M represent the de Rham cohomology of M by the Hodge theorem,

$$\begin{aligned}\text{ind}(d + \delta) &= \dim H^{\text{even}}(M; \mathbb{R}) - \dim H^{\text{odd}}(M; \mathbb{R}) \\ &= \sum_{k \geq 0} (-1)^k \dim H^k(M; \mathbb{R})\end{aligned}$$

But by definition the alternating sum of the Betti numbers of M is the Euler characteristic of M [Gr: 20]; thus the index of $d + \delta$ (an analytical invariant of M) equals the Euler characteristic of M (a topological invariant).

4.11 If the metric on a manifold is changed, the operator δ changes with it; this in turn changes the Laplacian Δ and the space of harmonic forms on M. There is one special case where the harmonic forms do not change with a change of metric. Given a metric g on M and a positive C^∞ function f on M, the new metric $g_1 := f^2 g$ is said to be *conformally equivalent* to g (this will be discussed further in 5.14). Suppose that the dimension of M is $n = 2k$. If $\{\omega^i\}$ is a g-orthonormal basis for M_p^*, then $\{f(p)\omega^i\}$ is a g_1-orthonormal basis for M_p^*; assume it is positively oriented with respect to the chosen local orientation of M. Let $\bigstar$ and $\bigstar_1$ be the star operators of g and g_1, respectively. By 3.18, if $c = f(p) \in \mathbb{R}_{>0}$, then

$$\begin{aligned}\bigstar_1(c\omega^1 \wedge \cdots \wedge c\omega^k) &= c\omega^{k+1} \wedge \cdots \wedge c\omega^n = c^k\omega^{k+1} \wedge \cdots \wedge \omega^n \\ &= c^k\bigstar(\omega^1 \wedge \cdots \wedge \omega^k) = \bigstar(c\omega^1 \wedge \cdots \wedge c\omega^k)\end{aligned}$$

so on the forms of middle dimension, $\bigstar = \bigstar_1$. Thus the conditions $d\mu = 0$ and $d\bigstar\mu = 0$ which characterize a harmonic form on a compact manifold (4.7, 4.9iv) are invariant under a conformal change of metric if μ is a form on M of middle dimension $k = n/2$. Therefore we have the following theorem.

Theorem Let (M, g) be a compact Riemannian manifold of dimension $n = 2k$. The space of harmonic forms on M of degree k is invariant under a conformal change of metric.

4.12 **Theorem** (Hodge) Each de Rham cohomology class of a compact Riemannian manifold M has a unique harmonic representative, that is, $H^r(M; \mathbb{R})$ is isomorphic to the vector space of harmonic r-forms on M.

PROOF This is proved for orientable M in chap. 6 of Warner's book. The key analysis in his proof is entirely local in nature, so it is valid in the nonorientable case. Here the nonorientable case is derived from the orientable case.

Let $\psi: \tilde{M} \to M$ be the Riemannian orientation covering of M from the proof of 4.7. Given $c \in H^r(M; \mathbb{R})$, let $\tilde{c} := \psi^* c \in H^r(\tilde{M}; \mathbb{R})$; since $\tilde{M}$ is orientable there is a unique harmonic representative $\tilde{\eta} \in A^r\tilde{M}$ for $\tilde{c}$. The involution τ of $\tilde{M}$ which switches the two points in $\tilde{M}$ over each point of M is an isometry, so $\tau^*\tilde{\eta}$ is also harmonic. The identity $\psi = \psi \circ \tau$ implies that $\tilde{c}$ is also the cohomology class of $\tau^*\tilde{\eta}$: $[\tau^*\tilde{\eta}] = \tau^*[\tilde{\eta}] = \tau^*\psi^*c = \psi^*c = \tilde{c}$. By uniqueness of the harmonic representative, $\tilde{\eta} = \tau^*\tilde{\eta}$, so $\tilde{\eta}$ is τ-invariant, and is thus the lift $\psi^*\eta$ of some $\eta \in A^rM$. But $\psi^*\Delta\eta = 0$, and ψ is a local diffeomorphism, so η is harmonic.

To see that η represents c, let $\mu \in A^rM$ be a representative of c. Since $\psi^*\eta$ and $\psi^*\mu$ both represent $\tilde{c}$, there exists $\beta \in A^{r-1}\tilde{M}$ such that $\psi^*(\eta - \mu) = d\beta$. But $d\tau^*\beta = d\beta$, so $\psi^*(\eta - \mu) = (d\beta + d\tau^*\beta)/2$. Since $(\beta + \tau^*\beta)/2$ is τ-invariant, it is the lift $\psi^*\alpha$ of some $\alpha \in A^{r-1}M$. Thus $\psi^*(\eta - \mu) = d\psi^*\alpha = \psi^*d\alpha$, and so $\eta - \mu = d\alpha$ because ψ is a local diffeomorphism. Hence $[\eta] = c$ as required.

Uniqueness of the harmonic representative is trivial.

4.13 **Theorem** (Chern) Let M be a compact manifold. If there exists a Riemannian metric g on M and an orientable vector subbundle E of TM such that the Levi-Civita parallel transport in TM is reducible to E, then $H^k(M; \mathbb{R}) \neq 0$, where k is the rank of E.

PROOF By the definition of orientability (1.65), there is a nonvanishing section σ of $\Lambda^k E$; the section $X := \sigma/\|\sigma\|$ of $\Lambda^k E$ has constant length 1, so as in 3.27, X is parallel. The form $X^\flat \in A^k M$ is parallel, and hence harmonic. Now apply 4.12.

GREEN'S THEOREM AND SOME APPLICATIONS

4.14 **Theorem** (Green) If $X \in \mathfrak{X}M$, where (M, g) is a compact Riemannian manifold, then $\int_M \operatorname{div} X = 0$.

PROOF If M is orientable, 4.5 and Stokes' theorem imply

$$\int_M \operatorname{div} X = \int_M (\operatorname{div} X)\omega = \int_M d\iota_X \omega = \int_{\partial M} \iota_X \omega = 0$$

If M is nonorientable, let $\psi\colon \tilde{M} \to M$ be the Riemannian orientation covering of M from the proof of 3.7. Let $\tilde{X} \in \mathfrak{X}\tilde{M}$ be ψ-related to X; since ψ is a local isometry, $\operatorname{div} \tilde{X} = (\operatorname{div} X) \circ \psi$, and $\int_M \operatorname{div} X = \frac{1}{2} \int_{\tilde{M}} \operatorname{div} \tilde{X} = 0$.

4.15 **Corollary 1** If (M, g) is a compact Riemannian manifold, then $\int_M \Delta f = 0$ for $f \in C^\infty M$.

PROOF By 4.9v, $\Delta f = -\operatorname{div} \nabla f$.

4.16 **Corollary 2** (Hopf-Bochner) If (M, g) is a connected, compact Riemannian manifold, and if $\Delta f \geq 0$ for $f \in C^\infty M$, then f is constant. In particular, $f \in C^\infty M$ is harmonic if and only if f is constant.

PROOF Since $\Delta f \geq 0$ and $\int_M \Delta f = 0$, $\Delta f = 0$, which implies $df = 0$; f is then constant because M is connected.

Here is an alternative proof: since M is connected,

$$\mathbb{R} \cong H^0(M; \mathbb{R}) \cong \{\text{harmonic } f \in C^\infty M\} \supset \{\text{constant } f \text{ on } M\} \cong \mathbb{R}$$

4.17 **Definition** Let (M, g) be a Riemannian manifold. Given $X \in \mathfrak{X}M$, its covariant differential ∇X is a section of End $TM \cong A^0(M, \text{End } TM)$, and its *second covariant differential* is $\nabla^2 X = d^\nabla \nabla X \in A^1(M, \text{End } TM) \cong \Gamma \operatorname{Hom}(TM, \text{End } TM)$. Denote by $\operatorname{tr} \nabla^2 X \in \mathfrak{X}M$ the so-called *contraction of* $\nabla^2 X$ *with respect to* g: $(\operatorname{tr} \nabla^2 X)_p := \sum \nabla^2 X(\epsilon_j, \epsilon_j)$, where $\{\epsilon_j\}$ is an orthonormal basis for M_p.

4.18 **Theorem** (Bochner) On a Riemannian manifold (M, g),

$$2\langle \operatorname{tr} \nabla^2 X, X \rangle + 2\|\nabla X\|^2 + \Delta\|X\|^2 = 0 \qquad X \in \mathfrak{X}M$$

If M is compact, if B is a positive semidefinite symmetric bilinear form on M, and if $X \in \mathfrak{X}M$ satisfies $B(X, \cdot) = \langle \operatorname{tr} \nabla^2 X, \cdot \rangle$, then X is parallel and $B(X, X) = 0$; if B is positive definite, then X is zero.

PROOF Let $\{E_j\}$ be an adapted moving frame near $p \in M$ (4.4); then

$$\begin{aligned}(2\langle \operatorname{tr} \nabla^2 X, X \rangle &+ 2\|\nabla X\|^2 + \Delta\|X\|^2)(p)\\ &= 2 \sum (\langle \nabla_{\epsilon_j} \nabla_{E_j} X, X_p \rangle + \|\nabla_{\epsilon_j} X\|^2) - \operatorname{div}(\nabla\|X\|^2)(p)\\ &= 2 \sum \epsilon_j \langle \nabla_{E_j} X, X \rangle - \sum \epsilon_j \langle \nabla\|X\|^2, E_j \rangle = 0\end{aligned}$$

If B is positive semidefinite, and $B(X, Y) = \langle \operatorname{tr} \nabla^2 X, Y \rangle$ for all $Y \in \mathfrak{X}M$, then $\Delta\|X\|^2 \le 0$, so (4.16) $\Delta\|X\|^2 = 0$ if M is compact; hence $\|\nabla X\|^2 = B(X, X) = 0$.

4.19 **Corollary 1** If $\nabla^2 X = 0$ for a vector field X on a compact Riemannian manifold, then X is parallel.

PROOF Set $B = 0$ in 4.18.

4.20 **Corollary 2** Let $X \in \mathfrak{X}M$, where M is compact; let $\mathscr{R}$ denote the Ricci tensor of a metric g on M (3.43). If $\mathscr{R}(X, X) \ge 0$ and $\mathscr{R}(X, \cdot) = \langle \operatorname{tr} \nabla^2 X, \cdot \rangle$, or if $\mathscr{R}(X, X) \le 0$ and $\mathscr{R}(X, \cdot) + \langle \operatorname{tr} \nabla^2 X, \cdot \rangle = 0$, then X is parallel and $\mathscr{R}(X, X) = 0$. If M has positive or negative Ricci curvature, respectively, then $X = 0$.

PROOF Use $B = \pm\mathscr{R}$ in 4.18.

Applications of this corollary will appear in chap. 5. As an exercise, the reader can prove that a Riemannian homogeneous space M of negative Ricci curvature (3.45) is noncompact by applying the corollary to the Killing fields on M (see 5.8).

WEITZENBÖCK'S FORMULA FOR THE LAPLACIAN

4.21 **Definition** For each r-form μ on a Riemannian manifold (M, g), define a $(0, r)$-tensor field ρ_μ on M by

$$\rho_\mu(v_1, \ldots, v_r) := \sum_{i=1}^{n} \sum_{j=1}^{r} (R(\epsilon_i, v_j)\mu)(v_1, \ldots, v_{j-1}, \epsilon_i, v_{j+1}, \ldots, v_r)$$

where $v_j \in M_p$, and $\{\epsilon_i\}$ is an orthonormal basis for M_p.

This is easily seen to be well-defined.

4.22 Theorem Weitzenböck's formula: If μ is a differential form on a Riemannian manifold, then

$$\Delta\mu = -\operatorname{div} \nabla\mu + \rho_\mu$$

$$\langle \Delta\mu, \mu\rangle = \tfrac{1}{2}\Delta\|\mu\|^2 + \|\nabla\mu\|^2 + \langle \rho_\mu, \mu\rangle$$

If M is flat, then ρ_μ is zero, so $\Delta\mu = -\operatorname{div} \nabla\mu$; thus $-\operatorname{div} \nabla\mu$ is formally the flat Laplacian. It must be noted that the formal flat Laplacian $-\operatorname{div} \nabla\mu = \Delta\mu - \rho_\mu$ is in general only a $(0, r)$-tensor field, and not an r-form; for this reason it is also called the "rough Laplacian" [BE].

PROOF Given $\mu \in A^r M$ and $U, V_1, \ldots, V_r \in \mathfrak{X}M$,

$$\begin{aligned}(\nabla\mu - d\mu)(U, V_1, \ldots, V_r) &= (\nabla_U \mu)(V_1, \ldots, V_r) - d\mu(U, V_1, \ldots, V_r)\\ &= \sum_{j=1}^{r} (\nabla_{V_j}\mu)(V_1, \ldots, V_{j-1}, U, V_{j+1}, \ldots, V_r)\end{aligned}$$

by 4.1. Fix $p \in M$, and assume $(\nabla V_j)_p = 0$; set $v_j := V_j|_p \in M_p$. If $\{E_j\}$ is an adapted moving frame near p, with $\epsilon_j := E_j|_p$, then by 4.7,

$$\begin{aligned}(\operatorname{div} \nabla\mu + \delta d\mu)(v_1, \ldots, v_r) &= \operatorname{div}(\nabla\mu - d\mu)(v_1, \ldots. v_r)\\ &= \sum_{i=1}^{n} (\nabla_{\epsilon_i}(\nabla\mu - d\mu))(\epsilon_i, v_1, \ldots, v_r)\\ &= \sum_i \epsilon_i((\nabla\mu - d\mu)(E_i, V_1, \ldots, V_r))\\ &= \sum_i \sum_{j=1}^{r} \epsilon_i((\nabla_{V_j}\mu)(V_1, \ldots, V_{j-1}, E_i, V_{j+1}, \ldots, V_r))\\ &= \sum_{i,j} (\nabla_{\epsilon_i}\nabla_{V_j}\mu)(v_1, \ldots, v_{j-1}, \epsilon_i, v_{j+1}, \ldots, v_r)\end{aligned}$$

On the other hand,

$$\begin{aligned}d\delta\mu(v_1, \ldots, v_r) &= \sum_{j=1}^{r} (-1)^{j+1}(\nabla_{v_j}\delta\mu)(v_1, \ldots, \hat{v}_j, \ldots, v_r)\\ &= \sum_j (-1)^j v_j\left(\sum_{i=1}^{n} (\nabla_{E_i}\mu)(E_i, V_1, \ldots, \hat{V}_j, \ldots, V_r)\right)\\ &= -\sum_{i,j} (\nabla_{v_j}\nabla_{E_i}\mu)(v_1, \ldots, v_{j-1}, \epsilon_i, v_{j+1}, \ldots, v_r)\end{aligned}$$

Thus, since $[E_i, V_j]_p = (\nabla_{E_i} V_j - \nabla_{V_j} E_i)_p = 0$ by assumption,

$$(\Delta\mu + \operatorname{div} \nabla\mu)(v_1, \ldots, v_r) = \rho_\mu(v_1, \ldots, v_r)$$

Finally,

$$\begin{aligned}(\operatorname{div} \nabla\mu)(v_1, \ldots, v_r) &= \sum_i (\nabla_{\epsilon_i} \nabla\mu)(\epsilon_i, v_1, \ldots, v_r) \\ &= \sum_i \epsilon_i((\nabla_{E_i}\mu)(V_1, \ldots, V_r)) \\ &= \sum (\nabla_{\epsilon_i} \nabla_{E_i} \mu)(v_1, \ldots, v_r)\end{aligned}$$

since all the fields are parallel at p. Thus (see 4.18),

$$\begin{aligned}\langle -\operatorname{div} \nabla\mu, \mu\rangle(p) &= -\sum_i \langle \nabla_{\epsilon_i} \nabla_{E_i} \mu, \mu_p\rangle \\ &= -\sum (\epsilon_i \langle \nabla_{E_i} \mu, \mu\rangle - \langle \nabla_{\epsilon_i} \mu, \nabla_{\epsilon_i} \mu\rangle) \\ &= -\tfrac{1}{2} \sum \epsilon_i E_i \|\mu\|^2 + \|\nabla\mu\|^2(p) \\ &= (\tfrac{1}{2}\Delta\|\mu\|^2 + \|\nabla\mu\|^2)(p)\end{aligned}$$

by 4.3 and 4.9v.

For $f \in C^\infty M \cong A^0 M$, $\rho_f = 0$, so $\Delta f = -\operatorname{div} \nabla f$ as in 4.9v.

Exercise Show that if M is compact, then $-\operatorname{div} \nabla\mu = \nabla^*\nabla\mu$, where ∇^* is the adjoint of ∇ with respect to the integration inner product on the vector space of $(0, \cdot)$-tensor fields.

4.23 **Corollary** If a connected, compact manifold M of dimension n admits a flat Riemannian metric g, then the dimension of $H^r(M; \mathbb{R})$ is less than or equal to $\binom{n}{r}$; equality holds for all r if and only if the holonomy group of (M, g) is trivial.

Proof Every parallel form on M is known to be harmonic. Conversely, if μ is harmonic, then since $R = 0$,

$$\int_M \|\nabla\mu\|^2 = -\tfrac{1}{2} \int_M \Delta\|\mu\|^2 = 0$$

Thus μ is parallel. Fix $p \in M$. By the Hodge theorem (4.12),

$$\dim H^r(M; \mathbb{R}) = \dim\{\text{parallel } r\text{-forms on } M\} \leq \dim \Lambda^r M_p^* = \binom{n}{r}$$

with equality if and only if parallel translation in $\Lambda^r T^* M$ is trivial around each loop at p.

4.24 Proposition Let (M, g) be a Riemannian manifold of constant sectional curvature K. The Laplacian on (M, g) is given by

$$\Delta\mu = -\text{div}\,\nabla\mu + r(n-r)K\mu \qquad \mu \in A^rM$$

PROOF The curvature operator (3.36) of (M, g) is $R = KI \in \Gamma\,\text{End}(\Lambda^2 TM)$ [GKM: 3.6(17)], that is, $R(U \wedge V) = K \cdot U \wedge V$, U, $V \in \mathfrak{X}M$. Fix $p \in M$, $v_1, \ldots, v_r \in M_p$, and an orthonormal basis $\{\epsilon_j\}$ for M_p. Since $R(U, V)\mu = -\mu \circ R(U, V)$ and $R(U, V)$ acts on $\Lambda^r\mathfrak{X}M$ as a derivation (2.69, 2.70), Weitzenböck's formula implies

$$\begin{aligned}
(\Delta\mu + \text{div}\,\nabla\mu)(v_1, \ldots, v_r) &= \rho_\mu(v_1, \ldots, v_r) \\
&= -\sum_{i,j} \mu(v_1, \ldots, v_{j-1}, R(\epsilon_i, v_j)\epsilon_i, v_{j+1}, \ldots, v_r) \\
&\quad - \sum_i \sum_{k \neq j} \mu(v_1, \ldots, v_{k-1}, R(\epsilon_i, v_j)v_k, v_{k+1}, \ldots, v_{j-1}, \epsilon_i, v_{j+1}, \ldots, v_r) \\
&= K \sum_{i,j} \mu(v_1, \ldots, v_{j-1}, v_j - \langle v_j, \epsilon_i\rangle\epsilon_i, v_{j+1}, \ldots, v_r) \\
&\quad + K \sum_i \sum_{k \neq j} \mu(v_1, \ldots, v_{k-1}, \langle \epsilon_i, v_k\rangle v_j - \langle v_j, v_k\rangle\epsilon_i, \\
&\qquad\qquad v_{k+1}, \ldots, v_{j-1}, \epsilon_i, v_{j+1}, \ldots, v_r) \\
&= K \sum_j \Big((n-1)\mu(v_1, \ldots, v_r) - \sum_{k \neq j} \mu(v_1, \ldots, v_r)\Big) \\
&= Kr(n-r)\mu(v_1, \ldots, v_r)
\end{aligned}$$

Exercises

1. Prove that if a compact manifold M admits a metric of constant positive curvature, then $H^r(M; \mathbb{R}) = 0$, $r = 1, \ldots, n-1$.
2. Derive a formula for Δ which involves the Ricci tensor $\mathscr{R}$ (see 5.4 for the case of a 1-form).

CHERN'S FORMULA FOR THE LAPLACIAN

In the last section we saw an explicit formula for $\Delta\mu$ as the formal flat Laplacian ($-\text{div}\,\nabla\mu$) plus a curvature term. Since the curvature tensor was defined in terms of holonomy, and in fact generates the holonomy algebra in the sense of Ambrose-Singer (2.47), we ought to be able to express $\Delta\mu + \text{div}\,\nabla\mu$ in terms of the holonomy algebra. For this we use the holonomy algebra valued curvature operator $R \in \Gamma\,\text{End}\,\Lambda^2 TM$ from 3.36.

4.25 **Theorem** Chern's formula: Let $G \subset O(n)$ be the holonomy group of a connected Riemannian manifold (M, g). Fix $p \in M$, and define an inner product on the holonomy algebra $\mathfrak{g}(M_p)$ by the inclusion $\mathfrak{g}(M_p) \subset \mathfrak{o}(M_p) \cong \Lambda^2 M_p$ (3.13). Extend each $X \in \mathfrak{g}(M_p)$ to a derivation of ΛM_p:

$$X(v_1 \wedge \cdots \wedge v_r) := \sum_j v_1 \wedge \cdots \wedge Xv_j \wedge \cdots \wedge v_r;$$

denote by ${}^tX \in \operatorname{End} \Lambda M_p^*$ the transpose of X (see 2.70). If $\{X_\alpha\}$ is an orthonormal basis for $\mathfrak{g}(M_p)$,

$$(\Delta\mu)_p = -\operatorname{div}(\nabla\mu)_p - \sum_\alpha {}^tR(X_\alpha){}^tX_\alpha\mu \qquad \mu \in A(M)$$

$$\langle \Delta\mu, \mu\rangle(p) = (\tfrac{1}{2}\Delta\|\mu\|^2 + \|\nabla\mu\|^2)(p) + \sum_\alpha \langle {}^tX_\alpha\mu, {}^tR(X_\alpha)\mu\rangle$$

PROOF [Ch2] [P2] [Wl 2] For $u, v \in M_p$, express ${}^tR(u, v) \in \mathfrak{g}(M_p^*)$ with respect to the orthonormal basis $\{{}^tX_\alpha\}$: by the symmetries for R (3.35),

$$\begin{aligned} {}^tR(u, v) &= \sum_\alpha \langle {}^tR(u, v), {}^tX_\alpha\rangle {}^tX_\alpha = \sum \langle R(u, v), X_\alpha\rangle {}^tX_\alpha \\ &= \sum \langle R(X_\alpha)v, u\rangle {}^tX_\alpha; \end{aligned}$$

If ρ_μ denotes Weitzenböck's curvature tensor from 4.21, then

$$\begin{aligned} \rho_\mu(v_1, \ldots, v_r) &= \sum_{i=1}^n \sum_{j=1}^r (R(\epsilon_i, v_j)\mu)(v_1, \ldots, v_{j-1}, \epsilon_i, v_{j+1}, \ldots, v_r) \\ &= -\sum_{i,j} {}^tR(\epsilon_i, v_j)(\mu)(v_1, \ldots, v_{j-1}, \epsilon_i, v_{j+1}, \ldots, v_r) \\ &= -\sum_{i,j,\alpha} \langle R(X_\alpha)v_j, \epsilon_i\rangle {}^tX_\alpha(\mu)\,(v_1, \ldots, v_{j-1}, \epsilon_i, v_{j+1}, \ldots, v_r) \\ &= -\sum_{\alpha,j} {}^tX_\alpha(\mu)(v_1, \ldots, v_{j-1}, R(X_\alpha)v_j, v_{j+1}, \ldots, v_r) \\ &= -\sum_\alpha {}^tX_\alpha(\mu) \circ R(X_\alpha)(v_1, \ldots, v_r) \\ &= -\sum_\alpha {}^tR(X_\alpha){}^tX_\alpha(\mu)(v_1, \ldots, v_r) \end{aligned}$$

because ${}^tR(X_\alpha)$ acts on $\Lambda(M_p^*)$ as a derivation. By Weitzenböck's formula,

$$\begin{aligned} \langle \Delta\mu, \mu\rangle(p) &= (\tfrac{1}{2}\Delta\|\mu\|^2 + \|\nabla\mu\|^2)(p) - \sum_\alpha \langle {}^tR(X_\alpha){}^tX_\alpha\mu, \mu_p\rangle \\ &= (\tfrac{1}{2}\Delta\|\mu\|^2 + \|\nabla\mu\|^2)(p) + \sum_\alpha \langle {}^tX_\alpha\mu, {}^tR(X_\alpha)\mu\rangle \end{aligned}$$

since $R(X_\alpha)$ is skew-symmetric.

4.26 **Corollary 1** (Chern) Let (M, g) be a connected Riemannian manifold. Suppose there exists a vector subbundle E of the exterior bundle ΛT^*M which is invariant under Levi-Civita parallel transport. If $P_E\colon \Lambda T^*M \to E$ is orthogonal projection, then $\Delta \circ P_E = P_E \circ \Delta$.

PROOF The extension of the Levi-Civita connection to ΛT^*M is reducible to E since E is invariant under parallel transport (2.57). Similarly, since holonomy in ΛT^*M is defined in terms of parallel transport, for all $p \in M$ each element of the holonomy algebra $\mathfrak{g}(p)$ at p leaves the fiber $E_p \subset \Lambda M_p^*$ invariant. Thus if μ is a section of E, then so is $\Delta\mu$ by Chern's formula.

But the orthogonal complement $E^\perp$ of E is also invariant under parallel transport, so if $\eta = \eta_E + \eta_{E^\perp} \in \Gamma(E \oplus E^\perp) = \Gamma\Lambda T^*M$, then

$$\Delta \circ P_E(\eta) = \Delta\eta_E = P_E(\Delta\eta_E + \Delta\eta_{E^\perp}) = P_E \circ \Delta(\eta)$$

4.27 **Corollary 2** Let (M, g) be a compact Riemannian manifold for which all the eigenvalues of the curvature operator $R \in \Gamma \text{ End } \Lambda^2 TM$ are non-negative. A differential form on M is harmonic if and only if it is parallel.

PROOF By 4.9iv a parallel form is harmonic.

Fix $p \in M$. Since the curvature operator on $\Lambda^2 M_p$ is symmetric (3.36), there exists an eigenbasis $\{X_\alpha\}$ for $\Lambda^2 M_p$, with corresponding eigenvalues $\lambda_\alpha \geq 0$: $R(X_\alpha) = \lambda_\alpha X_\alpha$. As in 4.25, extend each $Y \in \Lambda^2 M_p \cong \mathfrak{o}(M_p)$ to a derivation of ΛM_p, and let ${}^tY \in \text{End } \Lambda M_p^*$ be its transpose. It follows that ${}^tR(X_\alpha) = \lambda_\alpha {}^tX_\alpha$, so by Chern's formula (4.25),

$$\begin{aligned}\langle \Delta\mu, \mu\rangle(p) &= (\tfrac{1}{2}\Delta\|\mu\|^2 + \|\nabla\mu\|^2)(p) + \sum \lambda_\alpha \|{}^tX_\alpha\mu\|^2 \\ &\geq (\tfrac{1}{2}\Delta\|\mu\|^2 + \|\nabla\mu\|^2)(p)\end{aligned}$$

for all $\mu \in A(M)$. Thus (4.15), the integral over M of $\langle \Delta\mu, \mu\rangle$ is greater than or equal to zero, so $\nabla\mu = 0$ if $\Delta\mu = 0$.

4.28 The eigenvalues of the curvature operator R on (M, g) are a generalization of the sectional curvature of M; in particular, if λ_0 and λ_1 are the minimum and maximum eigenvalues of R over M, respectively, then $\lambda_0 \leq K_\sigma \leq \lambda_1$ for every 2-plane σ on M.

Definition The curvature operator R of (M, g) is called *positive definite* if all the eigenvalues of R are greater than zero.

4.29 **Theorem** (Bochner-Yano, Berger, Meyer) If an n-dimensional compact manifold M admits a metric g with positive definite curvature operator

R, then $H^r(M; \mathbb{R}) = 0$, $0 < r < n$. If in addition M is orientable and connected, then $H^*(M; \mathbb{R})$ is isomorphic to $H^*(S^n; \mathbb{R})$; in this case M is often called a "rational homology sphere."

PROOF Since M is compact, all eigenvalues of R are greater than or equal to some $\lambda > 0$. Fix $p \in M$, and choose an orthonormal eigenbasis $\{X_\alpha\}$ for $\mathfrak{o}(M_p) \cong \Lambda^2 M_p$: $R(X_\alpha) = \lambda_\alpha X_\alpha$, $\lambda_\alpha \geq \lambda > 0$. Since the holonomy algebra $\mathfrak{g}(p) \subset \mathfrak{o}(M_p)$ is generated by the vectors $R(X_\alpha)$, $\mathfrak{g}(p) = \mathfrak{o}(M_p)$. This will allow us to calculate the curvature term in Chern's formula for Δ by comparing M with the standard n-sphere.

Fix a point $q \in S^n$ and an isometry between M_p and S^n_q; the basis $\{X_\alpha\}$ corresponds to an orthonormal basis $\{Y_\alpha\}$ for $\mathfrak{o}(S^n_q)$, the value μ_p at p of $\mu \in A^r M$ corresponds to the value η_q for some (nonunique) $\eta \in A^r S^n$, and $\|{}^t X_\alpha \mu_p\| = \|{}^t Y_\alpha \eta_q\|$. By Chern's formula and the expression (4.24) for Δ on S^n,

$$\sum \|{}^t X_\alpha \mu_p\|^2 = \sum \|{}^t Y_\alpha \eta_q\|^2 = \langle \Delta\eta + \operatorname{div} \nabla\eta, \eta\rangle(q)$$
$$= r(n-r)\|\eta_q\|^2 = r(n-r)\|\mu_p\|^2$$

Therefore, since $\sum \langle {}^t R(X_\alpha)\mu, {}^t X_\alpha \mu\rangle \geq \lambda \sum \|{}^t X_\alpha \mu\|^2$,

$$\langle \Delta\mu, \mu\rangle \geq \tfrac{1}{2}\Delta\|\mu\|^2 + \|\nabla\mu\|^2 + \lambda r(n-r)\|\mu\|^2 \qquad \mu \in A^r M$$

and therefore $\int_M \langle \Delta\mu, \mu\rangle \geq \int_M (\|\nabla\mu\|^2 + \lambda r(n-r)\|\mu\|^2) \geq 0$ for $0 < r < n$; hence a harmonic r-form on M of degree other than 0 or n must be zero.

The original version of the theorem [YB] used exercise 2 from 4.24 to prove that if the eigenvalues of R are $\frac{1}{2}$-pinched, that is, if all the eigenvalues are positive and the ratio between the smallest and largest eigenvalues at each point is greater than or equal to $\frac{1}{2}$, then $H^*(M; \mathbb{R}) \cong H^*(S^n; \mathbb{R})$ for connected orientable M. Berger [B3] saw how to use just the minimum of the eigenvalues for an estimate on the forms of degree 2, and D. Meyer [Me] proved the theorem for all degrees. All these proofs used Weitzenböck's formula. The proof here is from [P2]; for related work, see [BK], [Ku], and [Ma].

In chap. 7 we shall see a theorem of Gallot and Meyer which partially describes the universal covering of M if the minimum eigenvalue is zero.

4.30 The proof of the next theorem uses extremely careful estimates of the term ρ_μ in Weitzenböck's formula.

Theorem (Berger) Let M be a compact manifold of dimension $n = 2m + 1$. If M admits a metric such that the sectional curvature is pinched between $2(m - 1)/(8m - 5)$ and 1, then $H^2(M; \mathbb{R}) = 0$.

PROOF See [B2].

CHAPTER

FIVE

GEOMETRIC VECTOR FIELDS ON RIEMANNIAN MANIFOLDS

In 2.119 and 2.121 the study of the affine vector fields on a manifold M with linear connection ∇ gave information on the affine transformation group of (M, ∇). Similarly, in 3.53 study of the Killing fields on a Riemannian manifold told us something about the isometry group of the manifold. In this chapter we will look at some other "geometric" vector fields on manifolds.

5.1 We begin with some comments which will be applied to the covariant differential $\nabla X \in \Gamma \operatorname{End} TM$ of a vector field X with respect to a connection.

Given an endomorphism L of an inner product space $(\mathscr{V}, \langle\ ,\ \rangle)$, it is useful to split L into its symmetric and skew-symmetric parts:

$$\begin{aligned}\langle Lu, v\rangle &= \tfrac{1}{2}(\langle Lu, v\rangle + \langle u, Lv\rangle) + \tfrac{1}{2}(\langle Lu, v\rangle - \langle u, Lv\rangle) \qquad u, v \in \mathscr{V}\\ &=: \langle L_{\text{sym}}u, v\rangle + \langle L_{\text{sk}}u, v\rangle = \langle (L_{\text{sym}} + L_{\text{sk}})u, v\rangle\end{aligned}$$

This splitting is invariant under the adjoint action of the orthogonal group $O(\mathscr{V})$ on the Lie algebra $\mathrm{gl}(\mathscr{V})$, for gAg^{-1} is symmetric or skew-symmetric if A is symmetric or skew-symmetric, respectively.

This invariant decomposition is reducible: if $\dim \mathscr{V} = n$, write L_{sym} as the sum

$$\frac{\operatorname{tr} L}{n} I + \left(L_{\text{sym}} - \frac{\operatorname{tr} L}{n} I\right)$$

where I is the identity transformation of $\mathscr{V}$; denote the trace-free part $L_{\text{sym}} - [(\text{tr } L)/n]I$ of L_{sym} by L_o. Since

$$g\frac{\text{tr } L}{n}Ig^{-1} = \frac{\text{tr } L}{n}I$$

and $gL_o g^{-1}$ is symmetric with trace zero, $g \in O(\mathscr{V})$, the splitting

$$L = \frac{\text{tr } L}{n}I + L_o + L_{\text{sk}}$$

of L is invariant under the adjoint action of $O(\mathscr{V})$. It can be shown that this decomposition is irreducible with respect to invariance under $O(\mathscr{V})$ [Wy*3*: V.B].

5.2 **Proposition** Let ∇ be the Levi-Civita connection on a Riemannian manifold (M, g). For each vector field $X \in \mathfrak{X}M$, the decomposition of the covariant differential $\nabla X \in \Gamma \text{ End } TM$ into symmetric and skew-symmetric parts is given for all $U, V \in \mathfrak{X}M$ by

$$\langle(\nabla X)_{\text{sym}} U, V\rangle = \tfrac{1}{2}(\mathscr{L}_X g)(U, V)$$
$$\langle(\nabla X)_{\text{sk}} U, V\rangle = \tfrac{1}{2}\, dX^\flat(U, V)$$

where $X^\flat = \langle X, \cdot\rangle \in A^1M$ is the metric dual of X from 3.8. Furthermore, the diagonal part of ∇X is

$$\frac{\text{div } X}{n}I \qquad n = \dim M$$

PROOF As in the proof of 3.52,

$$(\mathscr{L}_X g)(U, V) = X\langle U, V\rangle - \langle[X, U], V\rangle - \langle U, [X, V]\rangle$$
$$= \langle\nabla_U X, V\rangle + \langle U, \nabla_V X\rangle$$

because ∇ is torsion-free. Similarly,

$$dX^\flat(U, V) = U\langle X, V\rangle - V\langle X, U\rangle - \langle X, [U, V]\rangle$$
$$= \langle\nabla_U X, V\rangle - \langle U, \nabla_V X\rangle$$

The divergence of X is by definition (4.2) the trace of ∇X.

HARMONIC FIELDS

5.3 **Definition** A vector field X on a Riemannian manifold (M, g) is called *harmonic* if the metric dual 1-form $X^\flat$ is harmonic.

If M is compact, then by the Hodge theorem (4.12), $H^1(M; \mathbb{R})$ is isomorphic to the vector space of harmonic vector fields on M.

5.4 **Proposition** (de Rham) A vector field X on (M, g) is harmonic if and only if $\langle \operatorname{tr} \nabla^2 X, \cdot \rangle = \mathcal{R}(X, \cdot)$, where $\mathcal{R}$ is the Ricci tensor of g. In addition, X is harmonic if div $X = 0$ and ∇X is self-adjoint with respect to g; the converse holds if M is compact. Thus if M is compact, X is harmonic if and only if ∇X equals its trace-zero symmetric part.

PROOF Let $\{E_j\}$ be an adapted moving frame near $p \in M$ (4.4). Let $V \in \mathfrak{X}M$ with $(\nabla V)_p = 0$. By 2.69 and Weitzenböck's formula (4.22),

$$\begin{aligned}\Delta X^\flat(V)(p) &= \left(\sum (R(E_j, V)X^\flat)(E_j) - \operatorname{div} \nabla X^\flat(V)\right)(p)\\ &= \sum (X^\flat(R(V, E_j)E_j) - (\nabla_{E_j}\nabla_{E_j}X^\flat)(V))(p)\\ &= \sum (\langle R(V, E_j)E_j, X\rangle - \langle \nabla_{E_j}\nabla_{E_j}X, V\rangle)(p)\\ &= (\mathcal{R}(V, X) - \langle \operatorname{tr} \nabla^2 X, V\rangle)(p)\end{aligned}$$

By 5.2, ∇X is self-adjoint if and only if $dX^\flat = 0$. Since $\delta X^\flat = -\operatorname{div} X$ (4.3, 4.7), the result follows from 4.9iv.

5.5 **Corollary** Suppose X is a harmonic vector field on a compact Riemannian manifold (M, g). If $\mathcal{R}(X, X) \geq 0$, then X is parallel and $\mathcal{R}(X, X) = 0$. Thus there exists no nonzero harmonic field on a compact Riemannian manifold of positive Ricci curvature, so if a compact manifold M admits a metric of positive Ricci curvature, then $H^1(M; \mathbb{R}) = 0$.

PROOF Apply corollary 4.20 and the Hodge theorem.

Since S^n has constant curvature 1 if $n \geq 2$, it has positive Ricci curvature. Projective space $\mathbb{R}P^n$ (3.4*e*) has the same properties since the projection map $S^n \to \mathbb{R}P^n$ is a Riemannian covering; related examples are the lens spaces (see 7.18), which admit S^3 as a Riemannian covering manifold.

In chap. 6 we shall see that every compact, connected Lie group G admits an invariant metric of nonnegative sectional curvature; if in addition $H^1(G; \mathbb{R}) = 0$, then the Ricci curvature is constant positive (it should be noted [Wal] that Lie groups generally do not admit invariant metrics of positive sectional curvature). Examples of compact manifolds which admit metrics of positive Ricci curvature but for which no known metric has positive sectional curvature are products of spheres of dimensions > 1, and many homogeneous spaces (see 6.62). Further examples are some exotic spheres in high dimensions [Cgr] [Her], and in dimensions 7 and 15 [Na] [P1]; the last examples were also found by L. Bérard Bergery (unpublished).

Exercise Find two linearly independent harmonic fields on the 2-torus with the usual embedding into $\mathbb{R}^3$.

KILLING FIELDS

5.6 A vector field X on a Riemannian manifold (M, g) is a Killing field if and only if (3.52) $\mathscr{L}_X g = 0$. In particular, by 5.2 a Killing field is divergence-free and ∇X is skew-symmetric, that is (Killing's equation), $\langle \nabla_U X, V \rangle + \langle U, \nabla_V X \rangle = 0$, $U, V \in \mathfrak{X}M$.

5.7 **Proposition** If X is a Killing field on a Riemannian manifold (M, g), then $\langle \mathrm{tr}\, \nabla^2 X, \cdot \rangle + \mathscr{R}(X, \cdot) = 0$. If in addition M is compact and $\mathscr{R}(X, X) \leq 0$, then X is parallel and $\mathscr{R}(X, X) = 0$.

PROOF Fix $p \in M$ and an adapted moving frame $\{E_j\}$ near p. Given $V \in \mathfrak{X}M$ with $(\nabla V)_p = 0$, at the point p

$$\begin{aligned}
&\langle \mathrm{tr}\, \nabla^2 X, V \rangle + \mathscr{R}(X, V) \\
&\quad = \sum E_j \langle \nabla_{E_j} X, V \rangle + \langle \nabla_{E_j} \nabla_V X - \nabla_V \nabla_{E_j} X, E_j \rangle \\
&\quad = \sum (-E_j \langle E_j, \nabla_V X \rangle + E_j \langle \nabla_V X, E_j \rangle) - V \,\mathrm{div}\, X = 0
\end{aligned}$$

Now apply 4.20, which assumes the equality $\langle \mathrm{tr}\, \nabla^2 X, \cdot \rangle + \mathscr{R}(X, \cdot) = 0$.

5.8 **Corollary** A compact Riemannian manifold (M, g) of negative Ricci curvature has a finite isometry group.

PROOF By the Myers-Steenrod theorem (3.50), the isometry group of (M, g) is compact; if it were infinite, it would contain a nontrivial 1-parameter group of isometries of M. This would generate, by definition 3.51, a nonzero Killing field on M.

In Riemann surface theory [Sp] it is seen that a compact Riemann surface of genus greater than 1 has the hyperbolic plane H^2 as its universal Riemann covering surface. It can be shown that the deck transformations act on H^2 by isometries, so the quotient space inherits a Riemannian metric of sectional curvature -1 (just as $\mathbb{R}P^n$ inherited a metric of curvature 1 from S^n in 3.4*e*). Similarly, for each n a compact quotient of hyperbolic space H^n by a discrete group of isometries admits a metric of sectional curvature -1. Therefore all these spaces have negative Ricci curvature.

A Riemannian homogeneous space of negative Ricci curvature is noncompact, for by definition (3.4*d*), G acts on G/H by isometries, so each element of $\mathfrak{g}$ determines a Killing field on G/H (see 1.63*c*).

5.9 **Lemma** Let X be a vector field on a Riemannian manifold (M, g). With respect to the extension of g to a metric on the vector bundle $T^*M \otimes T^*M$ as in 3.6 and 3.9,

$$\|\mathscr{L}_X g\|^2 = 2\|\nabla X\|^2 + 2\,\mathrm{tr}(\nabla X \circ \nabla X) \in C^\infty M$$

PROOF If $\{\epsilon_j\}$ is an orthonormal basis for M_p, $p \in M$, then by the Ricci identity and the fact that ∇ is torsion-free,

$$\begin{aligned}\|\mathscr{L}_X g\|^2(p) &= \sum_{i,j} ((\mathscr{L}_X g)(\epsilon_i, \epsilon_j))^2 = \sum_{i,j} (\langle \nabla_{\epsilon_i} X, \epsilon_j\rangle + \langle \epsilon_i, \nabla_{\epsilon_j} X\rangle)^2 \\ &= 2 \sum_i (\langle \nabla_{\epsilon_i} X, \nabla_{\epsilon_i} X\rangle + \langle \epsilon_i, \nabla_{\nabla_{\epsilon_i} X} X\rangle) \\ &= 2(\|\nabla X\|^2 + \mathrm{tr}(\nabla X \circ \nabla X))(p)\end{aligned}$$

Exercise Fill in the details in the following alternative proof of the lemma. By 5.2, $(\nabla_U X)^\flat = \iota_U(\frac{1}{2}\mathscr{L}_X g + \frac{1}{2}\, dX^\flat)$ for all $U \in \mathfrak{X}M$, so since symmetric transformations are orthogonal to skew-symmetric transformations (3.10),

$$\begin{aligned}\|\nabla X\|^2 &= \tfrac{1}{4}\|\mathscr{L}_X g\|^2 + \tfrac{1}{4}\|dX^\flat\|^2 \\ &= \tfrac{1}{4}\|\mathscr{L}_X g\|^2 + \tfrac{1}{2}\|\nabla X\|^2 - \tfrac{1}{2}\,\mathrm{tr}(\nabla X \circ \nabla X)\end{aligned}$$

5.10 **Proposition** For each vector field X on a compact Riemannian manifold (M, g),

(i) $\displaystyle\int_M (\mathscr{R}(X, X) + \mathrm{tr}(\nabla X \circ \nabla X) - (\mathrm{div}\, X)^2) = 0$

(ii) $\displaystyle\int_M (\mathscr{R}(X, X) + \langle \mathrm{tr}\, \nabla^2 X, X\rangle + \tfrac{1}{2}\|\mathscr{L}_X g\|^2 - (\mathrm{div}\, X)^2) = 0$

PROOF Direct calculation using an adapted moving frame near $p \in M$ (4.4) shows that at p,

$$\begin{aligned}\mathscr{R}(X, X) + \mathrm{tr}(\nabla X \circ \nabla X) &= \mathrm{div}\, \nabla_X X - X(\mathrm{div}\, X) \\ &= \mathrm{div}\, \nabla_X X - \mathrm{div}((\mathrm{div}\, X)X) + (\mathrm{div}\, X)^2 \\ &= \mathrm{div}(\nabla_X X - (\mathrm{div}\, X)X) + (\mathrm{div}\, X)^2\end{aligned}$$

since $\operatorname{div} fX = Xf + f \operatorname{div} X$ for $f \in C^\infty M$. Identity (i) then follows from Green's theorem, and (ii) follows from (i), 5.9, and 4.18.

5.11 **Corollary** A vector field X on a compact Riemannian manifold (M, g) is a Killing field if and only if

$$\operatorname{div} X = \mathcal{R}(X, \cdot) + \langle \operatorname{tr} \nabla^2 X, \cdot \rangle = 0$$

PROOF $\Rightarrow$: This was proved in 5.6 and 5.7.

$\Leftarrow$: By 5.10ii, $\mathcal{L}_X g = 0$, so X is Killing by 3.52.

5.12 **Proposition** A vector field X on a compact Riemannian manifold (M, g) is parallel if and only if it is Killing and harmonic.

PROOF By 5.2, $\langle \nabla_U X, V \rangle = \frac{1}{2}(\mathcal{L}_X g)(U, V) + \frac{1}{2}\, dX^\flat(U, V)$. But $\mathcal{L}_X g = 0$ if and only if X is Killing (3.52), while $dX^\flat = 0$ for divergence-free X if and only if (5.4) X is harmonic; finally, a Killing field has divergence zero.

Exercise Where was compactness used in the proof?

5.13 **Proposition** Let X be a Killing field on a compact Riemannian manifold (M, g). For each harmonic form μ on M, $\mathcal{L}_X \mu = 0$. In particular if μ is a harmonic 1-form and M is connected, then $\mu(X)$ is a constant function, while if Y is a harmonic vector field, then $\langle X, Y \rangle$ is constant.

PROOF If $\{\psi_t\}$ is the 1-parameter group of X, then since isometries and Δ commute (4.9ii), $\psi_t^* \mu$ is harmonic for all t, and therefore $\mathcal{L}_X \mu = d/dt|_0 \psi_t^* \mu$ is harmonic. But $\mathcal{L}_X \mu = d\iota_X \mu + \iota_X\, d\mu = d\iota_X \mu$ since a harmonic form is closed. By the Hodge theorem (4.12), the only harmonic form on M which is cohomologous to zero is zero.

If μ is a harmonic 1-form, then $d\iota_X \mu = \mathcal{L}_X \mu - \iota_X\, d\mu = 0$.

Exercise (Bochner) Show that if $X \in \mathfrak{X}M$ is Killing and $Y \in \mathfrak{X}M$ is harmonic, then $\langle X, Y \rangle$ is a harmonic function. Outline: $\Delta\langle X, Y \rangle = -\operatorname{div}(d\langle X, Y \rangle) = -2\langle \nabla X, \nabla Y \rangle = 0$ by the exercise in 3.10.

CONFORMAL FIELDS

5.14 **Definition** (cf. 4.11) Riemannian metrics g_1 and g_2 on a manifold M are said to be *conformally equivalent* if there exists $\varphi \in C^\infty M$ such that $g_1 = \varphi \cdot g_2$. A *conformal manifold* is a manifold endowed with a conformal equivalence class of Riemannian metrics.

Let g_1 and g_2 be conformally equivalent metrics on M. For each $p \in M$, g_1 and g_2 induce the same angle measure on M_p; conversely, an angle measure determines a conformal equivalence class of inner products on M_p. Thus a conformal structure on M is equivalent to the specification of an angle measure on each tangent space such that the angle measure depends differentiably on the basepoint in M: if X and Y are vector fields, then the angle between X and Y is C^∞.

5.15 **Proposition** If g and $\tilde{g}$ are conformally equivalent metrics on M, $\tilde{g} = e^{2\varphi}g$, $\varphi \in C^\infty M$, then for $X, Y \in \mathfrak{X}M$,

$$\tilde{\nabla}_X Y = \nabla_X Y + X(\varphi)Y + Y(\varphi)X - \langle X, Y\rangle \nabla\varphi$$
$$\widetilde{\operatorname{div}}\, X = \operatorname{div} X + nX\varphi$$

Proof The result follows by direct computation from formula 3.28(3-1) for the Levi-Civita connection, and the definition of div X (4.2).

5.16 **Definition** A map between conformal manifolds is *conformal* if it preserves the angle measure; if representative Riemannian metrics g_j are chosen on M_j, then $f\colon M_1 \to M_2$ is *conformal* if f^*g_2 is conformally equivalent to g_1. A conformal diffeomorphism of a conformal manifold M is called a *conformal transformation* (or *conformorphism*) of M. A *conformal vector field* is a vector field whose local 1-parameter group consists of local conformal maps.

An isometric map of Riemannian manifolds is a conformal map of the underlying conformal manifolds; a Killing field on a Riemannian manifold is a conformal field on the underlying conformal manifold.

By a theorem of Liouville [Kr: 68.2], a conformal transformation of $\mathbb{R}^n$ maps a hypersphere or a hyperplane to a hypersphere or a hyperplane if $n \geq 3$.

5.17 **Definition** The *linear conformal group of* $\mathbb{R}^n$ is the Lie group $CO(n) = O(n) \times \mathbb{R}_{>0}$. It is a Lie subgroup of the *affine conformal group* $\mathbb{R}^n \times CO(n)$ of $\mathbb{R}^n$ (cf. 2.111).

The Lie algebra $\mathfrak{co}(n) = \mathfrak{o}(n) \times \mathbb{R}$ of $CO(n)$ consists of the matrices $A + tI$, $A \in \mathfrak{o}(n)$ and $t \in \mathbb{R}$, where I is the identity matrix.

Exercise Calculate the Lie algebra of the affine conformal group.

5.18 **Proposition** The following statements are equivalent for a vector field X on a Riemannian manifold (M, g):

(i) X is a conformal field.

(ii) $\mathscr{L}_X g = 2hg$ for some $h \in C^\infty M$.

(iii) $\mathscr{L}_X g = \dfrac{2 \operatorname{div} X}{n} g$

(iv) $\nabla X - (\operatorname{div} X/n)I$ is a section of the orthogonal algebra bundle $\mathfrak{o}(TM)$.

(v) ∇X is a section of the *conformal algebra bundle*

$$\mathfrak{co}(TM) := \{L \in \mathfrak{gl}(TM) \mid L = A + tI, \text{ where } A \in \mathfrak{o}(TM) \text{ and } t \in \mathbb{R}\}$$

PROOF If X is conformal, its 1-parameter group is given by $(\varphi_t^* g)_p = e^{2f(t,p)} g_p$ for some $f \in C^\infty(\mathbb{R} \times M)$; set

$$h(p) := \frac{\partial}{\partial t}\bigg|_0 f(t, p)$$

Conversely, let $k(u, v) = \langle u, v\rangle/(\|u\| \; \|v\|)$ be the cosine of the angle between nonzero vectors u and v at each point of M. Fix $p \in M$ and $u, v \in M_p - \{0\}$. Extend u and v to vector fields U and V near p such that $U \circ \varphi_t(p) = \varphi_{t*}u$ and $V \circ \varphi_t(p) = \varphi_{t*}v$; it follows that $[X, U]_p = 0$, so $\nabla_U X = \nabla_X U$ at p, and similarly with V. Furthermore,

$$\begin{aligned}\frac{d}{dt}\bigg|_0 \langle \varphi_{t*}u, \varphi_{t*}v\rangle &= X_p\langle U, V\rangle \\ &= (\langle \nabla_X U, V\rangle + \langle U, \nabla_X V\rangle)(p) \\ &= (\langle \nabla_U X, V\rangle + \langle U, \nabla_V X\rangle)(p) \\ &= (\mathscr{L}_X g)(u, v)\end{aligned}$$

Therefore $d/dt|_0 \, k(\varphi_{t*}u, \varphi_{t*}v)$ equals

$$\frac{(\mathscr{L}_X g)(u, v)}{\|u\| \; \|v\|} - \frac{\langle u, v\rangle((\mathscr{L}_X g)(u, u)\|v\|/\|u\| + \|u\|(\mathscr{L}_X g)(v, v)/\|v\|)}{2\|u\|^2\|v\|^2}$$

This is zero if $\mathscr{L}_X g = 2hg$ for some $h \in C^\infty M$. Applying this result to the vectors $\varphi_{s*}u$ and $\varphi_{s*}v$ we see that $d/dt|_s \, k(\varphi_{t*}u, \varphi_{t*}v) = 0$ for all s. Thus $k(\varphi_{t*}u, \varphi_{t*}v)$ is constant, and the local 1 parameter group of X preserves angles.

Now suppose X is conformal; for all $U, V \in \mathfrak{X}M$,

$$\langle \nabla_U X, V\rangle + \langle U, \nabla_V X\rangle = (\mathscr{L}_X g)(U, V) = \langle hU, V\rangle + \langle U, hV\rangle$$

where $\mathscr{L}_X g = 2hg$; thus $\langle(\nabla X - hI)U, V\rangle + \langle U, (\nabla X - hI)V\rangle = 0$, so $\nabla X - hI \in \Gamma\mathfrak{o}(TM)$, and $\nabla X \in \Gamma\mathfrak{co}(TM)$. Furthermore, the trace-free symmetric part of ∇X is zero, and $\nabla X - (\text{div } X/n)I$ has trace zero by 5.2, so $(\text{div } X)/n = h$.

Conversely, if $\nabla X - (\text{div } X/n)I \in \Gamma\mathfrak{o}(TM)$, then

$$\mathscr{L}_X g = \frac{2 \text{ div } X}{n} g$$

by the identity $(\mathscr{L}_X g)(U, V) = \langle \nabla_U X, V \rangle + \langle U, \nabla_V X \rangle$.

5.19 It can be proved [KN: note 11] that the group of all conformal transformations of a connected Riemannian manifold of dimension n is a Lie group of dimension less than or equal to $\binom{n+2}{2}$; its Lie algebra is isomorphic to the Lie algebra of complete conformal fields on M. Here we will look at the case where the maximum dimension is attained.

The diffeomorphism $\sigma: S^n - \{e_0\} \to \mathbb{R}^n$ which sends $(x^0, \ldots, x^n)$ to $(1 - x^0)^{-1}(x^1, \ldots, x^n)$ is called *stereographic projection*. Let $\bar{g}$ be the usual Riemannian metric on $\mathbb{R}^n$, and g the usual Riemannian metric on S^n; it follows that

$$(\sigma^* \bar{g})_p = \frac{1}{(1 - x^0)^2} g_p \qquad p = (x^0, \ldots, x^n) \in S^n - \{e_0\}$$

Thus $\sigma^* \bar{g}$ is conformally equivalent to g, so $S^n - \{e_0\}$ and $\mathbb{R}^n$ are Riemannian representatives of the same conformal manifold. For each Killing field X on $\mathbb{R}^n$ there is a unique conformal field V on S^n such that the restriction of V to $S^n - \{e_0\}$ is σ-related to X, and $V_{e0} = 0$. More generally, every element of the affine conformal algebra $\mathbb{R}^n \rtimes \mathfrak{co}(n)$ determines a conformal field on $\mathbb{R}^n$, and therefore on S^n because stereographic projection allows us to define an action of the affine conformal group on S^n: $gp := \sigma^{-1} g \sigma p$ for $p \neq e_0$, and $ge_0 = e_0$. Obviously this action is not transitive. To get all the conformal fields on S^n we need a conformal action by a larger group. *Exercise:* The fields above are C^∞.

Let Φ be the quadratic form of signature $(n + 1, 1)$ such that

$$\Phi(x) := x^{1\,2} + \cdots + x^{n2} - 2x^0 x^{n+1} \qquad x \in \mathbb{R}^{n+2}$$

The matrix of Φ is

$$S = \begin{bmatrix} 0 & 0 & -1 \\ 0 & I & 0 \\ -1 & 0 & 0 \end{bmatrix}$$

By a rotation of $\mathbb{R}^{n+2}$, Φ is equivalent to the standard Lorentz inner product on $\mathbb{R}^{n+2}$ from 3.41, so the group $O(n + 1, 1) := \{h \in GL(n + 2, \mathbb{R}) \mid {}^t h S h = S\}$ which leaves Φ invariant will be called the *Lorentz group*.

Since Φ is homogeneous and of maximal rank away from $0 \in \mathbb{R}^{n+2}$, its zero set projects down to an embedded submanifold of $\mathbb{R}P^{n+1}$, which will be denoted by $\Phi^{-1}(0)$. The general linear group $GL(n+2, \mathbb{R})$ acts naturally on $\mathbb{R}P^{n+1}$ by matrix multiplication on the representative points in $\mathbb{R}^{n+2} - \{0\}$ (cf. [W: 3.65(c)]), that is, g times the class of $v \in \mathbb{R}^{n+1} - \{0\}$ is the class of the point gv; as a subgroup of $GL(n+2, \mathbb{R})$, $O(n+1, 1)$ also acts on $\mathbb{R}P^{n+1}$. By definition, it maps the set $\Phi^{-1}(0)$ into itself; this is a C^∞ action of $O(n+1, 1)$ on the manifold $\Phi^{-1}(0)$. Furthermore, this action on $\Phi^{-1}(0)$ is transitive; in particular, the map $\pi\colon O(n+1, 1) \to \Phi^{-1}(0)$, $h \mapsto h \cdot [1:0:\cdots:0]$, is surjective. *Proof:* Embed $\mathbb{R}^n$ into the Lie algebra $\mathfrak{o}(n+1, 1)$ of $O(n+1, 1)$ as the subspace

$$\mathfrak{g}_{-1} := \left\{ \begin{bmatrix} 0 & 0 & 0 \\ {}^t v & 0 & 0 \\ 0 & v & 0 \end{bmatrix} \;\middle|\; v \in \mathbb{R}^n \right\}$$

where $v \in \mathbb{R}^n$ is written as a row vector. Under the exponential map,

$$\exp v = \exp \begin{bmatrix} 0 & 0 & 0 \\ {}^t v & 0 & 0 \\ 0 & v & 0 \end{bmatrix} = \begin{bmatrix} 1 & 0 & 0 \\ {}^t v & I & 0 \\ c & v & 1 \end{bmatrix} \qquad c = \frac{\|v\|^2}{2}$$

Thus for all $v \in \mathbb{R}^n \cong \mathfrak{g}_{-1}$,

$$\pi \circ \exp v = \text{coset in } \mathbb{R}P^{n+1} \text{ of } \begin{bmatrix} 1 & 0 & 0 \\ v & I & 0 \\ c & v & 1 \end{bmatrix} \begin{bmatrix} 1 \\ 0 \\ \vdots \\ 0 \end{bmatrix} = [1 : v : c]$$

so $\pi \circ \exp$ maps $\mathfrak{g}_{-1}$ onto $\Phi^{-1}(0) - \{[0:\cdots:0:1]\}$. But $[0:\cdots:0:1]$ is $\pi(S)$, S being the matrix of Φ. *QED.*

If $\sigma\colon S^n - \{e_0\} \to \mathbb{R}^n$ is the stereographic projection as before, then $k := \pi \circ \exp \circ \sigma$ maps $S^n - \{e_0\}$ to $\Phi^{-1}(0) - \{[0:\cdots:0:1]\}$; set $k(e_0) := [0:\cdots:0:1]$. The formulas

$$k(x^0, \ldots, x^n) = [2 - 2x^0 : 2x^1 : \cdots : 2x^n : 1 + x^0] \in \Phi^{-1}(0)$$

$$k^{-1}[m : w : s] = \left(\frac{2s - m}{2s + m}, \frac{2w}{2s + m} \right) \in S^n \qquad \Phi(m, w, s) = 0$$

show that k maps S^n diffeomorphically onto $\Phi^{-1}(0) \subset \mathbb{R}P^{n+1}$. Embedded into $\mathbb{R}P^{n+1}$ this way, S^n is called the *Möbius space* [Ko2].

Thus $S^n = \Phi^{-1}(0)$ is a homogeneous space. The isotropy group of the point $-e_0 \cong [1:0:\cdots:0]$ is

$$H = \left\{ \begin{bmatrix} a^{-1} & u & b \\ 0 & h & \zeta \\ 0 & 0 & a \end{bmatrix} \in O(n+1, 1) \right\}$$

where $\zeta = ah^t u$ and $b = a\|u\|^2/2$; let $\mathfrak{h}$ be the Lie algebra of H.

Exercises Verify that the diagonal subgroup $O(n) \times \mathbb{R}_{>0} = CO(n)$ of H acts on $\Phi^{-1}(0)$ the same way that $CO(n)$ acts on S^n via stereographic projection. Show that the action of the affine conformal group on S^n induced by stereographic projection agrees with the action of

$$\left\{ \begin{bmatrix} a^{-1} & 0 & 0 \\ \zeta & h & 0 \\ b & v & a \end{bmatrix} \in O(n+1, 1) \right\} \cong \mathbb{R}^n \times CO(n)$$

on $\Phi^{-1}(0)$. Verify that $O(n+1, 1)$ acts conformally on S^n.

Each element $X \in \mathfrak{o}(n+1, 1)$ of the Lie algebra of $O(n+1, 1)$ determines a vector field X^* (1.63c) on $O(n+1, 1)/H \cong S^n$ by

$$X^*_{\not p g} := \frac{d}{dt}\bigg|_0 \not p(\exp(tX)g) \qquad g \in O(n+1, 1)$$

where $\not p := k^{-1} \circ \pi$ is the projection map from $O(n+1, 1)$ onto S^n. Let us now calculate the integral curves of these vector fields for typical $X \in \mathfrak{o}(n+1, 1)$.

If X belongs to $\mathfrak{o}(n) \subset \mathfrak{co}(n) \subset \mathfrak{h}$, then X^* is just a Killing field on S^n; its flow consists of rotations of S^n. A typical flow of this type is shown in fig. 5-1. For future reference, the values of X^*, ∇X^*, and div X^* at the point $p = e_0$ are also listed; here $X_*|_p$ is the restriction to $S^n_p \subset \mathbb{R}^{n+1}_p$ of the tangent map induced by $X \in \mathfrak{o}(n) \subset \mathfrak{o}(n+1)$.

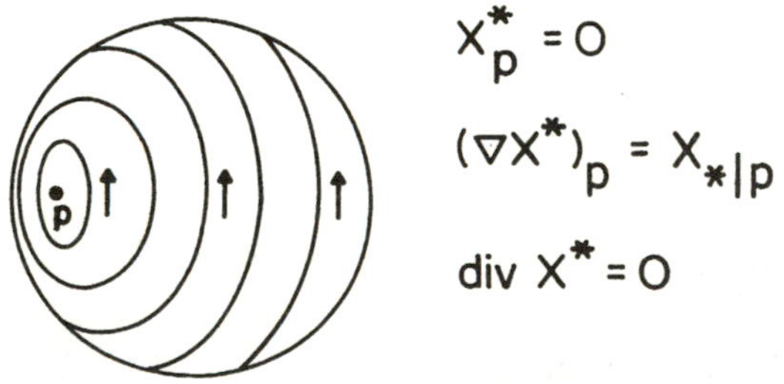

Figure 5-1 The integral curves of a rotation vector field X^* on S^2.

Next we look at the flow of the vector field X^* on S^n such that

$$X = \begin{bmatrix} -1 & 0 & 0 \\ 0 & 0 & 0 \\ 0 & 0 & 1 \end{bmatrix} \in \mathfrak{o}(n+1, 1)$$

Since X belongs to the isotropy algebra $\mathfrak{h}$ of the point $-e_0$ in S^n, the integral curve of X^* through $-e_0$ is constant. For this reason we look at the integral curve γ of X^* through a different point, say through a point $(0, v)$ on the "equator" in S^n, where $v \in \mathbb{R}^n$ has length 1. If v is written as a row vector, then

$$\gamma(s) = \exp \begin{bmatrix} -s & 0 & 0 \\ 0 & 0 & 0 \\ 0 & 0 & s \end{bmatrix} \cdot (0, v) = \begin{bmatrix} e^{-s} & 0 & 0 \\ 0 & I & 0 \\ 0 & 0 & e^s \end{bmatrix} \cdot \not{p} \begin{bmatrix} 1 & 0 & 0 \\ {}^t v & I & 0 \\ \frac{1}{2} & v & 1 \end{bmatrix}$$

$$= \not{p}\left(\begin{bmatrix} e^{-s} & 0 & 0 \\ 0 & I & 0 \\ 0 & 0 & e^s \end{bmatrix} \begin{bmatrix} 1 & 0 & 0 \\ {}^t v & I & 0 \\ \frac{1}{2} & v & 1 \end{bmatrix} \right) = k^{-1}\pi \begin{bmatrix} e^{-s} & 0 & 0 \\ {}^t v & I & 0 \\ \frac{1}{2}e^s & e^s v & e^s \end{bmatrix}$$

$$= k^{-1}[e^{-s} : v : \tfrac{1}{2}e^s] = \left(\frac{e^s - e^{-s}}{e^s + e^{-s}}, \frac{2v}{e^s + e^{-s}} \right)$$

The curve $\gamma(s) = (\tanh s, (\operatorname{sech} s)v)$ is a parametrization of the great circle in S^n through the points $-e_0 = \lim_{s \to -\infty} \gamma(s)$, $(0, v) = \gamma(0)$, and $e_0 = \lim_{s \to \infty} \gamma(s)$. The flow of X^* is shown in fig. 5-2.

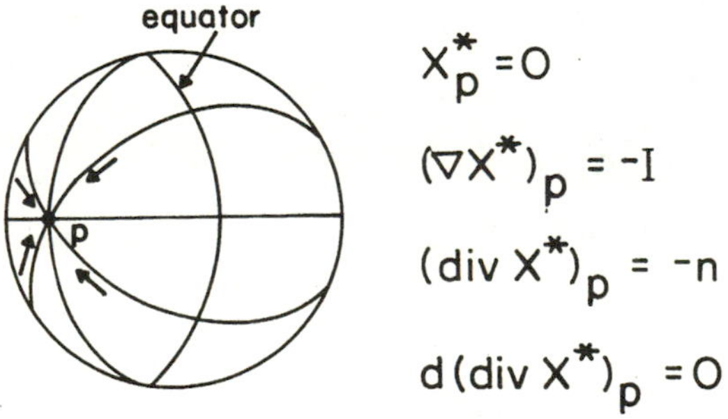

Figure 5-2 The integral curves of a great circle vector field on S^2.

To calculate the covariant differential of X^* near $p = e_0$, we use 5.15 and work in the conformally equivalent manifold $\mathbb{R}^n$. More precisely, let $\tau: S^n - \{-e_0\} \to \mathbb{R}^n$ be the stereographic projection from the point $-e_0$, $\tau(x^0, \ldots, x^n) := (1 + x^0)^{-1}(x^1, \ldots, x^n)$ (cf. 1.12j). For all

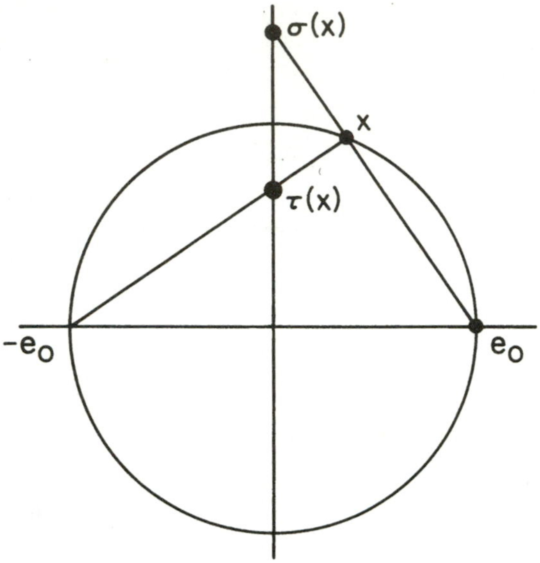

s, $\tau \circ \gamma(s) = e^{-s}v$, so the vector field on $\mathbb{R}^n$ which is τ-related to X^* on $S^n - \{-e_0\}$ is minus the identity, $-I_q = -q$. Let $\tilde{g}$ be the metric on $\mathbb{R}^n$ such that $\tau^*\tilde{g}$ is the usual metric g on $S^n - \{-e_0\}$. Since the manifolds $(S^n - \{-e_0\}, g)$ and $(\mathbb{R}^n, \tilde{g})$ are isometric, it suffices to calculate $\tilde{\nabla} I$ near 0 in $\mathbb{R}^n$. But $\tilde{g}$ is conformally equivalent to the standard metric $\bar{g}$ on $\mathbb{R}^n$, so we can calculate $\tilde{\nabla}$ using 5.15.

Specifically, for all $q \in \mathbb{R}^n$,

$$\tilde{g}_q = \left(\frac{2}{\|q\|^2 + 1}\right)^2 \bar{g}_q$$

so $\tilde{g} = e^{2\varphi}\bar{g}$, where $\varphi(q) := \ln 2 - \ln(\|q\|^2 + 1)$. The gradient of φ with respect to $\bar{g}$ is

$$(\nabla\varphi)_q = \frac{-2}{\|q\|^2 + 1} q$$

Hence, given a tangent vector (q, v) on $\mathbb{R}^n$,

$$\begin{aligned}\tilde{\nabla}_v I &= \nabla_v I + v(\varphi)I_q + I_q(\varphi)v - \langle v, I_q\rangle(\nabla\varphi)_q \\ &= v + v(\varphi)q + q(\varphi)v - \langle v, q\rangle(\nabla\varphi)_q \\ &= \frac{1 - \|q\|^2}{1 + \|q\|^2} v\end{aligned}$$

so that $\tilde{\nabla} I$ is symmetric. In particular, by 5.18 (or alternatively by 5.15), $(\operatorname{div} I)(q) = n(1 - \|q\|^2)/(1 + \|q\|^2)$.

Next let us look at a vector field Y^*, where

$$Y = \begin{bmatrix} 0 & 0 & 0 \\ {}^t w & 0 & 0 \\ 0 & w & 0 \end{bmatrix}$$

is in the abelian subalgebra $\mathfrak{g}_{-1}$ of $\mathfrak{o}(n+1, 1)$, $w \in \mathbb{R}^n$; for simplicity, assume $\|w\| = 1$. Let v be a unit vector in $\mathbb{R}^n$, and let γ be the integral curve of the vector field X^* associated with v above: $\gamma(t) = (\tanh t, (\operatorname{sech} t)v)$. Denote by β_t the integral curve of Y^* through the point $\gamma(t)$. As above,

$$\beta_t(s) = k^{-1}\left[e^{-t} : se^{-t}w + v : \frac{s^2 e^{-t} + 2s\langle v, w\rangle + e^t}{2}\right]$$

$$= \left(\frac{(s^2 - 1)e^{-t} + 2s\langle v, w\rangle + e^t}{(s^2 + 1)e^{-t} + 2s\langle v, w\rangle + e^t}, \frac{2se^{-t}w + 2v}{(s^2 + 1)e^{-t} + 2s\langle v, w\rangle + e^t}\right)$$

from which one obtains Y^* by differentiation. The integral curves of Y^* are shown in fig. 5-3. Each integral curve is a parametrization of the intersection of S^n with a hyperplane through the point e_0 which contains the tangent vector (e_0, w); in particular, each nonconstant integral curve parametrizes a circle in $\mathbb{R}^{n+1}$.

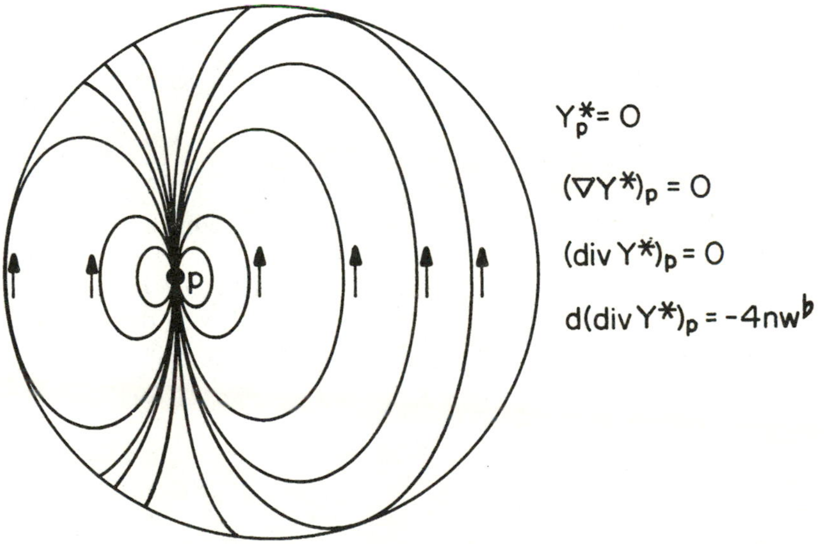

Figure 5-3 The integral curves of the vector field Y^* determined by w.

Exercise Show that Y^* is τ-related to the vector field U on $\mathbb{R}^n$ such that $U_q = \|q\|^2 w - 2\langle w, q\rangle q$, $q \in \mathbb{R}^n$. Study the vector field Z^* determined by the element

$$Z = \begin{bmatrix} 0 & z & 0 \\ 0 & 0 & {}^t z \\ 0 & 0 & 0 \end{bmatrix} \in \mathfrak{o}(n+1, 1) \qquad z \in \mathbb{R}^n$$

The dimension of $O(n+1, 1)$ is $\binom{n+2}{2}$. To specify a conformal vector field on S^n, the following linearly independent initial data at a point $p \in S^n$ are necessary and sufficient: $X_p \in S^n_p$, $(\nabla X - (\operatorname{div} X/n)I)_p \in \mathfrak{o}(S^n_p)$, $(\operatorname{div} X)(p) \in \mathbb{R}$, and $d(\operatorname{div} X)_p \in S^{n*}_p$. So far we have the following examples for $p = e_0 \in S^n$:

Fig. 5-1: $X_p = 0$, $(\nabla X - (\operatorname{div} X/n)I)_p \neq 0$, $\operatorname{div} X \equiv 0$

Fig. 5-2: $X_p = 0$, $(\nabla X - (\operatorname{div} X/n)I)_p = 0$, $(\operatorname{div} X)(p) \neq 0$, $d(\operatorname{div} X)_p = 0$

Fig. 5-3: $X_p = 0$, $(\nabla X - (\operatorname{div} X/n)I)_p = 0$, $(\operatorname{div} X)(p) = 0$, $d(\operatorname{div} X)_p \neq 0$

For an example with $X_p \neq 0$, and all the other data zero, rotate S^n about any axis perpendicular to e_0. *Exercise:* Show what happens for fields generated by the abelian subalgebra

$$\mathfrak{g}_1 = \left\{ \begin{bmatrix} 0 & z & 0 \\ 0 & 0 & {}^t z \\ 0 & 0 & 0 \end{bmatrix} \in \mathfrak{o}(n+1, 1) \,\middle|\, z \in \mathbb{R}^n \right\}$$

5.20 More generally, the conditions stated above are sufficient to specify a conformal vector field on any Riemannian manifold. In general, not all the data are necessary.

Theorem (Lelong-Ferrand, Obata) Let (M, g) be a compact Riemannian manifold of dimension $n \geq 3$. Denote by C the identity component of the Lie group of conformal transformations of (M, g). Either there exists a positive C^∞ function f on M such that C lies in the isometry group of the metric fg on M, or else (M, g) is conformally diffeomorphic to the standard n-sphere S^n.

Proof See [Le] or [Ob].

5.21 Corollary If the identity component of the conformal group of (M, g) is noncompact, then (M, g) is conformally S^n in dimensions $n \geq 3$.

Exercise (Bochner-Yano) Prove the following weaker version of 5.20: Let (M, g) be a compact Riemannian manifold of dimension $n \geq 2$. If X is a conformal field on M for which $\mathscr{R}(X, X) \leq 0$, then X is a Killing field. (Hint: Use 5.10, 5.12, and 5.18.)

AFFINE FIELDS

An affine field on a manifold (M, ∇) is a vector field whose local 1-parameter group consists of local affine maps. By 2.122, X is affine on (M, ∇) if and only if $[X, \nabla_U V] = \nabla_{[X, U]} V + \nabla_U [X, V]$, $U, V \in \mathfrak{X}M$; if ∇ is torsion-free, this is equivalent to the identity $\nabla_U \nabla X = R(U, X)$.

5.22 **Proposition** (Yano) If X is an affine field on a Riemannian manifold (M, g), then div X is locally constant, and $\mathscr{R}(X, Y) + \langle \operatorname{tr} \nabla^2 X, Y \rangle = 0$, $Y \in \mathfrak{X}M$. If M is compact, then X is a Killing field.

Proof Let $\{E_j\}$ be an adapted moving frame near $p \in M$. For $v \in M_p$,

$$v(\operatorname{div} X) = \sum \langle (\nabla_v \nabla X)(E_j), E_j \rangle = \sum \langle R(v, X)E_j, E_j \rangle = 0$$

so div X is closed. If M is compact, then div $X = 0$ by 4.14.

Next, since X is affine, $\nabla_{E_j} \nabla X = R(E_j, X)$ for all j; this implies that $\mathscr{R}(X, Y) + \langle \operatorname{tr} \nabla^2 X, Y \rangle = 0$ for all $Y \in \mathfrak{X}M$. Now apply 5.11.

Hano [Hn] replaced the compactness assumption by the assumption that the length of X is bounded; the proof is also found in [KN: VI.3.8].

Exercises

1. Construct an affine field on $\mathbb{R}^n$ which is not a Killing field.
2. Prove that the affine transformation group of a compact Riemannian manifold of negative Ricci curvature is discrete.

PROJECTIVE FIELDS

5.23 **Definition** A map $f: (\bar{M}, \bar{\nabla}) \to (M, \nabla)$ of manifolds with torsion-free connections is called *projective* if for each geodesic γ of $\bar{\nabla}$, $f \circ \gamma$ is a reparametrization of a geodesic of ∇: there must exist a strictly increasing C^∞ function h on some open interval such that $f \circ \gamma \circ h$ is a ∇-geodesic. Linear connections $\bar{\nabla}$ and ∇ on M are *projectively equivalent* if the identity map of M is projective. A *projective transformation of* (M, ∇) is a diffeomorphism which is projective.

For example, every affine map is projective. The terminology comes from classical projective geometry: a projective transformation of $\mathbb{R}P^n$ was defined to be a map which takes (unparametrized) projective lines to projective lines; the image of a geodesic in $\mathbb{R}P^n$ under such a map must usually be reparametrized before it becomes a geodesic.

5.24 **Proposition** (Weyl) Torsion-free linear connections $\bar{\nabla}$ and ∇ on M are projectively equivalent if and only if for each ∇-geodesic γ there is a C^∞ function f on the domain of γ such that $\bar{\nabla}_D \dot{\gamma} = f \cdot \dot{\gamma}$; this holds if and only if there is a 1-form α on M such that

$$\bar{\nabla}_U V - \nabla_U V = \alpha(U)V + \alpha(V)U \qquad U, V \in \mathfrak{X}M$$

Proof Let ∇ and $\bar{\nabla}$ be projectively equivalent. Let γ be a ∇-geodesic and h a function such that $\gamma \circ h$ is a $\bar{\nabla}$-geodesic. Since

$$\bar{\nabla}_D \dot{\widehat{\gamma \circ h}} = h'' \cdot \dot{\gamma} \circ h + (h')^2 \cdot \bar{\nabla}_{D \circ h} \dot{\gamma}$$

it suffices to set $f := -[h''/(h')^2] \circ h^{-1}$.

Conversely, given a C^∞ function f on an open interval, the differential equation $f \circ h = -h''/(h')^2$ has a solution: if F is a function such that $F' = f$ and C_1 and C_2 are appropriate constants, h can be obtained by solving the equation

$$\int_0^h e^{F(s) - C_1}\, ds = t + C_2$$

The result is C^∞ and strictly increasing as required.

Now suppose that such an f is given for each ∇-geodesic γ. Since $\bar{\nabla}$ and ∇ are torsion-free, the connection difference tensor field $\mathscr{D}(U, V) = \bar{\nabla}_U V - \nabla_U V$ is symmetric by the proof of 2.101. By hypothesis, $\mathscr{D}(\dot{\gamma}, \dot{\gamma}) = \bar{\nabla}_D \dot{\gamma} = f \cdot \dot{\gamma}$ for each ∇-geodesic γ; in particular, $\mathscr{D}(v, v)$ is a multiple of v for each $v \in TM$. Define a C^∞ function α on TM by $\alpha(v)v := \mathscr{D}(v, v)$ for $v \neq 0$, and $\alpha(0_p) := 0$ for all $p \in M$. Since α is fiberwise homogeneous, it is fiberwise linear by the proof of 2.33(3); thus $\alpha \in A^1 M$, and $\mathscr{D}(U, V) = \alpha(U)V + \alpha(V)U$.

The converse is obvious.

Exercise Show that a complete linear connection can be projectively equivalent to an incomplete one.

5.25 **Corollary** A diffeomorphism φ of M is a projective transformation of (M, ∇) if and only if there exists a 1-form α on M such that

$$\nabla_U V = \varphi_*^{-1} \nabla_U \varphi_* V + \alpha(U)V + \alpha(V)U \qquad U, V \in \mathfrak{X}M$$

5.26 **Definition** A vector field X on (M, ∇) is *projective* if its local 1-parameter group consists of local projective maps.

Exercise Affine fields are projective; construct projective fields on $\mathbb{R}^n$, S^n, and $\mathbb{R}P^n$ which are not affine.

5.27 **Proposition** A vector field X on (M, ∇), where ∇ is torsion-free, is projective if and only if there exists $\mu \in A^1M$ such that

$$(\mathscr{L}_X\nabla)(U, V) = \mu(U)V + \mu(V)U \qquad U, V \in \mathfrak{X}M$$

In this case, $(\nabla_U \nabla X)(V) = R(U, X)V + \mu(U)V + \mu(V)U$.

PROOF If the local 1-parameter group of X is $\{\varphi_t\}$, then by 5.25, there are 1-forms α_t on M such that

$$\nabla_U V = \varphi_{-t*}\nabla_U \varphi_{t*} V + \alpha_t(U)V + \alpha_t(V)U$$

As in the proof of 2.123 one obtains $\mu(v) = \lim_{t\to 0} \alpha_t(v)/t$ for all $v \in TM$.

5.28 **Proposition** Let X be a projective vector field on a Riemannian manifold (M, g), and let μ be the 1-form for X as in 5.5; then

$$\mathscr{R}(X, \cdot) + \langle \operatorname{tr} \nabla^2 X, \cdot \rangle = 2\mu \qquad \text{and} \qquad d(\operatorname{div} X) = (n+1)\mu$$

PROOF The proof is similar to the proof of Yano's theorem (5.22).

5.29 **Theorem** (Couty [Co]) If X is a projective vector field on a compact Riemannian manifold (M, g) such that $\mathscr{R}(X, X) \le 0$, then $\mathscr{R}(X, X) = \operatorname{div} X = 0$, and X is parallel.

PROOF By 5.28, $\mathscr{R}(X, Y) + \langle \operatorname{tr} \nabla^2 X, Y \rangle = 2Y(\operatorname{div} X)/(n+1)$ for $Y \in \mathfrak{X}M$. The metric dual $X^\flat = \langle X, \cdot \rangle$ is a 1-form on M; its exterior derivative $dX^\flat$ can be thought of either as a section of $\Lambda^2 T^*M$ or as a section of $\otimes^2 T^*M$. The norm squared of $dX^\flat$ as a section of $\otimes^2 T^*M$ can be calculated with respect to an adapted moving frame near $p \in M$; as in the exercise in 5.9, this yields

$$\tfrac{1}{2}\|dX^\flat\|^2 = \tfrac{1}{2}\sum_{i,j} dX^\flat(E_i, E_j)^2 = \|\nabla X\|^2 - \operatorname{tr}(\nabla X \circ \nabla X)$$

Thus by 4.18,

$$\tfrac{1}{2}\Delta\|X\|^2 + \frac{2}{n+1}\operatorname{div}(\operatorname{div}(X)X)$$

$$= -\langle \operatorname{tr} \nabla^2 X, X\rangle - \|\nabla X\|^2 + \frac{2}{n+1}\operatorname{div}(\operatorname{div}(X)X)$$

$$= \mathscr{R}(X, X) + \frac{2}{n+1}(\operatorname{div} X)^2 - \tfrac{1}{2}\|dX^\flat\|^2 - \operatorname{tr}(\nabla X \circ \nabla X)$$

By Green's theorem (4.14, 4.15), the integral over M of this expression is zero. Add to this the integral from 5.10i; the result is the equation

$$\int_M \left(2\mathscr{R}(X, X) - \tfrac{1}{2}\|dX^\flat\|^2 - \frac{n-1}{n+1}(\operatorname{div} X)^2\right) = 0$$

Since $\mathscr{R}(X, X) \leq 0$, each term in the integrand is zero. Thus ∇X is symmetric by 5.2, so since $\operatorname{div} X = 0$, X is harmonic by 5.4.

The identities $\operatorname{div} X = 0$ and $\mathscr{R}(X, \cdot) + \langle \operatorname{tr} \nabla^2 X, \cdot\rangle = 0$ imply X is a Killing field by 5.11. But by 5.12, a harmonic Killing field on a compact manifold is parallel.

5.30 To close the chapter, here is a theorem of Weyl which implies that a Riemannian manifold is equivalent to a fixed projective structure on a conformal manifold.

Theorem (Weyl [Wy2]) The conformal and projective structures of a Riemannian manifold completely determine the metric.

5.31 For an interesting extension of the Bochner technique [YB], which was the main tool used in this chapter, see [Wu].

CHAPTER

SIX

LIE GROUPS

The reader will already have noticed that Lie groups provide convenient examples of manifolds whose geometry can be studied relatively easily; this makes Lie groups (and their natural generalization—homogeneous spaces) useful as spaces on which to test many geometric conjectures. This chapter will present more of the elementary geometry of Lie groups and homogeneous spaces. For further information the reader is referred to [Che*1*], [DaZ], [He], [HS], [KN], [Mi*4*], and [Sw].

The following notation will be used throughout: if a Lie algebra $\mathfrak{g}$ is the vector space direct sum of subspaces $\mathfrak{h}_j$, then write $\mathfrak{g} = \mathfrak{h}_1 \oplus \cdots \oplus \mathfrak{h}_k$; if in addition the $\mathfrak{h}_j$ are known to be ideals of $\mathfrak{g}$, then write $\mathfrak{g} = \mathfrak{h}_1 \times \cdots \times \mathfrak{h}_k$.

A NEGATIVE CURVATURE EXAMPLE

In 3.38*b*, the reader saw that the sectional curvature of a bi-invariant metric on a Lie group is nonnegative. Before studying bi-invariant metrics in more detail we shall see, by way of contrast, an example of a Lie group with negative sectional curvature.

6.1 Fix an integer $n > 1$, and define

$$G := \left\{ \begin{bmatrix} 1 & 0 \\ v & sI \end{bmatrix} \in GL(n, \mathbb{R}) \,\middle|\, v \in \mathbb{R}^{n-1}, s > 0 \right\}$$

Here v is written as a column vector, and I is the $(n-1) \times (n-1)$ identity matrix.

The Lie algebra $\mathfrak{g}$ of G consists of the $n \times n$ matrices of the form

$$\begin{bmatrix} 0 & 0 \\ v & sI \end{bmatrix} \qquad v \in \mathbb{R}^{n-1},\ s \in \mathbb{R}$$

Set

$$E_j := \begin{bmatrix} 0 & 0 \\ e_j & 0 \end{bmatrix} \qquad j < n, \qquad E_n := \begin{bmatrix} 0 & 0 \\ 0 & I \end{bmatrix}$$

where $\{e_j\}$ is the usual orthonormal basis for $\mathbb{R}^{n-1}$. For $i, j < n$, $[E_i, E_j] = 0$, and $[E_n, E_j] = E_j$.

Exercise Show that for $n = 2$, this is (up to isomorphism) the only nonabelian 2-dimensional Lie algebra over $\mathbb{R}$.

Fix the left-invariant metric on G such that $\{E_j\}$ is an orthonormal basis for $\mathfrak{g}$ (3.4*b*). This metric is not bi-invariant on G because ad_{E_j} is not skew-symmetric for $j < n$. By 3.34*b*, the Levi-Civita connection satisfies the identity

$$\langle \nabla_X Y, Z \rangle = \tfrac{1}{2}(\langle [X, Y], Z \rangle - \langle [Y, Z], X \rangle + \langle [Z, X], Y \rangle)$$

for $X, Y, Z \in \mathfrak{g}$. Thus for $i, j < n$, and for all k,

$$\nabla_{E_i} E_j = \delta_{ij} E_n \qquad \nabla_{E_i} E_n = -E_i \qquad \nabla_{E_n} E_k = 0$$

Hence $K_{E_i \wedge E_j} = \langle R(E_i, E_j)E_j, E_i \rangle = -1$, so $(G, \langle\ ,\ \rangle)$ is a space of constant negative sectional curvature [Mi4].

Let us look at the geodesics of G. If f is a curve in $\mathbb{R}^{n-1}$ and h is a positive function, then

$$\gamma := \begin{bmatrix} 1 & 0 \\ f & hI \end{bmatrix}$$

is a curve in G; if the jth component of f is the real-valued function f_j, then

$$\dot{\gamma} = \sum_{j=1}^{n-1} \frac{f'_j}{h} E_j|_\gamma + \frac{h'}{h} E_n|_\gamma$$

so
$$\nabla_D \dot{\gamma} = \sum_{j=1}^{n-1} \left(\left(\frac{f'_j}{h}\right)' E_j|_\gamma + \frac{f'_j}{h} \nabla_{\dot{\gamma}} E_j \right) + \left(\frac{h'}{h}\right)' E_n|_\gamma + \frac{h'}{h} \nabla_{\dot{\gamma}} E_n$$

$$= \sum_j \left(\left(\frac{f'_j}{h}\right)' - \frac{f'_j h'}{h^2} \right) E_j|_\gamma + \left(\sum_j \left(\frac{f'_j}{h}\right)^2 + \left(\frac{h'}{h}\right)' \right) E_n|_\gamma$$

Thus γ is a geodesic if and only if $h > 0$ and

$$hf''_j - 2f'_j h' = 0 \qquad j = 1, \ldots, n-1$$

$$hh'' - (h')^2 + \sum_{j=1}^{n-1} (f'_j)^2 = 0$$

Given $r \in \mathbb{R}_{>0}$, $s \in \mathbb{R}$, $u \in \mathbb{R}^{n-1}$, and $v \in \mathbb{R}^{n-1} - \{0\}$, the geodesic γ in G such that

$$\gamma(0) = \begin{bmatrix} 1 & 0 \\ u & rI \end{bmatrix} \qquad \dot{\gamma}(0) = \begin{bmatrix} 0 & 0 \\ v & sI \end{bmatrix}$$

is

$$\gamma(t) := \begin{bmatrix} 1 & 0 \\ u + \dfrac{r}{\|v\|^2}[s + ar \cdot \tanh(at + \log b)]v & \dfrac{r^2 a}{\|v\|} \operatorname{sech}(at + \log b)I \end{bmatrix}$$

where $\qquad a = \dfrac{\sqrt{s^2 + \|v\|^2}}{r} \qquad b = \dfrac{\sqrt{s^2 + \|v\|^2} - s}{\|v\|}$

Exercise Find γ if $v = 0$. (Hint: Look at E_n.)

Since all the geodesics of G are defined for all time, G is complete (actually, completeness already follows from 3.55*d*). Thus G is a complete, simply connected, Riemannian manifold of constant sectional curvature -1. Such a manifold must be isometric to hyperbolic space H^n [BC: 9, prob. 27]. In fact, the Levi-Civita exponential map $f: G_e \to G$ (which does not equal the Lie group exponential map exp: $\mathfrak{g} \to G$) is a diffeomorphism, so if we fix $p \in H^n$ and an isometry $h: G_e \to H^n_p$, then $\exp_p \circ h \circ f^{-1}$ maps G to H^n. This map is onto since H^n is complete, and in fact it is an isometry (see the proof of 3.49). In particular, for all n, hyperbolic space H^n is isometric to a Lie group with a left-invariant metric.

We shall see in 6.24 that the situation is completely different for spheres: only S^1 and S^3 can be realized as Lie groups (this is already true when the spheres are considered just as C^∞ manifolds, regardless of any question about the metrics).

6.2 The Lie algebra $\mathfrak{g}$ of the Lie group G from 6.1 is not abelian, but the subalgebra of $\mathfrak{g}$ spanned by all $[U, V]$, $U, V \in \mathfrak{g}$, is abelian.

Definition The *derived algebra* of a Lie algebra $\mathfrak{g}$ is the subalgebra $[\mathfrak{g}, \mathfrak{g}]$ of $\mathfrak{g}$ spanned by all $[U, V]$, $U, V \in \mathfrak{g}$. If some term in the sequence $\mathfrak{g}$, $[\mathfrak{g}, \mathfrak{g}]$, $[[\mathfrak{g}, \mathfrak{g}], [\mathfrak{g}, \mathfrak{g}]]$, $\ldots$, is zero, then $\mathfrak{g}$ is called a *solvable Lie algebra*.

For example, every abelian Lie algebra is solvable, and the Lie algebra $\mathfrak{g}$ from 6.1 is a nonabelian solvable Lie algebra. More generally, Heintze [Hz] has shown that if a connected Lie group admits an invariant metric of nonpositive curvature, then its Lie algebra is solvable (cf. [Je*1*]).

BI-INVARIANT METRICS

6.3 Denote the center [W: 3.49] of a Lie group G by $Z(G)$, and the center of $\mathfrak{g}$ by $\mathfrak{z}(\mathfrak{g})$; when no confusion can ensue, simply write Z and $\mathfrak{z}$.

6.4 If $\tilde{G}$ is the universal covering space of a connected Lie group G, then [W: 3.25] $\tilde{G}$ can be made into a Lie group such that the covering map is a Lie group homomorphism.

Proposition If $\varphi: \hat{G} \to G$ is a covering homomorphism of connected Lie groups, then ker φ is a normal subgroup of the center $Z(\hat{G})$. In particular, $\pi_1(G)$ is commutative, and $\hat{G}/Z(\hat{G})$ is isomorphic to $G/Z(G)$.

PROOF If $g \in \ker \varphi$ and $h \in G$, then since φ is a homomorphism, $\varphi(gh) = \varphi(h) = \varphi(hg)$. Thus $gh = hg$ for all h near the identity element e of G because φ is a covering map. Since G is generated by an open neighborhood of e, $gh = hg$ for all $h \in G$.

In case $\hat{G}$ is the universal covering group $\tilde{G}$ of G, ker φ is the deck group, which is isomorphic to $\pi_1(G)$; hence $\pi_1(G)$ is abelian.

Finally, for all groups G covered by $\tilde{G}$,

$$G/Z(G) \cong \frac{\tilde{G}/\pi_1(G)}{Z(\tilde{G})/\pi_1(G)} \cong \tilde{G}/Z(\tilde{G})$$

6.5 By [W: 3.50], the kernel of the adjoint representation Ad of G on $\mathfrak{g}$ is $Z(G)$, so the image $\mathrm{Ad}(G) \subset GL(\mathfrak{g})$ is isomorphic to $G/Z(G)$. It follows from 6.4 that the subgroup $\mathrm{Ad}(G)$ of $GL(\mathfrak{g})$ is independent of the choice of connected Lie group G with prescribed Lie algebra $\mathfrak{g}$; hence $\mathrm{Ad}(G)$ depends only on the intrinsic structure of $\mathfrak{g}$, which motivates the notation in the following definition.

Definition The *adjoint group* of a Lie algebra $\mathfrak{g}$ is $\mathrm{Int}(\mathfrak{g}) := \mathrm{Ad}(G) \subset GL(\mathfrak{g})$, where G is any connected Lie group with Lie algebra $\mathfrak{g}$.

It follows that the Lie algebra of $\mathrm{Int}(\mathfrak{g})$ is the Lie subalgebra $\mathrm{ad}(\mathfrak{g})$ of $\mathfrak{gl}(\mathfrak{g})$; this is sometimes called the *adjoint algebra of* $\mathfrak{g}$.

6.6 **Definition** The *automorphism group* of a Lie algebra $\mathfrak{g}$ is $\mathrm{Aut}(\mathfrak{g}) := \{g \in GL(\mathfrak{g}) \mid g$ is a Lie algebra homomorphism: $g[U, V] = [gU, gV]$, $U, V \in \mathfrak{g}\}$. A *derivation* of $\mathfrak{g}$ is an element $L \in \mathfrak{gl}(\mathfrak{g})$ such that $L[U, V] = [LU, V] + [U, LV]$ for $U, V \in \mathfrak{g}$; let $\partial(\mathfrak{g})$ be the Lie algebra of derivations of $\mathfrak{g}$ under commutation.

6.7 **Proposition** The Lie algebra of $\mathrm{Aut}(\mathfrak{g})$ is $\partial(\mathfrak{g})$, and $\mathrm{Aut}(\mathfrak{g})$ is a closed subgroup of $GL(\mathfrak{g})$. Furthermore, $\mathrm{Int}(\mathfrak{g})$ is a Lie subgroup of $\mathrm{Aut}(\mathfrak{g})$.

PROOF A point $L \in \mathfrak{gl}(\mathfrak{g})$ belongs to the Lie algebra of $\mathrm{Aut}(\mathfrak{g})$ if and only if $\exp tL \in \mathrm{Aut}(\mathfrak{g})$ for all $t \in \mathbb{R}$, that is, if and only if $(\exp tL) \cdot [U, V] = [(\exp tL)U, (\exp tL)V]$, $t \in \mathbb{R}$, $U, V \in \mathfrak{g}$. Differentiation yields $L[U, V] = [LU, V] + [U, LV]$, so $L \in \partial(\mathfrak{g})$. Conversely, if $L \in \partial(\mathfrak{g})$, then $L^2[U, V] = L([LU, V] + [U, LV]) = [L^2U, V] + 2[LU, LV] + [U, L^2V]$; by induction, $L^n[U, V] = \sum_{j=0}^{n} \binom{n}{j}[L^jU, L^{n-j}V]$. Thus

$$
\begin{aligned}
(\exp tL)[U, V] &= \sum_{n=0}^{\infty} \frac{t^n L^n[U, V]}{n!} \\
&= \sum_{n=0}^{\infty} \sum_{k=0}^{n} [t^k L^k U/k!, t^{n-k} L^{n-k} V/(n-k)!] \\
&= \sum_{k=0}^{n} \sum_{n=k}^{\infty} [t^k L^k U/k!, t^{n-k} L^{n-k} V/(n-k)!] \\
&= \sum_{k, m=0}^{\infty} [t^k L^k U/k!, t^m L^m V/m!] \\
&= [(\exp tL)U, (\exp tL)V]
\end{aligned}
$$

Given $U, V \in \mathfrak{g}$, the inequality $g[U, V] \neq [gU, gV]$ is an open condition on $g \in GL(\mathfrak{g})$; thus $\mathrm{Aut}(\mathfrak{g}) = \bigcap_{U, V \in \mathfrak{g}} \{g \in GL(\mathfrak{g}) \mid g[U, V] = [gU, gV]\}$ is the intersection of closed subsets of $GL(\mathfrak{g})$, and is therefore closed in $GL(\mathfrak{g})$.

Finally, by the Jacobi identity, $\mathrm{ad}_U[V, W] = [\mathrm{ad}_U V, W] + [U, \mathrm{ad}_V W]$, $U, V, W \in \mathfrak{g}$, so $\mathrm{ad}(\mathfrak{g}) \subset \partial(\mathfrak{g})$.

The work in the last few sections is summarized by the following representation diagram for a connected Lie group G with Lie algebra $\mathfrak{g}$:

$$\begin{array}{ccccccc}
\mathfrak{g} & \xrightarrow{\text{ad}} & \text{ad}(\mathfrak{g}) & \longrightarrow & \partial(\mathfrak{g}) & \longrightarrow & \mathfrak{gl}(\mathfrak{g}) \\
\downarrow \text{exp} & & \downarrow \text{exp} & & \downarrow \text{exp} & & \downarrow \text{exp} \\
G & \xrightarrow[\text{Ad}]{} & \text{Int}(\mathfrak{g}) & \longrightarrow & \text{Aut}(\mathfrak{g}) & \longrightarrow & GL(\mathfrak{g})
\end{array}$$

In addition, if $g \in \text{Aut}(\mathfrak{g})$ and $U \in \mathfrak{g}$, then $[gU, V] = [gU, gg^{-1}V] = g[U, g^{-1}V]$, $V \in \mathfrak{g}$, so $\text{ad}_{gU} = g \circ \text{ad}_U \circ g^{-1}$ for all $g \in \text{Aut}(\mathfrak{g})$.

6.8 Now we return to the study of bi-invariant metrics on Lie groups.

Proposition A Lie group G admits a bi-invariant metric if and only if $\text{Int}(\mathfrak{g}) = \text{Ad}(G)$ has compact closure in $GL(\mathfrak{g})$.

PROOF $\Rightarrow$: By 3.4*b*, $\text{Int}(\mathfrak{g})$ lies in the compact subgroup $O(\mathfrak{g}, \langle\ ,\ \rangle)$ of $GL(\mathfrak{g})$, where $\langle\ ,\ \rangle$ is a bi-invariant metric on G.

$\Leftarrow$: Let H be the closure of $\text{Int}(\mathfrak{g})$ in $GL(\mathfrak{g})$, and fix a left-invariant volume form ω on H [W: 4.11]. Fix an inner product $\langle\ ,\ \rangle$ on $\mathfrak{g}$, and define $\widetilde{\langle U, V\rangle} := \int_H \langle hU, hV\rangle \omega_h$, $U, V \in \mathfrak{g}$. The bilinear form $\widetilde{\langle\ ,\ \rangle}$ is an inner product on $\mathfrak{g}$, and hence by 3.4*b* is a left-invariant metric on G. It suffices to show (3.4*b*) that $\widetilde{\langle\ ,\ \rangle}$ is invariant under $\text{Int}(\mathfrak{g}) \subset GL(\mathfrak{g})$. Fix $U, V \in \mathfrak{g}$, and define $f \in C^\infty H$ by $f(h) := \langle hU, hV\rangle$. The integral of f over H is right-invariant since H is compact [W: 4.11]. Thus for all $k \in H$,

$$\widetilde{\langle kU, kV\rangle} = \int_H \langle hkU, hkV\rangle \omega_h = \int_H f(hk)\omega_h$$
$$= \int_H f \circ R_k = \int_H f = \widetilde{\langle U, V\rangle}$$

For example, every compact Lie group admits a bi-invariant metric because $\text{Int}(\mathfrak{g}) = \text{Ad}(G)$ is compact; compactness of G is not necessary for the existence of a bi-invariant metric, as the abelian Lie group $\mathbb{R}^n$ shows.

In 6.17 we shall see that the existence of a bi-invariant metric actually implies that $\text{Int}(\mathfrak{g})$ is compact. The direct attempt at a proof using the proposition and the closure of $\text{Aut}(\mathfrak{g})$ in $GL(\mathfrak{g})$ from 6.7 fails because $\text{Int}(\mathfrak{g})$ is in general not closed in $\text{Aut}(\mathfrak{g})$.

Exercise Show that the conformal group $CO(n)$ from 5.17 admits a bi-invariant metric.

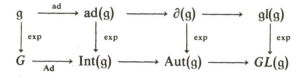

6.9 **Definitions** A Lie algebra is *simple* if it is nonabelian and has no proper ideals. A Lie algebra is *semisimple* if it is the direct sum of simple ideals. A Lie group is *simple* or *semisimple* if its Lie algebra is simple or semisimple, respectively.

A simple Lie algebra is clearly semisimple.

It can be shown [Ho: XVI.2.1] that every simply connected Lie group is the semidirect product of a maximal solvable Lie subgroup and a maximal semisimple Lie subgroup.

6.10 **Proposition** If $\mathfrak{g} = \mathfrak{g}_1 \times \cdots \times \mathfrak{g}_k$ is a simple ideal decomposition of a semisimple Lie algebra $\mathfrak{g}$, then:

(i) This ideal decomposition is unique (up to the order of the factors).
(ii) Every ideal of $\mathfrak{g}$ is the direct sum (unique up to order) of some of the $\mathfrak{g}_j$.
(iii) Every nonzero ideal of $\mathfrak{g}$ is the direct sum complement of an ideal of $\mathfrak{g}$.
(iv) The derived algebra $[\mathfrak{g}, \mathfrak{g}] = \mathfrak{g}$.
(v) The center $\mathfrak{z}(\mathfrak{g}) = 0$.
(vi) $\dim \mathfrak{g} \geq 3$.

PROOF Properties (i) to (v) are left as exercises. The only Lie algebras of dimension less than 3 are the abelian 1-dimensional Lie algebra, the abelian 2-dimensional Lie algebra, and the solvable 2-dimensional Lie algebra from 6.1 (up to isomorphism) [Ja: I.4].

6.11 Given U and V in a Lie algebra $\mathfrak{g}$, ad_U and ad_V are endomorphisms of $\mathfrak{g}$, and therefore so is $\mathrm{ad}_U \circ \mathrm{ad}_V$. From elementary linear algebra, $\mathrm{tr}(\mathrm{ad}_U \circ \mathrm{ad}_V)$ is well-defined, and $\mathrm{tr}(\mathrm{ad}_U \circ \mathrm{ad}_V) = \mathrm{tr}(\mathrm{ad}_V \circ \mathrm{ad}_U)$.

Definition The *Killing form of a Lie algebra* $\mathfrak{g}$ is the symmetric bilinear form $B(U, V) := \mathrm{tr}(\mathrm{ad}_U \circ \mathrm{ad}_V)$, $U, V \in \mathfrak{g}$.

6.12 **Lemma** If B is the Killing form of a Lie algebra $\mathfrak{g}$, then:

(i) $B([U, V], W) = B(U, [V, W])$, and therefore B is $\mathrm{Int}(\mathfrak{g})$-invariant.
(ii) More generally, B is $\mathrm{Aut}(\mathfrak{g})$-invariant.
(iii) If $\mathfrak{h}$ is an ideal in $\mathfrak{g}$, then so is $\mathfrak{h}^\perp := \{U \in \mathfrak{g} \mid B(U, V) = 0,\ V \in \mathfrak{h}\}$.
(iv) The restriction of B to an ideal $\mathfrak{h}$ of $\mathfrak{g}$ is the Killing form of the Lie algebra $\mathfrak{h}$.

PROOF (i): Since ad is a Lie algebra homomorphism,

$$\begin{aligned}\operatorname{tr}(\operatorname{ad}_{[U,V]} \circ \operatorname{ad}_W) &= \operatorname{tr}(\operatorname{ad}_U \circ \operatorname{ad}_V \circ \operatorname{ad}_W - \operatorname{ad}_V \circ \operatorname{ad}_U \circ \operatorname{ad}_W)\\ &= \operatorname{tr}(\operatorname{ad}_U \circ \operatorname{ad}_V \circ \operatorname{ad}_W - \operatorname{ad}_U \circ \operatorname{ad}_W \circ \operatorname{ad}_V)\\ &= \operatorname{tr}(\operatorname{ad}_U \circ \operatorname{ad}_{[V,W]})\end{aligned}$$

The Killing form B is then $\operatorname{Int}(\mathfrak{g})$-invariant by an argument similar to that in 3.4*b*; this also follows from (iii).

(ii): By the comment at the end of 6.7, for $g \in \operatorname{Aut}(\mathfrak{g})$, $U, V \in \mathfrak{g}$,

$$\begin{aligned}\operatorname{tr}(\operatorname{ad}_{gU} \circ \operatorname{ad}_V) &= \operatorname{tr}(g \circ \operatorname{ad}_U \circ g^{-1} \circ \operatorname{ad}_V) = \operatorname{tr}(g^{-1} \circ \operatorname{ad}_V \circ g \circ \operatorname{ad}_U)\\ &= \operatorname{tr}(\operatorname{ad}_{g^{-1}V} \circ \operatorname{ad}_U).\end{aligned}$$

(iii): For $U \in \mathfrak{h}$, $V \in \mathfrak{g}$, and $W \in \mathfrak{h}^\perp$, $B(U, [V, W]) = 0$ by (i).

(iv): Complete a basis for $\mathfrak{h}$ to a basis for $\mathfrak{g}$, so that

$$\operatorname{ad}_U \circ \operatorname{ad}_V = \begin{bmatrix} A & B \\ 0 & 0 \end{bmatrix} \begin{bmatrix} C & D \\ 0 & 0 \end{bmatrix} = \begin{bmatrix} AC & AD \\ 0 & 0 \end{bmatrix}$$

which has trace equal to $\operatorname{tr}(AC) = \operatorname{tr}(\overline{\operatorname{ad}}_U \circ \overline{\operatorname{ad}}_V)$, $\overline{\operatorname{ad}}$ being the adjoint representation of $\mathfrak{h}$ on $\mathfrak{h}$.

6.13 Examples

(*a*) The Killing form of $\mathfrak{gl}(n, \mathbb{R})$ is $B(U, V) = 2n \operatorname{tr}(UV) - 2 \operatorname{tr}(U) \operatorname{tr}(V)$. In this case, B is *degenerate*, that is, there is a nonzero U (for instance, $U = I$) in $\mathfrak{gl}(n, \mathbb{R})$ such that $B(U, V) = 0$ for all $V \in \mathfrak{gl}(n, \mathbb{R})$; furthermore, U spans an abelian ideal of $\mathfrak{gl}(n, \mathbb{R})$.

(*b*) The *special linear algebra* $\mathfrak{sl}(n, \mathbb{R}) := \{U \in \mathfrak{gl}(n, \mathbb{R}) \mid \operatorname{tr} A = 0\}$ is easily seen to be an ideal in $\mathfrak{gl}(n, \mathbb{R})$; in fact, $\mathfrak{sl}(n, \mathbb{R})$ is just the derived algebra of $\mathfrak{gl}(n, \mathbb{R})$. By 6.12iv, the Killing form of $\mathfrak{sl}(n, \mathbb{R})$ is therefore $B(U, V) = 2n \operatorname{tr}(UV)$, $U, V \in \mathfrak{sl}(n, \mathbb{R})$; it will be shown in 6.26 that B is nondegenerate.

6.14 Theorem Cartan's criterion for semisimplicity: A Lie algebra $\mathfrak{g}$ is semisimple if and only if its Killing form B is nondegenerate.

PROOF $\Rightarrow$: By 6.12iii, $\mathfrak{g}^\perp := \{U \in \mathfrak{g} \mid B(U, V) = 0, V \in \mathfrak{g}\}$ is an ideal of $\mathfrak{g}$. Since $\mathfrak{g}^\perp$ is abelian, it is zero by 6.10iii.

$\Leftarrow$: If $\mathfrak{g}$ is not simple, it contains a proper ideal $\mathfrak{h}$. The orthogonal complement $\mathfrak{h}^\perp$ of $\mathfrak{h}$, which is an ideal by 6.12iii, is also proper. It suffices to show that $\mathfrak{g} = \mathfrak{h} \times \mathfrak{h}^\perp$, and that $B|_{\mathfrak{h}}$ is nondegenerate, for then the result follows by induction.

Define $\psi\colon \mathfrak{g} \to \mathfrak{g}^*$ by $\psi U(V) := B(U, V)$; ψ is an isomorphism because B is nondegenerate (cf. the map $\flat$ from 3.8). But then $\psi(\mathfrak{h}) \subset \mathfrak{g}^*$ kills a subspace of $\mathfrak{g}$ of dimension equal to $\dim \mathfrak{g} -$

dim $\mathfrak{h}$, and $\mathfrak{h}^\perp = \{U \in \mathfrak{g} \mid \psi(\mathfrak{h})(U) = 0\}$ is the maximal subspace of $\mathfrak{g}$ killed by $\psi(\mathfrak{h})$; thus dim $\mathfrak{g} =$ dim $\mathfrak{h}$ + dim $\mathfrak{h}^\perp$.

The ideal $\mathfrak{h} \cap \mathfrak{h}^\perp$ of $\mathfrak{g}$ is abelian by the definition of $\mathfrak{h}^\perp$, the nondegeneracy of B, and 6.12i. Complete a basis for $\mathfrak{h} \cap \mathfrak{h}^\perp$ to a basis for $\mathfrak{g}$; given $U \in \mathfrak{h} \cap \mathfrak{h}^\perp$ and $V \in \mathfrak{g}$,

$$\mathrm{ad}_U \circ \mathrm{ad}_V = \begin{bmatrix} 0 & A \\ 0 & 0 \end{bmatrix} \begin{bmatrix} B & C \\ 0 & D \end{bmatrix} = \begin{bmatrix} 0 & AD \\ 0 & 0 \end{bmatrix}$$

which has trace zero for all $V \in \mathfrak{g}$. Since B is nondegenerate, U is then zero, and $\mathfrak{h} \cap \mathfrak{h}^\perp = \{0\}$.

Thus $\mathfrak{g} = \mathfrak{h} \times \mathfrak{h}^\perp$. By 6.12iv, the Killing form B' of $\mathfrak{h}$ is the restriction to $\mathfrak{h}$ of B, so if $B'(U, V) = 0$ for all $U, V \in \mathfrak{h}$, then for $U \in \mathfrak{h}$, $B(U, W) = 0$ for all $W \in \mathfrak{g}$, and $U = 0$; hence the Killing form of $\mathfrak{h}$ is nondegenerate, and similarly for that of $\mathfrak{h}^\perp$.

6.15 **Definition** A Lie algebra is said to be *compact* if it is the Lie algebra of some compact Lie group.

6.16 **Theorem** A Lie algebra $\mathfrak{g}$ is compact if and only if $\mathfrak{g} = \mathfrak{z} \times [\mathfrak{g}, \mathfrak{g}]$, where the ideal $[\mathfrak{g}, \mathfrak{g}]$ is semisimple (if nonzero) and compact.

PROOF $\Rightarrow$: Since $\mathfrak{g}$ is the Lie algebra of a compact Lie group G, $\mathrm{Int}(\mathfrak{g}) = \mathrm{Ad}(G)$ is compact, and therefore $\mathfrak{g}$ admits an $\mathrm{Int}(\mathfrak{g})$-invariant inner product $\langle\ ,\ \rangle$ by 6.8. Elementary linear algebra then implies that on the vector space level, $\mathfrak{g}$ is the direct sum of $\mathfrak{z}$ and its orthogonal complement $\mathfrak{z}^\perp$; since $\langle\ ,\ \rangle$ is invariant under $\mathrm{Int}(\mathfrak{g})$, $\mathfrak{z}^\perp$ is an ideal of $\mathfrak{g}$ as in 6.12iii, so $\mathfrak{g} = \mathfrak{z} \times \mathfrak{z}^\perp$.

Similarly, $\mathfrak{z}^\perp = \mathfrak{g}_1 \times \cdots \times \mathfrak{g}_k$, where each ideal $\mathfrak{g}_j$ of $\mathfrak{g}$ has no proper ideals. Since every element of $\mathfrak{z}^\perp$ commutes with $\mathfrak{z}$, the center of $\mathfrak{z}^\perp$ is trivial; this implies that each $\mathfrak{g}_j$ is simple, for if $\mathfrak{g}_j$ were not simple, then by 6.9 it would be abelian, and therefore lie in the center of $\mathfrak{z}^\perp$. Thus $\mathfrak{z}^\perp = \mathfrak{g}_1 \times \cdots \times \mathfrak{g}_k$ is the direct sum of simple ideals, and is therefore semisimple by definition. By the definition of $\mathfrak{z}$, $[\mathfrak{g}, \mathfrak{g}] = [\mathfrak{z}^\perp, \mathfrak{z}^\perp] = \mathfrak{z}^\perp$ (6.10iv), so $\mathfrak{g} = \mathfrak{z} \times [\mathfrak{g}, \mathfrak{g}]$, and $[\mathfrak{g}, \mathfrak{g}]$ is semisimple.

Finally, $[\mathfrak{g}, \mathfrak{g}] \cong \mathfrak{g}/\mathfrak{z} \cong \mathrm{ad}(\mathfrak{g})$ is compact by 6.5.

$\Leftarrow$: The center $\mathfrak{z}$ of $\mathfrak{g}$ is the Lie algebra of a torus, so $\mathfrak{g}$ is the Lie algebra of a product of compact Lie groups, and is therefore compact.

6.17 The next theorem summarizes the relationships among the various possible properties of a connected Lie group studied so far.

Theorem Let G be a connected Lie group with universal covering group $\tilde{G}$ and Lie algebra $\mathfrak{g}$. The following statements are equivalent:

(i) The Lie algebra $\mathfrak{g}$ is compact.

(ii) $\mathfrak{g}$ admits an $\mathrm{Int}(\mathfrak{g})$-invariant ($=\mathrm{Ad}(G)$-invariant) inner product.

(iii) $\mathfrak{g}$ admits an inner product for which $\mathrm{ad}_{\mathfrak{g}}$ is skew-symmetric.

(iv) $\mathrm{Int}(\mathfrak{g}) = \mathrm{Ad}(G)$ is relatively compact in $GL(\mathfrak{g})$.

(v) $\mathrm{Int}(\mathfrak{g}) = \mathrm{Ad}(G)$ is compact.

(vi) G (or equivalently, $\tilde{G}$) admits a bi-invariant metric.

(vii) G is the product of a vector group and a compact group.

(viii) $\tilde{G}$ is the product of a vector group and a semisimple compact group.

(ix) $G/Z(G)$ (or equivalently, $\tilde{G}/Z(\tilde{G})$) is compact.

(x) $\mathfrak{g} = \mathfrak{z} \times [\mathfrak{g}, \mathfrak{g}]$, where the ideal $[\mathfrak{g}, \mathfrak{g}]$ is compact.

(xi) $\mathfrak{g} = \mathfrak{z} \times [\mathfrak{g}, \mathfrak{g}]$, where $[\mathfrak{g}, \mathfrak{g}]$ is compact and semisimple (if nonzero).

Proof This follows from 6.4, 6.8, 6.16, and 3.4*b*.

A direct consequence of this theorem goes back to H. Weyl: if G is compact and semisimple, then so is $\tilde{G}$.

Exercise If $\mathfrak{g}$ is semisimple and compact, then $\mathrm{Int}(\mathfrak{g})$ is a compact subgroup of $GL(\mathfrak{g})$, and is therefore closed in $GL(\mathfrak{g})$. Prove that $\mathrm{Int}(\mathfrak{g})$ is closed in $GL(\mathfrak{g})$ for semisimple $\mathfrak{g}$ even without the compactness assumption. Proof outline: Since the Killing form B of $\mathfrak{g}$ is nondegenerate, for each derivation $D \in \partial(\mathfrak{g})$ the equality $B(\bar{D}, U) := \mathrm{tr}(D \circ \mathrm{ad}_U)$, $U \in \mathfrak{g}$, determines an element $\bar{D}$ of $\mathfrak{g}$. Since $D \in \partial(\mathfrak{g})$, $B(DU, V) = \mathrm{tr}(D \circ \mathrm{ad}_{[U,V]}) = B([\bar{D}, U], V)$ for all $U, V \in \mathfrak{g}$; therefore D equals $\mathrm{ad}_{\bar{D}}$, and $\partial(\mathfrak{g}) = \mathrm{ad}(\mathfrak{g})$. Now apply 6.7.

6.18 **Corollary** Assume G is connected; the Lie algebra $\mathfrak{g}$ is compact and semisimple if and only if the Killing form B of $\mathfrak{g}$ is negative definite. In this case, $-B$ is a bi-invariant metric on G, and G is compact.

Proof $\Rightarrow$: Let $\langle\ ,\ \rangle$ be an inner product on $\mathfrak{g}$ with respect to which ad_U is skew-symmetric for each $U \in \mathfrak{g}$; with respect to an orthonormal basis for $\mathfrak{g}$, each ad_U is a skew-symmetric matrix $[U_{ij}]$, so $B(U, U) = -\sum_{i,j} U_{ij}^2 \leq 0$, and equals zero if and only if $[U_{ij}] = 0$. In this case, $U \in \mathfrak{z}(\mathfrak{g})$; but $\mathfrak{z}(\mathfrak{g}) = 0$ since $\mathfrak{g}$ is semisimple.

$\Leftarrow$: Since $-B$ is an $\mathrm{Int}(\mathfrak{g})$-invariant inner product on $\mathfrak{g}$, $\mathfrak{g}$ is compact; semisimplicity of $\mathfrak{g}$ follows from the Cartan criterion.

6.19 **Definition** The *canonical metric* on a connected, compact, semisimple Lie group is minus the Killing form of its Lie algebra.

6.20 In 3.38b we saw that the sectional curvature of a bi-invariant metric on a Lie group is nonnegative. Now we calculate the Ricci curvature and estimate the eigenvalues of the curvature operator of the canonical metric on a compact, connected, semisimple Lie group.

Proposition With the canonical metric, a compact, connected, semisimple Lie group G is an Einstein manifold of Ricci curvature $\frac{1}{4}$.

PROOF Fix an orthonormal basis $\{E_j\}$ for $\mathfrak{g}$, and let $U, V \in \mathfrak{g}$. By 3.45 and 3.34b,

$$\mathscr{R}(U, V) = \sum \langle R(E_j, U)V, E_j\rangle = -\tfrac{1}{4}\sum \langle[[E_j, U], V], E_j\rangle$$
$$= -\tfrac{1}{4}\,\mathrm{tr}(\mathrm{ad}_U \circ \mathrm{ad}_V) = \tfrac{1}{4}\langle U, V\rangle$$

6.21 **Proposition** The eigenvalues of the curvature operator of a connected, compact, semisimple Lie group G lie between 0 and $\frac{1}{8}$.

PROOF Define $\alpha \in \mathrm{Hom}(\Lambda^2\mathfrak{g}, \mathfrak{g})$ to be the linear extension of the multiplication map $U \wedge V \mapsto [U, V]$. By 3.36 and 3.34$b$, $\langle R\sigma, \sigma\rangle$ equals $\|\alpha\sigma\|^2/4$ for $\sigma \in \Lambda^2\mathfrak{g}$. Let $\beta\colon \mathfrak{g} \to \Lambda^2\mathfrak{g}$ be the adjoint of α: $\langle \beta U, V \wedge W\rangle := \langle U, \alpha(V \wedge W)\rangle = \langle U, [V, W]\rangle$. With respect to an orthonormal basis $\{E_j\}$ for $\mathfrak{g}$, $\beta U = \sum_{i<j} \langle U, [E_i, E_j]\rangle \cdot E_i \wedge E_j$, $U \in \mathfrak{g}$. Thus

$$2\langle U, \alpha\beta V\rangle = 2\langle \beta U, \beta V\rangle = 2\sum_{i<j} \langle U, [E_i, E_j]\rangle\langle V, [E_i, E_j]\rangle$$
$$= \sum_i \langle [U, E_i], [V, E_i]\rangle = \langle U, V\rangle$$

so $\alpha\beta U = \frac{1}{2}U$, $U \in \mathfrak{g}$. Hence, if $\sigma = \beta U \in \Lambda^2\mathfrak{g}$, $U \in \mathfrak{g}$, then $\|\alpha\sigma\|^2 = \frac{1}{2}\langle U, \alpha\beta U\rangle = \frac{1}{2}\langle \beta U, \beta U\rangle = \frac{1}{2}\|\sigma\|^2$; since $\alpha|_{\beta(\mathfrak{g})^\perp} = 0$, it follows that $\langle R\sigma, \sigma\rangle = \frac{1}{4}\|\alpha\sigma\|^2 \leq \frac{1}{8}\|\sigma\|^2$.

6.22 **Theorem** The following statements are equivalent for a connected, compact Lie group G:

(i) G is semisimple.
(ii) $H_1(G; \mathbb{R}) = 0$.
(iii) $\pi_1(G)$ is finite.
(iv) The universal covering group $\tilde{G}$ is compact.
(v) The Lie algebra $\mathfrak{g}$ of G has trivial center.
(vi) The Killing form of $\mathfrak{g}$ is negative definite.

PROOF Statements (i), (iv), (v), and (vi) are equivalent by 6.17 and 6.18, while (iii) and (iv) are equivalent for arbitrary compact manifolds G, not just for Lie groups.

(ii) $\Leftrightarrow$ (iii): By the fundamental theorem of finitely generated abelian groups, $H_1(G; \mathbb{Z}) \cong k\mathbb{Z} \oplus \Gamma$, where k is a nonnegative integer, and Γ is a finite group. Since tensoring with $\mathbb{R}$ kills finite groups, $H_1(G; \mathbb{R}) \cong H_1(G; \mathbb{Z}) \otimes \mathbb{R} \cong k\mathbb{Z}$. Thus $H_1(G; \mathbb{Z})$ is finite if and only if $H_1(G; \mathbb{R}) = 0$. But $\pi_1(G)$ is abelian (6.4), so [Gr: 12.2] $H_1(G; \mathbb{Z}) \cong \pi_1(G)/[\pi_1(G), \pi_1(G)] = \pi_1(G)$.

Exercises Prove geometrically that (i) $\Rightarrow$ (iii): outline of proof—use 6.20 and Myers' theorem [GKM: 7.3(i)]. Prove geometrically that (i) $\Rightarrow$ (ii): outline of proof (Yano and Bochner)—if $\{E_j\}$ is an orthonormal basis for $\mathfrak{g}$, then $\sum R(U, E_j)E_j = U/4$, $U \in \mathfrak{g}$; it follows for $\mu \in A^1G$ that $\langle \rho_\mu, \mu \rangle = \|\mu\|^2/4$ (4.21).

6.23 É. Cartan showed that $\pi_2(G) = 0$ for a compact Lie group G [Ca6]. This theorem has been generalized by W. Browder to H-spaces [Br]. Applying the Hurewicz isomorphism [Hu: 2.9.1] we see that $H_2(G; \mathbb{Z}) = 0$ for every simply connected compact Lie group G, and therefore $H^2(G; \mathbb{R}) = 0$. Application of corollary 4.23 to the flat torus T^n, $n \geq 2$, shows that $H^2(G; \mathbb{R})$ is not always zero for a compact Lie group.

Theorem If G is a compact, semisimple Lie group, then $H^2(G; \mathbb{R}) = 0$.

PROOF Let $\mu \in A^2G$ be left-invariant, and let $\mu^\# \in \Lambda^2\mathfrak{g}$ be the metric dual of μ from 3.8: $\langle \mu^\#, U \wedge V \rangle = \mu(U, V)$, $U, V \in \mathfrak{g}$. With respect to an orthonormal basis $\{E_j\}$ for $\mathfrak{g}$, $\mu^\# = \sum_{i<j} \mu(E_i, E_j)E_i \wedge E_j$.

Given $U, V \in \mathfrak{g}$, the definition of ρ_μ (4.21), the symmetry relations for R and $\mathscr{R}$ (3.35, 3.44), and 6.20 imply

$$\begin{aligned}
\rho_\mu(U, V) &= \sum (\mu(R(U, E_i)E_i, V) - \mu(E_i, R(E_i, U)V) \\
&\qquad + \mu(U, R(V, E_i)E_i) - \mu(R(E_i, V)U, E_i)) \\
&= \sum (\mathscr{R}(U, E_i)\mu(E_i, V) + \mathscr{R}(V, E_i)\mu(U, E_i) \\
&\qquad - \mu(E_i, R(E_i, U)V + R(V, E_i)U)) \\
&= \sum \left(\frac{\langle U, E_i \rangle}{4} \mu(E_i, V) + \frac{\langle V, E_i \rangle}{4} \mu(U, E_i) + \mu(E_i, R(U, V)E_i) \right) \\
&= \frac{\mu(U, V)}{2} + \sum_{i, j} \langle R(U, V)E_i, E_j \rangle \mu(E_i, E_j) \\
&= \frac{\mu(U, V)}{2} - 2 \sum_{i<j} \mu(E_i, E_j) \langle R(U \wedge V), E_i \wedge E_j \rangle \\
&= \frac{\mu(U, V)}{2} - 2\langle R(\mu^\#), U \wedge V \rangle
\end{aligned}$$

where in the last two lines, R is the curvature operator. Thus by the definition of $\langle\ ,\ \rangle$ on T^*M (3.9) and 6.21,

$$\langle \rho_\mu, \mu \rangle = \frac{\|\mu\|^2}{2} - 2 \sum_{i<j} \langle R(\mu^*), E_i \wedge E_j \rangle \mu(E_i \wedge E_j)$$
$$= \frac{\|\mu\|^2}{2} - 2\langle R(\mu^*), \mu^* \rangle \geq \frac{\|\mu\|^2}{2} - \frac{\|\mu^*\|^2}{4} = \frac{\|\mu\|^2}{4}$$

If μ is harmonic, it is then zero by Weitzenböck's formula (4.22) and integration over M.

6.24 If $\tilde{G}$ is the universal covering group of a Lie group G, then [Gr: 7.12] $\pi_3(G) = \pi_3(\tilde{G})$. By the Hurewicz isomorphism, $\pi_3(\tilde{G}) \cong H_3(\tilde{G}; \mathbb{Z})$ since $\pi_2(G) = \pi_1(G) = 0$.

Theorem If G is a compact Lie group of dimension greater than 2, then $H_3(G; \mathbb{R})$ is nonzero [as is $\pi_3(G)$ if $\tilde{G}$ is compact].

PROOF Assume G is nonabelian, for otherwise G is a torus, and the theorem follows from 4.23. By 6.16, $[\mathfrak{g}, \mathfrak{g}]$ is semisimple and compact; its dimension is greater than or equal to 3 by 6.10.

Let $\langle\ ,\ \rangle$ be the bi-invariant metric on G which is the product of a Euclidean inner product on $\mathfrak{z}$ and minus the Killing form on $[\mathfrak{g}, \mathfrak{g}]$. Define $\mu \in A^3G$ by $\mu(U, V, W) := \langle [U, V], W \rangle$; μ is nonzero because $[\mathfrak{g}, \mathfrak{g}]$ is semisimple. Since $\nabla_U V = \frac{1}{2}[U, V]$, $U, V \in \mathfrak{g}$, μ is parallel by the Jacobi identity. But a parallel form is harmonic (4.9iv), so μ represents a nonzero element of $H^3(G; \mathbb{R})$.

6.25 **Corollary** The only spheres which are Lie groups are S^1 and S^3.

PROOF S^1 is the Lie group of complex numbers of absolute value 1 under multiplication. S^2 has no Lie group structure because a Lie group is parallelizable, while S^2 does not admit even one nonvanishing continuous vector field. S^3 is the Lie group of quaternions of modulus 1 under multiplication (details will appear in 6.45). Since $H^3(S^n; \mathbb{R}) = 0$ for $n > 3$, the proof is complete.

The last two results were known to É. Cartan [Ca6].

6.26 Nonzero eigenvalues for the Laplacian are generally not easy to compute; therefore the next theorem provides another example of how Lie groups are relatively easy to work with.

Theorem (Beers-Millman) If G is a compact, connected semisimple Lie group with its canonical metric, then $\frac{1}{2}$ is an eigenvalue of the Laplacian of multiplicity at least $2n$ on A^1G and A^2G, $n = \dim G$.

PROOF Fix an orthonormal basis $\{E_i\}$ for $\mathfrak{g}$. By the definition of $\langle\ ,\ \rangle$, $-\sum_i \mathrm{ad}_{E_i} \circ \mathrm{ad}_{E_i}$ is the identity map of $\mathfrak{g}$ (this was essentially proved in 6.21). Let $\mu \in \mathfrak{g}^*$ and $U \in \mathfrak{g}$. By Weitzenböck's formula,

$$\begin{aligned}\Delta\mu(U) &= -\sum (\nabla_{E_i}\nabla_{E_i}\mu)(U) + \sum \mu(R(U, E_i)E_i)) \\ &= -\tfrac{1}{4}\sum \mu([E_i, [E_i, U]]) - \tfrac{1}{4}\sum \mu([[U, E_i], E_i]) \\ &= -\tfrac{1}{2}\mu(\sum [E_i, [E_i, U]]) = \tfrac{1}{2}\mu(U)\end{aligned}$$

Similarly, $\Delta\eta = \frac{1}{2}\eta$ if η is a right-invariant 1-form on G.
Now given $\mu \in \mathfrak{g}^*$,

$$\Delta d\mu = (d\delta + \delta d)d\mu = d(d\delta + \delta d)\mu = d\Delta\mu = \tfrac{1}{2}d\mu$$

Since G is nonabelian, the identity $d\mu(U, V) = -\mu([U, V])$, U, $V \in \mathfrak{g}$, implies that $d\mu$ is nonzero if $\mu \neq 0$. Thus $d\mu$ is an eigenvector of Δ on A^2G for the eigenvalue $\frac{1}{2}$. Similarly, $d\eta$ is an eigenvector of Δ on A^2G for each nonzero right-invariant 1-form η on G.

Far more information is available in [BM].

SOME SIMPLE EXAMPLES

Sections 6.27 to 6.52 consist of a brief study of some classical simple Lie algebras and Lie groups.

6.27 Each $U \in \mathfrak{gl}(n, \mathbb{R})$ can be written uniquely as a skew-symmetric matrix plus a symmetric matrix (5.1): $U = (U - {}^tU)/2 + (U + {}^tU)/2$. If we set $\mathfrak{sym}(n) := \{\text{symmetric } U \in \mathfrak{gl}(n, \mathbb{R})\}$, then

$$\mathfrak{gl}(n, \mathbb{R}) = \mathfrak{o}(n) \oplus \mathfrak{sym}(n) \qquad \text{with dimensions } n^2 = \binom{n}{2} + \binom{n+1}{2}$$

where

$$[\mathfrak{o}(n), \mathfrak{o}(n)] \subset \mathfrak{o}(n) \qquad [\mathfrak{o}(n), \mathfrak{sym}(n)] \subset \mathfrak{sym}(n)$$
$$[\mathfrak{sym}(n), \mathfrak{sym}(n)] \subset \mathfrak{o}(n)$$

If $\mathfrak{sym}_o(n) := \mathfrak{sym}(n) \cap \mathfrak{sl}(n, \mathbb{R})$ and $\mathfrak{r} := \{tI \in \mathfrak{gl}(n, \mathbb{R}) \mid t \in \mathbb{R}\}$, then (see 5.1) $\mathfrak{gl}(n, \mathbb{R}) = \mathfrak{o}(n) \oplus \mathfrak{sym}_o(n) \oplus \mathfrak{r} = \mathfrak{sl}(n, \mathbb{R}) \oplus \mathfrak{r}$. The same bracket relations hold for $\mathfrak{sl}(n, \mathbb{R}) = \mathfrak{o}(n) \oplus \mathfrak{sym}_o(n)$ as for $\mathfrak{gl}(n, \mathbb{R})$ (with $\mathfrak{sym}_o$ substituted for $\mathfrak{sym}$).

Given $U = [U_{jk}] \in \mathfrak{o}(n)$ and $V = [V_{jk}] \in \mathfrak{sym}_o(n)$, the Killing form of $\mathfrak{sl}(n, \mathbb{R})$ from 6.13 satisfies the identities

$$B(U, U) = 2n \operatorname{tr}(UU) = -2n \operatorname{tr}(U^t U) = -2n \sum_{j,k} U_{jk}^2 \leq 0$$

$$B(V, V) = 2n \operatorname{tr}(VV) = 2n \operatorname{tr}(V^t V) = 2n \sum_{j,k} V_{jk}^2 \geq 0$$

$$B(U, V) = 2n \operatorname{tr}(UV) = -2n \operatorname{tr}({}^t U^t V)$$
$$= -2n \operatorname{tr}^t(VU) = -B(U, V) = 0$$

Thus B is negative definite on $\mathfrak{o}(n)$ and positive definite on $\mathfrak{sym}_o(n)$, and $\mathfrak{o}(n)$ and $\mathfrak{sym}_o(n)$ are orthogonal with respect to B. Hence B is indefinite but nondegenerate, so $\mathfrak{sl}(n, \mathbb{R})$ is a noncompact, semisimple Lie algebra for $n \geq 2$. Actually, $\mathfrak{sl}(n, \mathscr{R})$ is simple, as can be seen by checking the multiplication table.

Exercise Check directly that $\operatorname{Int}(\mathfrak{sl}(n, \mathbb{R}))$ is noncompact (see 6.17).

6.28 The *unitary algebra* $\mathfrak{u}(n) := \{U \in \mathfrak{gl}(n, \mathbb{C}) \mid {}^t U + \bar{U} = 0\}$ is a Lie algebra of dimension n^2 [W: 3.37]. The matrices of the form ciI, where $c \in \mathbb{R}$, $i^2 = -1$, and I is the identity matrix, are a nonzero abelian ideal of $\mathfrak{u}(n)$, so $\mathfrak{u}(n)$ is not a simple algebra.

The *special unitary algebra* $\mathfrak{su}(n) := \{\text{trace-free } U \in \mathfrak{u}(n)\}$ has dimension $n^2 - 1$. Its Killing form is $B(U, V) = 2n \operatorname{tr}(UV)$; since

$$B(U, U) = 2n \operatorname{tr}(UU) = -2n \operatorname{tr}(U\,{}^t\bar{U}) = -2n \sum_{j,k} |U_{jk}|^2$$

B is negative definite, so $\mathfrak{su}(n)$ is semisimple and compact (in fact, it is simple), and is therefore the Lie algebra of a compact Lie group. By [W: 3.37], the *special unitary group* $SU(n) := \{g \in GL(n, \mathbb{C}) \mid {}^t\bar{g} = g^{-1}, \det g = 1\}$ is a compact Lie group with Lie algebra $\mathfrak{su}(n)$. It will be proved in 6.47 that $SU(n)$ is simply connected.

6.29 It should be noted that if $U \in \mathfrak{u}(n)$, then the matrix iU is symmetric. Therefore, corresponding to the decomposition (6.27) of $\mathfrak{gl}(n, \mathbb{R})$ into skew-symmetric and symmetric parts there is a decomposition of $\mathfrak{gl}(n, \mathbb{C})$ into skew-Hermitian and symmetric parts:

$$\mathfrak{gl}(n, \mathbb{C}) = \mathfrak{u}(n) \oplus i\mathfrak{u}(n) \qquad \text{with real dimensions} \qquad 2n^2 = n^2 + n^2$$

Comparison of these dimensions with the corresponding dimensions in the real case emphasizes the difference in structure between the real Lie algebras $\mathfrak{gl}(n, \mathbb{R})$ and $\mathfrak{gl}(n, \mathbb{C})$.

6.30 There is a very important relationship between $\mathfrak{sl}(n, \mathbb{R})$ and $\mathfrak{su}(n)$. Given $U \in \mathfrak{su}(n)$, let $\operatorname{Re} U$ and $\operatorname{Im} U \in \mathfrak{gl}(n, \mathbb{R})$ denote the real and imaginary parts of the matrix U, respectively: $U = \operatorname{Re} U + i \operatorname{Im} U$. The conditions ${}^t U + \bar{U} = 0$ and $\operatorname{tr} U = 0$ imply that $\operatorname{Re} U \in \mathfrak{o}(n)$ and $\operatorname{Im} U \in \mathfrak{sym}_o(n)$. With the notation $i\mathfrak{sym}_o(n) := \{iV \mid V \in \mathfrak{sym}_o(n)\}$,

$$\mathfrak{su}(n) = \mathfrak{o}(n) \oplus i\mathfrak{sym}_o(n) \qquad \text{where} \qquad [\mathfrak{o}(n), \mathfrak{o}(n)] \subset \mathfrak{o}(n)$$
$$[\mathfrak{o}(n), i\mathfrak{sym}_o(n)] \subset i\mathfrak{sym}_o(n) \qquad [i\mathfrak{sym}_o(n), i\mathfrak{sym}_o(n)] \subset \mathfrak{o}(n)$$

Except for the factor i, this is the same situation as for $\mathfrak{sl}(n, \mathbb{R})$. In particular, the real linear map from $\mathfrak{su}(n)$ to $\mathfrak{sl}(n, \mathbb{R})$ which sends U to $\operatorname{Re} U + \operatorname{Im} U$ is a real vector space isomorphism, even though it is not a Lie algebra isomorphism (no Lie algebra isomorphism can exist between a compact and a noncompact Lie algebra). To see the exact Lie algebra relationship between $\mathfrak{sl}(n, \mathbb{R})$ and $\mathfrak{su}(n)$ it is necessary to work with complex Lie algebras.

6.31 **Definition** A *complex Lie algebra* is a complex vector space $\mathfrak{g}$ together with an anticommutative complex bilinear map $[\ ,\ ]: \mathfrak{g} \times \mathfrak{g} \to \mathfrak{g}$ such that $[U, [V, W]] + [V, [W, U]] + [W, [U, V]] = 0$. The *complexification of a real Lie algebra* $\mathfrak{h}$ is the complex vector space $\mathfrak{h}^{\mathbb{C}} = \mathfrak{h} \otimes \mathbb{C}$ (1.22), with the complex bilinear extension of the Lie bracket in $\mathfrak{h}$: given $U, V, X, Y \in \mathfrak{h}$,

$$[U + iV, X + iY] := [U, X] - [V, Y] + i([V, X] + [U, Y])$$

A real Lie algebra $\mathfrak{h}$ is called a *real form of a complex Lie algebra* $\mathfrak{g}$ if $\mathfrak{h}^{\mathbb{C}}$ is complex Lie algebra isomorphic to $\mathfrak{g}$.

For example, $\mathfrak{gl}(n, \mathbb{C})$ is naturally a complex Lie algebra as well as a real one; $\dim_{\mathbb{C}} \mathfrak{gl}(n, \mathbb{C}) = n^2$, whereas $\dim_{\mathbb{R}} \mathfrak{gl}(n, \mathbb{C}) = 2n^2$. Similarly, the complex special linear algebra $\mathfrak{sl}(n, \mathbb{C}) := \{U \in \mathfrak{gl}(n, \mathbb{C}) \mid \operatorname{tr} U = 0\}$ is a complex Lie algebra of complex dimension $n^2 - 1$ [W: 3.37]. The real Lie algebras $\mathfrak{u}(n)$ and $\mathfrak{su}(n)$, however, cannot be complex Lie algebras because they are not closed under multiplication by i.

It is easily checked that $\mathfrak{gl}(n, \mathbb{R})$ is a real form of $\mathfrak{gl}(n, \mathbb{C})$.

6.32 **Proposition** The real Lie algebras $\mathfrak{sl}(n, \mathbb{R})$ and $\mathfrak{su}(n)$ are both real forms of $\mathfrak{sl}(n, \mathbb{C})$: $\mathfrak{sl}(n, \mathbb{R})^{\mathbb{C}} \cong \mathfrak{sl}(n, \mathbb{C}) \cong \mathfrak{su}(n)^{\mathbb{C}}$.

Proof Given elements U and V in $\mathfrak{g} = \mathfrak{sl}(n, \mathbb{R})$ or $\mathfrak{su}(n)$, temporarily denote by $U +' iV$ the formal sum of U and iV in $\mathfrak{g}^{\mathbb{C}}$. The map $\mathfrak{g}^{\mathbb{C}} \to \mathfrak{sl}(n, \mathbb{C})$ which sends $U +' iV$ to the actual sum $U + iV$ of complex matrices is a complex linear map, since by 1.22, given $a + ib \in \mathbb{C}$,

$$\begin{aligned}(a + ib) \cdot (U +' iV) &= (aU - bV) +' i(bU + aV) \\ &\mapsto (aU - bV) + i(bU + aV) \\ &= (a + ib)(U + iV)\end{aligned}$$

Since the map is clearly bijective, it suffices to check the Lie brackets; for this it suffices to observe that the Lie brackets in both $\mathfrak{gl}(n, \mathbb{R})$ and $\mathfrak{gl}(n, \mathbb{C})$ are given by the commutator of the matrices involved, and for $U, V \in \mathfrak{gl}(n, \mathbb{R})$, the commutator of U and V in $\mathfrak{gl}(n, \mathbb{R})$ equals their commutator in $\mathfrak{gl}(n, \mathbb{C})$.

6.33 **Definition** Just as in the real case, a complex Lie algebra is called *simple* if it is nonabelian and has no proper ideals, and is called *semisimple* if it is the direct sum of simple ideals.

For example, $\mathfrak{sl}(n, \mathbb{C})$ is simple.

6.34 **Proposition** Let $\mathfrak{g}$ be a real Lie algebra; if $\mathfrak{g}^{\mathbb{C}}$ is simple, then $\mathfrak{g}$ is simple. More generally, $\mathfrak{g}$ is semisimple if and only if $\mathfrak{g}^{\mathbb{C}}$ is semisimple.

PROOF If $\mathfrak{g}$ is abelian, then so is $\mathfrak{g}^{\mathbb{C}}$. Similarly, if $\mathfrak{h}$ is a proper ideal of $\mathfrak{g}$, then $\mathfrak{h}^{\mathbb{C}}$ is a proper ideal of $\mathfrak{g}^{\mathbb{C}}$. The Killing form $B^{\mathbb{C}}$ of $\mathfrak{g}^{\mathbb{C}}$ is defined by $B^{\mathbb{C}}(U, V) := \operatorname{tr}(\mathrm{ad}_U^{\mathbb{C}} \circ \mathrm{ad}_V^{\mathbb{C}})$, where $\mathrm{ad}^{\mathbb{C}}$ is the adjoint representation of $\mathfrak{g}^{\mathbb{C}}$ on $\mathfrak{g}^{\mathbb{C}}$: $\mathrm{ad}_U^{\mathbb{C}}(V) = [U, V]$, $U, V \in \mathfrak{g}^{\mathbb{C}}$. It is easily checked that the proof of Cartan's criterion (6.14) goes through for complex Lie algebras, so $\mathfrak{g}^{\mathbb{C}}$ is semisimple if and only if $B^{\mathbb{C}}$ is nondegenerate. But $B^{\mathbb{C}}(U, V) = B(U, V)$ for all $U, V \in \mathfrak{g} \subset \mathfrak{g}^{\mathbb{C}}$.

6.35 **Definition** A real Lie algebra $\mathfrak{g}$ is called *absolutely simple* if $\mathfrak{g}^{\mathbb{C}}$ is simple.

A real form of a simple complex Lie algebra is therefore absolutely simple, and conversely.

Exercise Prove that if a simple real Lie algebra $\mathfrak{g}$ is not absolutely simple, then it is isomorphic to the *realification* of a simple ideal $\mathfrak{h}$ of $\mathfrak{g}^{\mathbb{C}}$, that is (cf. 6.32, 1.22, and 1.20), $\mathfrak{g}$ is isomorphic to the real Lie algebra underlying $\mathfrak{h}$.

6.36 Up to isomorphism, the absolutely simple algebra $\mathfrak{su}(n)$ is the only compact real form of $\mathfrak{sl}(n, \mathbb{C})$. More generally,

Theorem Every semisimple complex Lie algebra has a unique compact real form.

PROOF The complete proof [He: 3.6.3] demands too much preparation to be presented here; here is a partial proof which is sufficient for our purposes.

Let $\mathfrak{g}$ be a semisimple complex Lie algebra, and assume that a real form $\mathfrak{h}$ of $\mathfrak{g}$ is known. We may assume that $\mathfrak{h}$ is not compact. Since $\mathfrak{h}$ is semisimple, its Killing form B can be diagonalized, that is, there exists a basis $\{E_j, F_k\}$ for $\mathfrak{h}$, with dual basis $\{\eta_j, \mu_k\}$ for $\mathfrak{h}^*$, such that $B = -\sum \eta_j^2 + \sum \mu_k^2$; the number of negative terms is independent of the choice of basis. The basis $\{E_j, F_k\}$ may be chosen so that $\{E_j, iF_k\}$ is also a basis for a real form of $\mathfrak{g}$, and this real form is compact since its Killing form is negative definite.

The reader should verify that this process, which is called the *Weyl unitary trick*, could have been used to construct the compact real form $\mathfrak{su}(n)$ of $\mathfrak{sl}(n, \mathbb{C})$ from the noncompact real form $\mathfrak{sl}(n, \mathbb{R})$.

6.37 By 6.36, to classify the compact semisimple Lie algebras over $\mathbb{R}$ it suffices to classify the semisimple Lie algebras over $\mathbb{C}$, then to find a real form for each, and finally, to find their compact real forms by the Weyl unitary trick.

Killing and Cartan [HS] [Sw] classified the semisimple Lie algebras over $\mathbb{C}$ and $\mathbb{R}$. Over $\mathbb{C}$ their study yielded four classical families of simple Lie algebras, labeled A_n, B_n, C_n, and D_n, and five exceptional Lie algebras, labeled E_6, E_7, E_8, F_4, and G_2; the simple Lie algebras over $\mathbb{R}$ were then obtained either as real forms or as realifications of the simple Lie algebras over $\mathbb{C}$.

Sections 6.38 to 6.52 will consist of a brief look at the classical simple Lie algebras.

6.38 For all $n \geq 1$, A_n is the simple Lie algebra $\mathfrak{sl}(n+1, \mathbb{C})$, whose compact real form $\mathfrak{su}(n+1)$ we have already seen. Before we proceed to B_n, C_n, and D_n, a comment is in order about the nonsemisimple Lie algebra $\mathfrak{u}(n)$. The unitary group $U(n) = \{g \in GL(n, \mathbb{C}) \mid g^{-1} = {}^t\bar{g}\}$ is compact [W: 3.37], and therefore admits a bi-invariant metric, even though the Killing form of $\mathfrak{u}(n)$ is degenerate. Since each $U \in \mathfrak{u}(n)$ satisfies ${}^tU + \bar{U} = 0$, the symmetric bilinear form $\hat{B}(U, V) := 2n\,\mathrm{tr}(UV)$ is negative definite on $\mathfrak{u}(n)$; $\hat{B}$ is the formal extension to $\mathfrak{u}(n)$ of the Killing form of $\mathfrak{su}(n)$, and does not equal $\mathrm{tr}(ad \cdot \circ \mathrm{ad} \cdot)$ on $\mathfrak{u}(n)$. Because $\hat{B}$ is $\mathrm{Int}(\mathfrak{u}(n))$-invariant, $-\hat{B}$ determines a bi-invariant metric on $U(n)$.

6.39 The matrix of the so-called *symplectic 2-form* $\sum_{j=1}^{n} dx^j \wedge dx^{n+j}$ on $\mathbb{R}^{2n}$ (which will appear again in 8.2) is

$$\Omega = \begin{bmatrix} 0 & I \\ -I & 0 \end{bmatrix}$$

where I is the $n \times n$ identity matrix.

Definition For $\mathbb{F} = \mathbb{R}$ or $\mathbb{C}$, the *symplectic algebra over* $\mathbb{F}$ is $\mathfrak{sp}(n, \mathbb{F}) := \{U \in \mathfrak{gl}(2n, \mathbb{F}) \mid \Omega U + {}^tU\Omega = 0\}$; the *symplectic group over* $\mathbb{F}$ is $Sp(n, \mathbb{F}) := \{g \in GL(2n, \mathbb{F}) \mid {}^tg\Omega g = \Omega\}$.

Thus $Sp(n, \mathbb{R})$ preserves the symplectic 2-form $\sum dx^j \wedge dx^{n+j}$ on $\mathbb{R}^{2n}$, and $\mathfrak{sp}(n, \mathbb{R})$ preserves the form infinitesimally. The notation $Sp(2n, \mathbb{F})$ [and $\mathfrak{sp}(2n, \mathbb{F})$] is also in use.

6.40 **Proposition** The Lie algebra of $Sp(n, \mathbb{F})$ is $\mathfrak{sp}(n, \mathbb{F})$.

PROOF Given a 1-parameter subgroup γ of $Sp(n, \mathbb{F})$, differentiating the identity ${}^t\gamma(s)\Omega\gamma(s) = \Omega$, $s \in \mathbb{R}$, yields ${}^t\dot{\gamma}(0)\Omega + \Omega\dot{\gamma}(0) = 0$, which is the defining relation for $\mathfrak{sp}(n, \mathbb{F})$.

Conversely, given $U \in \mathfrak{sp}(n, \mathbb{F}) \subset \mathfrak{gl}(2n, \mathbb{F})$, let $\gamma(s) := \exp sU$, where exp is the exponential map of $GL(2n, \mathbb{F})$. Since ${}^tU = -\Omega U \Omega^{-1}$ ${}^t\gamma(s)\Omega\gamma(s) = \exp(-s\Omega U\Omega^{-1})\Omega\gamma(s) = \Omega\gamma(-s)\Omega^{-1}\Omega\gamma(s) = \Omega$.

6.41 Direct computation shows that

$$\mathfrak{sp}(n, \mathbb{F}) = \left\{ U = \begin{bmatrix} A & B \\ C & -{}^tA \end{bmatrix} \middle| \; A, B, C \in \mathfrak{gl}(n, \mathbb{F}),\ B, C \text{ symmetric} \right\} \quad (6\text{-}1)$$

so $\dim_{\mathbb{F}} \mathfrak{sp}(n, \mathbb{F}) = 2n^2 + n$. The Killing form of $\mathfrak{sp}(n, \mathbb{F})$ is given by $B(U, V) = (2n + 2)\operatorname{tr}(UV)$; this is nondegenerate, so $\mathfrak{sp}(n, \mathbb{F})$ is semisimple. Actually, $\mathfrak{sp}(n, \mathbb{F})$ is simple, and in fact, $\mathfrak{sp}(n, \mathbb{C})$ is the simple Lie algebra C_n from 6.37. Finally, $\mathfrak{sp}(n, \mathbb{R})$ is a real form of $\mathfrak{sp}(n, \mathbb{C})$; we will now see that it is noncompact, and then use the Weyl unitary trick (6.36) to find the unique compact real form of $\mathfrak{sp}(n, \mathbb{C})$.

By the decomposition of $\mathfrak{sl}(n, \mathbb{R})$ in 6.27, $\mathfrak{sp}(n, \mathbb{R})$ [which lies in $\mathfrak{sl}(2n, \mathbb{R})$ by eq. (6-1)] is the direct sum $(\mathfrak{o}(2n) \cap \mathfrak{sp}(n, \mathbb{R})) \oplus (\mathfrak{sym}_o(2n) \cap \mathfrak{sp}(n, \mathbb{R}))$. If $U = A + W \in \mathfrak{sp}(n, \mathbb{R})$, where $A \in \mathfrak{o}(2n)$ and $W \in \mathfrak{sym}_o(n)$, then as in 6.27, $B(A, A) \leq 0$ and $B(W, W) \geq 0$, with equality in each case only if the matrix is zero; similarly, $B(A, W) = 0$. By the Weyl unitary trick the compact real form $\mathfrak{g}$ of $\mathfrak{sp}(n, \mathbb{C})$ consists of all matrices of the form $A + iW$, where A and W are as above.

This can be broken down a bit further using eq. (6-1) and the decomposition of $\mathfrak{gl}(n, \mathbb{R})$ from 6.27. Given $C \in \mathfrak{o}(n)$ and $X, Y, Z \in \mathfrak{sym}(n)$,

$$\begin{bmatrix} C & X \\ -X & C \end{bmatrix} + \begin{bmatrix} Y & Z \\ Z & -Y \end{bmatrix} \in \mathfrak{sp}(n, \mathbb{R}) \qquad \begin{bmatrix} C & X \\ -X & C \end{bmatrix} + i\begin{bmatrix} Y & Z \\ Z & -Y \end{bmatrix} \in \mathfrak{g}$$

To find a more convenient description of $\mathfrak{g}$, observe that for C, X, Y, Z as above, ${}^t(C + iY) + \overline{C + iY} = 0$, so $C + iY \in \mathfrak{u}(n)$; similarly, $-X + iZ = -(\overline{X + iZ})$. Therefore,

$$\mathfrak{g} = \left\{ U = \begin{bmatrix} V & K \\ -\bar{K} & \bar{V} \end{bmatrix} \in \mathfrak{gl}(2n, \mathbb{C}) \;\middle|\; V \in \mathfrak{u}(n), K \in \mathfrak{sym}(n, \mathbb{C}) \right\}$$

$$= \mathfrak{sp}(n, \mathbb{C}) \cap \mathfrak{u}(2n) = \mathfrak{sp}(n, \mathbb{C}) \cap \mathfrak{su}(2n)$$

6.42 **Definition** The *symplectic algebra* is the real Lie algebra $\mathfrak{sp}(n) := \mathfrak{sp}(n, \mathbb{C}) \cap \mathfrak{su}(2n)$; the *symplectic group* is $Sp(n) := Sp(n, \mathbb{C}) \cap SU(2n)$.

Exercise Prove that $C_1 \cong A_1$; verify the corresponding Lie group isomorphisms $Sp(1, \mathbb{C}) \cong SL(2, \mathbb{C})$, $Sp(1, \mathbb{R}) \cong SL(2, \mathbb{R})$, $Sp(1) \cong SU(2)$.

6.43 Let $\mathbb{H}$ be the algebra of quaternions (1.12j). The $4n$-dimensional real vector space $\mathbb{H}^n$ is naturally an n-dimensional right module over the ring $\mathbb{H}$, with scalar multiplication $(v^1, \ldots, v^n) \cdot z := (v^1 z, \ldots, v^n z)$.

Definition The *symplectic inner product on* $\mathbb{H}^n$ is the map

$$\mathbb{H}^n \times \mathbb{H}^n \to \mathbb{H}, \ \langle (u^1, \ldots, u^n), (v^1, \ldots, v^n) \rangle := \bar{u}^1 v^1 + \cdots + \bar{u}^n v^n$$

The *quaternionic linear group* $GL(n, \mathbb{H})$ is the group of all invertible $\mathbb{H}$-linear transformations of $\mathbb{H}^n$: $g(uz) = g(u)z$, $g(u + v) = gu + gv$.

We shall see that $Sp(n)$ is isomorphic to the subgroup of $GL(n, \mathbb{H})$ which preserves the symplectic inner product.

6.44 Embed $\mathbb{C}$ into $\mathbb{H}$ by $a + bi \mapsto a + bi + 0j + 0k$; the restriction of the multiplication $\mathbb{H} \times \mathbb{H} \to \mathbb{H}$ to a map $\mathbb{H} \times \mathbb{C} \to \mathbb{H}$ exhibits $\mathbb{H}$ as a 2-dimensional complex vector space; the choice of basis $\{1, j\}$ for $\mathbb{H}$ over $\mathbb{C}$ fixes the isomorphism

$$\mathbb{C}^2 \cong \mathbb{H}, \qquad (z^1, z^2) \leftrightarrow z^1 + jz^2 \qquad z^1, z^2 \in \mathbb{C}$$

More generally, $\mathbb{H}^n$ is a $2n$-dimensional complex vector space; fix the isomorphism

$$\mathbb{C}^{2n} \cong \mathbb{H}^n, \qquad (u, v) \leftrightarrow u + jv \qquad u, v \in \mathbb{C}^n$$

where if $v = (v^1, \ldots, v^n) \in \mathbb{C}^n$, then $jv := (jv^1, \ldots, jv^n) \in \mathbb{H}^n$. This isomorphism determines a matrix representation of $GL(n, \mathbb{H})$ in $GL(2n, \mathbb{C})$: each $g \in GL(n, \mathbb{H})$ can be written in the form $g = A + jB$, where A, $B \in \mathfrak{gl}(n, \mathbb{C})$, and then as an element of $GL(2n, \mathbb{C})$, g takes the form

$$g = \begin{bmatrix} A & -\bar{B} \\ B & \bar{A} \end{bmatrix} \in GL(2n, \mathbb{C})$$

Proof: Given $u + jv \in \mathbb{H}^n$,

$$\begin{aligned}(A + jB)(u + jv) &= Au + Ajv + jBu + jBjv \\ &= Au + j^2\bar{B}v + jBu + j\bar{A}v \\ &= Au - \bar{B}v + j(Bu + \bar{A}v)\end{aligned}$$

which corresponds to the vector

$$\begin{bmatrix} Au - \bar{B}v \\ Bu + \bar{A}v \end{bmatrix} = \begin{bmatrix} A & -\bar{B} \\ B & \bar{A} \end{bmatrix} \begin{bmatrix} u \\ v \end{bmatrix} \in \mathbb{C}^{2n} \qquad \textit{QED}$$

Proposition The subgroup of $GL(n, \mathbb{H}) \subset GL(2n, \mathbb{C})$ which preserves the symplectic inner product on $\mathbb{H}^n$ is $Sp(n) = Sp(n, \mathbb{C}) \cap SU(2n)$.

PROOF Fix

$$g = \begin{bmatrix} A & -\bar{B} \\ B & \bar{A} \end{bmatrix} \in GL(n, \mathbb{H}) \subset GL(2n, \mathbb{C});$$

g belongs to $Sp(n)$ if and only if

$${}^t g\bar{g} = I \qquad \text{and} \qquad {}^t g\Omega g = \Omega = \begin{bmatrix} 0 & I \\ -I & 0 \end{bmatrix} \qquad (6.39,\ 6.41(6\text{-}1))$$

$\Leftrightarrow$ ${}^t\bar{A}A + {}^t\bar{B}B = I$ and ${}^tBA = {}^tAB$
$\Leftrightarrow$ ${}^t\overline{(A + jB)}(A + jB) = I$
$\Leftrightarrow$ $A + jB$ preserves the symplectic inner product on $\mathbb{H}^n$.

6.45 For example, $Sp(1)$ is the group of all quaternions z such that $\langle zu, zv\rangle = \langle u, v\rangle$ for all $u, v \in \mathbb{H}$; but $\langle zu, zv\rangle = \overline{zu}\, zv = \bar{u}\bar{z}zv$, and $\langle u, v\rangle = \bar{u}v$, so since $\bar{z}z$ belongs to the center $\mathbb{R}$ of $\mathbb{H}$, $z \in Sp(1)$ if and only if $\bar{z}z = 1$. Thus $Sp(1)$ is the unit sphere in $\mathbb{H} = \mathbb{R}^4$. In particular, the unit sphere $S^3 \subset \mathbb{R}^4$ is the multiplicative Lie group of unit quaternions; combining this with the exercises in 6.42 we obtain the isomorphisms $SU(2) \cong Sp(1) \cong S^3$, which imply that $SU(2) \cong Sp(1)$ is simply connected.

6.46 Proposition The symplectic group $Sp(1)$ is the manifold S^3; for $n > 1$, the homogeneous space $Sp(n)/Sp(n-1)$ is the sphere S^{4n-1}.

PROOF Since $Sp(n)$ acts differentiably on $\mathbb{H}^n = \mathbb{R}^{4n}$ and preserves the symplectic inner product, it acts differentiably on S^{4n-1}; this action is transitive, for given $v \in S^{4n-1}$, v can be completed to a basis $\{v_1, \ldots, v_n = v\}$ for $\mathbb{H}^n$ which is orthonormal with respect to the symplectic inner product, and then the matrix with the vectors $v_1, \ldots, v_n = v$ as columns belongs to $Sp(n)$ and maps the nth canonical basis vector e_n in $S^{4n-1} \subset \mathbb{H}^n$ to v. The isotropy subgroup in $Sp(n)$ of the point $e_n \in S^{4n-1}$ is $\begin{bmatrix} Sp(n-1) & 0 \\ 0 & 1 \end{bmatrix}$.

6.47 Proposition The compact simple groups $SU(n)$ and $Sp(n)$ are simply connected.

PROOF Let H be a closed subgroup of a Lie group G; denote the projection map of G onto $G/H =: M$ by $\wp$. Let $\gamma\colon S^1 \to G$ be a continuous curve mapping the basepoint $e^{i0} =: 1 \in S^1$ to the identity element e of G. Assume $\wp \circ \gamma$ is null-homotopic with fixed point e, and let $f\colon S^1 \times [0, 1] \to M$ be a fixed-point homotopy between $\wp \circ \gamma$ and the constant map $\wp(e)$: $f(s, 1) = \wp \circ \gamma(s)$ and $f(s, 0) = \wp(e)$, $s \in S^1$, and $f(1, t) = \wp(e)$, $0 \leq t \leq 1$. By the homotopy covering property of fiber bundles [S: 11.7], f can be lifted to a homotopy F of γ in G; $F\colon S^1 \times [0, 1] \to G$ such that $\wp \circ F = f$, $F(s, 1) = \gamma(s)$, and $F(1, t) = e$. Since $\wp \circ F(s, 0) = f(s, 0) = \wp(e)$ for all $s \in S^1$, $F(s, 0) \in H$. Thus F is a homotopy between γ and a curve in H.

It follows that if M and H are simply connected, then so is G. But $SU(2) \cong Sp(1) \cong S^3$ is simply connected, and the homogeneous spaces $SU(n)/SU(n-1)$ [W: 3.65(b)] and $Sp(n)/Sp(n-1)$ (6.46) are spheres of dimension greater than 1, and are therefore simply connected. By induction, $SU(n)$ and $Sp(n)$ are simply connected.

Exercises

1. The unitary group $U(n)$ is homomorphic to $SU(n) \times S^1$, and therefore $\pi_1 U(n) \cong \mathbb{Z}$. The homomorphism $S^1 \times SU(n) \to U(n)$, $(e^{it}, g) \mapsto e^{it}g$, is an n-fold covering map.
2. The standard identification of $\mathbb{C}^n$ with $\mathbb{R}^{2n}$ (1.20) determines a matrix representation of $GL(n, \mathbb{C})$ in $GL(2n, \mathbb{R})$ [cf. the representation of $GL(n, \mathbb{H})$ in $GL(2n, \mathbb{C})$ from 6.44]; given $g = A + iB \in GL(n, \mathbb{C})$, with $A, B \in \mathfrak{gl}(n, \mathbb{R})$, g is represented in $GL(2n, \mathbb{R})$

by the matrix $\begin{bmatrix} A & -B \\ B & A \end{bmatrix}$. Prove that $Sp(n, \mathbb{R}) \cap O(2n) \cong U(n)$, and therefore $\pi_1 Sp(n, \mathbb{R}) \cong \pi_1 U(n) \cong \mathbb{Z}$.

6.48 The *orthogonal algebra over* $\mathbb{C}$ [W: 3.37] is $\mathfrak{o}(n, \mathbb{C}) := \{U \in \mathfrak{gl}(n, \mathbb{C}) \mid {}^tU + U = 0\}$; it is the Lie algebra of the *complex orthogonal group* $O(n, \mathbb{C}) := \{g \in GL(n, \mathbb{C}) \mid {}^tgg = I\}$. The Killing form of $\mathfrak{o}(n, \mathbb{C})$ is $B(U, V) = (n-2)\operatorname{tr}(UV)$, so $\mathfrak{o}(n, \mathbb{C})$ is semisimple for $n > 2$; we shall see in 6.50 that $\mathfrak{o}(4, \mathbb{C})$ is not simple, but with this exception, $\mathfrak{o}(n, \mathbb{C})$ is in fact simple for $n > 2$. The algebra $\mathfrak{o}(2n+1, \mathbb{C})$ is denoted by B_n, $n \geq 1$ (6.37), and $\mathfrak{o}(2n, \mathbb{C})$ is denoted by D_n, $n \geq 2$.

The obvious real forms of the algebras $\mathfrak{sl}(n, \mathbb{C})$ and $\mathfrak{sp}(n, \mathbb{C})$ are noncompact, and the Weyl unitary trick is necessary to obtain the compact real forms; once obtained, the associated compact matrix groups are seen to be simply connected. The situation is quite different for $\mathfrak{o}(n, \mathbb{C})$. The obvious real form for $\mathfrak{o}(n, \mathbb{C})$ is the real orthogonal algebra $\mathfrak{o}(n)$ of skew-symmetric $n \times n$ matrices; since

$$B(U, U) = (n-2)\operatorname{tr}(UU) = -(n-2)\sum_{i,j} U_{ij}^2 \qquad U \in \mathfrak{o}(n)$$

which is negative for $U \neq 0$, $n > 2$, $\mathfrak{o}(n)$ is compact. The difficulty is that the Lie group $SO(n)$ (which is obtained by exponentiating $\mathfrak{o}(n) \subset \mathfrak{gl}(n, \mathbb{R})$) is not simply connected. For example, the fundamental group for $SO(2) \cong S^1$ is $\mathbb{Z}$, and it will be proved in 6.50 that $\pi_1 SO(3) \cong \mathbb{Z}_2 \cong \pi_1 SO(4)$; by arguments similar to those in 6.47 it can then be shown that $\pi_1 SO(n) \cong \mathbb{Z}_2$ for all $n > 2$ [S: 22.9].

6.49 **Definition** The 2-fold covering group of $SO(n)$ is denoted by $\operatorname{Spin}(n)$, $n \geq 2$, and is called the *spinor group*.

Thus $\operatorname{Spin}(n)$ is the universal covering group of $SO(n)$ for $n > 2$.

6.50 **Examples**

(*a*) $\operatorname{Spin}(3) \cong S^3 \cong Sp(1) \cong SU(2)$, that is, $C_1 \cong B_1 \cong A_1$.

Proof: It suffices to construct a 2-fold covering map $S^3 \to SO(3)$. Let $\mathcal{A}d: S^3 \to SO(4)$ be defined by $\mathcal{A}d_z v := zvz^{-1}$, $z \in S^3$, $v \in \mathbb{H}$. Identify $\mathbb{R}$ with the center of $\mathbb{H}$, which is the subspace of $\mathbb{H}$ spanned by the element 1; identify $\mathbb{R}^3$ with the span of i, j, and k in $\mathbb{H}$ in the obvious way. Since $\mathbb{R}$ is acted upon trivially by S^3, its orthogonal complement $\mathbb{R}^3$ is invariant under the action of S^3. Therefore the action of S^3 on $\mathbb{H}$ is the direct sum of the trivial action on $\mathbb{R}$ and the action of $SO(3)$ on $\mathbb{R}^3$:

$$\mathcal{A}d: S^3 \to SO(3) \cong \begin{bmatrix} 1 & 0 \\ 0 & SO(3) \end{bmatrix} \subset SO(4)$$

The kernel of $\mathcal{A}d$ is $\{1, -1\} \subset S^3$, and $\mathcal{A}d$ maps S^3 onto $SO(3)$, so $\mathcal{A}d$ is a covering homomorphism, which is universal because S^3 is simply connected. Hence $\pi_1 SO(3) \cong \ker \mathcal{A}d \cong \mathbb{Z}_2$, and $\text{Spin}(3) \cong S^3$. *QED.*

Exercises

1. Explicitly write down the covering map $SU(2) \to SO(3)$.
2. By 1.45*d* and [W: 3.65(a)], $SO(3)$ is the total space of a principal $SO(2)$-bundle over S^2 [called the *orthonormal frame bundle of* S^2 (9.5*b*)]. The composite map $S^3 \xrightarrow{\mathcal{A}d} SO(3) \xrightarrow{\not{p}} S^2$ is called the *Hopf fibration of* S^3 *over* S^2 (with fiber S^1). Show that $\not{p} \circ \mathcal{A}d$ is homotopically nontrivial, and therefore $\pi_3 S^2 \neq 0$.

(*b*) $\text{Spin}(4) \cong S^3 \times S^3 \cong \text{Spin}(3) \times \text{Spin}(3)$, that is, $C_2 \cong A_1 \times A_1$. *Proof:* Define $\varphi: S^3 \times S^3 \to SO(4)$ by $\varphi(z_1, z_2)(v) := z_1 v z_2^{-1}$, $v \in \mathbb{H}$. The homomorphism φ is surjective and has kernel $\{(1, 1), (-1, -1)\}$, and is therefore a covering homomorphism. Since $S^3 \times S^3$ is simply connected, it is the universal covering group of $SO(4)$. This being so, $\pi_1 SO(4) \cong \ker \varphi \cong \mathbb{Z}_2$, and $\text{Spin}(4) \cong S^3 \times S^3$. *QED.*

For the rest of example (*b*), φ will denote the map defined in the proof.

The manifolds $S^3 \times SO(3)$ and $SO(4)$ are diffeomorphic. *Proof:* By the identity $S^3 = SO(4)/SO(3)$, 1.45*d* implies that $SO(4)$ is a principal $SO(3)$-bundle over S^3; since the left-multiplication map $\varphi|_{S^3 \times \{1\}}$ is a section of this principal bundle, 1.52*e* implies that $SO(4)$ is the trivial principal bundle $S^3 \times SO(3)$. *QED.* The Lie groups $S^3 \times SO(3)$ and $SO(4)$, however, are not isomorphic. Instead, $SO(4)$ is isomorphic to the semidirect product $S^3 \rtimes SO(3)$, where multiplication in $S^3 \rtimes SO(3)$ is defined by $(u, g) \cdot (v, h) := (u \cdot a_g v, gh)$, $u, v \in S^3$, $g, h \in SO(3)$; here $SO(3)$ acts on S^3 by

$$a_g v := \begin{bmatrix} 1 & 0 \\ 0 & g \end{bmatrix} \varphi(v, 1) \begin{bmatrix} 1 & 0 \\ 0 & g^{-1} \end{bmatrix} \qquad g \in SO(3),\ v \in S^3$$

The isomorphism $S^3 \rtimes SO(3) \cong SO(4)$ is the map determined above by the section $\varphi|_{S^3 \times \{1\}}$ of the principal bundle $SO(4)$ over S^3.

Define a covering homomorphism f from $SO(4)$ to $SO(3) \times SO(3)$ by $f(g) := (\mathcal{A}d_h, \mathcal{A}d_k)$, where $g = \varphi(h, k) \in SO(4)$, $(h, k) \in S^3 \times S^3$, and $\mathcal{A}d$ is as defined in (*a*). We now have a commutative diagram of Lie group covering homomorphisms; the groups $SO(4)$ and $S^3 \times SO(3)$ are nonisomorphic 2-fold covering groups of $SO(3) \times SO(3)$.

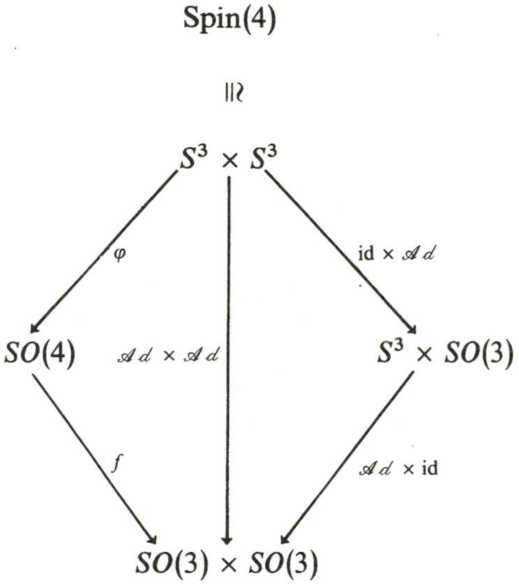

Exercise Recall the vector space isomorphism $\Lambda^2\mathbb{R}^n \cong \mathfrak{o}(n)$ from 3.13; under commutation, $\Lambda^2\mathbb{R}^n$ is a Lie algebra isomorphic to $\mathfrak{o}(n)$. Interpret the eigenspace decomposition of ★ on $\Lambda^2\mathbb{R}^4$ in terms of $\mathfrak{o}(4)$.

(*c*) $\text{Spin}(5) \cong Sp(2)$, that is, $C_2 \cong B_2$.
(*d*) $\text{Spin}(6) \cong SU(4)$, that is, $D_3 \cong A_3$.
Proof: It is expedient to prove (*d*) first. Define a quadratic form α on $\Lambda^2\mathbb{C}^4$ by $\alpha(u, v) := u \wedge v \in \Lambda^4\mathbb{C}^4 \cong \mathbb{C}$ (the exterior product is taken over $\mathbb{C}$, not $\mathbb{R}$); α is nondegenerate, and therefore an appropriate basis $\{\omega_j\}$ for $(\Lambda^2\mathbb{C}^4)^*$ puts α into the diagonal form $\alpha = \sum_{j=1}^{6} \omega_j^2$. Hence the subgroup of $GL(\Lambda^2\mathbb{C}^4)$ which leaves α invariant is isomorphic to the complex orthogonal group $O(\Lambda^2\mathbb{C}^4) \cong O(6, \mathbb{C})$ (6.48). The action of $SL(4, \mathbb{C})$ on $\mathbb{C}^4$ induces an action on $\Lambda\mathbb{C}^4$ (as in 1.35*c*); given $g \in SL(4, \mathbb{C})$ and $\sigma \in \Lambda^4\mathbb{C}^4$, $g\sigma = (\det g)\sigma = \sigma$, so $SL(4, \mathbb{C})$ leaves α invariant. Thus there is induced a homomorphism from $SL(4, \mathbb{C})$ to $O(\Lambda^2\mathbb{C}^4)$; the kernel is $\{\pm I\}$, so since $\dim SL(4, \mathbb{C}) = 30 = \dim O(6, \mathbb{C})$, this is a covering homomorphism, and $\mathfrak{sl}(4, \mathbb{C}) \cong \mathfrak{o}(\Lambda^2\mathbb{C}^4) \cong \mathfrak{o}(6, \mathbb{C})$. It follows that the compact real forms $\mathfrak{su}(4)$ and $\mathfrak{o}(6)$ are isomorphic, and therefore

so are the simply connected compact groups $SU(4)$ and Spin(6). *QED*. *Exercise:* Explicitly construct a covering map $SU(4) \to SO(6)$. (c): The skew-symmetric bilinear form

$$\Omega = \begin{bmatrix} 0 & I \\ -I & 0 \end{bmatrix}$$

on $\mathbb{C}^4$ which is preserved by $Sp(2, \mathbb{C})$ can be considered as a complex linear functional on $\Lambda^2\mathbb{C}^4$; its kernel is a subspace of $\Lambda^2\mathbb{C}^4$ of complex dimension 5. Since $SU(4)$ preserves the quadratic form α on $\Lambda^2\mathbb{C}^4 \supset \ker \Omega$, there is determined a homomorphism from $Sp(2) = SU(4) \cap Sp(2, \mathbb{C})$ to a compact subgroup of $O(\ker \Omega) \cong O(5, \mathbb{C})$; its kernel is $\{\pm I\} = \mathbb{Z}_2$, so this is a covering homomorphism. But the only connected compact subgroup of $O(5, \mathbb{C})$ of dimension $10 = \dim Sp(2)$ is $SO(5)$. Thus $Sp(2)$ is a 2-fold covering group of $SO(5)$, and $\text{Spin}(5) \cong Sp(2)$. *QED*.

For another proof of these isomorphisms, see [Pt: 13.60, 13.61].

6.51 For the exceptional Lie algebras, see [HS].

6.52 **Definition** The *Lorentz group* $O(k, n-k)$ is the subgroup of $GL(n, \mathbb{R})$ which leaves invariant the *Lorentz inner product* $\sum_1^k dx_j^2 - \sum_{k+1}^n dx_j^2$ *of signature* $(k, n-k)$ *on* $\mathbb{R}^n$ (see 3.39, 3.55c, 5.19).

It follows that the Lie algebra of $O(k, n-k)$ is

$$\mathfrak{o}(k, n-k) = \{U \in \mathfrak{gl}(n, \mathbb{R}) \mid {}^tUK + KU = 0\} \qquad K = \begin{bmatrix} I_k & 0 \\ 0 & -I_{n-k} \end{bmatrix}$$

where I_k is the $k \times k$ identity matrix.

Exercise Show that $\mathfrak{o}(k, n-k)$ is a noncompact real form of $\mathfrak{o}(n, \mathbb{C})$, and that $\mathfrak{o}(n)$ is related to $\mathfrak{o}(k, n-k)$ via the Weyl unitary trick.

HOMOGENEOUS SPACES

6.53 **Definition** An action of a Lie group G on a manifold M is called *effective* if the identity element e in G is the only element of G which acts trivially on M: if $gp = p$ for all $p \in M$, then $g = e$.

6.54 Proposition (Bourguignon, Yau) The only connected, compact Lie group which can act effectively on the torus T^n is a torus T^k, $k \leq n$.

PROOF Suppose that a connected compact group G acts effectively on T^n. It suffices to show that the Lie algebra $\mathfrak{g}$ of G is abelian. By averaging a metric on T^n over the action of G by integration (see the proof of 6.8) we may assume that G acts on T^n by isometries. By 5.13, $\mu(X)$ is constant on T^n if μ is a harmonic 1-form and X is a Killing field on T^n. Therefore if X and Y are Killing fields on T^n, then for each harmonic 1-form μ on T^n,

$$\mu([X, Y]) = X\mu(Y) - Y\mu(X) - d\mu(X, Y) = 0 \qquad (d\mu = 0)$$

If $\{\mu^i\}$ is a basis for the space of harmonic 1-forms on T^n, $i = 1, \ldots, n$, then $\mu^1 \wedge \cdots \wedge \mu^n$ is cohomologous to the volume form on T^n, and is therefore nonzero on some open subset of T^n. Therefore $[X, Y] = 0$ on this open set; since $[X, Y]$ is a Killing field on T^n, it is zero everywhere on T^n. But by hypothesis, G is a subgroup of the isometry group of T^n; thus (3.53) $\mathfrak{g}$ is isomorphic to a Lie subalgebra of the Killing fields on T^n, and is therefore abelian.

For an alternative proof, see [Y].

Exercise In the proof above, instead of averaging a metric on T^n over the action of G, it suffices to average a linear connection on T^n over the action of G.

6.55 Proposition Let $M = G/H$ be a homogeneous space. The subgroup N of G which acts trivially on M is the largest normal subgroup of G which lies in H. In particular,

(i) G/N acts effectively on M,
(ii) $M \cong (G/N)/(H/N)$, and
(iii) G acts effectively on M if and only if N is trivial.

PROOF It is clear that N lies in H; N is normal in G because given $n \in N$ and $g, h \in G$, $gng^{-1}\wp(h) = gn\wp(g^{-1}h) = g\wp(g^{-1}h) = \wp(h)$. Conversely, every normal subgroup of G acts trivially on M if it lies in H. Now let $n \in N$ and $g_1 \in G$; by normality there exists $n_1 \in N$ such that $ng_1 = g_1 n_1$, so for all $g \in G$, $gn\wp(g_1) = g\wp(g_1 n_1) = g\wp(g_1)$. Thus there is a well-defined action of G/N on M; this action is C^∞, effective, and transitive, and the isotropy subgroup of $\wp(e) \in M$ is H/N.

6.56 Example: The Berger spheres [B4] The sphere S^n is a homogeneous space $O(n+1)/O(n)$; the fact that S^3 is a Lie group lets us consider it as a homogeneous space in a different way. Let X be a nonzero left-invariant vector field on S^3, and let c be the 1-parameter subgroup of S^3 generated by X. Fix $b \in (0, 1]$, and let a be the square root of $1 - b^2$; define an action of $S^3 \times \mathbb{R}$ on S^3 by

$$(S^3 \times \mathbb{R}) \times S^3 \to S^3 \qquad ((g, s), v) \mapsto gvc(-as/b)$$

This action is obviously transitive, and the isotropy subgroup in $S^3 \times \mathbb{R}$ of the identity element $e \in S^3$ is $H := \{(c(at), bt) \mid t \in \mathbb{R}\} \cong \mathbb{R}$. Thus S^3 is $S^3 \times \mathbb{R}$ modulo a skew action of $\mathbb{R}$ on $S^3 \times \mathbb{R}$.

The action of $S^3 \times \mathbb{R}$ on S^3 is not effective: if $a = 0$, then the normal subgroup $\{e\} \times \mathbb{R}$ acts trivially, while if $a > 0$, then the normal subgroup $N := \{(e, 2\pi bj/a) \mid j \in \mathbb{Z}\}$ acts trivially. For $a = 0$, the quotient group $(S^3 \times \mathbb{R})/(\{e\} \times \mathbb{R}) \cong S^3$ acts effectively on S^3, while for $a > 0$, the quotient group $(S^3 \times \mathbb{R})/N \cong S^3 \times S^1$ acts effectively.

The Berger spheres will appear again in 6.61*c*.

Exercise The group $GL(n+1, \mathbb{F})$ acts transitively on projective space $\mathbb{F}P^n$. Find a group which acts effectively and transitively on $\mathbb{F}P^n$.

6.57 **Proposition** Let $M = G/H$ be a homogeneous space, where H is the isotropy subgroup of $p \in M$. If the linear isotropy representation $\lambda: H \to GL(M_p)$, $h \mapsto h_*|_p$, is *faithful*, that is, injective, then G acts effectively on M. Conversely, if M is connected and admits a G-invariant linear connection, and if the action of G on M is effective, then λ is faithful.

PROOF $\Rightarrow$: If $g \in G$ acts trivially on M, then $g \in H$, and $\lambda g = g_*|_p$ is the identity on M_p; thus $g = e \in H \subset G$.

$\Leftarrow$: Let $\lambda h = id$, $h \in H$. Since G acts on (M, ∇) by affine transformations, h is completely determined by $\lambda h = h_*|_p$ by 2.116. Thus h acts trivially on M, and therefore equals e.

6.58 The following result adds to the condition for the existence of an invariant metric on a homogeneous space from 3.4*d*.

Proposition The following conditions are equivalent for a homogeneous space $M = G/H$ (set $p = \wp(e) \in M$):

(i) G/H admits a G-invariant Riemannian metric, that is, the structure of a Riemannian homogeneous space.

(ii) M_p admits an inner product invariant under the subgroup λH of $GL(M_p)$, where λ is the linear isotropy representation of H on M_p.
(iii) $\mathfrak{g}/\mathfrak{h}$ admits an Ad_H-invariant inner product.
(iv) The closure of the subgroup λH of $GL(M_p)$ is compact.
(v) The closure of the subgroup Ad_H of $GL(\mathfrak{g}/\mathfrak{h})$ is compact.

PROOF The equivalence of (i) and (ii) was proved in 3.4*d*.

The adjoint action of H on $\mathfrak{g}$ is the restriction to H of the usual adjoint action of G on $\mathfrak{g}$; since $\mathrm{Ad}_h X \in \mathfrak{h}$ for $h \in H$ and $X \in \mathfrak{h}$, this induces an "adjoint" action of H on $\mathfrak{g}/\mathfrak{h}$. The representations Ad of H on $\mathfrak{g}/\mathfrak{h}$ and λ of H on M_p are equivalent under the isomorphism $\not{p}_\#: GL(\mathfrak{g}/\mathfrak{h}) \to GL(M_p)$ induced by the projection map $\not{p}$ from G to M, for since $\not{p}(hg) = \not{p}(hgh^{-1})$ for $h \in H$ and $g \in G$, $\lambda(h)\not{p}_*(X + \mathfrak{h}) = \not{p}_* \mathrm{Ad}_h(X + \mathfrak{h}) = (\not{p}_\# \mathrm{Ad}_h)\not{p}_*(X + \mathfrak{h})$ for all $h \in H$ and $X + \mathfrak{h} \in \mathfrak{g}/\mathfrak{h}$. This implies the equivalence of (ii) and (iii), and of (iv) and (v).

The equivalence of (ii) and (iv) follows by the argument in 6.8.

6.59 If G admits a left-invariant metric which is bi-invariant under the multiplication of the subgroup H, then G/H admits a G-invariant metric by 6.58, for an invariant metric on G which is bi-invariant under H is nothing but an inner product on $\mathfrak{g}$ that is invariant under the subgroup $\mathrm{Ad}(H) \subset \mathrm{Ad}(G) = \mathrm{Int}(\mathfrak{g})$, and therefore determines an Ad_H-invariant inner product on $\mathfrak{g}/\mathfrak{h}$. The converse is false in general—a G-invariant metric on G/H does not necessarily come from an invariant metric on G which is bi-invariant under H unless G acts effectively on G/H.

Proposition The following conditions are equivalent if G acts effectively on a connected homogeneous space $M = G/H$:
(i) M admits a G-invariant Riemannian metric.
(ii) G admits a left-invariant metric which is bi-invariant under H.
(iii) $\mathfrak{g}$ admits an Ad_H-invariant inner product.
(iv) $\mathrm{Ad}(H)$ has compact closure in $GL(\mathfrak{g})$.

PROOF The equivalence of (ii), (iii), and (iv) is similar to the corresponding proofs in 6.17 and 6.58; these conditions imply (i) by the discussion before the proposition.

(i) $\Rightarrow$ (iv): Let K be the full isometry group of the Riemannian homogeneous space M; since G acts effectively on M, G is a subgroup of K. The subgroup H of G is then a subgroup of the full isotropy subgroup $I \subset K$ of the point $\not{p}e \in M$; I is known to be compact [He: 4.2.5], so its image $\mathrm{Ad}_K I \subset GL(\mathfrak{k})$ under the adjoint representation Ad_K of K on $\mathfrak{k}$ is compact. The closure L of $\mathrm{Ad}_K H$ is then compact in $GL(\mathfrak{k})$. But $GL(\mathfrak{g})$ is a closed subgroup of $GL(\mathfrak{k})$, and $L \cap GL(\mathfrak{g})$ is the closure of $\mathrm{Ad}_G H \subset GL(\mathfrak{g})$; thus $\mathrm{Ad}_G H$ has compact closure in $GL(\mathfrak{g})$.

Exercise Does $\mathrm{Ad}_G H$ have compact closure in $\mathrm{Int}(\mathfrak{g})$? Prove that H and $\mathrm{Ad}_G H$ are isomorphic if G acts effectively on G/H.

6.60 An important special case is where G admits a bi-invariant metric.

Definition A Riemannian homogeneous space $M = G/H$ is called *normal homogeneous* if there exists a bi-invariant metric on G such that $p_*|_e$ maps the orthogonal complement $\mathfrak{h}^\perp$ of $\mathfrak{h}$ in $\mathfrak{g}$ isometrically to M_{pe}.

Notice that in the definition it is not assumed that G acts effectively on G/H.

6.61 **Examples**

(*a*) Define a bi-invariant metric on $SO(n+1)$ by $\langle U, V\rangle := \frac{1}{2}\,\mathrm{tr}(U\,{}^tV) = -B(U, V)/(2n-2)$, $n \geq 2$, where B is the Killing form of $\mathfrak{o}(n+1)$; since B is bi-invariant, so is $\langle\ ,\ \rangle$. Thus there is a normal homogeneous metric on the homogeneous space $S^n = SO(n+1)/SO(n)$. The isotropy subgroup and subalgebra of the point $e_0 \in S^n \subset \mathbb{R}^{n+1}$ are

$$H = \begin{bmatrix} 1 & 0 \\ 0 & SO(n) \end{bmatrix} \qquad \mathfrak{h} = \begin{bmatrix} 0 & 0 \\ 0 & \mathfrak{o}(n) \end{bmatrix}$$

The orthogonal complement of $\mathfrak{h}$ is the subspace

$$\mathfrak{m} := \left\{ \tilde{v} = \begin{bmatrix} 0 & -{}^tv \\ v & 0 \end{bmatrix} \in \mathfrak{o}(n+1) \,\middle|\, v \in \mathbb{R}^n \text{ (as a column vector)} \right\}$$

The metric on S^n is defined by the requirement that p_* map $\mathfrak{m} \subset \mathfrak{o}(n+1) \cong SO(n+1)_I$ isometrically to the tangent space on S^n at $e_0 = pI$. It follows that for each nonzero $v \in \mathbb{R}^n$, $p \circ \exp(t\tilde{v}) = \cos(t\|v\|)e_0 + \sin(t\|v\|)v/\|v\| \in S^n$; by 2.91c, this is a geodesic of the standard metric on S^n. In fact, the inner product on $S^n_{e_0}$ induced by p_* is the usual inner product from $\mathbb{R}^{n+1}$, so the normal homogeneous metric on $S^n = SO(n+1)/SO(n)$ is the usual metric on S^n.

(*b*) Similarly, the inner product $\langle U, V\rangle = \mathrm{tr}(U\,{}^t\bar{V}) = -B(U, V)/2n$ on the Lie algebra $\mathfrak{su}(n)$ determines a normal homogeneous metric on the homogeneous space $S^{2n-1} = SU(n)/SU(n-1)$. The standard embedding (6.47ii)

$$\varphi(A + iB) := \begin{bmatrix} A & -B \\ B & A \end{bmatrix}$$

of $\mathfrak{su}(n) \subset \mathfrak{gl}(n, \mathbb{C})$ into $\mathfrak{o}(2n) \subset \mathfrak{gl}(2n, \mathbb{R})$ is isometric with respect to the inner product defined here for $\mathfrak{su}(n)$ and the inner product on $\mathfrak{o}(2n)$ from (*a*). Nevertheless, $SU(n)$ does not induce the usual Riemannian metric on S^{2n-1} if $n > 2$. To see why, observe that the isotropy group of $e_1 \in \mathbb{C}^n$ is

$$K = \begin{bmatrix} 1 & 0 \\ 0 & SU(n-1) \end{bmatrix}$$

with Lie algebra

$$\mathfrak{k} = \left\{ \begin{bmatrix} 0 & 0 \\ 0 & A \end{bmatrix} \,\middle|\, A \in \mathfrak{su}(n-1) \right\}$$

The orthogonal complement of $\mathfrak{k}$ is

$$\mathfrak{n} = \left\{ \begin{bmatrix} (1-n)ai & -{}^t\bar{v} \\ v & aiI \end{bmatrix} \,\middle|\, v \in \mathbb{C}^{n-1}, a \in \mathbb{R} \right\}$$

The image $\varphi_* \mathfrak{n} \subset \mathfrak{o}(2n)$ does not equal the orthogonal complement $\mathfrak{m}$ of the subspace $\mathfrak{h}$ from (*a*). Therefore, the two groups induce different Riemannian structures on S^{2n-1}.

(*c*) The Berger spheres (continued from 6.56): Fix a nonconstant 1-parameter subgroup c in the Lie group S^3; fix $b \in (0, 1]$, and let $a = \sqrt{1 - b^2}$. Write S^3 as the homogeneous space $(S^3 \times \mathbb{R})/H$, where $H = \{(c(at), bt) \in S^3 \times \mathbb{R} \mid t \in \mathbb{R}\}$. The Lie group $S^3 \times \mathbb{R}$ has a natural bi-invariant metric—the product of the usual metrics on the factors; this determines a normal homogeneous metric on the homogeneous space $S^3 = (S^3 \times \mathbb{R})/H$.

Berger [B4] constructed these metrics as counterexamples to a conjecture of Klingenberg about closed geodesics ([*Kl1*], [GKM: 7.5A(iii)]). Since then they have served as counterexamples to conjectures on the first eigenvalue of the Laplacian on spheres ([Ur], [BB]). For other geometric points of view on the Berger spheres and related objects, see [We2], [Je3], and [Zi1].

More generally [Cha], define an action of $SU(n) \times \mathbb{R}$ on S^{2n-1} by $(g, t) \cdot v := e^{-iat/b} gv$, $v \in S^{2n-1} \subset \mathbb{C}^n$. If I is the $(n-1) \times (n-1)$ identity matrix, set

$$c(t) := \begin{bmatrix} e^{it} & 0 \\ 0 & e^{-it/n} I \end{bmatrix} \in SU(n)$$

and for a and b as above, let $H := \{(c(at), bt) \in SU(n) \times \mathbb{R} \mid t \in \mathbb{R}\}$. The usual bi-invariant metric on $SU(n) \times \mathbb{R}$ then determines a normal homogeneous metric on the homogeneous space $(SU(n) \times \mathbb{R})/H \cong S^{2n-1}$ [W: 3.65(b)].

Exercise Construct a family of bi-invariant metrics on $U(n)$ such that the associated normal homogeneous metrics on the homogeneous space $U(n)/U(n-1) = S^{2n-1}$ are the metrics constructed above.

6.62 Let $\mathfrak{m}$ be the orthogonal complement in $\mathfrak{g}$ of $\mathfrak{h}$, where $M = G/H$ is a normal homogeneous space; given $U \in \mathfrak{g}$, let $U = U^\top + U^\perp$ be the orthogonal decomposition according to $\mathfrak{g} = \mathfrak{h} \oplus \mathfrak{m}$.

Proposition If $M = G/H$ is a normal homogeneous space with Levi-Civita connection ∇, then $\nabla_U \wp_* V = \frac{1}{2}\wp_*[U, V]^\perp = \frac{1}{2}\wp_*[U, V]$, U, $V \in \mathfrak{m}$. The geodesics of ∇ are the curves of the form $\wp \circ \gamma$, where γ is an integral curve in G of some $U \in \mathfrak{m}$. The curvature tensor of ∇ is

$$R(\wp_* U, \wp_* V)\wp_* W = \tfrac{1}{4}\wp_*([U, [V, W]^\perp] + [V, [W, U]^\perp]) + \tfrac{1}{2}\wp_*[W, [U, V]^\perp] - \wp_*[[U, V]^\top, W]$$

for $U, V, W \in \mathfrak{m}$. If U and $V \in \mathfrak{m}$ have length 1, then

$$K_{\wp_* U \wedge \wp_* V} = \tfrac{1}{4}\|[U, V]^\perp\|^2 + \|[U, V]^\top\|^2$$

In particular, a normal homogeneous space has nonnegative curvature.

Proof The proof for ∇ is similar to 3.22*b* and 3.34*b*.

If γ is an integral curve of $U \in \mathfrak{m}$, then

$$\nabla_D \dot{\widehat{\wp \circ \gamma}} = \nabla_D \wp_* U \circ \gamma = \nabla_{\dot\gamma} \wp_* U = \nabla_{U \circ \gamma} \wp_* U = \tfrac{1}{2}\wp_*[U, U]_\gamma = 0$$

These are all the geodesics of M since $\wp_*$ is surjective.

By the structural equations from 2.103, for $U, V, W \in \mathfrak{m}$

$$\begin{aligned} R(\wp_* U, \wp_* V)\wp_* W &= \nabla_U \nabla_V \wp_* W - \nabla_V \nabla_U \wp_* W \\ &\quad - \nabla_W \wp_*[U, V] - \wp_*[[U, V], W] \\ &= \tfrac{1}{2}(\nabla_U \wp_*[V, W]^\perp - \nabla_V \wp_*[U, W]^\perp) \\ &\quad - \nabla_W \wp_*[U, V]^\perp - \wp_*[[U, V], W] \\ &= \tfrac{1}{4}\wp_*([U, [V, W]^\perp] - [V, [U, W]^\perp]) \\ &\quad - \tfrac{1}{2}\wp_*[W, [U, V]^\perp] - \wp_*[[U, V], W] \\ &= \tfrac{1}{4}\wp_*([U, [V, W]^\perp] + [V, [W, U]^\perp]) \\ &\quad + \tfrac{1}{2}\wp_*[W, [U, V]^\perp] - \wp_*[[U, V]^\top, W] \end{aligned}$$

Finally, for $U, V \in \mathfrak{m}$,

$$\begin{aligned}
\langle R(p_* U, p_* V)p_* V, p_* U\rangle
&= \tfrac{1}{4}\langle p_*[V, [V, U]^\perp], p_* U\rangle + \tfrac{1}{2}\langle p_*[V, [U, V]^\perp], p_* U\rangle \\
&\quad - \langle p_*[[U, V]^\top, V], p_* U\rangle \\
&= \tfrac{1}{4}\langle [V, [V, U]^\perp], U\rangle + \tfrac{1}{2}\langle [V, [U, V]^\perp], U\rangle - \langle [[U, V]^\top, V], U\rangle \\
&= \tfrac{1}{4}\|[U, V]^\perp\|^2 + \|[U, V]^\top\|^2
\end{aligned}$$

by the bi-invariance of the metric $\langle\ ,\ \rangle$ on G.

Exercise Let G be a compact Lie group. If H is a closed subgroup of G containing the *maximal torus of* G (that is, the identity component of $Z(G)$), then G/H admits an invariant metric of positive Ricci curvature.

6.63 More generally,

Definition A map $f: M \to N$ of Riemannian manifolds is called a *Riemannian submersion* if f is a submersion (1.5) and if for each $p \in M$, the *horizontal subspace of* M_p [orthogonal to the fiber over $f(p)$ in M] is mapped isometrically by $f_*|_p$ to $N_{f(p)}$.

It can be proved [Ol] that given a 2-plane σ in the horizontal space at $p \in M$, the sectional curvature of N with respect to the 2-plane $f_*\sigma$ is not less than the sectional curvature of M with respect to σ. This has found many applications in the study of nonnegatively curved manifolds; see, for example, [Ber], [GM], [Je2], [LY], [Na], [P1].

6.64 Let H be a Lie subgroup of G, with Lie algebra $\mathfrak{h} \subset \mathfrak{g}$; assume that $\mathfrak{m}$ is a direct sum complement to $\mathfrak{h}$ in $\mathfrak{g}$. With respect to an appropriate basis for $\mathfrak{g} = \mathfrak{h} \oplus \mathfrak{m}$, the restriction to H of $\mathrm{Ad}: G \to \mathrm{Int}(\mathfrak{g})$ takes the form

$$\mathrm{Ad}_h = \begin{array}{c} \begin{matrix} \mathfrak{h} & \mathfrak{m} \end{matrix} \\ \begin{bmatrix} A & B \\ 0 & C \end{bmatrix} \end{array} \begin{matrix} \mathfrak{h} \\ \mathfrak{m} \end{matrix} \qquad h \in H$$

since H is a subgroup of G. The submatrix $B = 0$ for all $h \in H$ if and only if the adjoint action of H on $\mathfrak{g}$, which is already reducible to an action on the subspace $\mathfrak{h}$ of $\mathfrak{g}$, is also reducible to an action on $\mathfrak{m}$; thus B is zero for all h in H if and only if $\mathrm{Ad}|_H$ is reducible to the direct sum of representations of H on $\mathfrak{h}$ and $\mathfrak{m}$.

Definition A homogeneous space G/H is called *reductive* if there exists a vector space decomposition $\mathfrak{g} = \mathfrak{h} \oplus \mathfrak{m}$ such that

$\mathrm{Ad}_H(\mathfrak{m}) \subset \mathfrak{m}$, where Ad_H is the restriction to H of $\mathrm{Ad}: G \to \mathrm{Int}(\mathfrak{g})$. In this case, $\mathfrak{h} \oplus \mathfrak{m}$ is called a *reductive decomposition of* $\mathfrak{g}$.

It will be seen that a reductive decomposition simplifies the study of the geometry of G/H. Not all homogeneous spaces are reductive. For example, punctured Euclidean space $\mathbb{R}^n - \{0\}$ is homogeneous since $GL(n, \mathbb{R})$ acts transitively on it. The isotropy subgroup H of the point $e_1 \in \mathbb{R}^n - \{0\}$ consists of all matrices of the form

$$\begin{bmatrix} 1 & v \\ 0 & g \end{bmatrix} \qquad g \in GL(n-1, \mathbb{R}),\ v \in \mathbb{R}^{n-1}$$

A typical element of the Lie algebra $\mathfrak{h}$ has the form

$$\begin{bmatrix} 0 & v \\ 0 & A \end{bmatrix} \qquad A \in \mathfrak{gl}(n-1, \mathbb{R}),\ v \in \mathbb{R}^{n-1}$$

and there is no direct sum complement to $\mathfrak{h}$ in $\mathfrak{gl}(n, \mathbb{R})$ which is Ad_H-invariant.

A reductive decomposition exists trivially if H is discrete in G (since $\mathfrak{h} = 0$); similarly, G/H is reductive if Ad_H is completely reducible to the direct sum of irreducible representations, that is, if there is a direct sum decomposition $\mathfrak{g} = \mathfrak{g}_1 \oplus \cdots \oplus \mathfrak{g}_k$ and an irreducible representation φ_j of H on each $\mathfrak{g}_j$ such that $\mathrm{Ad}_H|_{\mathfrak{g}_j} = \varphi_j$ for each j. Later we shall see that G/H is reductive if H is compact.

Exercise If $\mathfrak{g} = \mathfrak{h} \oplus \mathfrak{m}$ is a reductive decomposition of $\mathfrak{g}$, then $[\mathfrak{h}, \mathfrak{m}] \subset \mathfrak{m}$; the converse holds if H is connected.

6.65 **Proposition** If G/H is reductive, with reductive decomposition $\mathfrak{g} = \mathfrak{h} \oplus \mathfrak{m}$, then the invariant linear connections of G/H correspond to the (1, 2)-tensors on $\mathfrak{m}$ which commute with Ad_H.

PROOF Fix a linear map $\beta: \mathfrak{m} \otimes \mathfrak{m} \to \mathfrak{m}$ such that $\mathrm{Ad}_h \beta(U, V) = \beta(\mathrm{Ad}_h U, \mathrm{Ad}_h V)$, $U, V \in \mathfrak{m}$. Given $g \in G$ and $U, V \in \mathfrak{g}$ such that $\mathrm{Ad}_{g^{-1}} U$ and $\mathrm{Ad}_{g^{-1}} V$ are in $\mathfrak{m}$, define

$$\nabla_{U^*_{\not p g}} V^* := g_* \not p_* \beta(\mathrm{Ad}_{g^{-1}} U, \mathrm{Ad}_{g^{-1}} V)$$

where for $W \in \mathfrak{g}$, $W^* \in \mathfrak{X}(G/H)$ is defined as in 1.63*c* by

$$W^*_p := \frac{d}{dt}\bigg|_0 (\exp tW)p \qquad p \in G/H$$

Extend ∇ to all pairs of vector fields on G/H so that it is linear over $C^\infty M$ in the first argument and a derivation in the second argument.

Let $h \in H$. If $\mathrm{Ad}_{g^{-1}} W \in \mathfrak{m}$, $W \in \mathfrak{g}$, then $\mathrm{Ad}_{h^{-1}g^{-1}} W \in \mathfrak{m}$, so

$$\begin{aligned}\nabla_{U^*_{\not p gh}} V^* &= g_* h_* \not p_* \beta(\mathrm{Ad}_{h^{-1}g^{-1}} U, \mathrm{Ad}_{h^{-1}g^{-1}} V) \\ &= g_* \not p_* \mathrm{Ad}_h \beta(\mathrm{Ad}_{h^{-1}} \mathrm{Ad}_{g^{-1}} U, \mathrm{Ad}_{h^{-1}} \mathrm{Ad}_{g^{-1}} V) \\ &= g_* \not p_* \beta(\mathrm{Ad}_{g^{-1}} U, \mathrm{Ad}_{g^{-1}} V) = \nabla_{U^*_{\not p g}} V^*\end{aligned}$$

so ∇ is well-defined.

For each $W \in \mathfrak{g}$ and $k \in G$, $k_* W^* = (\mathrm{Ad}_k W)^* \circ k$ as a vector field along the map $k: G/H \to G/H$, so if $\mathrm{Ad}_{g^{-1}} U$ and $\mathrm{Ad}_{g^{-1}} V$ are in $\mathfrak{m}$, then

$$\begin{aligned}\nabla_{U^*_{\not p g}} k_* V^* &= \nabla_{U^*_{\not p g}} (\mathrm{Ad}_k V)^* \circ k \\ &= \nabla_{k_* U^*_{\not p g}} (\mathrm{Ad}_k V)^* = \nabla_{(\mathrm{Ad}_k U)^*_{\not p kg}} (\mathrm{Ad}_k V)^* \\ &= (kg)_* \not p_* \beta(\mathrm{Ad}_{g^{-1}k^{-1}} \mathrm{Ad}_k U, \mathrm{Ad}_{g^{-1}k^{-1}} \mathrm{Ad}_k V) \\ &= k_* \nabla_{U^*_{\not p g}} V^*\end{aligned}$$

Thus each $k \in G$ is an affine transformation of G/H by 2.114, and $(G/H, \nabla)$ is an affine homogeneous space (2.123iv).

Conversely, if a linear connection ∇ on G/H is invariant, then the (1, 2)-tensor β on $\mathfrak{m}$ defined by $\beta(U, V) := \not p_*|_{\mathfrak{m}}^{-1} \nabla_{U^*_{\not p e}} V^*$ is Ad_H-invariant.

6.66 **Corollary** A reductive homogeneous space admits an invariant connection.

Three natural examples of invariant connections related to a reductive decomposition $\mathfrak{g} = \mathfrak{h} \oplus \mathfrak{m}$ are determined by $\beta = 0$, $\beta = [\cdot, \cdot]_{\mathfrak{m}}$, and $\beta = -[\cdot, \cdot]_{\mathfrak{m}}$, where $[U, V]_{\mathfrak{m}}$ is the component of $[U, V]$ in the subspace $\mathfrak{m}$ of $\mathfrak{g}$. The *canonical connection* on a reductive homogeneous space G/H with respect to a reductive decomposition $\mathfrak{g} = \mathfrak{h} \oplus \mathfrak{m}$ is the one with $\beta = 0$.

Reductivity is not a necessary condition for the existence of an invariant connection; for example, the nonreductive homogeneous space $\mathbb{R}^n - \{0\}$ is naturally an affine homogeneous space.

6.67 Although $\mathbb{R}^n - \{0\}$ (which is not reductive) is naturally an affine homogeneous space, the natural invariant connection cannot be the Levi-Civita connection of a homogeneous Riemannian metric on $\mathbb{R}^n - \{0\}$, because by 3.55*d* every Riemannian homogeneous space is complete, while the standard connection on $\mathbb{R}^n - \{0\}$ is incomplete.

Proposition If M is a Riemannian homogeneous space G/H, where G acts effectively on M, then G/H is reductive.

PROOF By 6.59, $\mathfrak{g}$ admits an Ad_H-invariant inner product; set $\mathfrak{m} := \mathfrak{h}^\perp$. Given $U \in \mathfrak{h}$ and $V \in \mathfrak{m}$, $\langle U, \mathrm{Ad}_h V\rangle = \langle \mathrm{Ad}_{h^{-1}} U, V\rangle = 0$, so $\mathfrak{m}$ is invariant under Ad_H.

Conversely, a reductive homogeneous space admits an invariant Riemannian metric if Ad_H has compact closure in $GL(\mathfrak{m})$ (cf. 6.58); this is true, in particular, if H is compact.

6.68 Examples

(*a*) By 6.27, $\mathfrak{gl}(n, \mathbb{R}) = \mathfrak{o}(n) \oplus \mathfrak{sym}(n)$ is a reductive decomposition, so $GL(n, \mathbb{R})/SO(n)$ is a reductive homogeneous space; it admits an invariant Riemannian metric since $SO(n)$ is compact. Similarly, $SL(n, \mathbb{R})/SO(n)$, $U(n)/SO(n)$, $SU(n)/SO(n)$, $Sp(n, \mathbb{R})/SU(2n)$, and $Sp(n)/SU(2n)$ are reductive Riemannian homogeneous spaces.

(*b*) If $\mathfrak{h}$ is a Lie subalgebra of a Lie algebra $\mathfrak{g}$ and $\mathfrak{m}$ is a direct sum complement of $\mathfrak{h}$ in $\mathfrak{g}$, write $X_\mathfrak{m}$ for the component of $X \in \mathfrak{g}$ in $\mathfrak{m}$. A reductive Riemannian homogeneous space G/H is called *naturally reductive* if there exists a reductive decomposition $\mathfrak{g} = \mathfrak{h} \oplus \mathfrak{m}$ such that $\mathrm{ad}_X \colon \mathfrak{m} \to \mathfrak{m}$ is skew-symmetric for all $X \in \mathfrak{m}$:

$$\langle [X, Y]_\mathfrak{m}, Z\rangle + \langle Y, [X, Z]_\mathfrak{m}\rangle = 0 \qquad X, Y, Z \in \mathfrak{m}$$

Every normal Riemannian homogeneous space is naturally reductive, but the converse is false; for example [Zi2], the sphere S^{2n-1} admits a family of naturally reductive homogeneous metrics, of which the normal homogeneous Berger metrics from 6.61*c* are a proper subfamily.

As in 6.62, the geodesics of a naturally reductive homogeneous metric on G/H are the projections of the integral curves of the vector fields on G which belong to $\mathfrak{m}$. For more information about naturally reductive Riemannian homogeneous spaces, see [DaZ], where a complete classification of the naturally reductive metrics is given for simple Lie groups.

CHAPTER

SEVEN

SYMMETRIC SPACES

AFFINE SYMMETRIC SPACES

7.1 **Definition** Let ∇ be a linear connection on a manifold M; fix $p \in M$. If there exists an affine map $\mathscr{s}: M \to M$ such that $\mathscr{s} \circ \gamma(t) = \gamma(-t)$ for each geodesic γ with $\gamma(0) = p$, then $\mathscr{s}$ is called an *affine symmetry* (or *geodesic symmetry*) *of* (M, ∇) *about* p. If there exists an affine symmetry of (M, ∇) about each $p \in M$, then (M, ∇) is called an *affine symmetric space.*

It follows immediately from the definition that an affine symmetry $\mathscr{s}$ of (M, ∇) about $p \in M$ (M connected) is an involution of M, that is, $\mathscr{s} \neq \mathrm{id}_M$, but $\mathscr{s}^2 = \mathrm{id}_M$; in particular, $\mathscr{s}$ is a diffeomorphism of M. Furthermore, $\mathscr{s}_*|_p = -\mathrm{id}_{M_p}$; hence by 2.116 there is at most one affine symmetry about each $p \in M$ if M is connected.

7.2 **Proposition** If M is affine symmetric with respect to linear connections $\bar{\nabla}$ and ∇, and if the affine symmetries of $\bar{\nabla}$ and ∇ about each $p \in M$ agree, then $\bar{\nabla} = \nabla$.

PROOF As in 2.99, let $\mathscr{D} := \bar{\nabla} - \nabla$ be the connection difference tensor field; by hypothesis, if $\mathscr{s}$ is the affine symmetry about $p \in M$, then $\mathscr{s}_*|_p$ commutes with $\mathscr{D}$. Hence for $u, v \in M_p$,

$$\mathscr{D}(u, v) = \mathscr{D}(\mathscr{s}_* u, \mathscr{s}_* v) = \mathscr{s}_* \mathscr{D}(u, v) = -\mathscr{D}(u, v) = 0$$

Exercise If (M, ∇) is an affine symmetric space, prove that $T = \nabla R = 0$ (this will be proved in more generality in 7.16).

7.3 **Proposition** An affine symmetric space is affinely complete, and a connected affine symmetric space is affine homogeneous.

PROOF A geodesic segment $\gamma: [0, \tau] \to M$ can be extended to $[0, 2\tau]$ by reflecting γ in the point $\gamma(\tau)$:

$$\tilde{\gamma}(s) := \begin{cases} \gamma(s) & 0 \le s < \tau \\ \mathscr{s}_{\gamma(\tau)} \circ \gamma(2\tau - s) & \tau \le s \le 2\tau \end{cases}$$

Iteration yields an extension of γ to a geodesic defined on $[0, \infty)$, which can then be extended similarly to $\mathbb{R}$.

Now assume M is connected. It suffices to show that the affine transformation group $A(M, \nabla)$ (2.117) acts transitively on M. Fix p and $q \in M$, and let $\gamma = \gamma_k * \gamma_{k-1} * \cdots * \gamma_1$ be a broken geodesic joining p to q; denote by $\mathscr{s}_i$ the affine symmetry about the midpoint of γ_i. The composite map $\mathscr{s} := \mathscr{s}_k \circ \mathscr{s}_{k-1} \circ \cdots \circ \mathscr{s}_1$ is an affine transformation of (M, ∇) which moves p to q.

7.4 Unless otherwise stated, every symmetric space referred to from now on will be assumed to be connected.

Let (M, ∇) be an affine symmetric space; denote by G the identity component $A(M, \nabla)_o$ of the affine transformation group of M. For each $p \in M$, the orbit $G \cdot p$ of p under G is a connected set in M. Furthermore $G \cdot p$ is open. *Proof:* If H is the isotropy subgroup of p in $A(M, \nabla)$, let $\mathscr{p}$ be the projection map from $A(M, \nabla)$ to $A(M, \nabla)/H = M$. We must show that $\mathscr{p}^{-1}(G \cdot p)$ is open in $A(M, \nabla)$. But $G \cdot p = \mathscr{p}(G) \subset M$, so $\mathscr{p}^{-1}(G \cdot p) = \bigcup_{h \in H} R_h(G)$; this is open because G is open and each right translation R_h of $A(M, \nabla)$ is a diffeomorphism. *QED*. Given $q \in M$, $G \cdot p$ and $G \cdot q$ are either equal or disjoint. Since M is connected, it equals $G \cdot p$; thus G acts transitively on M.

Thus we can write $M = G/H$, where H is the isotropy subgroup of a point p_o in M; denote by $\mathscr{s} = \mathscr{s}_{p_o}$ the affine symmetry of (M, ∇) about p_o. Define a homomorphism $\sigma: G \to A(M, \nabla)$ by $\sigma(g) := \mathscr{s} g \mathscr{s} = \mathscr{s} g \mathscr{s}^{-1}$. Since $\sigma e = e$ and G is connected, σ maps G into itself; in fact, σ is an involutive automorphism of G: $\sigma^2 = \mathrm{id}_G$, $\sigma \neq \mathrm{id}_G$, and $\sigma(gh) = \sigma g \cdot \sigma h$, $g, h \in G$. Let G_σ be the group of fixed points of σ in G.

Proposition The identity component $(G_\sigma)_o$ of G_σ equals the identity component H_o of H, and H is a subgroup of G_σ.

Proof Given $h \in H$ and $v \in M_{p_o}$,

$$(\varsigma \circ h)_* v = \varsigma_* h_* v = -h_* v = h_*(-v) = (h \circ \varsigma)_* v$$

so $(\varsigma h \varsigma)_* |_{p_o} = h_* |_{p_o}$. Since M is connected, $\varsigma h \varsigma = h$ by 2.116. Hence $H \subset G_\sigma$, and therefore $H_o \subset (G_\sigma)_o$.

Now let c be a 1-parameter subgroup of G_σ; c lies in $(G_\sigma)_o$, and in particular, $\varsigma c \varsigma = c$, so $\varsigma c(t) p_o = c(t) \varsigma p_o = c(t) p_o$ for all t. Thus $c(t)p_o = p_o$ for all t because $p_o = c(0)p_o$ is an isolated fixed point of $\varsigma \in A(M, \nabla)$. Therefore c lies in H, and $(G_\sigma)_o \subset H_o$.

7.5 **Definition** A pair (G, H) of Lie groups, where H is a closed subgroup of G, is called a *symmetric pair* if there exists an involutive automorphism σ of G such that $(G_\sigma)_o \subset H \subset G_\sigma$.

Every affine symmetric space therefore determines a symmetric pair by 7.4; next we shall see that the converse holds.

7.6 **Theorem** The following statements are equivalent for a connected manifold M:

(i) M admits the structure of an affine symmetric space.

(ii) M is a homogeneous space G/H, where (G, H) is a symmetric pair.

(iii) There exists a C^∞ map μ: $M \times M \to M$ such that for all $p, q, r \in M$, $\mu(p, p) = p$, $\mu(p, \mu(p, q)) = q$, $\mu(p, \mu(q, r)) = \mu(\mu(p, q), \mu(p, r))$, and there exists a neighborhood U of p such that if $q \in U$ and $\mu(p, q) = q$, then $q = p$.

In this case, G can be chosen to act effectively on M.

Proof (i) $\Rightarrow$ (ii): This was proved in 7.3 to 7.5.

(ii) $\Rightarrow$ (iii): Let $\wp$: $G \to M$ be the projection. Define μ by

$$\mu(\wp g, \wp h) := \wp(g\sigma(g^{-1}h)) \qquad g, h \in G$$

This is well-defined since $H \subset G_\sigma$. For all g, h, and $k \in G$,

$$\mu(\wp g, \wp g) = \wp(g\sigma(g^{-1}g)) = \wp g$$

$$\mu(\wp g, \mu(\wp g, \wp h)) = \mu(\wp g, \wp(g\sigma(g^{-1}h)))$$
$$= \wp(g\sigma(g^{-1}g\sigma(g^{-1}h))) = \wp h$$

and

$$\mu(\mu(\wp g, \wp h), \mu(\wp g, \wp k)) = \wp(g\sigma(g^{-1}h)\sigma(\sigma(g^{-1}h)^{-1}g^{-1}g\sigma(g^{-1}k)))$$
$$= \wp(g\sigma(g^{-1}h)\sigma(\sigma h^{-1}\sigma k))$$
$$= \wp(g\sigma(g^{-1}h\sigma(h^{-1}k))$$
$$= \mu(\wp g, \mu(\wp h, \wp k))$$

Finally, since exp is injective on a neighborhood of 0 in $\mathfrak{g}$, there is a neighborhood W of e in G such that e is the only solution to the equation $g^2 = e$, $g \in W$. Let V be a connected neighborhood of e in G such that $h^{-1}\sigma(h) \in W$ for all $h \in V$.

Now let $h \in V$ such that $\mu(\wp e, \wp h) = \wp h$. By the definition of μ, $\wp \circ \sigma(h) = \wp h$, so $h^{-1}\sigma(h) \in H \subset G_\sigma$. Thus

$$(h^{-1}\sigma(h))^2 = h^{-1}\sigma(h)\sigma(h^{-1}\sigma(h)) = h^{-1}\sigma(h)\sigma(h^{-1})h = e$$

so $h^{-1}\sigma(h) = e$; since V is connected, $h \in (G_\sigma)_o \subset H$, and $\wp h = \wp e$. Similarly, for each $g \in G$, the identity $\mu(\wp g, \wp gh) = \wp gh$, $h \in V$, implies $\wp gh = \wp g$ as required.

(iii) $\Rightarrow$ (i): Define $\mathscr{s}_p q := \mu(p, q)$ for all $p, q \in M$. It suffices to construct a linear connection ∇ such that each $\mathscr{s}_p$ is an affine transformation of (M, ∇). By the Ambrose-Palais-Singer theorem on sprays (2.98, 2.104) it is sufficient to construct a spray which is invariant.

If γ is a C^∞ curve in M with $\gamma(0) = p$, then $\mathscr{s}_p^2 \gamma = \gamma$, so $\mathscr{s}_{p*}^2 \dot\gamma(0) = \dot\gamma(0)$; as a result, $\mathscr{s}_{p*} v = \pm v$ for each $v \in M_p$. We will now see that -1 is the only possible eigenvalue. Let $\bar\nabla$ be a linear connection on M, and define $\nabla_X Y := \frac{1}{2}(\bar\nabla_X Y + \mathscr{s}_{p*} \bar\nabla_X \mathscr{s}_{p*} Y)$ for all vector fields X and Y. Since $\mathscr{s}_p$ is an affine map with respect to ∇, $\mathscr{s}_p \circ \exp_p = \exp_p \circ \mathscr{s}_{p*}$. Hence, if $\gamma(t) := \exp_p tv$, $v \in M_p$, then $\mathscr{s}_p \circ \gamma(t) = \exp_p(t\mathscr{s}_{p*} v)$, so $\mathscr{s}_p \circ \gamma = \gamma$ if $\mathscr{s}_{p*} v = v$; in this case γ is constant since p is an isolated fixed point of $\mathscr{s}_p$, and therefore $v = 0$. Therefore $\mathscr{s}_{p*} v = -v$ for all $v \in M_p$.

Define $f: TM \times M \to TM$ by $f(v, q) := -\mathscr{s}_{q*} v$; since μ is C^∞, so is f. For each $v \in TM$, set $\hat v(q) := f(v, q)$, and define $S_v := \frac{1}{2}\hat v_* v \in TTM$; the vector S_v is tangent to TM at v, for if $\pi(v) = p \in M$, then $\hat v(p) = -\mathscr{s}_{p*} v = v$. The vector field S is C^∞ because f is C^∞.

Now to show that S is a spray on TM. Let γ be a curve in M with $\dot\gamma(0) = v \in M_p$. Since $\mathscr{s}_{p*} v = -v$, $\mathscr{s}_p \circ \gamma(t) = \gamma(-t) + o(t)$ as $t \to 0$ [to deal with the expression $\gamma(-t) + o(t)$ rigorously, one can calculate everything in $\mathbb{R}^n$ with respect to a chart x about p]. More generally, $\mathscr{s}_{\gamma(\sigma)} \gamma(\sigma + \tau) = \gamma(\sigma - \tau) + o(\tau)$ as $\tau \to 0$. In particular, $\mathscr{s}_{\gamma(t/2)} p = \mathscr{s}_{\gamma(t/2)} \gamma(t/2 - t/2) = \gamma(t) + o(t/2)$, so

$$\pi_* S_v = (\pi \circ \hat v)_* \frac{v}{2} = \frac{d}{dt}\Big|_0 \pi \circ \hat v \circ \gamma\left(\frac{t}{2}\right) = \frac{d}{dt}\Big|_0 \pi(-\mathscr{s}_{\gamma(t/2)*} v)$$

$$= \frac{d}{dt}\Big|_0 \mathscr{s}_{\gamma(t/2)} p = \frac{d}{dt}\Big|_0 \left(\gamma(t) + o\left(\frac{t}{2}\right)\right) = \dot\gamma(0) = v$$

Homogeneity: Given $c \in \mathbb{R}$, let $m_c(v) := cv$ for all $v \in TM$. For $v \in TM$ and $q \in M$,

$$\widehat{cv}(q) = -\sigma_{q*} cv = -c\sigma_{q*} v = -m_c \circ \sigma_{q*} v = m_c \circ \hat{v}(q)$$

so

$$S_{cv} = \tfrac{1}{2}\widehat{cv}_* cv = \frac{c}{2}\widehat{cv}_* v = \frac{c}{2} m_{c*} \hat{v}_* v = c \cdot m_{c*} S_v$$

Now we must see why the spray S is invariant under all the maps σ_{p**}, $p \in M$. If $v = \dot{\gamma}(0) \in M_q$ and $r \in M$, then

$$\sigma_p \sigma_r \gamma(t) = \mu(p, \mu(r, \gamma(t))) = \mu(\sigma_p r, \sigma_p \gamma(t)) = \sigma_{\sigma_p r} \sigma_p \gamma(t)$$

so

$$\begin{aligned}\sigma_{p*} \hat{v}(r) &= -\sigma_{p*} \sigma_{r*} v = -\dot{\widehat{\sigma_p \sigma_r \gamma}}(0) = -\dot{\widehat{\sigma_{\sigma_p r} \sigma_p \gamma}}(0) \\ &= -\sigma_{\sigma_p r *} \sigma_{p*} v = \widehat{\sigma_{p*} v} \circ \sigma_p(r)\end{aligned}$$

Therefore,

$$\sigma_{p**} S_v = \tfrac{1}{2}(\sigma_{p*} \hat{v})_* v = \tfrac{1}{2}\widehat{\sigma_{p*} v}_* \sigma_{p*} v = S_{\sigma_{p*} v}$$

The implication (iii) ⇒ (i) in the theorem is due to Loos [Lo].

7.7 **Corollary 1** The linear connection on an affine symmetric space is torsion-free.

PROOF The torsion-free linear connection associated with the spray S constructed in the proof of 7.6 (cf. 2.98, 2.104) is invariant under the maps σ_p, and by 7.2 is the unique invariant connection.

7.8 **Corollary 2** The tangent bundle of an affine symmetric space is naturally an affine symmetric space.

PROOF If μ is the multiplication map $M \times M \to M$ associated with the affine symmetric structure of M, then the map $\mu_*: TM \times TM \to TM$ satisfies the same properties as μ (see 7.6iii), so (TM, μ_*) is an affine symmetric space.

Exercise By 7.3, the affine symmetric space TM must be affine homogeneous. More generally, verify that if $M = G/H$ is homogeneous, then TM is the homogeneous space TG/TH.

7.9 **Definition** Let (M, ∇, μ) be an affine symmetric space. Define

$\mathrm{Aut}(M, \mu) := \{g \in \mathrm{Diff}(M) \mid g\mu(p, q) = \mu(gp, gq) \text{ for all } p, q \in M\}$; define $\mathfrak{aut}(M, \mu) := \{\text{complete } X \in \mathfrak{X}M \mid \mu_*(X, X) = X \circ \mu\}$.

Exercise Prove that $\mathfrak{aut}(M, \mu)$ is a Lie algebra.

7.10 Proposition Let (M, ∇, μ) be a connected affine symmetric space. The group $\mathrm{Aut}(M, \mu)$ is equal to the affine transformation group $A(M, \nabla)$ of the linear connection ∇; in particular, $\mathrm{Aut}(M, \mu)$ is a Lie group of dimension not greater than $n^2 + n$, where $n = \dim M$. The Lie algebra of $\mathrm{Aut}(M, \mu)$ is $\mathfrak{aut}(M, \mu) = \mathfrak{a}(M, \nabla)$, that is, the infinitesimal automorphisms of μ are just the complete affine vector fields on (M, ∇).

Proof Let $g \in \mathrm{Aut}(M, \mu)$. If γ is a geodesic, with $v := \dot\gamma(0)$, and if S is the geodesic spray from the proof of 7.6, then by 2.93,

$$\begin{aligned} g_{**} S_v = g_{**}\ddot\gamma(0) &= \frac{d}{dt}\bigg|_0 g_*\dot\gamma(t) = \frac{d}{dt}\bigg|_0 g_* \sigma_{\gamma(t/2)*}(-v) \\ &= \frac{\partial^2}{\partial t\,\partial s}\bigg|_0 g\mu(\gamma(t/2), \gamma(-s)) = \frac{\partial^2}{\partial t\,\partial s}\bigg|_0 \mu(g\gamma(t/2), g\gamma(-s)) \\ &= \frac{d}{dt}\bigg|_0 \sigma_{g\gamma(t/2)*}(-g_* v) = \frac{d}{dt}\bigg|_0 \widehat{g_* v} \circ g\gamma(t/2) \\ &= \tfrac{1}{2}\frac{d}{dt}\bigg|_0 \widehat{g_* v} \circ g\gamma(t) = \tfrac{1}{2}\widehat{g_* v}_* g_* v = S_{g_* v} \end{aligned}$$

Therefore, $\mathrm{Aut}(M, \mu) \subset A(M, \nabla)$.

Now suppose that $g \in A(M, \nabla)$; fix $p \in M$. Both $\sigma_{gp} \circ g$ and $g \circ \sigma_p$ are affine maps; they are equal by 2.116 because for all $v \in M_p$,

$$(\sigma_{gp} \circ g)_* v = \sigma_{gp*} g_* v = -g_* v = g_*(-v) = g_* \sigma_{p*} v = (g \circ \sigma_p)_* v$$

Thus, $A(M, \nabla) \subset \mathrm{Aut}(M, \mu)$.

The proof for the Lie algebras is similar to 2.119 and 2.123.

Fix $p \in M$. It will be proved in 7.25 that if $X \in \mathfrak{aut}(M, \mu)$ satisfies the identity $\sigma_{p*} X + X \circ \sigma_p = 0$, then $(\nabla X)_p = 0$.

7.11 Examples

(*a*) ***Lie groups*** Let ∇ be the torsion-free bi-invariant connection on a Lie group G defined by $\nabla_U V = \frac{1}{2}[U, V]$, $U, V \in \mathfrak{g}$, as in 2.102. Define μ on $G \times G$ by $\mu(g, h) := gh^{-1}g$, and set $\sigma_g h := \mu(g, h)$ as usual. It is easily verified that the axioms for a symmetric space multiplication from 7.6iii hold for μ; thus G is a symmetric space with respect to some linear connection. To prove that ∇ is in fact this connection it is sufficient to check the geodesics of ∇.

Fix $X \in \mathfrak{g}$, and let γ be the integral curve of X through $g \in G$, so that γ is the geodesic with initial tangent vector X_g; $\gamma(t) = g \exp tX$, where exp is the Lie group map from $\mathfrak{g}$ to G, not the connection-dependent map from G_g to G. For all t,

$$-s_{\gamma(t/2)*}\dot{\gamma}(0) = -\frac{d}{ds}\bigg|_0 \mu\left(g \exp\left(\frac{tX}{2}\right), g \exp(sX)\right)$$

$$= -\frac{d}{ds}\bigg|_0 g \exp\left(\frac{tX}{2}\right) \exp(-sX)g^{-1}g \exp\left(\frac{tX}{2}\right)$$

$$= -\frac{d}{ds}\bigg|_0 g \exp(tX - sX) = \dot{\gamma}(t)$$

Therefore $S_{\dot{\gamma}(0)} = \ddot{\gamma}(0)$. More generally, $S_{\dot{\gamma}(t)} = \ddot{\gamma}(t)$ for all t. Therefore $\dot{\gamma}$ is an integral curve of the spray determined by the multiplication map μ, so γ is a geodesic of the affine symmetric connection on (G, μ).

Exercise Check directly that the maps s_g are affine with respect to ∇. Hint: Check s_e first using bi-invariance.

Now back to the example. It is useful to consider G as the homogeneous space $(G \times G)/H$, where $G \times G$ acts on G by $(g, h) \cdot k := gkh^{-1}$, and the isotropy subgroup H of e is the diagonal $\Delta(G) = \{(g, g) \in G \times G\}$, which is itself isomorphic to G. The projection map p from $G \times G$ to G takes (g, h) to gh^{-1}.

Define an involutive automorphism σ of $G \times G$ by $\sigma(g, h) := (h, g)$. The fixed-point group $(G \times G)_\sigma$ equals $\Delta(G) = H$. Thus $(G \times G, H)$ is a symmetric pair. Given $(g_1, g_2), (h_1, h_2) \in G \times G$,

$$p((g_1, g_2)\sigma((g_1, g_2)^{-1}(h_1, h_2))) = (g_1, g_2)p(g_2^{-1}h_2, g_1^{-1}h_1)$$

$$= p(g_1, g_2)p(h_1, h_2)^{-1}p(g_1, g_2)$$

$$= \mu(p(g_1, g_2), p(h_1, h_2))$$

Therefore (G, ∇, μ) is the affine symmetric space associated with the symmetric pair $(G \times G, H)$ as in 7.6. Notice that in this case the group $G \times G$ might not act effectively on G.

(b) ***The sphere*** $S^n = SO(n+1)/SO(n)$. For each $p, q \in S^n$, define $s_p q := 2\langle p, q\rangle p - q$, where $\langle p, q\rangle$ is the usual inner product of p and q in $\mathbb{R}^{n+1}$. If $r \in S^n$ is perpendicular to p, and if $q = \cos(t)p +$

$\sin(t)r$ for some $t \in \mathbb{R}$, then $s_p q = \cos(t)p - \sin(t)r$; each map s_p is the restriction to S^n of an element of $O(n+1)$, and therefore acts on S^n isometrically, and hence affinely. This shows that (S^n, ∇, μ) is an affine symmetric space, where ∇ is the usual connection on S^n, and $\mu(p, q) = s_p q$.

Exercise Show that the isotropy subgroup $SO(n)$ of the point e_0 is the fixed point subgroup $SO(n+1)_\sigma$, where $\sigma g = s_{e_0} g s_{e_0}$.

(c) ***The space*** $SU(n)/SO(n)$ Complex conjugation of the entries of a matrix determines an involutive automorphism of $SU(n)$: $\sigma[g_{\alpha\beta}] := [\overline{g_{\alpha\beta}}]$. The fixed-point subgroup of σ is the closed subgroup $SO(n)$ of $SU(n)$. Thus the homogeneous space $SU(n)/SO(n)$ admits the structure of an affine symmetric space. The affine symmetry about a point $\wp g$, $g \in SU(n)$, is $\mu(\wp g, \wp h) = g\wp\sigma(g^{-1}h) = g\wp(\bar{g}^{-1}\bar{h}) = g\wp({}^t g\bar{h})$.

Exercises

1. Show that the Levi-Civita connection of the normal homogeneous metric on $SU(n)/SO(n)$ determined by minus the Killing form as an inner product on $\mathfrak{su}(n)$ is invariant under all the maps $s_{\wp g}$.
2. Find involutive automorphisms of $SL(n, \mathbb{R})$, $U(n)$, $Sp(n)$, and $Sp(n, \mathbb{R})$ which make the homogeneous spaces $SL(n, \mathbb{R})/SO(n)$, $U(n+1)/(U(1) \times U(n)) \cong \mathbb{C}P^n$, $Sp(n)/U(n)$, and $Sp(n, \mathbb{R})/U(n)$ into affine symmetric spaces.

LOCALLY AFFINE SYMMETRIC SPACES

Let ∇ be a linear connection on M. Fix $p \in M$, and let U be a neighborhood of 0 in M_p which is mapped diffeomorphically by $\exp_p$ to a neighborhood V of p in M. The map from V into M which sends $\exp_p u$ to $\exp_p(-u)$ sends every geodesic through p to the geodesic through p with the opposite initial tangent vector; in general, however, this map is not affine, since the image of a geodesic which does not pass through p need not be a geodesic. If this map is affine, then $V \cap \exp V$ is affine symmetric about p; if this happens for all $p \in M$, then locally M has the structure of an affine symmetric space.

The exposition in the following sections was influenced by Borel's notes [Bo3].

7.12 **Definition** Let ∇ be a linear connection on M. If for each $p \in M$ there is a neighborhood V of p in M and an affine map s_p: $V \to M$

such that $\mathscr{s}_{p*} v = -v$ for all $v \in M_p$, then (M, ∇) is called a *locally affine symmetric space* (or an *affine locally symmetric space*); each map $\mathscr{s}_p$ is called a *local affine symmetry of* (M, ∇).

For example, every open submanifold of an affine symmetric space is locally affine symmetric. Just as in 7.2, if M is locally affine symmetric with respect to linear connections $\bar{\nabla}$ and ∇, and if the local affine symmetries of $\bar{\nabla}$ and ∇ about each $p \in M$ agree, then $\bar{\nabla} = \nabla$.

7.13 Let ∇ be a linear connection on M; fix $p \in M$, and define an involution of a neighborhood of p by $\mathscr{s} \circ \exp_p(v) := \exp_p(-v)$. Define $\bar{\nabla}_U X := \mathscr{s}_* \nabla_U \mathscr{s}_* X$; it is easily checked that $\bar{\nabla}$ is a linear connection on M near p, and that $\mathscr{s}$ is affine if and only if $\bar{\nabla} = \nabla$.

Lemma The linear connections ∇ and $\bar{\nabla} := \mathscr{s}_* \nabla \mathscr{s}_*$ have the same exponential maps at p (the isolated fixed point of the involution $\mathscr{s}$). The torsion tensor $\bar{T}$ of $\bar{\nabla}$ is given by $\bar{T}(U, V) = \mathscr{s}_* T(\mathscr{s}_* U, \mathscr{s}_* V)$ for vector fields U and V near p; the corresponding identity also holds for the curvature tensors $\bar{R}$ and R. In particular, $\bar{T}_p = -T_p$, and $\bar{R}_p = R_p$.

PROOF If γ is a geodesic with $\gamma(0) = p$, then $\mathscr{s} \circ \gamma(t) = \gamma(-t)$, so $\mathscr{s}_* \dot{\gamma}(t) = -\dot{\gamma}(-t)$; this implies $\bar{\nabla}_D \dot{\gamma} = 0$, so $\overline{\exp}_p = \exp_p$.

By the structure equation (2.103), for vector fields U, V near p,

$$\bar{T}(U, V) = \mathscr{s}_*(\nabla_U \mathscr{s}_* V - \nabla_V \mathscr{s}_* U - \mathscr{s}_*[U, V]) = \mathscr{s}_* T(\mathscr{s}_* U, \mathscr{s}_* V)$$

and similarly, $\bar{R}(U, V)W = \mathscr{s}_* R(\mathscr{s}_* U, \mathscr{s}_* V)\mathscr{s}_* W$. At the point p then,

$$\bar{T}(u, v) = \mathscr{s}_* T(\mathscr{s}_* u, \mathscr{s}_* v) = -T(u, v) \qquad \bar{R}(u, v)w = R(u, v)w$$

7.14 **Definition** An affine transformation τ of a manifold M with linear connection ∇ is called a *transvection* if there exists a geodesic γ in M and a number t such that for all s, parallel translation along γ from $M_{\gamma(s)}$ to $M_{\gamma(t+s)}$ equals $\tau_*|_{\gamma(s)}$ (in particular, τ maps the image of γ to itself). An *infinitesimal transvection along a geodesic* γ is a vector field U on M whose local 1-parameter group consists of local transvections along γ.

7.15 **Proposition** Let ∇ be a torsion-free linear connection on M. The following statements are equivalent for an integral curve γ of a vector field $U \in \mathfrak{X}M$:

(i) $\nabla U = 0$ at each point $\gamma(t)$ (and in particular, γ is a geodesic).
(ii) For all $X \in \mathfrak{X}M$, $\nabla_U X = [U, X]$ along γ.
(iii) The vector field U is an infinitesimal transvection along γ.

PROOF (i)$\Leftrightarrow$(ii): Apply the identity $\nabla_X U = \nabla_U X - [U, X]$.

(ii)$\Rightarrow$(iii): Let X be a parallel field along γ; it is sufficient to show that $X_{\gamma(s)} = \varphi_{-t*} X_{\gamma(s+t)}$ for all s and t, where $\{\varphi_t\}$ is the 1-parameter group of U. For each t define a vector field $W(t)$ (cf. the proof of 2.123) on M by $W(t)_p := \varphi_{-t*} X_{\varphi_t p}$, $p \in M$. For each p, $W(\cdot)_p$ is a curve in M_p; we need to show that $W(t)_{\gamma(s)} = W(0)_{\gamma(s)}$ for all t and s. Using (ii) we obtain

$$\dot{W}(0)_{\gamma(s)} = \frac{\partial}{\partial t}\bigg|_0 \varphi_{-t*} X_{\varphi_t \gamma(s)} = [U, X]_{\gamma(s)} = (\nabla_U X)_{\gamma(s)} = 0$$

and

$$\dot{W}(t)_{\gamma(s)} = \frac{\partial}{\partial r}\bigg|_0 \varphi_{-(r+t)*} X_{\varphi_{t+r}\gamma(s)} = \varphi_{-t*} \frac{\partial}{\partial r}\bigg|_0 \varphi_{-r*} X_{\varphi_r \gamma(t+s)}$$
$$= \varphi_{-t*} \dot{W}(0)_{\gamma(s+t)} = 0$$

Hence

$$\varphi_{-t*} X_{\gamma(s+t)} = W(t)_{\gamma(s)} = W(0)_{\gamma(s)} = X_{\gamma(s)} = \mathbb{P}_\gamma^{-1} X_{\gamma(s+t)}$$

(iii)$\Rightarrow$(ii): For all s,

$$(\nabla_U X)_{\gamma(s)} = \frac{\partial}{\partial t}\bigg|_0 \mathbb{P}_\gamma^{-1} X_{\gamma(s+t)} = \frac{\partial}{\partial t}\bigg|_0 \varphi_{-t*} X_{\gamma(s+t)} = [U, X]_{\gamma(s)}$$

7.16 Theorem The following statements are equivalent for (M, ∇):

(i) (M, ∇) is locally affine symmetric.

(ii) The torsion T of ∇ is 0, and there exists an infinitesimal transvection along each geodesic segment in M.

(iii) $T = 0$, and the curvature tensor R is parallel.

PROOF (i)$\Rightarrow$(ii): If $\mathscr{s} = \mathscr{s}_p$, $p \in M$, then $\nabla = \mathscr{s}_* \nabla \mathscr{s}_*$, so by 7.13, $T_p = -T_p = 0$.

If $\gamma(t) = \exp tv$, $v \in M_p$, then for sufficiently small t, define $\tau_t := \mathscr{s}_{\gamma(t/2)} \circ \mathscr{s}_{\gamma(0)}$; each τ_t is a local affine map of (M, ∇). For a field U which is radially parallel about p (with respect to $\exp_p$), $\mathscr{s}_{\gamma(0)*} U = -U \circ \mathscr{s}_{\gamma(0)}$ because $\mathscr{s}_{\gamma(0)*} U$ is parallel along $\mathscr{s}_{\gamma(0)} \circ \gamma$ and $\mathscr{s}_{\gamma(0)*} U_{\gamma(0)} = -U_{\gamma(0)}$. Therefore

$$\tau_{t*} U_{\gamma(s)} = -\mathscr{s}_{\gamma(t/2)*} U \circ \mathscr{s}_{\gamma(0)} \gamma(s) = -\mathscr{s}_{\gamma(t/2)*} U_{\gamma(-s)} = U_{\gamma(s+t)}$$

Since U is parallel along γ, τ_t is a local transvection along γ.

(ii)$\Rightarrow$(iii): By hypothesis, parallel transport along a geodesic is induced by an affine map, and therefore preserves R by 2.115.

(iii) ⇒ (i): Fix $p \in M$; let N be a starlike neighborhood of 0 in M_p which is mapped diffeomorphically by $\exp_p$ to a neighborhood of p in M, and such that $-v$ is in N if v is. For $v \in N$, define $\mathscr{s} \circ \exp_p v := \exp(-v)$. Set $\bar{\nabla} := \mathscr{s}_* \nabla \mathscr{s}_*$ as in 7.13; it suffices to show that $\bar{\nabla} = \nabla$ on the neighborhood of p where it is defined.

If $X \in \mathfrak{X}M$ is ∇-parallel, then $\mathscr{s}_* X$ is $\bar{\nabla}$-parallel along $\mathscr{s}$; by 7.13, the curvature tensor $\bar{R}$ of $\bar{\nabla}$ is therefore $\bar{\nabla}$-parallel, and the torsion tensor of $\bar{\nabla}$ is zero. Finally, $\bar{R}_p = R_p$.

Set $A := [0, 1] \times N \subset \mathbb{R} \times M_p$, and define $\psi\colon A \to M$ by $\psi(t, v) := \exp_p tv$. Let $D = \partial/\partial t \in \mathfrak{X}A$, and let U, V be time-independent vector fields on A, that is, $dt(U) = dt(V) = 0$. Since $[D, U] = [D, V] = 0$, the structure equations (2.103) for ∇ along the map ψ imply

$$\nabla_D \psi_* U = \nabla_U \psi_* D$$

$$\nabla_D \nabla_U \psi_* V = R(\psi_* D, \psi_* U)\psi_* V + \nabla_U \nabla_D \psi_* V$$

By writing these equations in "polar coordinates" we will see that exactly the same equations hold for $\bar{\nabla}$ and $\bar{R}$; uniqueness of the solution to an initial value problem for a linear system of equations will then imply that $\bar{\nabla} = \nabla$ near p.

Let $\{E_j\}$ be a moving frame near p which is ∇-parallel along the geodesics emanating from p, and let $\{\omega^i\}$ be the dual basis field. The *local connection 1-forms* ω^i_j [Hi: 5.2] near p are then defined by $\omega^i_j(v) := \omega^i \nabla_v E_j$; since $\nabla_v E_j = \sum \omega^i_j(v) E_i|_{\pi v}$, the ω^i_j completely describe ∇ near p. Similarly, let $\{\bar{E}_j\}$ be a moving frame near p which is $\bar{\nabla}$-parallel along the geodesics emanating from p, with dual basis field $\{\bar{\omega}^i\}$. Assume that $\bar{E}_j|_p = E_j|_p$, in which case $\bar{\omega}^i|_p = \omega^i|_p$. As above, define $\bar{\omega}^i_j := \bar{\omega}^i \bar{\nabla} \bar{E}_j$.

It now suffices to show that $\bar{\omega}^i = \omega^i$ and $\bar{\omega}^i_j = \omega^i_j$ near p, for then $\bar{E}_j = E_j$ near p, and $\bar{\nabla}_v \bar{E}_j = \sum_i \bar{\omega}^i_j(v)\bar{E}_i = \sum_i \omega^i_j(v) E_i = \nabla_v E_j$. To do this we prove that the pullbacks of the forms by ψ agree on A in $\mathbb{R} \times M_p$. First, given $v \in N$, let $\gamma(t) = \psi(t, v) = \exp_p tv$, $0 \le t \le 1$. Since $\psi_* D_{(t, v)} = \dot{\gamma}(t)$,

$$\psi^* \bar{\omega}^i(D_{(t, v)}) = \bar{\omega}^i \dot{\gamma}(t) = \bar{\omega}^i v = \omega^i v = \omega^i \dot{\gamma}(t) = \psi^* \omega^i(D_{(t, v)})$$

Also, since the $\bar{E}_j$ and E_j are radially parallel,

$$\psi^* \bar{\omega}^i_j(D_{(t, v)}) = \bar{\omega}^i_j \dot{\gamma}(t) = \bar{\omega}^i \bar{\nabla}_{\dot{\gamma}(t)} \bar{E}_j = 0 = \cdots = \psi^* \omega^i_j(D_{(t, v)})$$

Next we will use the structure equations to prove that the forms agree on tangent vectors of the form $(0, u)$ on A. Define $z^i(t, u) := \omega^i u = \bar{\omega}^i u$, $(t, u) \in A$. Fix $u \in M_p$, and let U be the vector field on A

such that $U_{(t,v)} = (0, \mathscr{I}_v u)$; thus U is time-independent, and $dz^i U = \omega^i u = \bar{\omega}^i u$. Since the E_j are radially parallel about p,

$$\nabla_D \psi_* U = \sum_i \nabla_D \omega^i(\psi_* U) E_i|_\psi = \sum_i \frac{\partial \omega^i(\psi_* U)}{\partial t} E_i|_\psi.$$

$$\nabla_U \psi_* D = \sum_i \nabla_U z^i E_i|_\psi = \sum_i \left(dz^i U + \sum_j z^j \omega^i_j(\psi_* U)\right) E_i|_\psi$$

Thus, if we fix $v \in N$ and define

$$f^i(t) := \omega^i(\psi_* U_{(t,v)}) \qquad g^i_j(t) := \omega^i_j(\psi_* U_{(t,v)}) \qquad 0 \le t \le 1$$

then the first structure equation implies $\partial f^i/\partial t = \omega^i(u) + \sum_j \omega^j(v) g^i_j$, $i = 1, \ldots, n$; furthermore, $f^i(0) = \omega^i \psi_* U_{(0,v)} = 0$ because $\psi(0, v + tu) = \exp(0 \cdot (v + tu)) = p$.

Since R is parallel and the E_j and ω^i are radially parallel about p, the coefficients $R^i_{jkl} := \omega^i \circ R(E_k, E_l)E_j$ of R relative to $\{E_j\}$ are all constant: $R^i_{jkl} = R^i_{jkl}(p)$. If V is another time-independent vector field on A, then with U as above,

$$\begin{aligned}
\nabla_D \nabla_U \psi_* V &= \sum_i \Big(DU\omega^i(\psi_* V) + \sum_j (\omega^i_j(\psi_* U) D\omega^j(\psi_* V) \\
&\qquad + \omega^j(\psi_* V) D\omega^i_j(\psi_* U)\Big) E_i|_\psi \\
\nabla_U \nabla_D \psi_* V &= \sum_i \Big(UD\omega^i(\psi_* V) + \sum_j \omega^i_j(\psi_* U) D\omega^j(\psi_* V)\Big) E_i|_\psi
\end{aligned}$$

$$R(\psi_* D, \psi_* U)\psi_* V = \sum_{i,j,k,l} R^i_{jkl} z^k \omega^l(\psi_* U)\omega^j(\psi_* V) E_i|_\psi$$

Since $[U, D] = 0$, the second structure equation implies $\partial g^i_j/\partial t = \sum_{k,l} R^i_{jkl} \omega^k(v) f^l$, where v, f^i, and g^i_j are as above; furthermore, $g^i_j(0) = \omega^i_j \psi_* U_{(0,v)} = 0$.

Therefore the functions f^i, g^i_j satisfy the initial value problem

$$\frac{\partial f^i}{\partial t} = \sum_j \omega^j(v) g^i_j + \omega^i(u) \qquad f^i(0) = 0$$

$$\frac{\partial g^i_j}{\partial t} = \sum_{k,l} R^i_{jkl} \omega^k(v) f^l \qquad g^i_j(0) = 0$$

This is a first-order, constant-coefficient, linear system.

Similarly, define $\bar{f}^i(t) := \bar{\omega}^i \psi_* U_{(t,v)}$ and $\bar{g}^i_j(t) := \bar{\omega}^i_j \psi_* U_{(t,v)}$. Since $\bar{R}_p = R_p$, $\bar{E}_j|_p = E_j|_p$, and $\bar{\omega}^i|_p = \omega^i|_p$, the coefficients $\bar{R}^i_{jkl}$ of R with respect to $\{\bar{E}_j\}$ are just the coefficients R^i_{jkl} of R with respect to $\{E_j\}$. This implies that the functions $\bar{f}^i$, $\bar{g}^i_j$ satisfy the same initial value problem as the f^i, g^i_j; by the uniqueness of the solution to such a problem, $\bar{f}^i = f^i$ and $\bar{g}^i_j = g^i_j$ on the interval $[0, 1]$. Therefore $\psi^*\bar{\omega}^i = \psi^*\omega^i$ and $\psi^*\bar{\omega}^i_j = \psi^*\omega^i_j$, so $\bar{\omega}^i = \omega^i$ and $\bar{\omega}^i_j = \omega^i_j$. Thus $\bar{E}_j = E_j$, $\bar{\nabla}\bar{E}_j = \nabla E_j$, and $\bar{\nabla} = \nabla$ near p.

7.17 **Corollary** The curvature tensor of a locally affine symmetric space (M, ∇) is given by $R(u, v)w = -[[U, V], W]_p$, where U, V, and W are infinitesimal transvections along geodesics through p with initial tangent vectors $U_p = u$, $V_p = v$, and $W_p = w$.

PROOF Since $\mathscr{L}_U \nabla = 0$ (2.123), 7.15 and 7.16 imply

$$\nabla_u \nabla_V W = [U, \nabla_V W]_p = \nabla_{[U, V]_p} W + \nabla_v [U, W] = [V, [U, W]]_p$$

because $[U, V]_p = \nabla_u V - \nabla_v U = 0$, and similarly for $\nabla_v \nabla_U W$; hence

$$R(u, v)w = [V, [U, W]]_p - [U, [V, W]]_p = -[[U, V], W]_p$$

Exercise Give another proof using 2.123ii.

7.18 **Example: The lens spaces** Let p and q be relatively prime positive integers. For $v = 0, 1, \ldots, q-1$, let $r_v(q)$ be the rotation of $\mathbb{R}^2$ through the angle $2\pi v/q$, and let $r_v(p, q)$ be the rotation of $\mathbb{R}^2$ through the angle $2\pi vp/q$. Define an action of the group $\mathbb{Z}_q$ on $S^3 \subset \mathbb{R}^4$ by $(v, (u, v)) \mapsto (r_v(q)u, r_v(p, q)v)$, $u, v \in \mathbb{R}^2$. The quotient $L_{(q, p)}$ of S^3 modulo this action is a manifold called a *lens space;* its fundamental group is $\mathbb{Z}_q$. For example, $L_{(2, 1)}$ is the projective space $\mathbb{R}P^3$. Let $\bar{\nabla}$ be the standard linear connection on S^3. Define a connection ∇ on $L_{(q, p)}$ by the requirement that the projection map $\pi: (S^3, \bar{\nabla}) \to (L_{(q, p)}, \nabla)$ be an affine covering map. For each $x \in S^3$, the restriction of the affine symmetry $\bar{\jmath}_x$ of S^3 about x to a sufficiently small neighborhood of x can be pushed down to an involution $\jmath_{\pi x}$ of a neighborhood of $\pi x \in L_{(q, p)}$, $\jmath_{\pi x}\pi y := \pi \bar{\jmath}_x y$. Each involution $\jmath_{\pi x}$ is a local affine symmetry of $L_{(q, p)}$ about πx, so $L_{(q, p)}$ is locally an affine symmetric space (but in general, not globally [Wo]).

Exercise Study the Möbius strip with universal affine covering $\mathbb{R} \times (-1, 1)$, the "infinite" Möbius strip covered affinely by $\mathbb{R} \times \mathbb{R}$, and the Klein bottle covered affinely by $\mathbb{R}^2$.

7.19 **Theorem** If a locally affine symmetric space (M, ∇) is affinely complete, connected, and simply connected, then it is globally an affine symmetric space.

PROOF See [KN: XI.1.2].

The affine symmetric space $\mathbb{R}P^n$ shows that simple connectivity is not necessary for a symmetric space, but the lens spaces show that the theorem is false without the assumption. Completeness is necessary by 7.3.

7.20 **Corollary** The universal affine covering of an affinely complete, connected, locally affine symmetric space is affine symmetric.

SYMMETRIC LIE ALGEBRAS

7.21 **Definition** A triple $(\mathfrak{g}, \mathfrak{h}, \tau)$ is called a *symmetric* (or *involutive*) *Lie algebra* if $\mathfrak{g}$ is a Lie algebra, τ is an involutive automorphism of $\mathfrak{g}$, and $\mathfrak{h}$ is the eigenspace of the eigenvalue 1 of τ.

It follows immediately that $\mathfrak{h}$ is a subalgebra of $\mathfrak{g}$.

7.22 **Proposition** If (G, H) is a symmetric pair for an affine symmetric space (7.5, 7.6), then $(\mathfrak{g}, \mathfrak{h}, \sigma_*|_e)$ is a symmetric Lie algebra. Conversely, if $(\mathfrak{g}, \mathfrak{h}, \tau)$ is a symmetric Lie algebra, then there exists a symmetric pair (G, H) such that the Lie algebra of G is isomorphic to $\mathfrak{g}$, the Lie algebra of H is isomorphic to $\mathfrak{h}$, and the automorphisms $\sigma_*|_e$ and τ are equivalent.

PROOF $\Rightarrow$: By definition (7.5), σ acts trivially on H, so $\sigma_*|_e$ acts trivially on $H_e \cong \mathfrak{h}$.

$\Leftarrow$: If G is the connected, simply connected Lie group with Lie algebra $\mathfrak{g}$, then there is a unique homomorphism σ from G to G with $\sigma_*|_e = \tau$; since τ is an involutive automorphism of $\mathfrak{g}$, σ is an automorphism of G, and is involutive. Let H be the Lie subgroup of G with Lie algebra $\mathfrak{h}$.

It follows from 7.6 that a symmetric Lie algebra determines an affine symmetric space.

7.23 **Definition** Let $(\mathfrak{g}, \mathfrak{h}, \tau)$ be a symmetric Lie algebra. The *canonical decomposition of* $\mathfrak{g}$ is $\mathfrak{g} = \mathfrak{h} \oplus \mathfrak{m}$, where $\mathfrak{m}$ is defined as the (-1)-eigenspace of τ in $\mathfrak{g}$.

The following relations hold:

$$[\mathfrak{h}, \mathfrak{h}] \subset \mathfrak{h} \qquad [\mathfrak{h}, \mathfrak{m}] \subset \mathfrak{m} \qquad [\mathfrak{m}, \mathfrak{m}] \subset \mathfrak{h}$$

For example, if $X \in \mathfrak{h}$ and $U \in \mathfrak{m}$, then $\tau[X, U] = [\tau X, \tau U] = [X, -U] = -[X, U]$. In particular, the canonical decomposition of a symmetric Lie algebra is reductive, so a symmetric space is reductive homogeneous.

Exercise Calculate the symmetric Lie algebras for the examples in 7.11.

7.24 **Proposition** Let (G, H) be a symmetric pair for a connected, affine symmetric space (M, ∇), where G is the identity component $A(M, \nabla)_o$ of the affine transformation group of (M, ∇), and H is the isotropy subgroup of a point $p \in M$. Let $\mathfrak{g} = \mathfrak{h} \oplus \mathfrak{m}$ be the canonical reductive decomposition of $\mathfrak{g}$, and let $\tilde{\nabla}$ be the canonical connection on the reductive homogeneous space G/H from 6.66. The connections ∇ and $\tilde{\nabla}$ are equal.

PROOF Since ∇ is the unique linear connection on M which is invariant under all the symmetries s_q (7.2), it suffices to show that $\tilde{\nabla}$ is also s_q-invariant for all $q \in M$; because of the affine invariance of ∇ and $\tilde{\nabla}$ and the equality $s_{\wp g} = g s_p g^{-1}$, $g \in G$, from 7.6, it suffices to check this for $s_p = s_{\wp e} =: s$.
By definition (6.66), $\tilde{\nabla}_{X^*} Y^* = 0$ at p for X, $Y \in \mathfrak{m}$, so we must check that $\tilde{\nabla}_{X^*} s_* Y^*$ also vanishes at p. For all $g \in G$ (1.63c),

$$s_* Y^*_{\wp g} = \frac{d}{dt}\bigg|_0 \wp\sigma(\exp(tY)g) = \wp_*(\widetilde{\sigma_* Y})_{\sigma g} = (\sigma_* Y)^*_{s_p \wp g}$$

so the fields $(\sigma_* Y)^*$ and Y^* are s-related. Therefore,

$$(\tilde{\nabla}_{X^*} s_* Y^*)_p = (\tilde{\nabla}_{s_* X^*}(\sigma_* Y)^*)_p = (\tilde{\nabla}_{(\sigma_* X)^*}(\sigma_* Y)^*)_{sp} = 0$$

because $\sigma_* X$ and $\sigma_* Y$ belong to $\mathfrak{m}$ and $sp = p$.

Exercise Show directly that ∇ satisfies the definition of $\tilde{\nabla}$.

7.25 **Theorem** Let (G, H) be a symmetric pair for a connected, affine symmetric space (M, ∇), where $G = A(M, \nabla)_o$ and H is the isotropy subgroup of a point $p \in M$. Let $\mathfrak{g} = \mathfrak{h} \oplus \mathfrak{m}$ be the canonical reductive decomposition of $\mathfrak{g}$. Identify $\mathfrak{g} = \mathfrak{a}(M, \nabla)$ with the Lie algebra of complete affine vector fields on (M, ∇) as in 2.119. If $X \in \mathfrak{h} \subset \mathfrak{X}M$, then $X_p = 0 \in M_p$. For a fixed $U \in \mathfrak{X}M$, the following conditions are equivalent:

(i) $U \in \mathfrak{m} \subset \mathfrak{X}M$.
(ii) U is affine, and $s_{p*} U = -U \circ s_p$.
(iii) $\gamma(t) := \exp(tU)p$ is a geodesic of ∇, and U is an infinitesimal transvection along γ (7.14).

PROOF If $X \in \mathfrak{h}$, then $\exp tX$ fixes p for all t, so $X_p = 0 \in M_p$.

(i)$\Leftrightarrow$(ii): This follows from the identity $\sigma_* |_e U = \mathfrak{s}_{p*} U \circ \mathfrak{s}_p$.

(i)$\Rightarrow$(iii): For $U \in \mathfrak{m}$, set $\gamma(t) := \exp tU \in G$; since $\mathrm{Ad}_{\gamma(-t)} U = U$, by the definition of the canonical connection on G/H, $\nabla_X U = 0$ at $\gamma(t)$ for all $X \in \mathfrak{g}$ such that $\mathrm{Ad}_{\gamma(-t)} X \in \mathfrak{m}$. Thus $(\nabla U)_{\gamma(t)} = 0$, so 7.15 applies.

(iii)$\Rightarrow$(i): The subspace $\mathfrak{m}$ of $\mathfrak{g}$ is isomorphic to M_p; thus there exists $V \in \mathfrak{m}$ such that $V_p = U_p \in M_p$. By the work above and 7.15, $(\nabla V)_p = 0$; but by the proof of 2.121, two affine vector fields on M which have the same value and the same covariant differential at a point of M are equal. Hence $U = V \in \mathfrak{m}$.

Just as in 7.15, if U is an infinitesimal transvection along a geodesic γ through $p \in M$, then $(\nabla_U X)_p = [U, X]_p$ for all $X \in \mathfrak{X}M$; similarly, if U, V, and W are infinitesimal transvections along geodesics through $p \in M$, then $R(U, V)W = -[[U, V], W]$ at p. These identities hold because an affine symmetric space is locally symmetric. On the other hand, much of the work on symmetric Lie algebras can be carried out locally on affine symmetric spaces, or more generally, on spaces with a connection invariant under its parallel transport [Bo3] [N2].

RIEMANNIAN SYMMETRIC SPACES

7.26 **Definition** A *Riemannian symmetric space* is a Riemannian manifold which is affine symmetric with respect to the Levi-Civita connection. A *locally Riemannian symmetric space* is a Riemannian manifold which is locally affine symmetric with respect to the Levi-Civita connection.

The affine symmetries of a Riemannian symmetric space are isometries. Furthermore, a Riemannian homogeneous space M is Riemannian symmetric if there exists an affine symmetry about some point p in M. Finally, the curvature tensor and sectional curvature of a locally Riemannian symmetric space are invariant under parallel transport; conversely [He: IV.1.3], if the sectional curvature of a Riemannian manifold is invariant under parallel transport, then the manifold is locally a Riemannian symmetric space.

7.27 **Proposition** Let (M, ∇) be an affine symmetric space, let $G = A(M, \nabla)_o$, and let H be the isotropy subgroup in G of a point $p \in M$. There exists a Riemannian metric on M, with respect to which (M, ∇) is a Riemannian symmetric space, if and only if Ad_H has compact closure, where $\mathrm{Ad}: G \to GL(\mathfrak{g})$.

PROOF $\Rightarrow$: This is a special case of 6.59.

$\Leftarrow$: Let $\mathfrak{g} = \mathfrak{h} \oplus \mathfrak{m}$ be the canonical decomposition of $\mathfrak{g}$ from 7.23: $\sigma_* |_e X = -X$ for all $X \in \mathfrak{m}$. Let $\langle\ ,\ \rangle$ be an Ad_H-invariant inner product on $\mathfrak{g}$ such that $\mathfrak{h}$ and $\mathfrak{m}$ are orthogonal to each other; push $\langle\ ,\ \rangle$ down to an invariant metric on $M = G/H$. Fix $U \in \mathfrak{m}$, and set $\gamma(t) := (\exp tU)p$; by 7.25iii, γ is a geodesic, and parallel translation along γ from M_p to $M_{\gamma(t)}$ is given by $(\exp tU)_* |_p$. Given $v, w \in M_p$, let V and W be the parallel fields along γ with initial values v and w, respectively. Since the metric on M is invariant, $\langle V, W\rangle(t) = \langle (\exp tU)_* v, (\exp tU)_* w\rangle = \langle v, w\rangle$ for all t. Thus $D\langle V, W\rangle = 0 = \langle \nabla_D V, W\rangle + \langle V, \nabla_D W\rangle$ along γ, so the Ricci identity holds along γ. Since both $\langle\ ,\ \rangle$ and ∇ are invariant, the Ricci identity holds everywhere on M, and ∇ is a Riemannian connection with respect to $\langle\ ,\ \rangle$. But ∇ is torsion-free, so it is the Levi-Civita connection of $\langle\ ,\ \rangle$.

7.28 **Definition** A symmetric Lie algebra $(\mathfrak{g}, \mathfrak{h}, \tau)$ is called an *orthogonal symmetric Lie algebra* if the Lie subgroup of $GL(\mathfrak{g})$ with Lie algebra $\mathrm{ad}(\mathfrak{h})$ is compact.

7.29 **Examples**

(*a*) Let G be the connected isometry group of a Riemannian symmetric space M; the isotropy subgroup H of a point of M is compact, so the symmetric Lie algebra $(\mathfrak{g}, \mathfrak{h}, \sigma_* |_e)$ associated with the symmetric pair (G, H) is orthogonal symmetric.

(*b*) Let G be a connected Lie group. As in 7.11*a*, let σ be the automorphism of $G \times G$ such that $\sigma(g, h) := (h, g)$. The fixed-point subgroup of σ is $H := \Delta(G) \subset G \times G$. The $(+1)$-eigenspace of $\sigma_*|_e$ is $\mathfrak{h} = \Delta_* \mathfrak{g} \subset \mathfrak{g} \times \mathfrak{g}$, while the (-1)-eigenspace is $\mathfrak{m} = \{(X, -X) \in \mathfrak{g} \times \mathfrak{g}\}$. Hence $(\mathfrak{g} \times \mathfrak{g}, \Delta_* \mathfrak{g}, \sigma_*|_e)$ is a symmetric Lie algebra. The image in $GL(\mathfrak{g} \times \mathfrak{g})$ of $\mathrm{Ad}|_H$ is $\{(\mathrm{Ad}_g, \mathrm{Ad}_g) |\ g \in G\} \cong \mathrm{Ad}(G) \subset GL(\mathfrak{g})$. Since $\Delta(GL(\mathfrak{g}))$ is closed in $GL(\mathfrak{g} \times \mathfrak{g})$, Ad_H is relatively compact in $GL(\mathfrak{g} \times \mathfrak{g})$ if and only if Ad_G is relatively compact in $GL(\mathfrak{g})$. By 6.17, this holds if and only if G is the product of a compact group and a vector group; in this case G admits a bi-invariant metric, and in fact G is Riemannian symmetric with respect to each bi-invariant metric.

Exercises Verify directly that the Lie algebras $(\mathfrak{o}(3) \times \mathfrak{o}(3), \Delta_* \mathfrak{o}(3), \sigma_* |_e)$ and $(\mathfrak{o}(4), \mathfrak{o}(3), \varphi_* |_e)$ are orthogonal symmetric Lie algebras, where $SO(3)$ is embedded into $SO(4)$ as the subgroup (6.50)

$$\begin{bmatrix} 1 & 0 \\ 0 & SO(3) \end{bmatrix} \qquad \text{and} \qquad \varphi g := \begin{bmatrix} -1 & 0 \\ 0 & I \end{bmatrix} g \begin{bmatrix} -1 & 0 \\ 0 & I \end{bmatrix} \qquad g \in SO(4)$$

where I is the 3×3 identity matrix. Prove that these orthogonal symmetric Lie algebras are isomorphic.

7.30 At this point the study of Riemannian symmetric spaces has been made fairly algebraic. The next result is quite similar to the work done for Lie groups in chap. 6.

Theorem Let G be the identity component $I(M, \langle\ ,\ \rangle)_o$ of the isometry group of a connected Riemannian symmetric space, and let H be the isotropy subgroup of a point $p \in M$, with associated orthogonal symmetric Lie algebra $(\mathfrak{g}, \mathfrak{h}, \sigma_* |_e)$. Let $\mathfrak{m}$ be the (-1)-eigenspace of $\sigma_* |_e$ in $\mathfrak{g}$, and let B be the Killing form of $\mathfrak{g}$.

(i) If $\mathfrak{g}$ is semisimple and B is negative definite on $\mathfrak{m}$, then M is a compact Riemannian symmetric space with nonnegative sectional curvature and positive Ricci curvature.

(ii) If $\mathfrak{g}$ is semisimple and B is positive definite on $\mathfrak{m}$, then M is diffeomorphic to Euclidean space (and, in particular, is simply connected) and has nonpositive sectional curvature and negative Ricci curvature.

(iii) If $\mathfrak{m}$ is an abelian ideal of $\mathfrak{g}$, then M is flat and its universal Riemannian covering space is Euclidean space.

Proof (i) and (ii): Assume $\mathfrak{g}$ is semisimple. The inner product $\langle\ ,\ \rangle$ on $\mathfrak{m}$ is Ad_H-invariant, so just as in the proof of 6.18, ad_X is skew-symmetric with respect to $\langle\ ,\ \rangle$ for each $X \in \mathfrak{h}$, and then $B(X, X) \leq 0$; furthermore, B is negative definite on $\mathfrak{h}$ because $\mathfrak{g}$ is semisimple (Cartan's criterion, 6.14).

Define $L \in \mathrm{End}(\mathfrak{m})$ by $\langle LU, V\rangle := B(U, V)$, $U, V \in \mathfrak{m}$. Because both B and the inner product $\langle\ ,\ \rangle$ on $\mathfrak{m}$ are Ad_H-invariant, L commutes with the adjoint action of $\mathfrak{h}$ on $\mathfrak{m}$: given $X \in \mathfrak{h}$ and U, $V \in \mathfrak{m}$,

$$\begin{aligned} \langle L[X, U], V\rangle &= B([X, U], V) = -B(U, [X, V]) \\ &= -\langle LU, [X, V]\rangle = \langle [X, LU], V\rangle \end{aligned}$$

Since L is self-adjoint, its eigenvalues $\lambda_1, \ldots, \lambda_k$ are real; in case (i) all λ_j are negative, and in case (ii) all λ_j are positive. Let $\mathfrak{m} = \mathfrak{m}_1 \oplus \cdots \oplus \mathfrak{m}_k$ be the associated eigenspace decomposition; $L|_{\mathfrak{m}_j}$ is multiplication by λ_j, so $B(\mathfrak{m}_i, \mathfrak{m}_j) = 0$ for $i \neq j$ since $\langle \mathfrak{m}_i, \mathfrak{m}_j\rangle = 0$

in this case. Let $X \in \mathfrak{h}$, $U \in \mathfrak{m}_i$, and $V \in \mathfrak{m}_j$, $i \neq j$. Since ad_X and L commute, $[X, U] \in [\mathfrak{h}, \mathfrak{m}_i] \subset \mathfrak{m}_i$, so $B(X, [U, V]) = B([X, U], V) = 0$; but $[U, V] \in \mathfrak{h}$, and B is negative definite on $\mathfrak{h}$, so $[U, V] = 0$. Therefore, $[\mathfrak{m}_i, \mathfrak{m}_j] = 0$ for $i \neq j$.

Let $U = U_1 + \cdots + U_k$ and $V = V_1 + \cdots + V_k$, with $U_i, V_i \in \mathfrak{m}_i$. Identify $\mathfrak{m}$ with M_p; by 7.25 and 7.17, at the point p in M

$$\begin{aligned}\langle R(U, V)V, U\rangle &= -\langle[[U, V], V], U\rangle = -\sum \langle[[U_i, V_i], V_i], U_i\rangle \\ &= -\sum \frac{1}{\lambda_i}\langle\lambda_i[[U_i, V_i], V_i], U_i\rangle \\ &= -\sum \frac{1}{\lambda_i} B([[U_i, V_i], V_i], U_i) \\ &= \sum \frac{1}{\lambda_i} B([U_i, V_i], [U_i, V_i])\end{aligned}$$

which has the opposite sign from the λ_i since $[\mathfrak{m}, \mathfrak{m}] \subset \mathfrak{h}$ and B is negative definite on $\mathfrak{h}$.

Fix an orthonormal basis $\{E_{ij}\}$ for $\mathfrak{m}_i$ and an orthonormal basis $\{F_k\}$ for $\mathfrak{h}$; for each $U \in \mathfrak{m}$,

$$\mathscr{R}(U, U) = \sum_{i,j} \langle R(E_{ij}, U)U, E_{ij}\rangle = \sum \frac{1}{\lambda_i} B([E_{ij}, U], [E_{ij}, U])$$

Since all the λ_i are nonzero and have the same sign, $\mathscr{R}(U, U)$ is zero if and only if $[E_{ij}, U] = 0$ for all i and j. Hence $\mathscr{R}(U, U) = 0$ if and only if $[\mathfrak{m}, U] = 0$. But

$$B(U, U) = \sum_{i,j} \langle[U, [U, E_{ij}]], E_{ij}\rangle + \sum_k \langle[U, [U, F_k]], F_k\rangle$$

which equals zero if $[\mathfrak{m}, U] = 0$. Thus, since B is nondegenerate, $U = 0$ if $\mathscr{R}(U, U) = 0$, so the Ricci curvature of M cannot vanish; $\mathscr{R}$ is therefore positive in case (i) and negative in case (ii) by the inequalities on the sectional curvature in those cases.

In case (i) M is compact by Myers' theorem [GKM: 7.3(i)]; in case (ii) it is diffeomorphic to Euclidean space by a theorem of Kobayashi [KN: VIII.8.3] (cf. the Hadamard-Cartan theorem [GKM: 7.2]).

In case (iii), $R = 0$ by 7.17, and $\mathbb{R}^n$ is the only complete, simply connected, flat Riemannian manifold of dimension n.

If $\mathfrak{g}$ is simple, then [KN: XI.8.6] M is an Einstein space, that is, (3.45), it has constant Ricci curvature.

7.31 The following definition is suggested by 7.30.

Definition An orthogonal symmetric Lie algebra $(\mathfrak{g}, \mathfrak{h}, \tau)$ with $\mathfrak{g}$ semisimple is *of compact type* if B is negative definite on $\mathfrak{m}$, and is *of noncompact type* if B is positive definite on $\mathfrak{m}$. An orthogonal symmetric Lie algebra $(\mathfrak{g}, \mathfrak{h}, \tau)$ is *of Euclidean type* if $\mathfrak{m}$ is an abelian ideal of $\mathfrak{g}$.

7.32 The next result, which will be presented without proof, shows how the various types of symmetric spaces fit together.

Theorem If $(M, \langle\ ,\ \rangle)$ is a Riemannian symmetric space, and if $(\mathfrak{g}, \mathfrak{h}, \tau = \sigma_*|_e)$ is the orthogonal symmetric Lie algebra associated with the isometry group of $(M, \langle\ ,\ \rangle)$; then there exists an ideal decomposition $\mathfrak{g} = \mathfrak{g}_o \times \mathfrak{g}_+ \times \mathfrak{g}_-$ such that (i) the ideals are invariant under τ, and are mutually perpendicular with respect to the Killing form of $\mathfrak{g}$; (ii) if τ_o, τ_+, and τ_- are the restrictions of τ to $\mathfrak{g}_o$, $\mathfrak{g}_+$, and $\mathfrak{g}_-$, respectively, with $(+1)$-eigenspaces $\mathfrak{h}_o$, $\mathfrak{h}_+$, and $\mathfrak{h}_-$, then $(\mathfrak{g}_o, \mathfrak{h}_o, \tau_o)$, $(\mathfrak{g}_+, \mathfrak{h}_+, \tau_+)$, and $(\mathfrak{g}_-, \mathfrak{h}_-, \tau_-)$ are orthogonal symmetric Lie algebras of Euclidean type, noncompact type, and compact type, respectively. Furthermore, if M is simply connected, then there exists an isometric product decomposition $M = M_o \times M_+ \times M_-$, where M_o is a Euclidean space with orthogonal symmetric Lie algebra $(\mathfrak{g}_o, \mathfrak{h}_o, \tau_o)$, M_+ is a simply connected Riemannian symmetric space with orthogonal symmetric Lie algebra $(\mathfrak{g}_+, \mathfrak{h}_+, \tau_+)$, and M_- is a simply connected Riemannian symmetric space with orthogonal symmetric Lie algebra $(\mathfrak{g}_-, \mathfrak{h}_-, \tau_-)$.

PROOF [He: V.1.1], [KN: XI.7.2, 7.5, 7.6].

7.33 We now take another look at the Weyl unitary trick from 6.30, 6.36, 6.41, and 6.48.

Definition Let $(\mathfrak{g}, \mathfrak{h}, \tau)$ be an orthogonal symmetric Lie algebra, with canonical reductive decomposition $\mathfrak{g} = \mathfrak{h} \oplus \mathfrak{m}$. Denote by $\mathfrak{g}^*$ the real linear subspace $\mathfrak{h} \oplus i\mathfrak{m}$ of the complexification $\mathfrak{g}^{\mathbb{C}}$ of $\mathfrak{g}$; $\mathfrak{g}^*$ is closed under $[\cdot, \cdot]$ in $\mathfrak{g}^{\mathbb{C}}$, and is therefore a real Lie algebra. Define τ^* on $\mathfrak{g}^*$ by $\tau^*(X + iY) := X - iY$, $X \in \mathfrak{h}$, $Y \in \mathfrak{m}$. The triple $(\mathfrak{g}^*, \mathfrak{h}, \tau^*)$ is called the *dual of* $(\mathfrak{g}, \mathfrak{h}, \tau)$.

Proposition The dual $(\mathfrak{g}^*, \mathfrak{h}, \tau^*)$ of an orthogonal symmetric Lie algebra $(\mathfrak{g}, \mathfrak{h}, \tau)$ is orthogonal symmetric; its dual is $(\mathfrak{g}, \mathfrak{h}, \tau)$. An orthogonal symmetric Lie algebra is of compact type if and only if its dual is of noncompact type.

PROOF The map τ^* is an involutive automorphism of $\mathfrak{g}^*$ with fixed-point set $\mathfrak{h}$, so it suffices to prove that the Lie subgroup of $GL(\mathfrak{g}^*)$ with Lie algebra $\mathrm{ad}(\mathfrak{h})$ is compact; this follows by the definition of $\mathfrak{g}^*$ because the Lie subgroup of $GL(\mathfrak{g})$ with Lie algebra $\mathrm{ad}(\mathfrak{h})$ is compact.

Just as in the proof of 6.36, the Killing forms of $\mathfrak{g}$ and $\mathfrak{g}^*$ have opposite signs on $\mathfrak{m}$ and $i\mathfrak{m}$, so the result follows by 7.40.

Exercise What is the dual of an orthogonal symmetric Lie algebra of Euclidean type?

7.34 **Examples**

(*a*) By 6.30, the dual of the orthogonal symmetric Lie algebra $(\mathfrak{su}(n), \mathfrak{o}(n), \sigma_*|_e)$ (with σ defined as in 7.11*c*) is $(\mathfrak{sl}(n, \mathbb{R}), \mathfrak{o}(n), \sigma_*|_e^*)$.

(*b*) By 6.41, $\mathfrak{sp}(n, \mathbb{R}) = (\mathfrak{o}(2n) \cap \mathfrak{sp}(n, \mathbb{R})) \oplus (\mathfrak{sym}(2n) \cap \mathfrak{sp}(n, \mathbb{R})) =: \mathfrak{h} \oplus \mathfrak{m}$. It is easily checked that $[\mathfrak{h}, \mathfrak{h}] \subset \mathfrak{h}$, $[\mathfrak{h}, \mathfrak{m}] \subset \mathfrak{m}$, and $[\mathfrak{m}, \mathfrak{m}] \subset \mathfrak{h}$; thus if τ is defined by $\tau(X + Y) := X - Y$, $X \in \mathfrak{h}$, $Y \in \mathfrak{m}$, then $(\mathfrak{sp}(n, \mathbb{R}), \mathfrak{h}, \tau)$ is a symmetric Lie algebra. It is orthogonal because $O(2n) \cap Sp(n, \mathbb{R})$ is closed in $GL(2n, \mathbb{R})$, and is therefore compact; furthermore it is of noncompact type by 6.41. Its dual is $(\mathfrak{sp}(n), \mathfrak{h}, \tau^*)$, which is of compact type.

Exercise What happens on the symmetric space level? What is the dual of the orthogonal symmetric Lie algebra $(\mathfrak{o}(n+1), \mathfrak{o}(n), \sigma_*|_e)$ associated with the Riemannian symmetric space S^n? What is the Riemannian symmetric space associated with $(\mathfrak{o}(n+1)^*, \mathfrak{o}(n), \sigma_*|_e^*)$? Discuss 6.52 in terms of symmetric spaces.

7.35 The final theorem in this chapter provides a good example of how symmetric spaces aid in the study of more general geometric questions.

Recall Meyer's theorem (4.29): if an n-dimensional, connected, compact oriented manifold M admits a metric g with positive definite curvature operator, then $H^*(M; \mathbb{R})$ is isomorphic to $H^*(S^n; \mathbb{R})$. Now we see what happens if the curvature operator is only nonnegative.

Theorem (Gallot-Meyer) If $\hat{M}$ is a compact, connected, simply connected Riemannian manifold with nonnegative curvature operator, then $\hat{M}$ is a product of manifolds for which the holonomy group acts irreducibly on each tangent space; for each factor M, either

(i) $\dim M = n$ and the holonomy group is isomorphic to $SO(n)$, in which case $H^*(M; \mathbb{R}) \cong H^*(S^n; \mathbb{R})$;

(ii) $\dim M = 2n$ and the holonomy group is isomorphic to $U(n)$, in which case $H^*(M; \mathbb{R}) \cong H^*(\mathbb{C}P^n; \mathbb{R})$; or

(iii) M is isometrically a symmetric space of compact type.

PROOF The proof uses theorems of Berger [B*1*], Borel [Bo*1*, *2*], Brown-Gray [BrG], Montgomery-Samelson [MoS], and Tachibana [T], not all of which could be proved here; therefore the reader is referred to [GaM].

Exercise Using the proof of Meyer's theorem from 4.29, prove that if the curvature operator of (M, g) is positive definite, where M is compact, connected, and orientable, then the holonomy group of g is $SO(n)$.

CHAPTER

EIGHT

SYMPLECTIC AND HERMITIAN VECTOR BUNDLES

In chap. 3 we studied the geometry of Riemannian vector bundles, that is, vector bundles for which the fibers are inner product spaces and the inner product depends differentiably on the basepoint; we also looked very briefly at Lorentz metrics on the tangent bundle. Both Riemannian and Lorentz metrics were nondegenerate, symmetric bilinear forms on the fibers. In this chapter we will look at other forms on the fibers of a vector bundle; they will also be nondegenerate, but will not be symmetric.

SYMPLECTIC VECTOR BUNDLES

8.1 **Definition** A *symplectic vector space* is a finite-dimensional real vector space $\mathscr{V}$ endowed with a skew-symmetric, nondegenerate bilinear form Ω, called a *symplectic form on* $\mathscr{V}$.

A symplectic form is thus a skew-symmetric, real bilinear map Ω from $\mathscr{V} \times \mathscr{V}$ to $\mathbb{R}$ with the property that if $\Omega(u, v) = 0$ for all $v \in \mathscr{V}$, then $u = 0$. It follows that the rank of Ω equals the dimension of $\mathscr{V}$, that is, if $\Omega = \sum A_{ij}\omega^i \wedge \omega^j$ with respect to a basis $\{\omega^i\}$ for the dual space $\mathscr{V}^*$, then the rank of the matrix A of Ω equals the dimension of $\mathscr{V}$. Furthermore, by an appropriate choice of basis, the matrix A of Ω can be put into a very simple form.

8.2 **Proposition** Let $(\mathscr{V}, \Omega)$ be a symplectic vector space. There exists a basis $\{\omega^i\}$ for $\mathscr{V}^*$ such that $\Omega = \sum \omega^i \wedge \omega^{n+i}$; with respect to the dual basis for $\mathscr{V}$ the matrix of Ω is

$$\begin{bmatrix} 0 & I \\ -I & 0 \end{bmatrix}$$

I being the $n \times n$ identity matrix. In particular, the dimension of $\mathscr{V}$ is even, $\dim \mathscr{V} = 2n$.

Proof Let ϵ_1 and ϵ_{n+1} be points in $\mathscr{V}$ such that $\Omega(\epsilon_1, \epsilon_{n+1})$ is nonzero; we may assume that it equals 1. Let P be the subspace of $\mathscr{V}$ spanned by ϵ_1 and ϵ_{n+1}; the matrix of the restriction of Ω to P is

$$\begin{bmatrix} 0 & 1 \\ -1 & 0 \end{bmatrix}$$

Let $P^\perp$ be the orthogonal complement of P with respect to Ω: $P^\perp := \{v \in \mathscr{V} \mid \Omega(v, w) = 0 \text{ for all } w \in P\}$. Given $v \in \mathscr{V}$, if $w := \Omega(v, \epsilon_{n+1})\epsilon_1 - \Omega(v, \epsilon_1)\epsilon_{n+1}$, then $v = w + (v - w)$ is a decomposition of v as the sum of an element of P plus an element of $P^\perp$; this decomposition is unique by the nondegeneracy of Ω. Hence $\mathscr{V} = P \oplus P^\perp$. Furthermore, the restriction of Ω to $P^\perp$ is a symplectic form on $P^\perp$, so we may continue by induction.

It follows that if $(\mathscr{V}, \Omega)$ is a $2n$-dimensional symplectic vector space, then $\Omega^n = \Omega \wedge \cdots \wedge \Omega = (-1)^{[n/2]} n! \omega^1 \wedge \cdots \wedge \omega^{2n}$.

As an example of a symplectic vector space we have $\mathbb{R}^{2n}$ with the canonical symplectic form $\Omega_o := dx^1 \wedge dx^{n+1} + \cdots + dx^n \wedge dx^{2n}$ from 6.39. For more information about symplectic vector spaces, see [AM: 3.1], [BG: 2.23], [GS: IV.2], or [We3]; for a much more classical treatment from first principles, see [Ar: III].

8.3 **Definition** A linear map f from a symplectic vector space $(\mathscr{V}_1, \Omega_1)$ to a symplectic vector space $(\mathscr{V}_2, \Omega_2)$ is called *symplectic* if ${}^t f \Omega_2 = \Omega_1$, that is, if $\Omega_2(fu, fv) = \Omega_1(u, v)$ for all $u, v \in \mathscr{V}_1$.

For example, the real symplectic group $Sp(n, \mathbb{R})$ from 6.39 is the subgroup of $GL(2n, \mathbb{R})$ consisting of symplectic self-maps of $(\mathbb{R}^{2n}, \Omega_o)$; since $Sp(n, \mathbb{R})$ preserves Ω_o, it also preserves the usual volume element on $\mathbb{R}^{2n}$; this also follows from the inclusion of $Sp(n, \mathbb{R})$ into $SL(2n, \mathbb{R})$ from 6.41.

8.4 **Definition** A *symplectic vector bundle over a manifold* M is a real vector bundle E over M endowed with a skew-symmetric form $\Omega \in \Gamma(E^* \otimes E^*)$ such that for all $p \in M$, (E_p, Ω_p) is a symplectic vector space; the form Ω is called a *symplectic 2-form on* E.

A symplectic vector bundle is automatically orientable, with the usual orientation being the one represented by the form $(-1)^{[m/2]}\Omega^m$, where $2m$ is the rank of E.

8.5 **Proposition** (Lichnerowicz [Li*1*]) Let (E, Ω) be a symplectic vector bundle over a manifold M; there exists a Riemannian metric on E which is compatible with Ω in the sense that the bundle isomorphism $E \to E^*$, $u \mapsto \Omega(u, \cdot)$, is a bundle isometry with respect to the dual metrics from 3.9.

PROOF Let $\langle\ ,\ \rangle_o$ be an arbitrary Riemannian metric on E; define a field $\hat{\Omega}$ of endomorphisms of E by $\langle U, \hat{\Omega}V\rangle_o := \Omega(U, V)$, $U, V \in \Gamma E$. Since $\hat{\Omega}$ is skew-symmetric with respect to $\langle\ ,\ \rangle_o$ and $-\hat{\Omega}^2$ is symmetric with respect to $\langle\ ,\ \rangle_o$, the eigenfunctions of $-\hat{\Omega}^2$ are nonnegative real functions on M; they are in fact positive because Ω is nondegenerate. For each eigenfunction λ of $-\hat{\Omega}^2$, let $E_\lambda \subset E$ be the eigenbundle of λ. It follows that $E = \oplus E_\lambda$.

Define a new metric $\langle\ ,\ \rangle$ on E by setting $\langle E_\lambda, E_\mu\rangle = 0$ for $\lambda \neq \mu$, and $\langle U, V\rangle := \sqrt{\lambda}\langle U, V\rangle_o$ for $U, V \in \Gamma E_\lambda$. For $U, V \in \Gamma E_\lambda$,

$$\begin{aligned}\langle\Omega(U, \cdot), \Omega(V, \cdot)\rangle_{E^*} &= \lambda^{-1/2}\langle\hat{\Omega}U, \hat{\Omega}V\rangle_o = \lambda^{-1/2}\langle -\hat{\Omega}^2 U, V\rangle_o \\ &= \lambda^{1/2}\langle U, V\rangle_o = \langle U, V\rangle\end{aligned}$$

Hence, since E_λ and E_μ are orthogonal for $\lambda \neq \mu$, $\langle\Omega(U, \cdot), \Omega(V, \cdot)\rangle = \langle U, V\rangle$ for all $U, V \in \Gamma E$.

Exercise Is the compatible Riemannian metric unique?

8.6 **Definition** A symplectic 2-form Ω on the tangent bundle TM of a manifold M is called an *almost symplectic structure on* M, and (M, Ω) is called an *almost symplectic* (or *almost Hamiltonian*) *manifold*. An almost symplectic structure Ω on M is called *symplectic* if it is closed, $d\Omega = 0$; in this case, (M, Ω) is called a *symplectic* (or *Hamiltonian*) *manifold*.

8.7 **Examples**

(*a*) The *canonical symplectic structure on* $\mathbb{R}^{2n}$ is the "extension" of the canonical symplectic form Ω_o on the vector space $\mathbb{R}^{2n}$ (see the process for extending an inner product on $\mathbb{R}^n$ to a metric on the manifold $\mathbb{R}^n$ from 3.4*a*): given tangent vectors (p, u) and (p, v) at $p \in \mathbb{R}^{2n}$, define $\Omega((p, u), (p, v)) := \Omega_o(u, v)$. *Exercise:* Show that Ω is a closed (and in fact, exact) 2-form on $\mathbb{R}^{2n}$.

(*b*) After $\mathbb{R}^{2n}$, the standard example of a manifold which admits a symplectic structure is the cotangent bundle of another manifold. Let $M = T^*N$, and define a 1-form θ on M by $\theta_\mu := \pi^*\mu$, $\mu \in M$, that is, for all $\mu \in T^*N$ and $v \in (T^*N)_\mu$, $\theta(v) := (\pi^*\mu)(v) = \mu(\pi_* v)$.

$$\begin{array}{ccc} TT^*N & \xrightarrow{\pi_*} & TN \\ \downarrow & & \downarrow \\ T^*N & \xrightarrow[\pi]{} & N \end{array}$$

The 1-form θ on $M = T^*N$ is called the *fundamental* (or *canonical*) *1-form on* T^*N. Let us look at θ in the usual bundle coordinates on T^*N. Let $(x \circ \pi, y)$ be the bundle chart on T^*N associated with a chart $x = (x^1, \ldots, x^n)$ on N; the coordinates y_j are defined by $y_j(\mu) := \mu(X_j|_{\pi(\mu)})$, $X_j := \partial/\partial x^j$. For $\mu \in N_p^*$ and $v \in (T^*N)_\mu$,

$$\pi_* v = \sum v(x^j \circ \pi) X_j|_p \qquad \mu = \sum y_i(\mu)\, dx_p^i$$

and $$\theta v = \sum_{i,j} y_i(\mu)\, dx^i(v(x^j \circ \pi) X_j) = \sum y_i(\mu) v(x^i \circ \pi)$$

Hence in local coordinates, $\theta = \sum y_i \cdot d(x^i \circ \pi)$, or as it is usually written, $\theta = \sum y_i\, dx^i$.

Define the *fundamental 2-form on* T^*N by $\Omega = -d\theta$. In local coordinates, $\Omega = \sum dx^i \wedge dy_i$, so Ω is clearly a symplectic 2-form on T^*N. In the light of this example, the canonical symplectic structure on $\mathbb{R}^{2n}$ from (*a*) is really a symplectic structure on $T^*\mathbb{R}^n = \mathbb{R}^n \times (\mathbb{R}^n)^*$. More generally, every symplectic manifold is locally of the form $T^*\mathbb{R}^n$ in the sense of the following theorem of Darboux: an almost symplectic manifold (M, Ω) is symplectic, that is, Ω is closed, if and only if there is a chart (x, U) around each point of M such that on U, $\Omega = \sum dx^i \wedge dx^{n+i}$. For the proof, see [We*3*] or [AM: 3.2.2].

(*c*) Unlike T^*N, the tangent bundle TN has no canonical symplectic structure; however, suppose that $\langle\ ,\ \rangle$ is a given Riemannian metric on N. Pull the canonical 1-form θ on T^*N back to a 1-form $\tilde{\theta} := \flat^*\theta$ on TN by means of the bundle isomorphism $\flat\colon TN \to T^*N$ from 3.8. Similarly, let $\tilde{\Omega} := \flat^*\Omega = -\flat^* d\theta = -d\tilde{\theta}$; $\tilde{\Omega}$ is also symplectic.

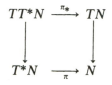

Let $\kappa\colon TTN \to TN$ be the connection map (2.49, 2.52) of the Levi-Civita connection of $\langle\ ,\ \rangle$: $\kappa(Y_* v) := \nabla_v Y$, $v \in TN$, $Y \in \mathfrak{X}N$. As in 2.50, $(\pi_*, \kappa)\colon TTN \to TN \oplus TN$ is a vector bundle isomorphism along the projection map π from TN to N, and determines an almost complex structure $J \in \Gamma$ End TTN on the manifold TN (1.58) such that if $(\pi_*, \kappa)(\xi) = (u, v) \in TN \oplus TN$, then $(\pi_*, \kappa) \cdot (J\xi) = (-v, u)$, that is, $\pi_* J = -\kappa$ and $\kappa J = \pi_*$. Let $\widetilde{\langle\ ,\ \rangle}$ be the Sasaki metric on the manifold TN determined by $\langle\ ,\ \rangle$ and κ (3.4f): $\widetilde{\langle \zeta, \xi \rangle} = \langle \pi_* \zeta, \pi_* \xi \rangle + \langle \kappa\zeta, \kappa\xi \rangle$.

We will now see that $\tilde{\theta}(V) = \widetilde{\langle S, V \rangle}$, $V \in \mathfrak{X}TN$, where S is the geodesic spray on TN of $\langle\ ,\ \rangle$, and that for all $U, V \in \mathfrak{X}TN$, $\tilde{\Omega}(U, V) = \widetilde{\langle JU, V \rangle}$. *Proof:* Let $\bar{\pi}$: $TT^*N \to T^*N$ be the tangent bundle projection map of the manifold T^*N. Given $z \in TN$ and $v \in (TN)_z$,

$$\begin{aligned} \tilde{\theta}(v) &= \theta(\flat_* v) = z'(\bar{\pi}_* \flat_* v) = z'(\pi_* v) \\ &= \langle z, \pi_* v \rangle = \langle \pi_* S_z, \pi_* v \rangle = \widetilde{\langle S_z, v \rangle} \end{aligned}$$

since the geodesic spray S is horizontal by definition (2.93). Now let ∇ be the pullback of the Levi-Civita connection of $\langle\ ,\ \rangle$ along π: $TN \to N$. The identity map I: $TN \to TN$ is a vector field along π, so its covariant differential is well-defined; for all $U \in \mathfrak{X}TN$, $\nabla_U I = \kappa(I_* U) = \kappa U$ (cf. 3.62). Thus since $\pi_* S = I$,

$$\begin{aligned} \tilde{\Omega}(U, V) &= -d\tilde{\theta}(U, V) \\ &= -U\langle \pi_* S, \pi_* V \rangle + V\langle \pi_* S, \pi_* U \rangle + \langle \pi_* S, \pi_*[U, V] \rangle \\ &= -\langle \nabla_U I, \pi_* V \rangle - \langle I, \nabla_U \pi_* V \rangle + \langle \nabla_V I, \pi_* U \rangle \\ &\quad + \langle I, \nabla_V \pi_* U \rangle + \langle I, \pi_*[U, V] \rangle \\ &= -\langle \kappa U, \pi_* V \rangle + \langle \kappa V, \pi_* U \rangle \\ &\quad - \langle I, \nabla_U \pi_* V - \nabla_V \pi_* U - \pi_*[U, V] \rangle \\ &= \langle \pi_* JU, \pi_* V \rangle + \langle \kappa JU, \kappa V \rangle = \widetilde{\langle JU, V \rangle} \end{aligned}$$

by the structure equation 2.103i. *QED.*

Exercises Prove that J acts isometrically with respect to the Sasaki metric, and that $\tilde{\Omega}$ and $\langle\ ,\ \rangle$ are compatible in the sense of 8.5. Show that the orientation induced on the manifold TN by J (1.67) is the same as that induced by the symplectic 2-form $\tilde{\Omega}$ (8.4).

8.8 Theorem (Hodge-Lichnerowicz) If M is a compact symplectic manifold of dimension $2n$, then $H^{2k}(M; \mathbb{R}) \neq 0$, $0 \leq k \leq n$.

PROOF If $\langle\ ,\ \rangle$ is a Riemannian metric on M compatible with Ω, then Ω^n equals $(-1)^{[n/2]}n!$ times the usual Riemannian volume element on M, so the integral over M of Ω^n is nonzero. For each k, Ω^k

is closed and therefore represents a cohomology class on M; this class is nonzero since the class of the top power Ω^n is nonzero.

Exercise More generally [Li*1*], show that if a compact Riemannian manifold M admits a parallel 2-form of rank $2r \leq n$, then $H^{2k}(M; \mathbb{R}) \neq 0$, $k \leq r$.

8.9 There is still more information available about the de Rham cohomology of a compact Riemannian manifold which admits a symplectic 2-form which is parallel in addition to being closed (cf. 4.9iv).

Theorem (Hodge-Lichnerowicz) Let M be a compact Riemannian manifold of dimension $2n$. If M admits a parallel symplectic 2-form, then the Betti numbers of M satisfy the inequalities

$$b_r(M) \leq b_{r+2}(M) \qquad 0 \leq r < n$$

and

$$b_r(M) \geq b_{r+2}(M) \qquad r \geq n$$

Proof By Poincaré duality it suffices to prove the inequalities just for $r < n$. Let Ω be a symplectic 2-form on M which is parallel with respect to the Riemannian metric. It is sufficient to prove that the map from A^rM to $A^{r+2}M$ which sends an r-form μ to the $(r+2)$-form $\Omega \wedge \mu$ is injective for $r < n$, and that $\Omega \wedge \mu$ is harmonic if μ is, for then the unique harmonic representative μ of each nonzero de Rham cohomology class of degree less than n is sent to the harmonic representative of a nonzero class of degree 2 higher.

Let (x, U) be a Darboux chart on M as in 8.7*b*, so that

$$\Omega|_U = \sum_1^n dx^i \wedge dx^{n+i}$$

If μ is an r-form on M with $r < n$, then by writing $\mu|_U$ in terms of the dx^i one checks that $\Omega \wedge \mu$ is nonzero if μ is.

Now we check that $\Omega \wedge \mu$ is closed and coclosed if μ is harmonic; the first follows from the identity $d(\Omega \wedge \mu) = d\Omega \wedge \mu + \Omega \wedge d\mu$.

Let $\mu \in A^rM$ be harmonic; assume $r > 0$ since a harmonic function is constant on each component of M. Let $\{E_j\}$ be an orthonormal moving frame over an open subset U of M; define an r-form $\tilde{\mu}$ on U by $\tilde{\mu} := \sum \iota_{E_j}\Omega \wedge \iota_{E_j}\mu$. This is well-defined independently of the choice of orthonormal moving frame, and therefore $\tilde{\mu}$ can be extended to an r-form on all of M.

Before calculating $d\tilde{\mu}$, we observe that if X is a vector field on M such that $(\nabla X)_p = 0$ for some $p \in M$, then at p, $\mathscr{L}_X\mu = \nabla_X\mu$, and

$\mathscr{L}_X\Omega = \nabla_X\Omega = 0$. For example, given vector fields $V_1, \ldots, V_r$,

$$\begin{aligned}(\mathscr{L}_X\mu)(V_1, \ldots, V_r) &= X\mu(V_1, \ldots, V_r) - \sum \mu(V_1, \ldots, [X, V_i], \ldots, V_r)\\ &= (\nabla_X\mu)(V_1, \ldots, V_r)\\ &\quad + \sum \mu(V_1, \ldots, \nabla_X V_j - [X, V_j], \ldots, V_r)\\ &= (\nabla_X\mu)(V_1, \ldots, V_r) + \sum \mu(V_1, \ldots, \nabla_{V_j}X, \ldots, V_r)\end{aligned}$$

so $\mathscr{L}_X\mu$ equals $\nabla_X\mu$ at p by the assumption on X.

Thus if $\{E_j\}$ is an adapted moving frame on M near p (4.4), then

$$\begin{aligned}d\tilde{\mu}_p &= \sum (d\iota_{E_j}\Omega \wedge \iota_{E_j}\mu - \iota_{E_j}\Omega \wedge d\iota_{E_j}\mu)_p\\ &= \sum (\mathscr{L}_{E_j}\Omega \wedge \iota_{E_j}\mu - \iota_{E_j}\Omega \wedge \mathscr{L}_{E_j}\mu)_p = -(\sum \iota_{E_j}\Omega \wedge \nabla_{E_j}\mu)_p\end{aligned}$$

because the E_j are radially parallel about p. By 4.7,

$$\begin{aligned}\delta(\Omega \wedge \mu)_p &= -(\sum \iota_{E_j}\nabla_{E_j}(\Omega \wedge \mu))_p = -(\sum \iota_{E_j}(\Omega \wedge \nabla_{E_j}\mu))_p\\ &= (\Omega \wedge \delta\mu - \sum \iota_{E_j}\Omega \wedge \nabla_{E_j}\mu)_p = d\tilde{\mu}_p\end{aligned}$$

The Hodge decomposition theorem [W: 6.8] states that the only form on M which is both exact and coexact is zero. Hence $\delta(\Omega \wedge \mu) = d\tilde{\mu} = 0$, and $\Omega \wedge \mu$ is coclosed; since it is closed, it is harmonic.

8.10 **Theorem** (Hodge-Lichnerowicz) Let M be a compact Riemannian manifold. If M admits a parallel symplectic 2-form, then the odd-dimensional Betti numbers of M are even.

PROOF Let $\langle\ ,\ \rangle_o$ be the Riemannian metric on M, and let Ω be the symplectic 2-form which is parallel with respect to $\langle\ ,\ \rangle_o$. As in the proof of 8.5, define a (1, 1)-tensor field $\hat{\Omega}$ on M by $(\hat{\Omega}V)^\flat := -\iota_V\Omega$ for all $V \in \mathfrak{X}M$, where $\flat$: $TM \to T^*M$ is as in 3.8. Since Ω is parallel, so are $\hat{\Omega}$ and $-\hat{\Omega}^2$; it follows that each eigenbundle E_λ of $-\hat{\Omega}^2$ in TM is parallel, and that each eigenfunction λ of $-\hat{\Omega}^2$ is constant. Hence the new metric $\langle\ ,\ \rangle$ defined on M as in 8.5 has the same Levi-Civita connection as $\langle\ ,\ \rangle_o$; consequently Ω is parallel with respect to $\langle\ ,\ \rangle$. From now on work with the new metric $\langle\ ,\ \rangle$.

Fix $r > 0$, and let φ be the map from A^rM to A^rM which sends an r-form μ to the r-form $\iota_{(\mu^\#)}\Omega^r/r!$, where $\#: \Lambda^rT^*M \to \Lambda^rTM$ is as in 3.8. As in the proof of 8.9, if μ is harmonic, so is $\varphi\mu$.

By writing $\Omega = \sum \omega^i \wedge v^i$ with respect to a local orthonormal basis field $\{\omega^i, v^i\}$ for T^*M, one sees that on A^rM, $\varphi^2 = (-1)^r$. In particular, for odd r, $\varphi^2 =$ minus the identity on A^rM, so (cf. 1.58) A^rM can be given a complex structure by defining $(a + bi)\mu := a\mu +$

$b\varphi\mu$, $a + bi \in \mathbb{C}$, $\mu \in A^r M$. The subspace of harmonic r-forms is a complex linear subspace, and therefore has even real dimension.

8.11 **Definition** A vector field X on a symplectic manifold (M, Ω) is called a *symplectic* (or *Hamiltonian*) *vector field* if $\iota_X \Omega = dh$ for some C^∞ function h on M; h is called the *Hamiltonian function*.

Since the symplectic 2-form of a symplectic manifold is closed, it follows that $\mathscr{L}_X \Omega = 0$ for a symplectic vector field; the converse is true only locally. In fact [Cal], by the de Rham theorem, $H^1(M; \mathbb{R})$ is isomorphic to $\mathfrak{g}/\mathfrak{h}$, where $\mathfrak{g}$ is the vector space of vector fields X on M such that $\mathscr{L}_X \Omega = 0$, and $\mathfrak{h} \subset \mathfrak{g}$ is the space of symplectic vector fields. A similar result [Li6] holds for the first cohomology of M with compact support.

8.12 **Theorem** Let $(N, \langle\ ,\ \rangle)$ be a Riemannian manifold, and let $\tilde{\Omega} = \flat^* \Omega$ be the symplectic structure on $M = TN$ induced by the metric $\langle\ ,\ \rangle$ and the canonical symplectic structure Ω on T^*N as in 8.7*c*. The geodesic spray of $\langle\ ,\ \rangle$ is a Hamiltonian vector field on $(M, \tilde{\Omega})$.

PROOF Let $\widetilde{\langle\ ,\ \rangle}$ be the Sasaki metric on M, with Levi-Civita connection $\tilde{\nabla}$. By the exercise in 3.29, if S is the geodesic spray of the metric $\langle\ ,\ \rangle$ on N, then $\tilde{\nabla}_S S = 0$. If $\tilde{\theta} = \flat^* \theta$ as in 8.7*c*, then for all $U \in \mathfrak{X}M$,

$$\begin{aligned}(\iota_S \tilde{\Omega})(U) &= -d\tilde{\theta}(S, U) = -S\tilde{\theta}(U) + U\tilde{\theta}(S) + \tilde{\theta}([S, U]) \\ &= -S\widetilde{\langle S, U\rangle} + U\widetilde{\langle S, S\rangle} + \widetilde{\langle S, [S, U]\rangle} \\ &= -\widetilde{\langle \tilde{\nabla}_S S, U\rangle} - \widetilde{\langle S, \tilde{\nabla}_S U - [S, U]\rangle} + U\widetilde{\langle S, S\rangle} \\ &= -\widetilde{\langle S, \tilde{\nabla}_U S\rangle} + U\widetilde{\langle S, S\rangle} = \tfrac{1}{2}U\widetilde{\langle S, S\rangle}\end{aligned}$$

Hence $\iota_S \tilde{\Omega} = dh$, where $h := \tfrac{1}{2}\widetilde{\|S\|^2} \in C^\infty M$.

Exercise Show that the Hamiltonian h of the geodesic spray (which is unique up to an additive constant) is the function $\tfrac{1}{2}\|v\|^2$ on TN; this is the so-called *energy function* from the calculus-of-variations approach to the study of geodesics on N (see [Be], [GKM: 4.1(ii)], [Kl2], [We*1*]).

HERMITIAN VECTOR BUNDLES

In secs. 8.13 to 8.21 we shall see that symplectic vector bundles are equivalent to complex vector bundles; when both attributes are considered together, the result is a so-called Hermitian vector bundle. This should be compared with example 8.7*c*.

8.13 **Theorem** The following conditions are equivalent for a real vector bundle E over a manifold M:

(i) E is the realification of a complex vector bundle over M, that is, there exists a complex structure on E in the form (1.57, 1.58) of a section J of End E such that $J^2 = -\text{id}$.

(ii) E admits a symplectic 2-form Ω (8.4).

Given either a complex structure J or a symplectic 2-form Ω on E, the other can be chosen so that the identity $\Omega(JX, JY) = \Omega(X, Y)$ is satisfied for $X, Y \in \Gamma E$; in this case, $\langle X, Y\rangle := \Omega(X, JY)$ defines a Riemannian metric on M with respect to which J acts by isometries. In addition, the orientations induced on E by J (1.67) and Ω (8.4) agree.

Proof Let J_o be the usual complex structure on the real vector space $\mathbb{R}^{2m}$ underlying $\mathbb{C}^m$ (1.20), and let Ω_o be the usual symplectic form on $\mathbb{R}^{2m}$; for $u, v \in \mathbb{R}^{2m}$ the identity $\Omega_o(J_o u, J_o v) = \Omega_o(u, v)$ is satisfied. In matrix terms,

$$J_o = \begin{bmatrix} 0 & -I \\ I & 0 \end{bmatrix} = -\Omega_o \qquad \text{and} \qquad {}^t J_o \Omega_o J_o = \Omega_o$$

An isomorphism between a complex vector space $\mathscr{V}$ and $\mathbb{C}^m$ then induces a symplectic structure on the real vector space $\mathscr{V}^{\mathbb{R}}$ underlying $\mathscr{V}$ which is compatible with the complex structure J on $\mathscr{V}^{\mathbb{R}}$.

Let $\mathscr{U} = \{U_\alpha\}$ be a locally finite open covering of M such that E is trivial over each set U_α. A bundle chart on E over U_α determines a symplectic 2-form Ω_α on E over U_α; if $\{f_\alpha\}$ is a partition of unity with respect to $\mathscr{U}$, then (see 3.3) $\Omega := \sum f_\alpha \cdot \Omega_\alpha$ defines a symplectic 2-form on E. Since each local form Ω_α is compatible with the complex structure J of E, so is Ω. Thus (i) implies (ii).

(ii) $\Rightarrow$ (i): In 8.5 we saw that associated with a symplectic 2-form Ω on E there is associated a Riemannian metric $\langle\ ,\ \rangle$. Define $J \in \Gamma$ End E by $\langle JX, Y\rangle := \Omega(X, Y)$, $X, Y \in \Gamma E$. By 3.9 and 8.5,

$$\langle JX, JY\rangle_E = \langle (\iota_X \Omega)^\#, (\iota_Y \Omega)^\#\rangle_E = \langle \iota_X \Omega, \iota_Y \Omega\rangle_{E^*} = \langle X, Y\rangle_E$$

so J acts isometrically. The identities $\langle JX, JY\rangle = \langle X, Y\rangle$ and $\langle JX, Y\rangle = \Omega(X, Y) = -\langle X, JY\rangle$ then imply that $J^2X = -X$, so $\Omega(JX, JY) = \langle J^2X, JY\rangle = \langle JX, Y\rangle = \Omega(X, Y)$.

Finally, let $\{X_j\}$ be a local orthonormal basis field for E such that $JX_j = X_{m+j}$, $j = 1, \ldots, m$ (such a basis was constructed iteratively in the proof of 1.57). By 1.65 and 1.67, the basis orientation for E induced by J is compatible with the orientation induced on E^* by the basis $\{\omega^i\}$ dual to $\{X_j\}$; but $\Omega = \sum \omega^i \wedge \omega^{m+i}$, so Ω induces the orientation on E^* given by the $2m$-form $(-1)^{[m/2]}\Omega^m = m!\,\omega^1 \wedge \cdots \wedge \omega^{2m}$ (8.4).

8.14 Let Ω be a symplectic 2-form on a complex vector bundle (E, J) over M such that $\Omega(JX, JY) = \Omega(X, Y)$ for all $X, Y \in \Gamma E$, that is, $\Omega(JX, Y) + \Omega(X, JY) = 0$. Let $\otimes$ denote the tensor product of complex vector bundles, and define $H: E^* \otimes E^* \to \mathbb{C}$ by $H(X, Y) := \Omega(X, JY) + i\Omega(X, Y)$. From 8.13 we know that the real part of H determines a Riemannian metric on E with respect to which J acts isometrically, $\langle X, Y\rangle := \Omega(X, JY)$. Furthermore, $\Omega(X, Y) = \langle JX, Y\rangle$, so H can also be written $H(X, Y) = \langle X, Y\rangle + i\langle JX, Y\rangle = \langle X, Y\rangle + i\Omega(X, Y)$. Finally, $H(Y, X) = \overline{H(X, Y)}$, and $H(X, JY) = iH(X, Y) = -H(JX, Y)$, $i = \sqrt{-1}$.

8.15 **Definition** A *Hermitian inner product* on a complex vector space $\mathscr{V}$ is a function $H: \mathscr{V} \times \mathscr{V} \to \mathbb{C}$ such that (i) H is bilinear over $\mathbb{R} \subset \mathbb{C}$; (ii) the real part of H is a real inner product on the real vector space underlying $\mathscr{V}$; (iii) $H(v, u) = \overline{H(u, v)}$ and $H(u, iv) = iH(u, v)$.†

For example, the usual Hermitian inner product on $\mathbb{C}^m$ is

$$H_o(u, v) = \langle u, v\rangle + i\Omega_o(u, v) = \sum \bar{u}^j v^j \qquad u = (u^j),\ v = (v^j) \in \mathbb{C}^m$$

where $\langle\ ,\ \rangle$ is the usual inner product on $\mathbb{R}^{2m}$, and Ω_o is the usual symplectic form on $\mathbb{R}^{2m}$. An element $g \in GL(m, \mathbb{C})$ leaves H invariant if and only if it leaves $\langle\ ,\ \rangle$ and Ω_o invariant, so the subgroup of $GL(m, \mathbb{C})$ which preserves H is $O(2m) \cap Sp(m, \mathbb{R}) = U(m)$ (by 6.47, exercise 2). *Exercise:* Prove this directly.

8.16 **Definition** A *Hermitian vector bundle* over a manifold M is a complex vector bundle E endowed with a Hermitian metric, that is, a section H of $E^* \otimes E^*$ such that H_p is a Hermitian inner product on E_p for each $p \in M$.

† The reader who prefers the alternative convention $H(iu, v) = iH(u, v)$ is invited to use it, but is cautioned that a number of signs will have to be changed throughout these sections.

8.17 Examples

(*a*) Theorem 8.13 implies that a Hermitian vector bundle is equivalent to a complex vector bundle with a compatible symplectic 2-form and a compatible Riemannian metric (such exist by 8.13).

(*b*) Let E be the universal line bundle over $\mathbb{C}P^n$ from 1.12*e*. Define a Hermitian metric on E by restricting the usual Hermitian inner product on $\mathbb{C}^{n+1}$ to each 1-dimensional subspace.

(*c*) If $M = TN$, where $(N, \langle\ ,\ \rangle)$ is a Riemannian manifold, then $\tilde{H} := \widetilde{\langle\ ,\ \rangle} + i\tilde{\Omega}$ is a Hermitian metric on TM, where $\widetilde{\langle\ ,\ \rangle}$ is the Sasaki metric and $\tilde{\Omega}$ is the associated symplectic 2-form from 8.7*c*.

(*d*) If $(E, \langle\ ,\ \rangle)$ is a Riemannian vector bundle, then $(E^{\mathbb{C}}, H)$ is a Hermitian vector bundle, where $H(aX, bY) := \bar{a}b\langle X, Y\rangle$, $a, b \in \mathbb{C}$, and $X, Y \in \Gamma E$. This bundle is Hermitian isometric to the Hermitian vector bundle $(E \oplus E, J, H)$, where $J(X, Y) := (-Y, X)$, and

$$H((W, X), (Y, Z)) := \langle W, Y\rangle - \langle X, Z\rangle + i(\langle W, Z\rangle + \langle X, Y\rangle)$$

The bundle isomorphism from $E \oplus E$ to $E^{\mathbb{C}}$ takes (X, Y) to $X + iY$.

Exercise Show that the map $(\pi_*, \kappa)\colon TTN \to TN \oplus TN = TN^{\mathbb{C}}$ is a Hermitian vector bundle isometry along $\pi\colon TN \to N$, where N is a Riemannian manifold, and the Hermitian structures are as in 8.17*c* and *d*.

8.18 Definition Let H be a Hermitian metric on a complex vector bundle E over a manifold M. A connection ∇ in E is called *Hermitian* if H is parallel, $\nabla H = 0$.

Equivalently, ∇ is Hermitian if and only if $H(X, Y)$ is constant for all parallel sections X and Y along every C^∞ curve in M.

Exercise (cf. 3.21) A Hermitian vector bundle admits a Hermitian connection.

8.19 Proposition Let J be a complex structure on a real vector bundle E over a manifold M, and let H be a Hermitian metric on the complex vector bundle (E, J). Let ∇ be a connection in the real bundle E. If any two of the fields J, Ω, and $\langle\ ,\ \rangle$ are parallel with respect to ∇, then so is the third; in this case, ∇ is Hermitian.

Proof First assume ∇ is Riemannian. The identities $\nabla\langle\ ,\ \rangle = 0$ and $\Omega(X, Y) = \langle JX, Y\rangle$ imply that $(\nabla_U\Omega)(X, Y) = \langle(\nabla_U J)X, Y\rangle$, $U \in \mathfrak{X}M$, $X, Y \in \Gamma E$. Thus Ω is parallel if and only if J is parallel.

Now assume ∇ is symplectic, $\nabla\Omega = 0$. The identities $\nabla\Omega = 0$ and $\langle X, Y\rangle = \Omega(X, JY)$ imply that $(\nabla_U\langle\ ,\ \rangle)(X, Y) = \Omega(X, (\nabla_U J)Y)$, so $\langle\ ,\ \rangle$ is parallel if and only if J is parallel.

Under the hypotheses of the proposition, ∇H is not defined unless J is parallel so that ∇ can be thought of as a connection in the complex vector bundle (E, J). If we define ∇H purely formally as the sum $\nabla\langle\ ,\ \rangle + i\nabla\Omega$, then the formal converse to the proposition holds.

8.20 **Proposition** The curvature tensor R of a Hermitian connection on a Hermitian vector bundle (E, H) is skew-Hermitian.

PROOF If $X, Y \in \Gamma E$ and if $U, V \in \mathfrak{X}M$ with $[U, V] = 0$, then as in 3.24,

$$\begin{aligned} H(R(U, V)X, Y) &= -UH(X, \nabla_V Y) - H(\nabla_V X, \nabla_U Y) \\ &\quad + VH(X, \nabla_U Y) + H(\nabla_U X, \nabla_V Y) \\ &= -H(X, R(U, V)Y) \end{aligned}$$

8.21 Now we specialize to the tangent bundle case. Assume that J is a complex structure on the tangent bundle TM of a manifold M; as in 1.58, (M, J) is called an *almost complex manifold*. Let $H = \langle\ ,\ \rangle + i\Omega$ be a Hermitian metric on the complex vector bundle (TM, J). If the Levi-Civita connection ∇ of $\langle\ ,\ \rangle$ is a complex connection in (TM, J), then by 8.19, it is a Hermitian connection. Furthermore, Ω is parallel, and therefore (4.1) it is closed, that is, (M, Ω) is a symplectic manifold (8.6). This imposes rather stringent requirements on M; for example, if M is compact, then theorems 8.8, 8.9, and 8.10 hold. In 8.41 this situation will be dealt with further, but first we need more structure on the manifold M itself.

COMPLEX MANIFOLDS

8.22 Recall the identification of $\mathbb{C}^n$ with its underlying real vector space $\mathbb{R}^{2n}$ from 1.20: $(u^1 + iv^1, \ldots, u^n + iv^n) \leftrightarrow (u^1, \ldots, u^n, v^1, \ldots, v^n)$, $u^j, v^j \in \mathbb{R}$; writing vectors as columns for the sake of matrix multiplication we have

$$u + iv \leftrightarrow \begin{bmatrix} u \\ v \end{bmatrix} \in \mathbb{R}^{2n} \qquad u, v \in \mathbb{R}^n$$

Multiplication by i in $\mathbb{C}^n$ then corresponds to left multiplication by the matrix

$$J_o = \begin{bmatrix} 0 & -I \\ I & 0 \end{bmatrix} \in GL(2n, \mathbb{R})$$

for

$$i(u + iv) = -v + iu \leftrightarrow \begin{bmatrix} -v \\ u \end{bmatrix} = \begin{bmatrix} 0 & -I \\ I & 0 \end{bmatrix} \begin{bmatrix} u \\ v \end{bmatrix}$$

It follows that $(\mathbb{R}^{2n}, J_o)$ is a complex vector space isomorphic to $\mathbb{C}^n$. This was used in 1.21*a* to induce a complex vector bundle structure $\mathbb{C}^n \times \mathbb{C}^n$ on the tangent bundle $T\mathbb{C}^n$ of $\mathbb{C}^n$, and has also been used many times since in other settings.

The group $G := \{g \in GL(2n, \mathbb{R}) \mid gJ_o = J_o g\}$ acts on $(\mathbb{R}^{2n}, J_o)$ by complex linear transformations, and this action is equivalent to the usual action of $GL(n, \mathbb{C})$ on $\mathbb{C}^n$; indeed (6.46ii), $GL(n, \mathbb{C})$ is isomorphic to G by the map from $GL(n, \mathbb{C})$ to $GL(2n, \mathbb{R})$ such that

$$A + iB \mapsto \begin{bmatrix} A & -B \\ B & A \end{bmatrix} \qquad A, B \in \mathrm{gl}(n, \mathbb{R})$$

and this isomorphism commutes with the actions of $GL(n, \mathbb{C})$ and G on $\mathbb{C}^n$ and $(\mathbb{R}^{2n}, J_o)$, respectively.

8.23 **Definition** Let U be an open subset of $\mathbb{C}^n$. A function $f\colon U \to \mathbb{C}$ is called *holomorphic* (or *complex analytic*) if for each $a = (a^1, \ldots, a^n)$ in U there is a convergent power series of the form

$$h_a(z) = \sum_{\substack{j_k = 0 \\ k = 1, \ldots, n}}^{\infty} c_{j_1 \cdots j_n}(z^1 - a^1)^{j_1} \cdots (z^n - a^n)^{j_n} \qquad z = (z^1, \ldots, z^n)$$

on a neighborhood of a such that $f(z) = h_a(z)$ for all z near a. A map $g = (g^1, \ldots, g^m)\colon U \to \mathbb{C}^m$ is called *holomorphic* if each component $g^j\colon U \to \mathbb{C}$ is holomorphic.

8.24 **Definition** The *canonical differential operators $\partial/\partial z^j$ on the holomorphic functions on* $\mathbb{C}^n$ are defined by

$$\frac{\partial f}{\partial z^j}(a) := \lim_{\zeta \to a^j} \frac{f(a^1, \ldots, a^{j-1}, \zeta, a^{j+1}, \ldots, a^n) - f(a^1, \ldots, a^n)}{\zeta - a^j}$$

where f is assumed to be holomorphic on a neighborhood of a in $\mathbb{C}^n$.

Write $z^j = x^j + y^j, j = 1, \ldots, n$. The map $(x, y) = (x^1, \ldots, x^n, y^1, \ldots, y^n)$ from $\mathbb{C}^n$ to $\mathbb{R}^{2n}$ is a C^∞ chart on the manifold $\mathbb{C}^n$ which realizes the standard identification of $\mathbb{C}^n$ with $\mathbb{R}^{2n}$. If $f = u + iv$ is a function from an open set $U \subset \mathbb{C}^n$ to $\mathbb{C}$ which is C^∞ as a map of the underlying real manifolds (actually C^1 is sufficient), then the usual differential operators $\partial/\partial x^j$ and $\partial/\partial y^j$ on $\mathbb{C}^n$ can be applied to u and v. Just as in elementary complex analysis, f is holomorphic if and only if the Cauchy-Riemann equations hold for each j:

$$\frac{\partial u}{\partial x^j} = \frac{\partial v}{\partial y^j} \qquad \frac{\partial u}{\partial y^j} = -\frac{\partial v}{\partial x^j}$$

Furthermore, if f is holomorphic, then

$$\frac{\partial f}{\partial z^j} = \frac{\partial u}{\partial x^j} + i\frac{\partial v}{\partial x^j} =: \frac{\partial}{\partial x^j}(u + iv) = \frac{\partial f}{\partial x^j}$$

and
$$i\frac{\partial f}{\partial z^j} = \frac{\partial u}{\partial y^j} + i\frac{\partial v}{\partial y^j} =: \frac{\partial}{\partial y^j}(u + iv) = \frac{\partial f}{\partial y^j}$$

Thus the differential operator $\partial/\partial z^j$ can be represented by $\partial/\partial x^j$ (extended to complex C^∞ functions), and the operator $i\partial/\partial z^j$ can be represented by $\partial/\partial y^j$.

8.25 Let $J_o \in \Gamma$ End $T\mathbb{C}^n$ be the complex structure induced on the real vector bundle $T\mathbb{C}^n$ by its identification with the complex vector bundle $\mathbb{C}^n \times \mathbb{C}^n$. It follows that for each j, $J_o(\partial/\partial x^j) = \partial/\partial y^j$, and $J_o(\partial/\partial y^j) = -\partial/\partial x^j$ by the identities at the end of 8.24.

Proposition Suppose that $f\colon U \to \mathbb{C}^m$ is a C^∞ map, where U is open in $\mathbb{C}^n$. The map f is holomorphic if and only if its differential $f_*\colon T\mathbb{C}^n \to T\mathbb{C}^m$ commutes with the complex structures on $T\mathbb{C}^n$ and $T\mathbb{C}^m$.

PROOF It suffices to check the result for $m = 1$.

If $f = u + iv\colon U \to \mathbb{C}$, then for simplicity write $\partial u/\partial x$ for the row vector $[\partial u/\partial x^1 \cdots \partial u/\partial x^n]$, and similarly for $\partial u/\partial y$, $\partial v/\partial x$, and $\partial v/\partial y$. By 8.22,

$$f_* \circ J_o = \begin{bmatrix} \partial u/\partial x & \partial u/\partial y \\ \partial v/\partial x & \partial v/\partial y \end{bmatrix} \begin{bmatrix} 0 & -I \\ I & 0 \end{bmatrix} = \begin{bmatrix} \partial u/\partial y & -\partial u/\partial x \\ \partial v/\partial y & -\partial v/\partial x \end{bmatrix}$$

and
$$J_o \circ f_* = \begin{bmatrix} 0 & -I \\ I & 0 \end{bmatrix} \begin{bmatrix} \partial u/\partial x & \partial u/\partial y \\ \partial v/\partial x & \partial v/\partial y \end{bmatrix} = \begin{bmatrix} -\partial v/\partial x & -\partial v/\partial y \\ \partial u/\partial x & \partial u/\partial y \end{bmatrix}$$

Hence $f_* \circ J_o = J_o \circ f_*$ if and only if the Cauchy-Riemann equations hold.

For example, $GL(n, \mathbb{C})$ acts holomorphically on $\mathbb{C}^n$.

8.26 Definition A *complex manifold of dimension n* is a topological manifold M together with a collection (called a *complex analytic atlas*) of homeomorphisms (called *charts*) from open subsets of M to open subsets of $\mathbb{C}^n$; this atlas is required to have the following properties: if $z_\alpha: U_\alpha \to V_\alpha$ and $z_\beta: U_\beta \to V_\beta$ are charts with $U_\alpha \cap U_\beta \neq \phi$, then the transition function $z_\alpha \circ z_\beta^{-1}$ is a holomorphic map from $z_\beta(U_\alpha \cap U_\beta)$ into $\mathbb{C}^n$. Furthermore, the union of all the domains U_α of the charts must equal M, and the atlas must be maximal with respect to the complex analytic compatibility condition.

As an alternative to the maximality condition, one could say that two complex analytic atlases on a given topological manifold are equivalent if their union is a complex analytic atlas; a complex manifold would then be a topological manifold together with an equivalence class of complex analytic atlases.

Let $z = x + iy: U \to \mathbb{C}^n$ be the decomposition of a chart on a complex manifold M into its real and imaginary parts with respect to the standard identification of $\mathbb{C}^n$ with $\mathbb{R}^{2n}$; the pair (x, y) is then a homeomorphism from U to an open subset of $\mathbb{R}^{2n}$. Given two such holomorphic charts on overlapping subsets of M the transition function between them is a real analytic map of open subsets of $\mathbb{R}^{2n}$, and is, in particular, C^∞. Thus the complex analytic atlas on a complex manifold induces a real analytic atlas and a C^∞ atlas on the underlying topological manifold; hence every complex manifold is simultaneously a real analytic and a C^∞ manifold in a canonical fashion.

The basic examples of complex manifolds are n-dimensional complex space $\mathbb{C}^n$, n-dimensional complex projective space $\mathbb{C}P^n$ (1.12*e*), Riemann surfaces, and open subsets of complex manifolds.

8.27 **Definition** If M and N are complex manifolds, then a map f from M to N is called *holomorphic* in case it is holomorphic with respect to the holomorphic charts on M and N. A *holomorphic function on* M is a holomorphic map from M to $\mathbb{C}$.

Holomorphic functions on a complex manifold satisfy the maximum modulus principle from complex analysis: if f is a holomorphic function on a connected open set U in a complex manifold such that f attains a maximum absolute value in U, then f is constant. This implies, among other things, that no compact complex manifold can be holomorphically embedded into $\mathbb{C}^n$ for any n (for each coordinate of such an embedding would have to attain a maximum absolute value somewhere on the compact manifold). Also familiar from complex analysis is the fact that if two holomorphic functions agree on an open neighborhood of a point p in a connected complex manifold M, then they agree on all of M.

8.28 **Definition** Let f and g be holomorphic functions defined on open sets in a complex manifold M; *f and g are equivalent at a point p* in the intersection of their domains if they agree on a neighborhood of p. An equivalence class of holomorphic functions at a point p in a complex manifold M is called a *germ of holomorphic functions at p*.

On a connected complex manifold a germ of holomorphic functions at a point p consists of exactly one function (up to the domain of definition); thus the previous definition does not seem to accomplish much. The reason for introducing germs of holomorphic functions at a point p is that holomorphic tangent vectors will act on the holomorphic functions defined near p; just as in the real C^∞ case, if tangent vectors were defined directly by their action on the functions, then all points in a manifold would have the same zero tangent vector. Using germs for the definition instead of the functions themselves is merely a technical device to obtain different zero tangent vectors at distinct points of the manifold.

The germs of holomorphic functions on a complex manifold M at a point $p \in M$ form an algebra over $\mathbb{C}$, which will be denoted by F_p; although we will not need this information, the fact that locally germs contain exactly one element says that the algebra F_p has no zero divisors.

8.29 **Definition** Let p be a point in a complex manifold M. A *holomorphic tangent vector on M at p* is a complex linear derivation on the algebra F_p of germs of holomorphic functions at p, that is, a holomorphic tangent vector at p is a complex linear map $v: F_p \to \mathbb{C}$ such that given f and $g \in F_p$, $v(fg) = v(f)g(p) + f(p)v(g)$. The set of holomorphic tangent vectors on M at p is called the *holomorphic tangent space on M at p*; temporarily the holomorphic tangent space on M at p will be denoted by the symbol $\mathscr{M}_p$.

With respect to the operations

$$(u + v)(f) := uf + vf \qquad (cv)(f) := c \cdot vf \qquad c \in \mathbb{C}$$

the holomorphic tangent space $\mathscr{M}_p$ on M at p is a complex vector space.

For example, the holomorphic tangent space on the complex manifold $\mathbb{C}^n$ at a point p is the n-dimensional complex vector space spanned by the canonical operators $\partial/\partial z^j|_p$ from 8.24.

8.30 **Definition** Let f be a holomorphic map from a complex manifold M to a complex manifold N. For each point $p \in M$, define the *induced map* $f_*|_p: \mathscr{M}_p \to \mathscr{N}_{f(p)}$ on the holomorphic tangent spaces by

$(f_* v)(\varphi) := v(\varphi \circ f)$ for all holomorphic functions φ defined on an open neighborhood of $f(p)$ in N.

The map from $\mathcal{M}_p$ to $\mathcal{N}_{f(p)}$ induced by a holomorphic map f from M to N is complex linear. Furthermore, the induced maps satisfy the chain rule: $f_* \circ g_* = (f \circ g)_*$ for g from L to M and f from M to N.

8.31 **Proposition** If M is an n-dimensional complex manifold, then for each $p \in M$ the holomorphic tangent space $\mathcal{M}_p$ is an n-dimensional complex vector space.

PROOF Let (z, U) be a holomorphic chart about a point $p \in M$. Since z is invertible, the induced map $z_*|_p$ is an isomorphism from the holomorphic tangent space $\mathcal{M}_p$ to the holomorphic tangent space on $\mathbb{C}^n$ at the point $z(p)$. But by the remark at the end of 8.29, the holomorphic tangent space on $\mathbb{C}^n$ at $z(p)$ is n-dimensional since it is spanned by the canonical operators $\partial/\partial z^j$ at $z(p)$.

8.32 **Definition** The *holomorphic tangent bundle of a complex manifold M* is the union (which will temporarily be denoted by the symbol τM) over M of the holomorphic tangent spaces $\mathcal{M}_p$. The *projection map* π *from* τM *to* M is defined by $\pi(\mathcal{M}_p) := p$, $p \in M$. Given a holomorphic chart (z, U) on M, the *associated bundle chart on* τM is the composite map $\pi^{-1}U \xrightarrow{z_*} T\mathbb{C}^n \xrightarrow{\cong} \mathbb{C}^n \times \mathbb{C}^n$, also to be denoted by z_*.

8.33 The holomorphic tangent bundle of a complex manifold is a complex vector bundle over the underlying real C^∞ manifold; in addition, the holomorphic tangent bundle is a complex manifold in its own right. As in the real C^∞ case, a holomorphic chart (z, U) on M and the canonical differential operators $\partial/\partial z^j$ on $\mathbb{C}^n$ induce local sections of τM of the form $z_*^{-1}(\partial/\partial z^j) \circ z$; again, as in the real C^∞ case, little confusion will ensue if we agree to denote these local sections of τM by the symbol $\partial/\partial z^j$.

8.34 Now we investigate the relationship between the holomorphic tangent bundle τM of a complex manifold M and the ordinary real tangent bundle TM of the underlying real C^∞ manifold.

Proposition The real vector bundle underlying the holomorphic tangent bundle τM of a complex manifold M is isomorphic to the real tangent bundle TM of the underlying real manifold M.

PROOF First of all, the result is true for the complex manifold $\mathbb{C}^n$ by the remark at the end of 8.29. In particular, the holomorphic tangent bundle chart z_* induced on τM by a holomorphic chart z on M is a real vector bundle chart on the real vector bundle underlying τM.

Given a holomorphic chart (z, U) on M, write $z^j = x^j + iy^j$ for each j, so that the map $(x, y) = (x^1, \ldots, x^n, y^1, \ldots, y^n)$ from U into $\mathbb{R}^{2n}$ is a real analytic (and hence C^∞) chart on the subordinate real manifold M; the induced map $(x, y)_*: TU \to T\mathbb{R}^{2n}$ is then one of the usual tangent bundle charts on TM. But $(x, y)_*$ is exactly z_* considered as a real vector bundle chart on the real vector bundle subordinate to τM.

To make this identification of τM (as a real vector bundle) and TM clearer, observe that just as in 8.24, each real tangent vector on M can be extended to an action on the holomorphic functions on M by the rule

$$v(f) := v(\operatorname{Re} f) + iv(\operatorname{Im} f) \qquad v \in TM$$

It follows that if (x, y) is the real chart on M corresponding to a holomorphic chart $z = x + iy$ on M, then when differentiating holomorphic functions on a neighborhood of a point $p \in M$, $\partial/\partial x^j = \partial/\partial z^j$ and $\partial/\partial y^j = i\,\partial/\partial z^j$.

8.35 To complete the identification of τM and TM, recall that the real vector bundle underlying a complex vector bundle admits (1.57) a complex structure in the form of a field J of endomorphisms of the fibers such that $J^2 = -\mathrm{id}$. By 8.24, if z_α and z_β are holomorphic charts on open sets U_α and U_β in M with $U := U_\alpha \cap U_\beta$ nonempty, then the expressions $z_{\alpha *}^{-1} \circ J_o \circ z_{\alpha *}$ and $z_{\beta *}^{-1} J_o \circ z_{\beta *}$ agree on U, where J_o is as in 8.25.

Definition The *canonical complex structure* J on the real tangent bundle TM of a complex manifold M is defined locally on M by the formula $J := z_*^{-1} \circ J_o \circ z_*$, where z is a holomorphic chart on M, and J_o is the canonical complex structure on $T\mathbb{C}^n$ (8.25). As in 1.58, J is also called the *canonical almost complex structure on M*.

From now on the holomorphic tangent space on M at a point p will be denoted by the pair (M_p, J_p), and the holomorphic tangent bundle of M will be denoted by (TM, J).

Exercises Prove the following: If (x, y) is the real C^∞ chart on M associated to a holomorphic chart z on an open set U in M, then $J(\partial/\partial x^j) = \partial/\partial y^j$ and $J(\partial/\partial y^j) = -\partial/\partial x^j$. A C^∞ map f from a complex manifold M to a complex manifold N is holomorphic if and only if f_* commutes with the almost complex structures on M and N: $f_* \circ J_M = J_N \circ f_*$. If J is the canonical almost complex structure on a complex manifold M, then J_* is the canonical almost complex structure on the complex manifold TM.

8.36 **Definition** Let (M, J) be an almost complex manifold. A *local automorphism of* J is a local diffeomorphism φ of M such that $\varphi_* \circ J = J \circ \varphi_*$. An *infinitesimal automorphism of* J is a vector field X on M whose local 1-parameter group consists of local automorphisms of J. The almost complex structure J on M is called *integrable* if it is the canonical almost complex structure induced by some complex manifold structure on M.

By analogy with 2.123 one can check that a vector field X on M is an infinitesimal automorphism of an almost complex structure J on M if and only if $\mathscr{L}_X J = 0$, that is, if and only if $[X, JY] = J[X, Y]$ for all $Y \in \mathfrak{X}M$.

8.37 **Theorem** An almost complex structure J on a C^∞ manifold M is integrable if and only if for all $X, Y \in \mathfrak{X}M$,

$$[JX, JY] = [X, Y] + J[JX, Y] + J[X, JY]$$

PROOF $\Rightarrow$: This follows directly from the exercise in 8.35 and the fact that the Lie bracket of any two coordinate vector fields on an open subset of M is zero [W: exercise 1.15].

$\Leftarrow$: This is a hard theorem of Newlander-Nirenberg [NN]. For a proof under the added assumption that M is real analytic, see [KN: appendix 8].

The expression

$$N(X, Y) := [X, Y] + J[JX, Y] + J[X, JY] - [JX, JY],$$

whose vanishing is the integrability condition for an almost complex structure J, is a tensor field, called the *Nijenhuis torsion tensor field of* J.

8.38 **Corollary** If X is an infinitesimal automorphism of the canonical almost complex structure J of a complex manifold M, then so is JX. Furthermore, each infinitesimal automorphism of J is a holomorphic map from M to TM in this case, and is therefore called a *holomorphic vector field on* M. Conversely, each holomorphic section of TM is an infinitesimal automorphism of J.

Proof The identity $[X, JY] = J[X, Y]$, $Y \in \mathfrak{X}M$, for an infinitesimal automorphism X of J (8.36) implies that $J[JX, Y] = [JX, JY] + N(X, Y)$, so JX is also an infinitesimal automorphism of J if $N = 0$.

If $\{\varphi_t\}$ is the 1-parameter group of X, then for $v \in TM$,

$$X_* Jv = \frac{\partial}{\partial t}\bigg|_0 \varphi_{t*} Jv = \frac{\partial}{\partial t}\bigg|_0 J_* \varphi_{t*} v = J_{**} X_* v$$

Therefore X_* commutes with the canonical almost complex structures on M and TM, and is therefore a holomorphic map by the exercises in 8.35. The proof of the converse is similar to 2.123.

For example, the coordinate fields $\partial/\partial z^j$ of a holomorphic chart on M are local holomorphic vector fields on M.

8.39 **Definition** A Hermitian metric $H = \langle\ ,\ \rangle + i\Omega$ on the tangent bundle TM of an almost complex manifold (M, J) is called a *Kähler metric* if the symplectic 2-form Ω is closed, that is, if Ω is a symplectic structure on M. In this case Ω is called the *Kähler form of H*, and the symplectic manifold (M, Ω) is called an *almost Kähler manifold*. If in addition J is integrable, so that (M, J) is a complex manifold, then (M, Ω) is called a *Kähler manifold*.

8.40 **Theorem** Let $H = \langle\ ,\ \rangle + i\Omega$ be a Hermitian metric on the tangent bundle of an almost complex manifold (M, J); suppose that ∇ is the Levi-Civita connection of the Riemannian metric $\langle\ ,\ \rangle = \operatorname{Re} H$. The following statements are equivalent:

(i) The almost complex structure J is parallel.
(ii) The symplectic 2-form Ω is parallel.
(iii) The connection ∇ is Hermitian.
(iv) The 2-form Ω is closed and J is integrable.
(v) The pair (M, Ω) is a Kähler manifold.

Proof The equivalence of (i), (ii), and (iii) is shown in 8.19, while by 4.1, Ω is closed if it is parallel. Now fix a point $p \in M$, and let X, Y, Z be vector fields on M such that $(\nabla X)_p = (\nabla Y)_p = (\nabla Z)_p = 0$. At p the identity $\Omega(X, Y) = \langle JX, Y\rangle$ from 8.14 implies

$$\begin{aligned} d\Omega(X, Y, Z) &- d\Omega(X, JY, JZ) \\ &= X\langle JY, Z\rangle - Y\langle JX, Z\rangle + Z\langle JX, Y\rangle + X\langle Y, JZ\rangle \\ &\quad + \langle [X, JY], Z\rangle - \langle [X, JZ], Y\rangle + \langle J[JY, JZ], X\rangle \\ &= 2\langle (\nabla_X J)Y, Z\rangle + \langle X, [Y, JZ] - [Z, JY] + [JY, JZ]\rangle \\ &= 2\langle (\nabla_X J)Y, Z\rangle - \langle JX, N(Y, Z)\rangle \end{aligned}$$

where N is the Nijenhuis torsion tensor from 8.37. Thus (i) and (ii) imply N is zero since $d\Omega$ already equals zero by (ii). Conversely, if $d\Omega$ and N are zero, then so is ∇J.

8.41 **Corollary** (Hodge) If (M, Ω) is a compact Kähler manifold of complex dimension n, then all the even-dimensional Betti numbers of M are nonzero, all the odd-dimensional Betti numbers of M are even, and $b_r(M) \leq b_{r+2}(M)$, $0 \leq r < n$, with the corresponding inequalities for $r \geq n$.

PROOF Theorems 8.40, 8.8, 8.9, and 8.10.

This proof is due to Lichnerowicz; for Chern's proof, which uses his formula for the Laplacian from 4.25, see [Ch2].

This corollary implies that a complex manifold must be very special topologically before the Levi-Civita connection of the real part of a Hermitian metric can be a Hermitian connection. For example, let $m, n > 0$. Calabi and Eckmann [CaE] showed that the product $S^{2m+1} \times S^{2n+1}$ is a complex manifold; for details, see [Ch4]. Since the Betti numbers of this manifold do not satisfy 8.41, $S^{2m+1} \times S^{2n+1}$ cannot be a Kähler manifold, so the Levi-Civita connection of a metric cannot be Hermitian.

Evidently, the normalization "torsion $= 0$" that is so useful in Riemannian geometry is not the right condition for Hermitian geometry. To express the proper normalization condition for Hermitian geometry we need to look at the complexification of the bundle TM.

8.42 **Definition** A *complex tangent vector on a real C^∞ manifold M* is a point in the complexified tangent bundle $TM^{\mathbb{C}}$; a section of $TM^{\mathbb{C}}$ is called a *complex vector field on M*. Complex tangent vectors act as complex linear derivations on the C^∞ complex functions on M by the rule $(u + iv)(f + ig) := uf - vg + i(ug + vf)$, $u, v \in M_p$, $p \in M$, and $f, g \in C^\infty M$. A *complex differential form on M* is an element $\mu = \eta + i\nu \in A(M)^{\mathbb{C}} := \mathbb{C} \otimes A(M)$, $\eta, \nu \in A(M)$.

8.43 **Proposition** The natural complex conjugations on the complex C^∞ functions, the complex vector fields, and the complex differential forms on a manifold M are related by the identities

$$\bar{V}f = \overline{V\bar{f}} \qquad \bar{\mu}(V_1, \ldots, V_r) = \overline{\mu(\bar{V}_1, \ldots, \bar{V}_r)}$$

where $V, V_1, \ldots, V_r$ are complex vector fields, f is a complex C^∞ function, and μ is a complex r-form on M.

PROOF If $V = X + iY$, where X and Y are real C^∞ vector fields, and if $f = g + ih$, $g, h \in C^\infty M$, then by definition, $\bar{V} = X - iY$ and $\bar{f} = g - ih$, so

$$\begin{aligned} \bar{V}\bar{f} &= (X - iY)(g + ih) = Xg + Yh + i(Xh - Yg) \\ &= \overline{(X + iY)(g - ih)} \end{aligned}$$

and similarly for the differential form case.

8.44 **Definition** The *complex exterior derivative operator on M* is the complex linear extension to $A(M)^{\mathbb{C}}$ of the usual exterior derivative operator d on $A(M)$: $d(\mu + i\eta) := d\mu + i\, d\eta$, $\mu, \eta \in A(M)$. Similarly extend the Lie bracket [,] to the complex vector fields on M by complex bilinearity. The natural extension to $TM^{\mathbb{C}}$ of a linear connection ∇ on TM is defined by $\nabla_U(X + iY) := \nabla_U X + i\nabla_U Y$. It is useful to extend ∇ by complex linearity in the first argument also; by abuse of language this extended operator will also be called a connection on TM.

By 8.43, given a complex function f and a complex vector field X, $\overline{df}(X) = \overline{df(\bar{X})} = \overline{\bar{X}f} = X\bar{f} = d\bar{f}(X)$, so $\overline{df} = d\bar{f}$.

8.45 Suppose that M is a $2n$-dimensional C^∞ manifold with an almost complex structure J. In addition to the real tangent bundle TM, M has two "complex tangent bundles": the bundle (TM, J), which has fiber $\mathbb{C}^n$, and the bundle $TM^{\mathbb{C}}$, which has fiber $\mathbb{C}^{2n}$. Furthermore, the real tangent bundle TM is naturally a real subbundle of $TM^{\mathbb{C}}$ in the sense that it is precisely the set of all $v \in TM^{\mathbb{C}}$ such that $\bar{v} = v$ (1.23).

Extend J complex linearly to $TM^{\mathbb{C}}$: $J(U + iV) := JU + iJV$, $U, V \in \mathfrak{X}M$. Since $J^2 = -\text{id}$, the eigenvalues of J are i and $-i$.

Definition Let $TM^{(1,0)}$ and $TM^{(0,1)}$ denote the eigenbundles of J for the eigenvalues i and $-i$, respectively. For $(r, s) = (1, 0)$ or $(0, 1)$, a section of $TM^{(r,s)}$ is called a *vector field of type* (r, s).

8.46 If J is an almost complex structure on M, then so is $-J$.

Proposition The complex vector bundles $TM^{(1,0)}$ and (TM, J) are isomorphic; similarly, the complex vector bundles $TM^{(0,1)}$ and $(TM, -J)$ are isomorphic. In particular, $TM^{\mathbb{C}} = TM^{(1,0)} \oplus TM^{(0,1)} \cong (TM, J) \oplus (TM, -J)$. Complex conjugation on $TM^{\mathbb{C}}$ is a conjugate linear automorphism of $TM^{\mathbb{C}}$ which switches the bundles $TM^{(1,0)}$ and $TM^{(0,1)}$.

PROOF The map from TM to $TM^{\mathbb{C}}$ which sends v to $\frac{1}{2}(v - iJv)$ is a real vector bundle homomorphism into the subbundle $TM^{(1,0)}$;

similarly, the map $v \mapsto \frac{1}{2}(v + iJv)$ is a homomorphism into $TM^{(0,1)}$. Both these homomorphisms are injective.

On the other hand, given $p \in M$ and $u, v \in M_p$,

$$u + iv = \frac{(u + Jv) - iJ(u + Jv)}{2} + \frac{(u - Jv) + iJ(u - Jv)}{2}$$

so $TM^{(1,0)} = \{v - iJv \mid v \in TM\}$, and $TM^{(0,1)} = \{v + iJv \mid v \in TM\}$. The rest of the proposition is now clear.

8.47 Let J be the canonical almost complex structure on a complex manifold M; by 8.35, the holomorphic tangent bundle of M is isomorphic to the complex vector bundle (TM, J), where TM is the C^∞ real tangent bundle of M. By 8.46, (TM, J) is isomorphic to the complex subbundle $TM^{(1,0)}$ of the complexified tangent bundle $TM^{\mathbb{C}}$. Thus if M is a complex manifold we can think of $TM^{(1,0)}$ as the holomorphic tangent bundle of M.

Given $v \in TM$, the element of $TM^{(1,0)}$ which corresponds to v as a holomorphic tangent vector on M is $\frac{1}{2}(v - iJv)$. For example, if z is a holomorphic chart on M, then the local holomorphic vector field $\partial/\partial z^j$ corresponds to the local section $\frac{1}{2}(\partial/\partial x^j - iJ\,\partial/\partial x^j)$ of $TM^{(1,0)}$. This correspondence is not just formal, for $J\,\partial/\partial x^j = \partial/\partial y^j$ (8.35), so

$$\frac{1}{2}\left(\frac{\partial}{\partial x^j} - iJ\frac{\partial}{\partial x^j}\right)(x^k + iy^k) = \frac{1}{2}\left(\frac{\partial}{\partial x^j} - i\frac{\partial}{\partial y^j}\right)(x^k + iy^k) = \delta_{jk} = \frac{\partial z^k}{\partial z^j}$$

hence by the chain rule, elements of $TM^{(1,0)}$ differentiate the holomorphic functions on M exactly the same way as the corresponding holomorphic tangent vectors.

On the other hand, $\frac{1}{2}(\partial/\partial x^j + iJ\,\partial/\partial x^j)(x^k + iy^k) = 0$; it follows from the chain rule that a complex C^∞ function f on M is holomorphic if and only if $\frac{1}{2}(v + iJv)f = 0$ for all v in the real tangent bundle TM. By analogy with elementary complex analysis [Car: II.2.2; VI.1.1], the elements of $TM^{(0,1)}$ are called *antiholomorphic tangent vectors*.

Similarly, $\frac{1}{2}(\partial/\partial x^j - i\,\partial/\partial y^j)(g + ih) = 0$ if and only if g and h satisfy the conjugate Cauchy-Riemann equations $\partial g/\partial x^j = -\partial h/\partial y^j$, $\partial g/\partial y^j = \partial h/\partial x^j$; hence a complex C^∞ function f is called *antiholomorphic* if and only if $vf = 0$ for all $v \in TM^{(1,0)}$. As in the exercise in 8.35, this can be expressed in terms of a complex structure. We know that $TM^{(0,1)}$ is isomorphic to the complex vector bundle $(TM, -J)$, so $-J$ is formally the complex structure conjugate to J; in fact, a complex C^∞ function is antiholomorphic if and only if $f_* \circ (-J) = J_o \circ f_*$, where J_o is the usual complex structure on $\mathbb{R}^2$ from 8.22, so $-J$ is really conjugate to J.

It is to be emphasized that by identifying holomorphic tangent vectors with elements of $TM^{\mathbb{C}}$ we have naturally extended holomorphic tangent vectors to complex derivations on all the complex C^∞ functions on M, whereas by the original definition they acted only on holomorphic functions.

8.48 **Definition** A complex (r, s)-tensor field L on an almost complex manifold M is said to be *of type* (p, q), where p and q are nonnegative integers with $p + q = s$, if $L(V_1, \ldots, V_s)$ is zero in case more than p of the V_j are of type $(1, 0)$ or more than q of the V_j are of type $(0, 1)$.

Exercises

1. A complex 1-form μ is of type $(1, 0)$ if and only if $\mu \circ J = i\mu$, and is of type $(0, 1)$ if and only if $\mu \circ J = -i\mu$.
2. A form μ is of type (p, q) if and only if $\bar{\mu}$ is of type (q, p).
3. If μ is of type (p, q) and η is of type (r, s), then $\mu \wedge \eta$ is of type $(p + r, q + s)$.
4. A (p, q)-form is zero if $p > n$ or $q > n$, $n = \dim M$.
5. Assume M is a complex manifold; a C^∞ complex function f is holomorphic if and only if df is of type $(1, 0)$, and is antiholomorphic if and only if df is of type $(0, 1)$.
6. Let H be a Hermitian metric on TM; extend H complex bilinearly to a complex $(0, 2)$-tensor field

$$H^{\mathbb{C}} \in \Gamma \operatorname{Hom}(TM^{\mathbb{C}}, M \times \mathbb{C})$$

on M. The extension $H^{\mathbb{C}}$ is of type $(1, 1)$; in particular, this is true for the complex bilinear extension of $\langle \ , \ \rangle = \operatorname{Re} H$, and for the complex bilinear extension of $\Omega = \operatorname{Im} H$.

8.49 **Proposition** If μ is a (p, q)-form on an almost complex manifold (M, J), then $d\mu$ is a sum of forms of types $(p + 2, q - 1)$, $(p + 1, q)$, $(p, q + 1)$, and $(p - 1, q + 2)$. The following conditions are equivalent for (M, J):

(i) If X and Y are vector fields of type $(1, 0)$, then so is $[X, Y]$.

(ii) If X and Y are vector fields of type $(0, 1)$, then so is $[X, Y]$.

(iii) If μ is of type (p, q), then $d\mu$ is a sum of forms of types $(p + 1, q)$ and $(p, q + 1)$.

(iv) The almost complex structure J is integrable.

PROOF (i)$\Leftrightarrow$(ii): Complex conjugation commutes with $[\ ,\ ]$.

For the first result, let μ be a (p, q)-form, $p + q = r$; given fields X_j,

$$d\mu(X_0, \ldots, X_r) = \sum (-1)^j X_j \mu(X_0, \ldots, \hat{X}_j, \ldots, X_r) + \sum_{i<j} (-1)^{i+j} \mu([X_i, X_j], X_0, \ldots, \hat{X}_i, \ldots, \hat{X}_j, \ldots, X_r)$$

By inspection, this vanishes if more than $p + 2$ of the X_j are of type $(1, 0)$ or if more than $q + 2$ of the fields are of type $(0, 1)$. Further inspection verifies that (i) and (ii) imply (iii).

Conversely, if X and Y are of type $(1, 0)$, then for each $(0, 1)$-form μ, $\mu[X, Y] = X\mu Y - Y\mu X - d\mu(X, Y) = 0$ if (iii) holds; hence $[X, Y]$ is of type $(1, 0)$.

Similarly, the Nijenhuis torsion tensor N from 8.37 vanishes if (i) and (ii) are satisfied. Conversely, given real vector fields X and Y on M, the complex fields $X - iJX$ and $Y - iJY$ are of type $(1, 0)$, hence the identity $J[X - iJX, Y - iJY] = i[X - iJX, Y - iJY] + J(N(X, Y) + iJN(X, Y))$ implies (i) if $N = 0$.

8.50 **Definition** Decompose the exterior derivative operator d on the complex forms of a complex manifold M as $d = \partial + \bar{\partial}$ (also written $d = d' + d''$), where if μ is a form of type (p, q), then $\partial\mu$ is the part of $d\mu$ of type $(p + 1, q)$, and $\bar{\partial}\mu$ is the part of $d\mu$ of type $(p, q + 1)$. A differential form μ on M is *holomorphic* if it is of type $(p, 0)$ and $\bar{\partial}\mu = 0$; an *antiholomorphic form* is a $(0, q)$-form such that $\partial\mu = 0$.

By exercise 8.48(5), it follows that a C^∞ complex function f on M is holomorphic if and only if $\bar{\partial}f = 0$.

Exercises Prove the following: If z is a holomorphic chart on a complex manifold M, then the forms dz^j are holomorphic. A $(p, 0)$-form μ is holomorphic if and only if $d\mu$ is of type $(p + 1, 0)$; this holds if and only if the C^∞ complex coefficients $\mu_{j_1 \cdots j_p}$ in the expansion $\mu = \sum \mu_{j_1 \cdots j_p} dz^{j_1} \wedge \cdots \wedge dz^{j_p}$ are holomorphic functions. A form μ is holomorphic if and only if $\bar{\mu}$ is antiholomorphic. The identity $d^2 = 0$ implies the identities $\partial^2 = \bar{\partial}^2 = \partial\bar{\partial} + \bar{\partial}\partial = 0$.

8.51 **Definition** A complex vector bundle E over a complex manifold M is called *holomorphic* if there is an atlas $\{(\pi, \varphi_\alpha): \pi^{-1}U_\alpha \to U_\alpha \times \mathbb{C}^m\}$ such that $\{U_\alpha\}$ is an open covering of M and such that if $U_\alpha \cap U_\beta \neq \varnothing$, then the bundle transition function $f_{\alpha,\beta}$ defined on $U_\alpha \cap U_\beta$ by $f_{\alpha,\beta}(p) := \varphi_\alpha \circ (\varphi_\beta|_{E_p})^{-1}$ is a holomorphic map from $\varphi_\beta(U_\alpha)$ to $\mathbb{C}^m$.

A holomorphic vector bundle E over a complex manifold M is automatically a complex manifold, so it makes sense to refer to a section X of the bundle as being holomorphic; simply require that the map X from M into E be a holomorphic map of complex manifolds. For instance, the holomorphic tangent bundle of M is a holomorphic vector bundle, as is the universal line bundle over $\mathbb{C}P^n$ (1.12e).

8.52 Now we return to the problem of a proper normalization for a Hermitian connection on a Hermitian vector bundle over a complex manifold which was mentioned in 8.41.

If H is a Hermitian metric on a complex vector bundle E over a manifold M, then the condition that a connection ∇ in E be Hermitian is the equation $\nabla H = 0$, where

$$(\nabla_V H)(X, Y) = VH(X, Y) - H(\nabla_V X, Y) - H(X, \nabla_V Y) \qquad V \in \mathfrak{X}M,\ X, Y \in \Gamma E$$

The natural extension of this equation which allows complex tangent vector fields is

$$(\nabla_V H)(X, Y) = VH(X, Y) - H(\nabla_{\bar{V}} X, Y) - H(X, \nabla_V Y) \qquad V \in \Gamma TM^{\mathbb{C}},\ X, Y \in \Gamma E$$

Theorem Let (E, H) be a Hermitian holomorphic vector bundle over a complex manifold M. There is a unique connection ∇ in E such that

(i) ∇ is Hermitian with respect to complex tangent vectors:

$$vH(X, Y) = H(\nabla_{\bar{v}} X, Y) + H(X, \nabla_v Y) \qquad v \in TM^{\mathbb{C}},\ X, Y \in \Gamma E$$

(ii) For each holomorphic section X of E, the section ∇X of $\operatorname{Hom}(TM^{\mathbb{C}}, E)$ is of type (1, 0): $\nabla_v X = 0$ for all $v \in TM^{(0,1)}$.

PROOF Uniqueness: If X and Y are local holomorphic sections of E, and if v is of type (1, 0), then $H(X, \nabla_v Y) = vH(X, Y)$ since $\bar{v}$ is of type (0, 1).

Existence: It is enough to show that such a connection exists locally, for by uniqueness, any two local connections must agree on any open set where they are both defined.

Let $\{X_j\}$ be a local holomorphic basis field for E over an open set U in M. Define $\nabla_v X_j := 0$ for v of type (0, 1), and set $H(X_j, \nabla_v X_k) := vH(X_j, X_k)$ for v of type (1, 0). Extend ∇ linearly in the first argument over the complex C^∞ functions on M, and extend ∇ to a complex derivation in the second argument: $\nabla_v(fX + Y) := v(f)X_p + f(p)\nabla_v X + \nabla_v Y$ for $v \in M_p^{\mathbb{C}}$ and $X, Y \in \Gamma E$. The operator ∇ is now a connection of the required type over the open set U in M.

Exercise If $\{Y_j\}$ is a local holomorphic basis field for E over an open set V in M with $U \cap V \neq \varnothing$, use the Cauchy-Riemann equations to check directly that the $\{X_j\}$-connection equals the $\{Y_j\}$-connection over $U \cap V$.

8.53 **Lemma** Let ∇ be a connection in the bundle (TM, J), where J is the canonical almost complex structure on a complex manifold M; assume that ∇ is a torsion-free connection in TM. A vector field X on M is a holomorphic vector field if and only if $J\nabla_V X = \nabla_{JV} X$ for all $V \in \mathfrak{X}M$.

PROOF By 8.36 and 8.38, X is a holomorphic vector field if and only if $[X, JV] = J[X, V]$ for all $V \in \mathfrak{X}M$; since J is parallel,

$$\begin{aligned}[X, JV] - J[X, V] &= \nabla_X JV - \nabla_{JV} X - J\nabla_X V + J\nabla_V X \\ &= J\nabla_V X - \nabla_{JV} X\end{aligned}$$

8.54 **Theorem** Let $H = \langle\ ,\ \rangle + i\Omega$ be a Hermitian metric on the tangent bundle of a complex manifold M; (M, Ω) is a Kähler manifold if and only if the unique Hermitian connection of type (1, 0) on TM is the Levi-Civita connection of $\langle\ ,\ \rangle$ extended complex linearly in the first argument.

PROOF $\Rightarrow$: Let us check that the extension ∇ of the Levi-Civita connection of $\langle\ ,\ \rangle$ is Hermitian with respect to complex tangent vectors and is of type (1, 0); the result follows by uniqueness.

For the first property it is sufficient to show that if W is a complex C^∞ vector field on M and if X and Y are local holomorphic vector fields on M, then $WH(X, Y) = H(\nabla_{\bar{W}} X, Y) + H(X, \nabla_W Y)$, for both sides of the equality behave properly when X and Y are multiplied by complex C^∞ functions. If $W = U + iV$ for $U, V \in \mathfrak{X}M$, then by 8.40,

$$\begin{aligned}WH(X, Y) &= U\langle X, Y\rangle + iV\langle X, Y\rangle + iU\Omega(X, Y) - V\Omega(X, Y) \\ &= \langle\nabla_U X, Y\rangle + \langle X, \nabla_U Y\rangle + i\langle\nabla_V X, Y\rangle + i\langle X, \nabla_V Y\rangle \\ &\quad + i\Omega(\nabla_U X, Y) + i\Omega(X, \nabla_U Y) - \Omega(\nabla_V X, Y) - \Omega(X, \nabla_V Y)\end{aligned}$$

By the lemma and the relations among J, Ω, and $\langle\ ,\ \rangle$ (8.13, 8.17),

$$\langle\nabla_V X, Y\rangle = \Omega(\nabla_V X, JY) = -\Omega(J\nabla_V X, Y) = \Omega(\nabla_{-JV} X, Y)$$

But in the complex vector bundle (TM, J), $JV = iV$. Thus $\langle\nabla_V X, Y\rangle = \Omega(\nabla_{-iV} X, Y)$. Similarly, $\langle X, \nabla_V Y\rangle = \Omega(X, \nabla_{iV} Y)$, $-\Omega(\nabla_V X, Y) = \langle\nabla_{-iV} X, Y\rangle$, and $-\Omega(X, \nabla_V Y) = \langle X, \nabla_{iV} Y\rangle$.

Consequently,

$$\begin{aligned} WH(X, Y) &= H(\nabla_U X, Y) + H(X, \nabla_U Y) \\ &\quad + H(\nabla_{-iV} X, Y) + H(X, \nabla_{iV} Y) \\ &= H(\nabla_{U-iV} X, Y) + H(X, \nabla_{U+iV} Y) \end{aligned}$$

Therefore the Levi-Civita connection is Hermitian.

Now let $V \in \mathfrak{X}M$, and let X be a holomorphic vector field; since $\nabla_V X$ is a section of the complex vector bundle (TM, J),

$$\nabla_{V+iJV} X = \nabla_V X + i\nabla_{JV} X = \nabla_V X + iJ\nabla_V X = \nabla_V X - \nabla_V X = 0$$

Thus the Levi-Civita connection is of type (1, 0).

$\Leftarrow$: If ∇ is Hermitian, then by 8.40, (M, Ω) is Kähler.

Exercise Use the lemma to prove that the torsion tensor T of a Hermitian connection on a Hermitian manifold satisfies the identity $T(JX, Y) = T(X, JY)$ for all $X, Y \in \mathfrak{X}M$.

THE CURVATURE OF KÄHLER MANIFOLDS

8.55 **Proposition** The following properties hold for the curvature tensor R on a Kähler manifold (M, Ω): for all $X, Y \in \mathfrak{X}M$,

(i) $R(X, Y) \circ J = J \circ R(X, Y)$, that is, the section $R(X, Y)$ of the bundle $\mathrm{End}(TM)$ lies in the subbundle $\mathrm{End}(TM, J)$ of endomorphisms of the holomorphic tangent bundle of M.

(ii) $R(JX, JY) = R(X, Y)$.

(iii) $R(X, Y)$ is skew-Hermitian with respect to the Hermitian metric $H = \Omega(\cdot, J\,\cdot) + i\Omega$.

(iv) The Ricci tensor satisfies

$$\mathscr{R}(JX, JY) = \mathscr{R}(X, Y) = \tfrac{1}{2}\,\mathrm{tr}\, J \circ R(X, JY)$$

Proof (i): This was proved in 2.44 (whenever J is parallel).

(ii): By the block permutation identity for the curvature tensor of the Levi-Civita connection of $\langle\ ,\ \rangle = \Omega(\cdot, J\,\cdot)$ (3.35),

$$\begin{aligned} \langle R(JX, JY)U, V\rangle &= \langle R(U, V)JX, JY\rangle = \langle JR(U, V)X, JY\rangle \\ &= \langle R(U, V)X, Y\rangle = \langle R(X, Y)U, V\rangle \end{aligned}$$

(iii): This was proved in 8.20 for every Hermitian connection.

(iv): If $\{E_j\}$ is a local orthonormal moving frame with respect to

$\langle\ ,\ \rangle$, then so is $\{JE_j\}$, so by (i) and (ii),

$$\begin{aligned}\mathscr{R}(JX, JY) &= \sum \langle R(E_j, JX)JY, E_j\rangle \\ &= \sum \langle R(JE_j, X)Y, JE_j\rangle = \mathscr{R}(X, Y)\end{aligned}$$

Finally, with the obvious notation for the maps involved,

$$\begin{aligned}\mathscr{R}(X, Y) &= \mathrm{tr}[V \mapsto R(V, X)Y] = \mathrm{tr}[JV \mapsto R(JV, X)Y] \\ &= -\mathrm{tr}[JV \mapsto R(V, JX)Y] \\ &= \mathrm{tr}[JV \mapsto R(JX, Y)V + R(Y, V)JX] \\ &= -\mathrm{tr}[JV \mapsto J \circ R(JX, Y)JV + R(V, Y)JX)] \\ &= \mathrm{tr}[JV \mapsto J \circ R(X, JY)JV - R(JV, JY)JX] \\ &= \mathrm{tr}\, J \circ R(X, JY) - \mathscr{R}(JY, JX) \\ &= \mathrm{tr}\, J \circ R(X, JY) - \mathscr{R}(X, Y)\end{aligned}$$

8.56 Recall from 3.37 that if $(M, \langle\ ,\ \rangle)$ is a Riemannian manifold, then the sectional curvature of M is defined with respect to each tangent 2-plane σ on M by $K_\sigma := \langle R(u, v)v, u\rangle$, where $\{u, v\}$ is an orthonormal basis for the 2-plane $\sigma \subset M_p$, $p = \pi(\sigma) \in M$. If (M, Ω) is a Kähler manifold, then not all the tangent 2-planes on M are relevant to the Kähler geometry of M.

Definition Let p be a point in a complex manifold M; a 2-plane $\sigma \subset M_p$ is called a *holomorphic 2-plane* if it is a complex subspace of the holomorphic tangent space (M_p, J_p).

Of course a holomorphic 2-plane has complex dimension 1. It follows immediately that a 2-plane σ in M_p is holomorphic if and only if it is invariant under J, or equivalently, if and only if there exists some $v \in \sigma$ such that $\sigma = v \wedge Jv$.

8.57 **Definition** The *holomorphic sectional curvature on a Kähler manifold* (M, Ω) is the restriction to the holomorphic 2-planes on M of the sectional curvature of the Riemannian metric $\langle\ ,\ \rangle = \Omega(\cdot, J\ \cdot)$.

8.58 An analog of Schur's theorem from 3.42ii holds for holomorphic sectional curvature. The reader is referred to [KN: IX.7.4, 7.5] for a proof.

Theorem Let M be a connected Kähler manifold of complex dimension ≥ 2. If for all $p \in M$ the holomorphic sectional curvature K_σ is independent of the choice of holomorphic 2-plane $\sigma \subset M_p$, then the holomorphic sectional curvature K of M is constant, $K = \kappa$. In this case, the sectional curvature of M with respect to an arbitrary real 2-plane σ on M is

$$K_\sigma = \frac{1 + 3\cos^2 \alpha(\sigma)}{4} \kappa$$

where $\alpha(\sigma)$ is the angle between the 2-planes σ and $J\sigma$.

8.59 **Corollary** If M is a connected Kähler manifold of complex dimension ≥ 2 with constant sectional curvature, then M is flat.

PROOF If σ is a 2-plane on M for which $J\sigma \perp \sigma$, then $K_{J\sigma} = \frac{1}{4}K_\sigma$.

8.60 Let ∇ be a connection with curvature tensor R in a complex vector bundle E over a manifold M. Recall from 2.85*a* that if we define expression $\hat{c}_j(R)$ on M by the identity

$$\sum_{j=0}^{m} \hat{c}_j(R)\lambda^{m-j} := \det\left(\lambda I - \frac{1}{2\pi i} R\right) \qquad \lambda \in \mathbb{R}$$

then the polarization $c_j(R)$ of $\hat{c}_j(R)$ is a closed $2j$-form, and that the cohomology classes of M represented by the forms $c_j(R)$ depend only on the complex vector bundle E, not on the choice of connection on E.

The class represented by the form $c_j(R)$ is called the jth Chern class of E, and is an integral cohomology class of M. The form $c_j(R)$ is called the jth Chern form of the connection ∇ on E.

Theorem If $\mathscr{R}$ is the Ricci curvature tensor of the Riemannian metric on a Kähler manifold M, then the first Chern form of the tangent bundle (TM, J) of M is given by

$$(c_1(R))(X, Y) = \frac{1}{2\pi} \mathscr{R}(JX, Y)$$

PROOF Denote $c_1(R)$ by $\mu \in A^2M$. By definition, for all $X, Y \in \mathfrak{X}M$,

$$\mu(X, Y) = -\frac{1}{2\pi i} \operatorname{tr}_{\mathbb{C}} R(X, Y) = \frac{i}{2\pi} \operatorname{tr}_{\mathbb{C}} R(X, Y)$$

where $\operatorname{tr}_{\mathbb{C}} R(X, Y)$ denotes the trace over $\mathbb{C}$ of the section $R(X, Y)$ of the complex algebra bundle $\mathfrak{gl}(TM, J)$. By 8.55iii, $R(X, Y)$ is actually a section of the unitary algebra bundle $\mathfrak{u}(TM, H) \subset \mathfrak{gl}(TM, J)$ of the Hermitian metric H on TM.

To relate $\mathrm{tr}_{\mathbb{C}}\, R(X, Y)$ to the real trace of $R(X, Y) \in \Gamma\mathfrak{gl}(TM)$, observe that if $\varphi\colon \mathfrak{gl}(n, \mathbb{C}) \to \mathfrak{gl}(2n, \mathbb{R})$ is the usual embedding from 8.22, then

$$\mathrm{tr}\, \varphi(A) = \mathrm{tr} \begin{bmatrix} B & -C \\ C & B \end{bmatrix} = 2\, \mathrm{tr}_{\mathbb{C}}\, B \qquad A = B + iC \in \mathfrak{gl}(n, \mathbb{C})$$

Furthermore, if $A \in \mathfrak{u}(n)$, then the diagonal elements of iA are real, so $\mathrm{tr}[J \circ \varphi(A)] = \mathrm{tr}\, \varphi(iA) = 2\, \mathrm{tr}_{\mathbb{C}}\, iA = 2i\, \mathrm{tr}_{\mathbb{C}}\, A$. Thus by 8.55,

$$\begin{aligned} \mu(X, Y) &= \frac{i}{2\pi}\, \mathrm{tr}_{\mathbb{C}}\, R(X, Y) = \frac{1}{4\pi}\, \mathrm{tr}[J \circ R(X, Y)] \\ &= \frac{1}{4\pi}\, \mathrm{tr}[J \circ R(JX, JY)] = \frac{1}{2\pi}\, \mathcal{R}(JX, Y) \end{aligned}$$

8.61 Example The Riemann sphere S^2 is a Riemann surface, and is therefore a 1-dimensional complex manifold (which is in fact holomorphically diffeomorphic to $\mathbb{C}P^1$). The holomorphic atlas on S^2 is generated by the following two charts: for p equal to either the north or south pole, stereographically project $S^2 - \{p\}$ from p onto the equatorial plane $\mathbb{R}^2 \cong \mathbb{C}$ (1.12*j*). The usual Riemannian metric on S^2 is invariant under the canonical almost complex structure J on S^2; in fact, if $\{u, v\}$ is a positively oriented orthonormal basis for the tangent space on S^2 at p, then (8.35, exercise) $Ju = v$ and $Jv = -u$.

Since the curvature tensor of S^2 is $R(X, Y) = X \wedge Y$ by 3.34*c*, it follows that the first Chern form on S^2 is $1/2\pi$ times the volume form of S^2. Thus when we evaluate the first Chern class of S^2 over the fundamental homology cycle by integrating the first Chern form $c_1(R)$ over S^2 [W: 4.17], we obtain

$$\int_{S^2} c_1(R) = \frac{1}{2\pi}\, \mathrm{vol}\, S^2 = 2$$

This is the Euler characteristic of S^2. In fact, the integral equation above is really an example of the Gauss-Bonnet theorem, for it can be shown [Mi3] that the top-dimensional Chern class of a complex vector bundle is the Euler class of the underlying oriented real vector bundle. For more details, see [KN: XII.5].

CHAPTER

NINE

OTHER DIFFERENTIAL GEOMETRIC STRUCTURES

Let N be a manifold with a fixed geometric structure; examples include a vector space (possibly with an inner product), an affine space, a Lie group (possibly with an invariant connection or metric), a homogeneous space (for instance the Möbius space $O(n + 1, 1)/H$ from 5.19), and a symmetric space. Let E be a C^∞ fiber bundle over a manifold M with standard fiber N such that the structure group G of E acts on N so as to preserve the given structure. The geometric structure of N can then be imposed on each fiber of E. Suppose there is given a system $\mathbb{P}$ of C^∞ lifts into E of the C^∞ curves of M such that for each curve γ, the induced map $E_{\gamma(a)} \to E_{\gamma(b)}$ along γ preserves the geometric structure of the fibers; if $\mathbb{P}$ satisfies axioms similar to those in 2.6 (this will be made precise later), then we may think of $\mathbb{P}$ as a parallelism structure in E; the pair $(E, \mathbb{P})$ is a generalization of the geometric structure of N over the base manifold M.

Examples already familiar to the reader are linear parallel transport in a vector bundle (2.7), isometric parallel transport in a Riemannian vector bundle (3.19) or a Lorentz vector bundle (3.39), and affine parallel transport in an affine bundle (2.108, 2.110). Similarly, parallel transport in the tensor product $E_1 \otimes E_2$ of vector bundles (each given a connection) should preserve the tensor product structure, not just the vector space structure, of the fibers.

For an example which has not been explored in this book, consider the theory of relativity. Lorentz, Einstein, and Minkowski realized that the proper geometric setting for a study of motion and electromagnetism was the so-called Minkowski space—$\mathbb{R}^4$ with a Lorentz inner product (3.39) [MTW]. Einstein then realized that a comprehensive treatment of gravitation should be based on the geometry of a 4-dimensional manifold M (called space-time) with a Lorentz metric on the tangent bundle; thus each fiber of TM was required to be Minkowski space, and the tensor calculus had to respect this Minkowski structure. When Levi-Civita invented parallel transport a few years later, it too respected the Minkowski structure of the fibers, that is, the Lorentz metric was parallel.

More recently, Yang and Mills [YM] generalized some ideas of Weyl to introduce a new geometric technique into quantum physics; the result was essentially the independent rediscovery of parallel transport in a fiber bundle. The reader will discover [Bl] [DM] [Re] ['t H] that a problem of terminology has arisen (see 2.110): while physicists generally refer to the Levi-Civita connection when studying general relativity, they prefer Weyl's term "gauge theory" (which came out of his attempt to express the theory of relativity in terms of conformal geometry [Bl]) when studying quantum theory via parallel transport.

For example, the so-called *isospin space* for the proton-neutron doublet [Fr: 2.14] is $\mathbb{C}^2$ with its usual Hermitian inner product H_o (8.15), so for global questions one can use a Hermitian vector bundle E of rank 2; actually, it is expedient to work not with $(\mathbb{C}^2, H_o)$, but with its "internal symmetry group" $SU(2)$ instead, so physicists find it more natural to look at the principal $SU(2)$-bundle of special unitary frames of E than at E itself. A far more significant application of this approach has led via the group $SU(2) \times U(1)$ to a highly successful unified theory of the electromagnetic and weak forces; it is hoped that a similar approach with an appropriate group may bring the strong force into the theory. The final step in this program would be to unify the fourth force—gravitation—with the other three.

All this has naturally inspired much work by differential geometers, of which only [AHS], [BL], and [BLS] are cited as examples.

In this chapter a number of important geometric structures will be studied in terms of parallel transport in fiber bundles. Principal fiber bundles will be dealt with first because of the relative simplicity of Lie groups, and because the work in 1.47 and 1.49 allows us to use principal fiber bundles as the main tool in the study of more general fiber bundles.

PARALLELISM IN PRINCIPAL FIBER BUNDLES

9.1 **Definition** A *parallelism structure in a principal fiber bundle* $\wp: B \to M$ with fiber (and group) G is a system $\mathbb{P}$ of C^∞ lifts to B of the C^∞ curves in M such that the following axioms hold:

(1) Existence: For each $b \in B$ and each C^∞ curve $\gamma: [a, c] \to M$ with $\gamma(a) = \wp(b)$, there is given a unique C^∞ lift $\mathbb{P}_\gamma b$ of γ to B (that is,

$\mathbb{P}_\gamma b$ is a C^∞ curve in B such that $\not{p} \circ \mathbb{P}_\gamma b = \gamma$) with initial value b; $\mathbb{P}_\gamma b$ is said to be *parallel along* γ and is also called a *horizontal curve in B*.

(2) Homogeneity and invertibility: Given a C^∞ curve γ: $[a, c] \to M$ the parallel transport map $B_{\gamma(a)} \to B_{\gamma(c)}$ along γ which sends $b = \mathbb{P}_\gamma b(a)$ to its parallel translate $\mathbb{P}_\gamma b(c)$ commutes with the right action of G on B: $\mathbb{P}_\gamma(bg)(c) = \mathbb{P}_\gamma b(c) \cdot g$, $g \in G$. This map is invertible, and its inverse is parallel transport along the reverse curve γ^- [2.6(2)].

(3) Parametrization independence: Let γ: $[a, c] \to M$ be a C^∞ curve, and let φ: $[d, e] \to [a, c]$ be a C^∞ function such that $\varphi(d) = a$ and $\varphi(e) = c$. It is required that $\mathbb{P}_{\gamma \circ \varphi} b(t) = (\mathbb{P}_\gamma b) \circ \varphi(t)$ for all b and t.

(4) C^∞ dependence on initial conditions: For each open set U in M and each C^∞ map f: $TU \to M$ such that $f(0_p) = p$, $p \in U$, the following map is required to be C^∞:

$$f\colon \Delta^*(TU \times \not{p}^{-1}U) \to B \qquad (v, b) \mapsto \mathbb{P}_\gamma b(1)$$

where $\Delta(p) := (p, p)$, $p \in U$, and $\gamma(t) := f(tv)$.

(5) Initial uniqueness: Fix a point $p \in M$. Let β and γ be C^∞ curves in M emanating from $p = \beta(0) = \gamma(0)$ such that $\dot\beta(0) = \dot\gamma(0)$. It is required that $\dot{\widehat{\mathbb{P}_\beta b}}(0) = \dot{\widehat{\mathbb{P}_\gamma b}}(0)$ for all $b \in \not{p}^{-1}(p)$.

9.2 Just as with vector bundles, parallel transport in a principal bundle can be expressed as a "horizontal" distribution on the bundle.

Definition A *connection on a principal fiber bundle* B over M with fiber G is a vector subbundle $\mathscr{H}$ of $TB \to B$ such that

(i) $\mathscr{H}$ is complementary to the vertical bundle: $TB = \mathscr{H} \oplus \mathscr{V}B$;

(ii) $\mathscr{H}$ is homogeneous: $R_{g*}\mathscr{H}_b = \mathscr{H}_{bg}$ for all $b \in B$, $g \in G$.

9.3 **Theorem** Let π: $E \to M$ be a vector bundle of rank m, and let $\not{p}$: $B \to M$ be the bundle of bases of E (1.45*e*). There is a one-to-one correspondence between the connections on E, the connections on B, and the parallelism structures in B.

Proof By 2.28 and 2.33, a connection $\mathscr{H}$ on E is equivalent to a system of parallel transport in E. This induces parallel transport in B as follows: let γ be a C^∞ curve in M, and let b be a basis $\{\epsilon_1, \ldots, \epsilon_m\}$ for the vector space $E_{\gamma(a)}$; if X_j is the $\mathscr{H}$-parallel section of E along γ such that $X_j(a) = \epsilon_j$, then $\mathbb{P}_\gamma b(t)$ is defined to be the basis $\{X_1(t), \ldots, X_m(t)\}$ for $E_{\gamma(t)}$. The required properties for $\mathbb{P}$ in B follow from those for $\mathbb{P}$ in E (2.6).

Conversely, a parallelism structure $\mathbb{P}$ in B determines parallel transport in E: if $X_1, \ldots, X_m$ are sections of E along a C^∞ curve in M such that $\{X_1(t), \ldots, X_m(t)\}$ is a basis for $E_{\gamma(t)}$ for each t, and if the section $\{X_1, \ldots, X_m\}$ of B along γ is parallel, then each X_j is defined to be parallel along γ; more generally, each constant-coefficient linear combination of the X_j is parallel along γ.

The correspondence between the parallelism structures in B and the connections on B is proved analogously to 2.28 and 2.33.

Exercise By analogy with 2.35, show that every principal fiber bundle admits a connection, and show (without reference to vector bundles) that there is a one-to-one correspondence between the connections and the parallelism structures on a principal fiber bundle.

9.4 It is worthwhile to see the explicit relationship between a connection $\mathscr{H}$ on a vector bundle E and the corresponding connection $\bar{\mathscr{H}}$ on the bundle B of bases of E. If the standard fiber of E is a vector space $\mathscr{V}$, then (1.46) E is naturally isomorphic to the bundle $B \times_{GL(\mathscr{V})} \mathscr{V}$. As in 1.47, denote by $\mathscr{P}: B \times \mathscr{V} \to E$ the projection map. According to the proof in 9.3, if a curve β in B is an $\bar{\mathscr{H}}$-horizontal lift of a curve γ in M, and if $v \in \mathscr{V}$, then the curve $\mathscr{P}(\beta, v)$ in E is $\mathscr{H}$-horizontal, that is, parallel, along γ. Since the tangent fields $\dot{\beta}$ and $\dot{\overline{\mathscr{P}(\beta, v)}}$ lie in the horizontal bundles $\bar{\mathscr{H}}$ and $\mathscr{H}$, respectively, $\mathscr{P}_*$ maps the bundle $\bar{\mathscr{H}} \times \{0\}$ on $B \times \mathscr{V}$ to the bundle $\mathscr{H}$ on E; this map is a bundle isomorphism along $\mathscr{P}$.

Much of the work from chap. 2 can be carried out for connections on principal fiber bundles; after a few examples of principal fiber bundles we shall see just a few of the important constructions.

9.5 **Examples** In these examples, assume for the sake of simplicity that the standard fiber for each real vector bundle is $\mathbb{R}^n$, and the standard fiber for each complex vector bundle is $\mathbb{C}^n$.

(*a*) The *frame bundle of a manifold M* is defined to be the bundle of bases of TM. Every linear connection on M can be represented as a connection on the frame bundle. Denote $B(TM)$ by LM.

(*b*) Let $(E, \langle\ ,\ \rangle)$ be a Riemannian vector bundle over M. Denote by $O(E)$ the subset of the bundle B of bases of E consisting of the orthonormal bases for the fibers of E. As in 1.45*e*, each point $b \in B$ can be thought of as an isomorphism from $\mathbb{R}^n$ to a fiber of E; the points of $O(E)$ are just the elements of B which map $\mathbb{R}^n$ isometrically to the corresponding fibers of E. It is easily checked that $O(E)$ is a principal subbundle of B with fiber $O(n)$, because the change of basis matrix between two orthonormal bases for an inner product space is an orthogonal matrix.

Conversely, a principal $O(n)$-subbundle Q of B determines a Riemannian metric in E by the requirement that each $b \in Q$ map $\mathbb{R}^n$ isometrically to the fiber $E_{\not p b}$.

If $\bar{\mathscr{H}}$ is the connection on B corresponding to a Riemannian connection on E as in 9.3, then $\bar{\mathscr{H}}$ can be reduced to $O(E)$ in the following sense. Given $b \in O(E)$, let β be the parallel curve in B with initial value b over a curve γ in M with $\gamma(0) = \not p(b)$; for all t, $\beta(t)$ is an orthonormal basis for $E_{\gamma(t)}$, so the curve β lies in the subbundle $O(E)$ of B. In particular, the horizontal space $\bar{\mathscr{H}}_b$ on B at b lies in the subspace $O(E)_b$ of the tangent space B_b for all $b \in O(E)$. Define $\mathscr{H} := \bar{\mathscr{H}} \cap TO(E) \subset TB$; it follows that $TO(E) = \mathscr{H} \oplus \mathscr{V}O(E)$, and that $\mathscr{H}$ is right-invariant under the action of $O(n)$ on $O(E)$. Thus $\mathscr{H}$ is a connection on $O(E)$, and by definition, it represents the original Riemannian connection on E.

The bundle $O(TM)$ associated with a Riemannian metric on M is called the *orthonormal frame bundle of M*.

9.6 **Definition** A principal subbundle P of a principal bundle B is called a *reduction of B*. A reduction of the frame bundle LM of TM to a principal G-subbundle P is called a *G-structure on M*. A connection $\mathscr{H}$ on a principal bundle B is said to be *reducible to a principal subbundle P* if $\mathscr{H} \cap TP$ is a connection on P.

9.7 **Examples** (continued from 9.5)

(*c*) A conformal structure on M (5.14) is equivalent to a $CO(n)$-structure P on M; a point $b \in LM$ belongs to P if and only if b is an orthonormal basis for $M_{\not p b}$ with respect to some Riemannian metric in the conformal class of metrics on TM. A linear connection on M preserves the conformal structure (that is, leaves some representative Riemannian metric invariant under parallel transport) if and only if it is reducible to P.

(*d*) A real vector bundle E over M is orientable (1.65) if and only if there is a principal $GL_+(n, \mathbb{R})$-subbundle of the bundle B of bases of E, where $GL_+(n, \mathbb{R})$ is the subgroup of $GL(n, \mathbb{R})$ of matrices with positive determinant; given an orientation of E, the so-called *orientation bundle* is the subbundle of B consisting of all the positively oriented bases for the fibers of E.

If E is oriented, every parallel translate of a positively oriented basis is positively oriented; thus a connection on B is automatically reducible to the orientation bundle of each orientation of E.

(*e*) If M is orientable, the corresponding $GL_+(n, \mathbb{R})$-structure $P \subset LM$ can be further reduced to an $SL(n, \mathbb{R})$-structure Q. In fact, let ω be a volume element on M representing the given orientation of M. The set of all volume-preserving $b: \mathbb{R}^n \to M_{\not p b}$ in P with respect to the standard volume element on $\mathbb{R}^n$ is a principal $SL(n, \mathbb{R})$-subbundle of P (and of LM).

Taking this one step further, we see that an $SL(n, \mathbb{R})$-structure on M can be reduced to an $SO(n)$-structure; this is equivalent to the choice of a Riemannian metric for which the given volume element is the Riemannian volume element.

In the first case a connection on LM is reducible to Q if and only if the volume element ω is parallel; the connection is further reducible to the $SO(n)$-structure if and only if both the volume element and the metric are parallel.

(*f*) If E is a complex vector bundle over M, the principal $GL(n, \mathbb{C})$-bundle $B(E)$ of complex bases of E can be embedded into the principal $GL(2n, \mathbb{R})$-bundle $B(E^{\mathbb{R}})$ of bases of the realification of E: each isomorphism $b: \mathbb{C}^n \to E_{\pi(b)}$ is a real isomorphism of the underlying real vector spaces. Conversely, if Q is a principal $GL(n, \mathbb{C})$-subbundle of the bundle B of bases of a real vector bundle E_1 of fiber dimension $2n$, then define a complex structure $J \in \Gamma\ \mathrm{End}(E_1)$ (as in 1.57) by $J_p := b \circ J_o \circ b^{-1}$ for some $b \in Q_p$, $p \in M$, where J_o is the usual complex structure on $\mathbb{R}^{2n}$ from 1.20 (cf. 8.35); J is well-defined since $GL(n, \mathbb{C})$ commutes with J on $\mathbb{R}^{2n}$.

In this case, a connection on the real vector bundle E_1 is a connection on the complex vector bundle (E_1, J) if and only if (2.15) J is parallel, or equivalently, if and only if the connection on $B(E_1)$ is reducible to a connection on Q.

(*g*) A real vector bundle E over a manifold M admits a symplectic 2-form (8.4) if and only if the basis bundle B can be reduced to a principal $Sp(n, \mathbb{R})$-subbundle (rank $E = 2n$).

Exercise Interpret 8.13 in terms of principal bundles. (Hint: see 6.47, exercise 2.)

(*h*) A complex vector bundle E over M always admits a Hermitian metric by 8.17; such a Hermitian metric is equivalent to a reduction of the principal $GL(n, \mathbb{C})$-bundle of bases of E to a principal $U(n)$-subbundle.

HOLONOMY AND CURVATURE IN PRINCIPAL FIBER BUNDLES

9.8 **Definition** Let $\mathbb{P}$ be a system of parallel transport in a principal fiber bundle B over a manifold M; fix $p \in M$. The *holonomy group of* $\mathbb{P}$ *at* p is the group $G(p)$ of diffeomorphisms of the fiber B_p of the form $\mathbb{P}_\gamma$, where γ is a piecewise C^∞ loop in M based at p; the group operation is composition along product paths. The *restricted holonomy group of* $\mathbb{P}$ *at* p is the subgroup $G_o(p)$ of the holonomy group $G(p)$ consisting of parallel translations around null-homotopic loops based at p.

9.9 **Proposition** Let $\mathbb{P}$ be a system of parallel transport in a principal fiber bundle $\not{p}: B \to M$ with group G. Each point $b \in B$ determines an isomorphism from the holonomy group $G(\not{p}b)$ of $\mathbb{P}$ at the basepoint $\not{p}b \in M$ to a subgroup of G, which we denote by $\Phi(b)$. The subgroups $\Phi(b)$ and $\Phi(b_1)$ of G determined by points b and b_1 in B over the same basepoint in M are conjugate subgroups of G. Furthermore, if points b and b' in B can be joined by a horizontal curve in B, then $\Phi(b) = \Phi(b')$.

PROOF Fix $b \in B$, $p := \not{p}b \in M$. Given a loop γ in M at p, map $\mathbb{P}_\gamma \in G(p)$ to the unique $g \in G$ such that $bg = \mathbb{P}_\gamma b$.

If $b_1 = bh$ for some $h \in G$, then $\mathbb{P}_\gamma b_1 = (\mathbb{P}_\gamma b) \cdot h = bgh = b_1 h^{-1}gh$, so $\Phi(b)$ and $\Phi(b_1)$ are conjugate in G.

Now let b' be a point in B such that $b' = \beta(1)$ for some horizontal curve β in B with $\beta(0) = b$; β is the horizontal lift to B of some curve α in M: $\beta = \mathbb{P}_\alpha$. If γ is a loop at p such that $\mathbb{P}_\gamma b = bg$, then $\alpha * \gamma * \alpha^-$ is a loop at b', and $\mathbb{P}_{\alpha * \gamma * \alpha^-} b' = \mathbb{P}_\alpha \mathbb{P}_\gamma b = \mathbb{P}_\alpha(bg) = (\mathbb{P}_\alpha b) \cdot g = b'g$. Thus $\Phi(b) \subset \Phi(b')$; the reverse inclusion follows by the same argument along α^-.

Similarly, each $b \in B$ determines an isomorphism from the restricted holonomy group $G_o(\not{p}b)$ to a subgroup $\Phi_o(b)$ of $\Phi(b) \subset G$; the other statements apply to $\Phi_o(b)$ also.

Exercise Let $\mathbb{P}$ be a parallelism structure in a vector bundle E over a connected manifold M, and let $\bar{\mathbb{P}}$ be the corresponding parallel transport in the bundle B of bases of E. Prove that $\Phi(b)$ is isomorphic to the holonomy group of $\mathbb{P}$ in E for all $b \in B$.

9.10 Just as in 2.24 and 2.25, the holonomy group $\Phi(b) \subset G$ of a connection $\mathscr{H}(\simeq \mathbb{P})$ on a principal G-bundle B, $b \in B$, is a Lie group, and its identity component is the restricted holonomy group $\Phi_o(b)$. Furthermore, $\Phi(b)$ is trivial for all $b \in B$ if and only if B is a trivial bundle and $\mathscr{H}$ is the product connection on B (cf. 2.37).

Theorem Let $\mathscr{H}$ be a connection on a principal bundle B over a connected manifold M. Fix a point $b_o \in B$, and set P equal to the set of all points in B which can be joined to b_o by an $\mathscr{H}$-horizontal curve in B. Then P is a principal subbundle of B with group $\Phi(b_o)$ and the connection $\mathscr{H}$ on B can be reduced to P.

PROOF Fix a point $p \in M$, and let (x, U) be a chart on M near p. For all $v \in M_p$, define $\gamma_v(t) := x^{-1}(x(p) + t\, dx(v))$ for t sufficiently small. Now let β be a horizontal curve in B joining b_o to some point b in P over p: $\beta(0) = b_o$, and $\beta(1) = b$. Define a section s of B over U by setting $s \circ \gamma_v(t) := (\mathbb{P}_{\gamma_v} b)(t)$; the section s is P-valued because the piecewise smooth curve $\mathbb{P}_{\gamma_v} * \beta$ is horizontal.

Define a bundle chart on $P|_U$ by $(\not{p}, \varphi)(s_q \cdot h) := (q, h) \in U \times \Phi(b)$. This imposes a C^∞ bundle structure on P. Since the action $(U \times \Phi(b)) \times \Phi(b) \to U \times \Phi(b)$, $((q, h), g) \mapsto (q, hg)$, is C^∞ and free, so is the right action of $\Phi(b_o) = \Phi(b) \subset G$ on P. Thus P is a principal bundle with group $\Phi(b_o)$; it is a subbundle of B because the chart above extends to a chart on B.

By definition, $\mathscr{H} \cap TP$ is a connection on P.

9.11 **Definition** The subbundle P of B with group $\Phi(b_o) \subset G$ constructed in 9.10 is called the *holonomy bundle through b_o of the connection $\mathscr{H}$ on B*.

Exercise Let $\mathscr{H}$ be a connection on the bundle B of bases of a vector bundle E over M; let $\mathfrak{g}$ be the Lie algebra of the holonomy group G of $\mathscr{H}$ on E, and let $\mathfrak{g}(E)$ be the holonomy algebra bundle of $\mathscr{H}$ as in 2.46. Let $P \subset B$ be a holonomy bundle of $\mathscr{H}$, and let $\mathscr{V}P$ be the vertical bundle of P as in 1.25. Define a map ψ: $\mathscr{V}P \to \mathfrak{g}(E)$ by $\psi\xi := b \circ A \circ b^{-1}$, where

$$\xi = \frac{d}{dt}\bigg|_0 b \cdot \exp(tA) \in \mathscr{V}P \qquad \text{at } b \in P,\ A \in \mathfrak{g}$$

and exp is the exponential map from $\mathfrak{g}$ to G. Prove that ψ is a Lie algebra bundle isomorphism along the map $\not{p}$: $P \to M$ (first it must be proved that $\mathscr{V}P$ is a Lie algebra bundle over P).

9.12 **Definition** A connection on a principal bundle is *flat* if it is involutive when considered as a C^∞ distribution.

With this definition, the principal bundle analogs of 2.39, 2.40, and 2.41 are valid. Furthermore, just as in 2.43 we can use the vertical component of the Lie bracket of two horizontal fields on B to measure the failure of $\mathscr{H}$ to be an involutive distribution on B. This leads to a provisional definition of the curvature form of $\mathscr{H}$ on B as the map which takes a pair (U, V) of horizontal fields on B to $\mathscr{V}[U, V] \in \Gamma\mathscr{V}B$.

In the vector bundle case, the map $\mathrm{pr}_2: \mathscr{V}E \to E$ allowed us to identify $\mathscr{V}[U, V]$ with a vector in E because by 1.27, $\mathscr{V}E$ is isomorphic to E along the projection $E \to M$. This is not true in the principal bundle case, since the fiber in $\mathscr{V}B$ over $b \in B$ is a Lie algebra, while the fiber in B through b is diffeomorphic to a Lie group. For many purposes it is sufficient to think of the curvature form on B as being $\mathscr{V}B$-valued, but in general another approach is necessary.

Exercise Continue with the notation from the exercise in 9.11. Show that if $\bar{\mathscr{H}}$-horizontal vector fields $\bar{U}$ and $\bar{V}$ on P are $\not{p}$-related to vector fields U and V on M, then $\psi(\mathscr{V}[\bar{U}, \bar{V}]) = -R(U, V) \circ \not{p}$ as a section of $\mathfrak{g}(E)$ along the map $\not{p}: P \to M$.

9.13 To express the curvature of a connection on a principal bundle in a useful form, we need something analogous to the connection map.

Definition Let $\mathscr{H}$ be a connection on a principal G-bundle B over a manifold M; let $\mathfrak{g}$ be the Lie algebra of G. The *connection form of* $\mathscr{H}$ is the $\mathfrak{g}$-valued 1-form ω on B (1.54, 1.56*b*) such that

$$\omega: TB \to \mathfrak{g}$$

$$v \mapsto \lambda_{b*}|_e^{-1}(\mathscr{V}v) \qquad v \in B_b,\ b \in B$$

where $\lambda_b: G \to B$ is the diffeomorphism onto the fiber through b defined by $\lambda_b g := bg = R_g b$. The *curvature form of* $\mathscr{H}$ is the $\mathfrak{g}$-valued 2-form $\Omega: TB \wedge TB \to \mathfrak{g}$ such that $\Omega(u, v) := -\omega[U, V]_b$, where U and V are horizontal fields on B which extend the horizontal vectors $\mathscr{H}u$ and $\mathscr{H}v$.

To check that Ω is well-defined, let X be any horizontal field on B near b, and let $f \in C^\infty B$ such that $f(b) = 0$; if U is a horizontal extension of $\mathscr{H}u \in \mathscr{H}_b$, then so is $U + fX$. Now let V be a horizontal field on B near b. Since $f(b) = 0$ and X is horizontal,

$$\mathscr{V}[U + fX, V]_b = \mathscr{V}([U, V]_b + f(b)[X, V]_b - V_b(f) \cdot X_b) = \mathscr{V}[U, V]_b$$

9.14 Now we look at some of the properties of the $\mathfrak{g}$-valued connection form on a principal G-bundle. Let $\lambda_b g := bg$, $b \in B$, $g \in G$.

Proposition The connection form ω of a connection $\mathscr{H}$ on a principal G-bundle B satisfies the identities (i) $v = \lambda_{b*}\omega v$ for $v \in \mathscr{V}B_b$, $b \in B$; (ii) $\omega v = 0$ for $v \in \mathscr{H}_b$, $b \in B$; (iii) $R_g^*\omega = \mathrm{Ad}_{g^{-1}} \circ \omega$, $g \in G$. Conversely, given a $\mathfrak{g}$-valued differential 1-form ω on B which satisfies identities (i) and (iii), the subbundle $\mathscr{H} := \ker \omega \subset TB$ is a connection on B. A curve β in B is horizontal if and only if $\omega\dot{\beta} = 0$.

PROOF $\Rightarrow$: Identities (i) and (ii) are clear by 9.13 and 9.2.

(iii): Given $b \in B$ and $g, h \in G$, $bhg = bgg^{-1}hg$, so $\lambda_{bg}^{-1} \circ R_g \circ \lambda_b(h) = L_{g^{-1}} \circ R_g(h) \in G$. Therefore given $v = \lambda_{b*}\omega v \in \mathscr{V}B_b$,

$$(R_g^*\omega)v = \omega R_{g*}v = \lambda_{bg}^{-1}{}_* R_{g*}\lambda_{b*}\omega v = L_{g^{-1}*}R_{g*}\omega v = \mathrm{Ad}_{g^{-1}} \circ \omega(v)$$

On the other hand, both $R_g^*\omega$ and $\mathrm{Ad}_{g^{-1}} \circ \omega$ are zero on $\mathscr{H}$.

$\Leftarrow$: The kernel of ω is a C^∞ vector bundle on B because ω is a maximal rank differential form on B; it is complementary to $\mathscr{V}B$ by (i), and is invariant under G by (iii). Thus it is a connection.

9.15 **Definition** Let B be a principal G-bundle over a manifold M. For each $X \in \mathfrak{g}$, the *fundamental vertical field* $\tilde{X}$ *on* B is defined by $\tilde{X}_b := \lambda_{b*}X \in \mathscr{V}B_b$, $b \in B$, where $\lambda_b g := bg \in B$, $g \in G$.

Condition (i) of 9.14 states that if ω is the connection form of a connection $\mathscr{H}$ on B, then $\omega\tilde{X} = X$ for all $X \in \mathfrak{g}$.

9.16 The next proposition should be compared with 1.63*c* and 6.54.

Proposition Let ω be the connection form of a connection $\mathscr{H}$ on a principal bundle B with group G. For $X \in \mathfrak{g}$ and $g \in G$, $R_{g*}\tilde{X} = \widetilde{\mathrm{Ad}_{g^{-1}}X} \circ R_g$. If $X, Y \in \mathfrak{g}$, $[\tilde{X}, \tilde{Y}] = \widetilde{[X, Y]}$. If $X \in \mathfrak{g}$ and if $\bar{U}$ is a horizontal field on B, then $[\tilde{X}, \bar{U}]$ is a horizontal field on B; furthermore, if $\bar{U}$ is the horizontal lift of some $U \in \mathfrak{X}M$, then $[\tilde{X}, \bar{U}] = 0$.

PROOF The first identity is essentially 9.14iii; in fact,

$$\tilde{X}_{bg} = \lambda_{bg*}X = \frac{d}{dt}\bigg|_0 bg \cdot \exp(tX)g^{-1}g = R_{g*}\lambda_{b*}\mathrm{Ad}_g X = R_{g*}\widetilde{\mathrm{Ad}_g X_b}$$

Next, for $X \in \mathfrak{g}$, the 1-parameter group of $\tilde{X} \in \Gamma\mathscr{V}B \subset \mathfrak{X}B$ is

$\{\varphi_t = R_{\exp(tX)}\}$. Therefore given $Y \in \mathfrak{g}$ and $b \in B$,

$$[\tilde{X}, \tilde{Y}]_b = \frac{d}{dt}\bigg|_0 \varphi_{-t*}\tilde{Y}_{\varphi_t b} = \frac{d}{dt}\bigg|_0 R_{\exp(-tX)*}\lambda_{\varphi_t b*}Y$$
$$= \frac{\partial^2}{\partial t\,\partial s}\bigg|_0 \varphi_t(b)\exp(sY)\exp(-tX)$$
$$= \frac{\partial^2}{\partial t\,\partial s}\bigg|_0 b\cdot\exp(tX)\exp(sY)\exp(-tX)$$
$$= \lambda_{b*}\frac{d}{dt}\bigg|_0 R_{\exp(-tX)*}Y_{\exp(tX)} = \lambda_{b*}[X, Y] = \widetilde{[X, Y]}_b$$

Similarly, if $\bar{U} \in \mathfrak{X}B$ is horizontal, then since $R_{\exp(-tX)*}\mathscr{H} = \mathscr{H}$,

$$[\tilde{X}, \bar{U}]_b = \frac{d}{dt}\bigg|_0 \varphi_{-t*}\bar{U}_{\varphi_t b}$$

is the initial tangent vector to a curve in the vector space $\mathscr{H}_b$, and is therefore horizontal. Finally, if $p_*\bar{U} = U \circ p$ for some $U \in \mathfrak{X}M$, then $p_*[\tilde{X}, \bar{U}] = [0, U] \circ p = 0$; hence $[\tilde{X}, \bar{U}]$ vanishes since p_* maps $\mathscr{H}_b$ isomorphically to M_{pb} for all $b \in B$.

9.17 The exterior derivative operator on real-valued differential forms on a manifold M can be extended to differential forms with values in a vector space $\mathscr{V}$. For our purposes the simplest approach is to let ∇ be the trivial connection in the trivial vector bundle $M \times \mathscr{V}$, and to set $d := d^\nabla$, the exterior covariant derivative operator in $A(M, M \times \mathscr{V})$ (2.75). *Exercise:* If $\mu \in A(M, M \times \mathscr{V})$ and $\eta \in \mathscr{V}^*$, then $d(\eta \circ \mu) = \eta \circ d\mu$, where the first d is the usual operator on $A(M)$.

Theorem The structure equation: Let ω and Ω be the connection form and curvature form, respectively, of a connection $\mathscr{H}$ on a principal G-bundle B over a manifold M. For all $Y, Z \in \mathfrak{X}B$,

$$\Omega(Y, Z) = d\omega(Y, Z) + [\omega Y, \omega Z]$$

In particular, if Y and Z are horizontal, then $d\omega(Y, Z) = \Omega(Y, Z)$.

PROOF Let $U, V \in \mathfrak{X}M$, and $W, X \in \mathfrak{g}$, with corresponding horizontal lifts $\bar{U}, \bar{V}$ and fundamental vertical fields $\tilde{W}, \tilde{X}$ on B. It suffices to check the various combinations of horizontal and vertical fields.

First, since $\mathscr{H}\tilde{W} = \mathscr{H}\tilde{X} = 0$, $\Omega(\tilde{W}, \tilde{X}) = 0$; but by 9.16,

$$d\omega(\tilde{W}, \tilde{X}) + [\omega\tilde{W}, \omega\tilde{X}] = \tilde{W}X - \tilde{X}W - \omega[\tilde{W}, \tilde{X}] + [W, X]$$
$$= -\omega\widetilde{[W, X]} + [W, X] = 0$$

Note: W and X should be considered as constant maps from B to $\mathfrak{g}$.

Next, $\Omega(\bar{U}, \tilde{X}) = 0$, while

$$d\omega(\bar{U}, \tilde{X}) + [\omega\bar{U}, \omega\tilde{X}] = \bar{U}X - \tilde{X}\omega\bar{U} - \omega[\bar{U}, \tilde{X}] + [0, X] = 0$$

Finally,

$$\Omega(\bar{U}, \bar{V}) = -\omega[\bar{U}, \bar{V}] = \bar{U}\omega\bar{V} - \bar{V}\omega\bar{U} - \omega[\bar{U}, \bar{V}] + [\omega\bar{U}, \omega\bar{V}]$$
$$= d\omega(\bar{U}, \bar{V}) + [\omega\bar{U}, \omega\bar{V}]$$

By means of 1.56*b* we can write the structure equation in the form

$$\Omega = d\omega + \tfrac{1}{2}[\omega, \omega] \in A^2(B, B \times \mathfrak{g})$$

The equation $d\omega = \Omega - \frac{1}{2}[\omega, \omega]$ gives a decomposition of $d\omega$ into horizontal and vertical parts.

Exercise Prove the Bianchi identity $d\Omega = [\Omega, \omega]$.

9.18 **Corollary** The curvature form Ω of a connection $\mathscr{H}$ on a principal G-bundle B satisfies the identity $R_g^*\Omega = \mathrm{Ad}_{g^{-1}} \circ \Omega$, $g \in G$.

PROOF For all $g \in G$,

$$R_g^*\Omega = R_g^*(d\omega + \tfrac{1}{2}[\omega, \omega]) = dR_g^*\omega + \tfrac{1}{2}[R_g^*\omega, R_g^*\omega]$$
$$= d(\mathrm{Ad}_{g^{-1}} \circ \omega) + \tfrac{1}{2}[\mathrm{Ad}_{g^{-1}}\, \omega, \mathrm{Ad}_{g^{-1}}\, \omega]$$
$$= \mathrm{Ad}_{g^{-1}}(d\omega + \tfrac{1}{2}[\omega, \omega])$$

A form μ on B such that $R_g^*\mu = \mathrm{Ad}_{g^{-1}} \circ \mu$ is called *equivariant*.

Exercise Interpret the corollary in terms of the exercise in 9.12.

9.19 By 2.52, 2.58, and 9.14, a connection ∇ in a vector bundle E over a manifold M is equivalent to a connection form ω on the bundle B of bases of E. We shall now see the direct relationship between ∇ and ω.

Proposition Let ∇ be a connection in a vector bundle E over a manifold M, and let ω be the corresponding connection form on the bundle B of bases of E. Fix $p \in M$, $b \in B_p$, and a tangent vector v at b. Given a C^∞ curve β in B such that $\dot{\beta}(0) = v$, set $\gamma := p \circ \beta$. For all $\xi \in \mathbb{F}^m$, define a section $\beta\xi$ of E along γ by $(\beta\xi)_t := \beta(t)(\xi) \in E_{\gamma(t)}$. The covariant derivative of $\beta\xi$ along γ at time zero is

$$\nabla_{D_0} \beta\xi = b(\omega(v)\xi)$$

Thus, by abuse of notation, $\nabla_D \beta = \beta \cdot \omega(\dot{\beta})$ for each curve β in B.

PROOF Let η be the horizontal lift of γ to B such that $\eta(0) = b = \beta(0)$. Let g be the C^∞ curve in $GL(m, \mathbb{F})$ such that $\beta(t) = \eta(t)g(t)$ for all t. By the product rule,

$$\dot\beta(0) = \dot{\widehat{\eta g}}(0) = R_{g(0)*}\dot\eta(0) + \lambda_{\eta(0)*}\dot g(0) = \dot\eta(0) + \lambda_{b*}\dot g(0)$$

since $g(0) = I \in GL(m, \mathbb{F})$, where $\lambda_b h = bh \in B$ for all $h \in GL(m. \mathbb{F})$. Thus $\omega v = \omega\dot\beta(0) = \dot g(0)$ by 9.14 since $\dot\eta(0)$ is horizontal.

For each t, $\beta(t)$ is a basis $\{X_j|_t\}$ for $E_{\gamma(t)}$, and $\eta(t) = \{Y_j|_t\}$, where $X_j|_t = \sum g_{ij}(t) Y_i|_t$, and therefore

$$\nabla_{D_0}\beta\xi = \sum_{i,j}\xi^j\nabla_{D_0}g_{ij}Y_i = \sum_{i,j}\xi^j g'_{ij}(0)Y_i|_0$$
$$= b\left(\sum_j g'_{1j}(0)\xi^j, \ldots, \sum_j g'_{mj}(0)\xi^j\right) = b\dot g(0)\xi = b\omega(v)\xi$$

9.20 Continuing with E, ∇, B, and ω as in 9.19, we now see the relationship between the curvature tensor of ∇ and the curvature form of ω (this formalizes the exercises in 9.11, 9.12, and 9.18).

Proposition Let E, ∇, B, and ω be as in 9.19; let R be the curvature tensor of ∇, and let Ω be the curvature form of ω. Given $b \in B$ and tangent vectors u, v on B at b, the curvature matrix $\Omega(u, v) \in \mathfrak{gl}(m, \mathbb{F})$ is the matrix with respect to the basis b for $E_{\wp b}$ of the curvature transformation $R(\wp_* u, \wp_* v)$ on $E_{\wp b}$, that is,

$$R(\wp_* u, \wp_* v)b\xi = b(\Omega(u, v)\xi) \qquad \xi \in \mathbb{F}^m$$

PROOF We may assume that u and v are horizontal, since both sides of the equality are zero if one of the vectors is vertical.

Let U and V be horizontal vector fields on B such that $U_b = u$, $V_b = v$, and U and V are $\wp$-related to vector fields W, X on M. Let Y and Z be the horizontal lifts of W and X, respectively, to fields on the vector bundle E. If $\mathscr{P}\colon B \times \mathbb{F}^m \to E$ is the projection map from 1.47, $\mathscr{P}(b, \xi) = b\xi \in E_{\wp b}$, then by 9.3 and the uniqueness of horizontal lifts, the vector fields $(U, 0)$ and $(V, 0)$ on $B \times \mathbb{F}^m$ are $\mathscr{P}$-related to the vector fields Y and Z on E. Hence $\mathscr{P}_*([U, V], 0) = [Y, Z] \circ \mathscr{P}$.

The map $\mathscr{P}$ is a section of E along the map $\wp \circ \mathrm{pr}_1\colon B \times \mathbb{F}^m \to M$; the covariant derivative operator in the next sequence of calculations is the pullback of ∇ in E along this map $\wp \circ \mathrm{pr}_1$. For all $\xi \in \mathbb{F}^m$, it follows from the definition of Ω (9.13), R (2.43), and κ

(2.49) that

$$R(\not{p}_* u, \not{p}_* v)b\xi = -\kappa[Y, Z]_{b\xi} = -\kappa\mathscr{P}_*([U, V], 0)_{(b, \xi)}$$
$$= -\nabla_{([U, V], 0)}\mathscr{P}$$

Let β be a curve in B such that $\dot{\beta}(0) = [U, V]_b$; then by the chain rule for ∇ (2.54) and 9.13,

$$R(\not{p}_* u, \not{p}_* v)b\xi = -\nabla_{\dot{\widehat{(\beta, \xi)}}(0)}\mathscr{P} = -\nabla_{D0}\mathscr{P} \circ (\beta, \xi) = -\nabla_{D0}\beta\xi$$
$$= -b\omega([U, V]_b) = b\Omega(u, v)\xi$$

Exercises

1. Give another proof of the proposition using the structure equations in 2.67 and 9.17.
2. Interpret 3.24 and 8.20 in terms of the proposition.
3. If ∇ is a Riemannian connection in a Riemannian vector bundle (E, g) over M, and if $b = \{\epsilon_j\}$ is an orthonormal basis for E_p, then given vectors u and v tangent to the orthonormal frame bundle of E at b, show that $\Omega(u, v)$ is the matrix $[\langle R(\not{p}_* u, \not{p}_* v)\epsilon_j, \epsilon_i\rangle] \in \mathfrak{o}(m)$.

9.21 A special case of the following theorem was stated in 2.47.

Theorem (Ambrose-Singer) Let B be a principal fiber bundle with group G over a connected manifold M. Given a connection $\mathscr{H}$ on B and a point $b_o \in B$, let $H = \Phi(b_o) \subset G$ be the holonomy group of $\mathscr{H}$ with respect to b_o, and let $P \subset B$ be the holonomy bundle of $\mathscr{H}$ through b_o (9.11). The Lie algebra of $\Phi(b_o)$ is the subalgebra of the Lie algebra $\mathfrak{g}$ of G spanned by all $\Omega(u, v)$, $u, v \in P_b$, $b \in P$.

PROOF The subspace $\mathfrak{h}$ of $\mathfrak{g}$ spanned by all $\Omega(u, v)$, $u \wedge v \in \Lambda^2 TP$, is contained in the Lie algebra of $\Phi(b_o)$. To prove that $\mathfrak{h}$ equals the Lie algebra of $\Phi(b_o)$, it suffices to show that $\mathfrak{h}$ is the Lie algebra of the group of P.

Define a distribution $\mathscr{D}$ on P by

$$\mathscr{D}_b := \mathscr{H}_b \oplus \lambda_{b*}\mathfrak{h} \subset P_b \qquad b \in P$$

where λ_b: $\Phi(b_o) \to P$ by $\lambda_b g := bg$. If $X_1, \ldots, X_n$ are vector fields on M which form a local basis for TM near some point $p \in M$, and if $\{U_j\}$ is a basis for $\mathfrak{h}$, then the horizontal lifts $\bar{X}_j$ to P of the X_j and the fundamental vertical fields $\tilde{U}_j$ from 9.15 form a local C^∞ basis for $\mathscr{D}$ on a neighborhood of the fiber $\not{p}^{-1}p$, so the distribution $\mathscr{D}$ is C^∞.

Furthermore, $\mathscr{D}$ is involutive because (i) the Lie bracket of any two vertical fields on P is a vertical field, (ii) each bracket $[\tilde{U}_i, \bar{X}_j]$ is zero by 9.16, and (iii) for all $b \in P$, $[\bar{X}_i, \bar{X}_j]_b = \overline{[X_i, X_j]}_b - \lambda_{b*}\Omega(\bar{X}_i, \bar{X}_j)_b$, which is in $\mathscr{D}_b$ by definition.

Now let β be a horizontal curve in P with $\beta(0) = b_o$; since $\dot{\beta}(t)$ is in $\mathscr{H}_{\beta(t)} \subset \mathscr{D}_{\beta(t)}$ for all t, β lies in the maximal integral manifold of $\mathscr{D}$ through b_o; consequently, the maximal integral manifold of $\mathscr{D}$ through b_o contains, and therefore equals P.

Thus $\mathfrak{h}$ is a Lie subalgebra of the group $\Phi(b_o)$ of P with the property that $\lambda_{b*}\mathfrak{h}$ equals the vertical space $\mathscr{V}P_b$ for all $b \in P$; hence $\mathfrak{h}$ equals the Lie algebra of $\Phi(b_o)$.

Exercise Derive theorem 2.47 as a corollary of this theorem.

CHARACTERISTIC CLASSES OF PRINCIPAL BUNDLES

In chap. 2 we saw how a connection in a vector bundle E over a manifold M induced various cohomology classes on M; these classes were characteristic of E in that they were independent of the choice of connection used to construct the differential forms representing the cohomology classes. Now we shall see how a connection on a principal bundle yields the same result.

9.22 Let B be a principal G-bundle over a manifold M. Suppose that Ω is the curvature form of a connection ω on B; by definition, Ω is a 2-form on B with values in the Lie algebra $\mathfrak{g}$ of G. In what follows we will essentially be using the tensor algebra bundle of the trivial vector bundle $B \times \mathfrak{g}$; since this bundle is trivial, the B will be suppressed. Nevertheless, unless otherwise stated, multiplication is always assumed to be the tensor product in the tensor algebra of the vector space $\mathfrak{g}$, not the Lie algebra multiplication in $\mathfrak{g}$ (cf. 1.56c, 2.83). Also since the vector bundle $B \times \mathfrak{g}$ is trivial, it has the usual trivial connection ∇, which means that d^∇ is the usual exterior derivative operator d on vector-valued differential forms as it was used in the structure equation in 9.17.

If f is an invariant polynomial of degree k on $\mathfrak{g}$, then f can be applied to the k-fold product of Ω with itself; the result is a differential $2k$-form $f(\Omega \otimes \cdots \otimes \Omega)$. We will see in 9.23 that $d(f(\Omega \otimes \cdots \otimes \Omega)) = 0$. Thus $f(\Omega \otimes \cdots \otimes \Omega)$ is a closed differential form on B, and (if f is real) therefore represents a de Rham cohomology class on B; the claim is that this is the lift to B of a cohomology class on M. The class downstairs on M is the cohomology class of the polynomial f and the bundle B

induced by the connection ω; just as in 2.84, this class on M depends only on B and f, not on the connection ω.

9.23 **Proposition** Let B be a principal G-bundle over M. A differential form μ on B is the lift of a form on M if and only if μ is horizontal (that is, $\iota_v \mu = 0$ if v is vertical) and invariant under the right action of G on B (that is, $R_g^* \mu = \mu$ for all $g \in G$).

PROOF $\Rightarrow$: Suppose that $\mu = \wp^* \eta$, $n \in A(M)$, where $\wp: B \to M$ is the bundle projection. For $g \in G$, $R_g^* \mu = R_g^* \wp^* \eta = (\wp \circ R_g)^* \eta = \wp^* \eta = \mu$ since $\wp \circ R_g = \wp$. Now let v be a vertical tangent vector on B. Since $\wp_* v = 0$, $\iota_v \mu = \iota_v \wp^* \eta = \wp^* \iota_{\wp_* v} \eta = 0$.

$\Leftarrow$: Given a horizontal invariant r-form μ on B, define $\eta \in A^r M$ as follows: for $p \in M$ and $v_1, \ldots, v_r \in M_p$, choose a point $b \in B_p$ and vectors $w_1, \ldots, w_r$ on B at b such that $\wp_* w_j = v_j$ for all j. Set $\eta(v_1, \ldots, v_r) := \mu(w_1, \ldots, w_r)$. This is independent of the choice of vectors w_j at b by the horizontality of μ, and is independent of the choice of $b \in \wp^{-1} p$ by the right invariance of μ. The form η on M is C^∞ because given a local vector field on M near p, it can be lifted to a local C^∞ vector field on B near b.

Given an invariant polynomial f on $\mathfrak{g}$ and the curvature form Ω of a connection ω on B, $f(\Omega \otimes \cdots \otimes \Omega)$ is horizontal and right-invariant because the curvature form Ω is horizontal by definition, and satisfies the equivariance identity $R_g^* \Omega = \mathrm{Ad}_{g^{-1}} \Omega$. Thus it is the lift of a differential form on M, which will be denoted by $f(R)$. As in 2.83,

$$\begin{aligned} \wp^* df(R) = df\underbrace{(\Omega \otimes \cdots \otimes \Omega)}_{k \text{ factors}} &= f(d^\nabla(\Omega \otimes \cdots \otimes \Omega)) \\ &= kf(d^\nabla \Omega \otimes \Omega \otimes \cdots \otimes \Omega) \\ &= kf([\Omega, \omega] \otimes \Omega \otimes \cdots \otimes \Omega) \end{aligned}$$

by the Bianchi identity from the exercise in 9.17. But $\iota_v \wp^* df(R)$ is 0 if v is vertical, while $f([\Omega, \omega] \otimes \Omega \otimes \cdots \otimes \Omega)(v_1, \ldots, v_{2k}) = 0$ if all the v_j are horizontal. This implies that $\wp^* df(R) = d\wp^* f(R) = 0$; thus $df(R) = 0$, so $f(R)$ is closed. Thus if f is a real invariant polynomial on $\mathfrak{g}$, then $f(R)$ represents a de Rham cohomology class on M.

Exercise Let B be the bundle of bases of a vector bundle E over M, and let ω be the connection form on B of a connection ∇ in E (9.19); let Ω be the curvature form of ω, and R the curvature tensor of ∇. If f is an invariant polynomial on the Lie algebra $\mathfrak{gl}(\mathscr{V})$, where $\mathscr{V}$ is the standard fiber of E, then the differential form $f(R)$ defined above is identical to the form $f(R \otimes \cdots \otimes R)$ defined in 2.84.

PARALLEL TRANSPORT IN FIBER BUNDLES

Suppose that $\wp: B \to M$ is a principal fiber bundle with group G. By 1.47, given a C^∞ action $\mu: G \times F \to F$ of G on a manifold F, there is an associated bundle $\pi: B \times_G F \to M$ with fiber F and structure group G. Each point of $B \times_G F$ is an equivalence class of pairs $(b, \xi) \in B \times F$, where $(bg, \xi) \sim (b, g\xi)$ for all $g \in G$; the equivalence class of (b, ξ) is just the orbit of (b, ξ) in $B \times F$ under the right action $(B \times F) \times G \to B \times F$, $(b, \xi) \cdot g := (bg, g^{-1}\xi)$. The projection $\mathscr{P}: B \times F \to B \times_G F$ is a principal G-bundle. The relations among the bundles are summarized by the diagram below.

$$\begin{array}{ccc} B \times F & \xrightarrow{\text{pr}_1} & B \\ \Big\downarrow{\scriptstyle \mathscr{P}} & & \Big\downarrow{\scriptstyle \wp} \\ B \times_G F & \xrightarrow[\pi]{} & M \end{array}$$

9.24 **Definition** Given a system $\mathbb{P}$ of parallel transport in the principal bundle B, define *parallel transport in* $B \times_G F$ so that if β is a horizontal lift to B of a curve γ in M, then for all $\xi \in F$, the curve $\mathscr{P}(\beta, \xi)$ in $B \times_G F$ is a parallel section of $B \times_G F$ along γ. If $\mathscr{H} \subset TB$ is the connection associated with $\mathbb{P}$, then [En] the *associated connection on* $B \times_G F$ is defined so that the horizontal space at $\mathscr{P}(b, \xi)$ is $\mathscr{P}_*(\mathscr{H}_b \times \{0_\xi\})$.

As in 9.3, a lift α to $B \times_G F$ of a curve γ in M is parallel along γ if and only if $\dot{\alpha}$ is a horizontal vector at each point.

By analogy with 2.18 and 9.8, the holonomy group of a connection on $B \times_G F$ can also be defined; it follows that if $\Phi \subset G$ is the holonomy group of $\mathbb{P}$ in B, then the holonomy group of the connection on $B \times_G F$ equals $\eta(\Phi) \subset \text{Diff}(F)$, where $\eta: G \to \text{Diff}(F)$ is the homomorphism induced by the action of G on F. For example, if G acts effectively on F, then the holonomy group in $B \times_G F$ is isomorphic to Φ, while if G acts trivially on F, then the holonomy group in $B \times_G F$ is trivial.

9.25 **Examples**

(*a*) The vector bundle case was done in 9.3.

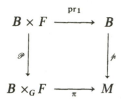

(*b*) Let $(E, \langle\ ,\ \rangle)$ be a Riemannian vector bundle over M; the *unit sphere bundle in* E is defined by $E_1 := \{v \in E \mid \|v\| = 1\}$ (see 3.61). If $O(E)$ is the bundle of orthonormal bases of $(E, \langle\ ,\ \rangle)$ from 9.5*b*, then $E_1 = O(E) \times_{O(m)} S^{m-1}$, where $m = \text{rank } E$. Fix a Riemannian connection on E, with the associated connection on $O(E)$. The parallel transport induced on E_1 as the associated bundle $O(E) \times_{O(m)} S^{m-1}$ by the Riemannian connection on $O(E)$ is identical to that induced by the inclusion of E_1 into E. In particular, parallel transport in E_1 preserves the natural Riemannian structure of the fibers of E_1.

(*c*) Recall from 5.17 that the linear conformal group of $\mathbb{R}^m$ is $CO(m) \cong O(m) \times \mathbb{R}_{>0}$. Let $(E, \langle\ ,\ \rangle)$ be a Riemannian vector bundle over M; the class of metrics on E which are conformally equivalent to $\langle\ ,\ \rangle$ (5.14, 9.7*c*) can be represented by a principal $CO(m)$-subbundle $CO(E)$ on the basis bundle B of E. The orthogonal group $O(m)$ is naturally a subgroup of $CO(m)$, and we can write $CO(E) = O(E) \times_{O(m)} CO(m)$, where $O(E)$ is the orthonormal frame bundle of E.

As in 5.19, $CO(m)$ acts conformally on S^m. It follows that each fiber of the bundle $CO(E) \times_{CO(m)} S^m$ admits a natural conformal structure. For example, fix a point $p \in M$; denote by N the fiber of $CO(E) \times_{CO(m)} S^m$ over p. Each $b \in CO(E)_p$ is identified with the map from S^m to N which sends ξ to $\mathscr{P}(b, \xi)$. Given tangent vectors u and v on N at $b\xi$, $\xi \in S^m$, define the angle $\sphericalangle(u, v)$ between u and v to be the angle $\sphericalangle(b_*^{-1}u, b_*^{-1}v)$ between their inverse images on S^m at ξ. This is independent of the choice of b over p, for if $b_1 = bh$, $h \in CO(m)$, then $\sphericalangle(b_{1*}^{-1}u, b_{1*}^{-1}v) = \sphericalangle(b_*^{-1}u, b_*^{-1}v)$ because $h \in CO(m)$ is a conformal transformation of S^m.

Similarly, the fact that $CO(m)$ fixes the points $(\pm 1, 0, \ldots, 0)$ in S^m singles out their images in each fiber of $CO(E) \times_{CO(m)} S^m$; in fact, each of the points $(\pm 1, 0, \ldots, 0)$ defines a natural section of the bundle by $p \mapsto \mathscr{P}(b, (\pm 1, 0, \ldots, 0))$ for any $b \in CO(E)_p$.

A connection on $CO(E)$ induces parallel transport in $CO(E) \times_{CO(m)} S^m$ which preserves the conformal structure of the fibers. Similarly, the sections induced by the points $(\pm 1, 0, \ldots, 0)$ are parallel. Later on we shall see a naturally related parallel transport in $CO(E) \times_{CO(m)} S^m$ for the special case $E = TM$; this related parallel transport will still be conformal, but will not leave the sections above parallel; this will rely on our study of the conformal sphere S^n in chap. 5.

(*d*) As in 5.19, $GL(m + 1, \mathbb{R})$ acts on $\mathbb{R}P^m$ by matrix multiplication on the representative points in $\mathbb{R}^{m+1} - \{0\}$. Embed $GL(m, \mathbb{R})$ into $GL(m + 1, \mathbb{R})$ as the subgroup

$$\begin{bmatrix} 1 & 0 \\ 0 & GL(m, \mathbb{R}) \end{bmatrix}$$

Since $GL(m + 1, \mathbb{R})$ maps projective lines in $\mathbb{R}P^m$ to projective lines, so does the subgroup $GL(m, \mathbb{R})$; in addition, $GL(m, \mathbb{R})$ fixes the equivalence class $[1:0:\cdots:0]$ of the point $(1, 0, \ldots, 0)$ in $\mathbb{R}^{m+1} - \{0\}$.

Let E be a vector bundle over M with standard fiber $\mathbb{R}^m$, and let B be the bundle of bases of E. Each fiber of the bundle $B \times_{GL(m,\mathbb{R})} \mathbb{R}P^m$ is naturally a projective space; given $p \in M$ and $b \in B_p$, a projective line in the fiber of $B \times_{GL(m,\mathbb{R})} \mathbb{R}P^m$ over p is defined to be the image under b of a projective line in $\mathbb{R}P^m$. This is independent of the choice of $b \in B_p$ because $GL(m, \mathbb{R})$ acts on $\mathbb{R}P^m$ by projective linear transformations [compare the discussion of $\nprec$ in (*c*)]. Just as in (*c*), the point $[1:0:\cdots:0]$ in $\mathbb{R}P^m$ determines a section of the bundle because it is invariant under the action of $GL(m, \mathbb{R})$.

For the case $E = TM$ there is a naturally related parallel transport which preserves the projective structure of the fibers, but does not leave the section above parallel.

Exercise We know that S^m is the 1-point compactification of $\mathbb{R}^m$; show that the bundle $CO(E) \times_{CO(m)} S^m$ from (*c*) is the bundle of 1-point compactifications of the fibers of E. Similarly, show that the bundle $B \times_{GL(m,\mathbb{R})} \mathbb{R}P^m$ from (*d*) is the bundle of projective completions of the fibers of E.

9.26 Given a connection on a principal G-bundle B and an action of G on a manifold F, the associated parallel transport in the bundle $B \times_G F$ satisfies all the axioms from 9.1 except for one change. Since in the general case there is no natural fiber-preserving action of G on $B \times_G F$, axiom 9.1(2) must be changed to:

(2′) Given a C^∞ curve $\gamma: [a, c] \to M$, the parallel transport map $\pi^{-1}\gamma(a) \to \pi^{-1}\gamma(c)$ along γ which sends ξ to its parallel translate is a diffeomorphism. This map is invertible, and its inverse is parallel transport along the reverse curve γ^-.

9.27 On the other hand, the given action of G on F may preserve more than just the C^∞ structure of F. For example, F may be a vector space and G may act linearly; in this case, $B \times_G F$ is a vector bundle, and parallel transport is the same as in chap. 2. Further examples were seen in 9.25.

As yet another example, suppose that F is a homogeneous space, $F = K/H$, and that the action of G on F commutes with the action of K. The group K then acts naturally on the manifold $B \times_G F$; this action preserves the fibers, and is transitive on each one. In this case, $B \times_G F$ is a bundle of homogeneous spaces, and parallel transport preserves the homogeneous space structure of the fibers. In addition, there may be given an invariant connection or metric on $F = K/H$; if the action of G on F preserves this extra structure, then each fiber of $B \times_G F$ also inherits the connection or metric. Parallel transport in $B \times_G F$ preserves all this specified structure on the fibers.

Thus to axiom (2′) we may add the statement that parallel transport in $B \times_G F$ preserves any structure on the fibers which they inherit from a G-invariant structure on F.

It is to be noted that in axiom 9.1(2), what is preserved by parallel transport in a principal G-bundle B is the homogeneous space structure of the fibers of the bundle; unless B is a trivial bundle, there is no group structure on the fibers of B, for specifying a group structure on each fiber would yield a section of B, $p \mapsto$ the identity element of the group B_p.

If G acts on a Lie group K by automorphisms, then each fiber of $B \times_G K$ is canonically a Lie group, and parallel translation is a Lie group isomorphism between fibers; the prototype for this is the vector group $\mathbb{R}^m$ and group $GL(m, \mathbb{R})$, which yield a vector bundle.

Exercise Let $\langle\ ,\ \rangle$ be a G-invariant metric (of Riemannian, Lorentz, or Hermitian type) on a manifold F. Construct the natural metric on each fiber of $B \times_G F$, and show that it is invariant under parallel transport induced by a connection on B. Given a metric g on M, define a metric on the manifold $B \times_G F$ which is the F-metric on vertical vectors and g on horizontal vectors (cf. 3.4*f*). Are the fibers in $B \times_G F$ *totally geodesic* (that is, is every geodesic in a fiber also a geodesic in $B \times_G F$)? If $B \times_G F$ has a natural section σ induced by a fixed point of G in F, is the submanifold $\sigma(M)$ of $B \times_G F$ totally geodesic?

CARTAN CONNECTIONS

Suppose that H is a closed subgroup of G, and that F is the homogeneous space G/H. In this case, only the center of G acts canonically on the fibers of $B \times_G F$. Thus although each fiber of $B \times_G F$ is diffeomorphic to G/H, it is not canonically a homogeneous space (as a fiber of the bundle) unless G is abelian. However, information of a different sort is available because of the fact that H acts naturally on B.

9.28 **Proposition** Let H be a closed subgroup of a Lie group G, and let B be a principal G-bundle over M. The associated bundle $E := B \times_G (G/H)$ is diffeomorphic as a bundle to the quotient bundle B/H. The bundle E admits a section if and only if B admits a principal H-subbundle Q, and in this case E is diffeomorphic as a bundle to $Q \times_H (G/H)$; if they exist, such sections and subbundles are in one-to-one correspondence.

PROOF Let $o \in F = G/H$ be the coset of H. Map $B \times_G F$ to B/H so that $\mathscr{P}(b, go)$ goes to the coset of bg in B modulo H; the inverse map takes the coset of b mod H to $\mathscr{P}(b, o)$.

If σ is a section of E, let Q be the set of all $b \in B$ such that the coset of b mod H equals $\sigma_{\not{p}b}$. Conversely, this equation defines a section of E for each subbundle Q of B with group H.

Fix such a bundle Q. For each $b \in B$ there exists $g \in G$ such that $bg \in Q$; the identity $\mathscr{P}(b, \xi) = \mathscr{P}(bg, g^{-1}\xi)$, $\xi \in F$, then shows that $E_{\not{p}b}$ equals the fiber in $Q \times_H F$ over $\not{p}b \in M$.

The section of E determined by an H-subbundle Q of B is just the section $\mathscr{P}(\cdot, o)$ of $Q \times_H F$ determined by the fixed point o of H in F.

As an example, consider the homogeneous space $GL(m, \mathbb{R})/GL_+(m, \mathbb{R}) \cong \mathbb{Z}_2$; a vector bundle E is orientable if and only if its basis bundle B admits a principal $GL_+(m, \mathbb{R})$-subbundle if and only if $B \times_{GL(m, \mathbb{R})} \mathbb{Z}_2$ admits a section. If E is orientable, an orientation is just a choice of section.

9.29 To the previous situation we now add a restriction on the dimension of the fiber G/H.

Definition (Cartan-Ehresmann) Let Q be a principal H-bundle over M, where H is a closed subgroup of a given Lie group G such that $\dim(G/H) = \dim M$. A *Cartan connection on Q* is a $\mathfrak{g}$-valued 1-form ϑ on Q such that (i) $\vartheta(\tilde{X}) = X$ for all $X \in \mathfrak{h}$, where $\tilde{X}$ is the fundamental vertical field on Q determined by X (9.15); (ii) $R_h^*\vartheta = \mathrm{Ad}_{h^{-1}} \circ \vartheta$ for $h \in H$; (iii) for all $b \in Q$, $\vartheta_b : Q_b \to \mathfrak{g}$ is an isomorphism.

Exercise Show that if Q admits a Cartan connection, then it is a parallelizable manifold.

9.30 Given M, H, G, and Q as in 9.29, let σ be the section $p \mapsto \mathscr{P}(b, o)$, $b \in Q_p$, of $E = Q \times_H (G/H)$ from 9.28. As in 1.25, let $\pi_{\mathscr{V}} : \mathscr{V}E \to E$ be the vertical bundle over E; the restriction of $\mathscr{V}E$ to the submanifold $\sigma(M)$ of E is essentially the composite bundle $\pi \circ \pi_{\mathscr{V}}: \mathscr{V}E|_{\sigma M} \to M$.

Proposition Let Q be a principal H-bundle over M, where H is a closed subgroup of a given Lie group G such that $\dim(G/H) = \dim M$. Set $E := Q \times_H (G/H)$, and let σ be the section $\mathscr{P}(\cdot, o)$ of E. If there exists a Cartan connection ϑ on Q, then the restricted vertical bundle $\mathscr{V}E|_{\sigma M}$ is isomorphic to TM. In this sense, for all $p \in M$ the manifold E_p is tangent to the manifold $M \cong \sigma M$ at the common point $p \cong \sigma p$.

PROOF Fix an isomorphism $(G/H)_o \cong \mathbb{R}^n$; the linear isotropy representation of H then induces a homomorphism $\lambda: H \to GL(n, \mathbb{R})$. Since $E = Q \times_H (G/H)$, the restricted bundle $\mathscr{V}E|_{\sigma M}$ is isomorphic to $Q \times_{\lambda H} \mathbb{R}^n$.

Let $\varphi: G \to G/H$ be the projection. Given $b \in Q$ and $v \in Q_b$, map the tangent vector $\not{p}_* v \in M_{\not{p}b}$ to $b_* \varphi_* \vartheta v$, where b is the diffeomorphism $\xi \mapsto \mathscr{P}(b, \xi)$ of G/H onto $E_{\not{p}b}$. If $\not{p}_* v = 0$, then $v = \tilde{X}_b$ for some $X \in \mathfrak{h}$, so $b_* \varphi_* \vartheta v = 0$, and conversely. Thus the image of $w = \not{p}_* v$ is independent of the choice of $v \in Q_b$. Now let $h \in H$. Since $hgo = hgh^{-1}o \in G/H$, $(bh)_* \varphi_* \vartheta R_{h*} v = b_* \varphi_* \mathrm{Ad}_h \mathrm{Ad}_{h^{-1}} \vartheta v = b_* \varphi_* \vartheta v$, so the choice of $b \in Q_p$ is irrelevant. The given map is a bundle isomorphism.

Exercise Prove the converse if G/H is reductive (6.64). (Hint: Extend an ordinary connection on Q.)

9.31 A Cartan connection ϑ: $TQ \to \mathfrak{g}$ on a principal H-bundle Q over M determines parallel transport in $E = Q \times_H (G/H)$ by the following roundabout procedure. Let $B := Q \times_H G$ be the natural extension of Q to a principal G-bundle; we know that $E \cong B \times_G (G/H) \cong B/H$. Let ω be the unique connection form (in the usual sense) on B such that $\iota^*\omega = \vartheta$, where ι: $Q \to B$ is the obvious embedding. Given $b \in Q$, $g \in G$, and $v \in B_{bg}$, choose $X \in \mathfrak{g}$ such that $R_{g^{-1}*} v - \tilde{X}_b \in Q_b$; then

$$\begin{aligned}\omega v &= (R_g^*\omega)(\tilde{X}_b) + (R_g^*\omega)(R_{g^{-1}*} v - \tilde{X}_b) \\ &= \mathrm{Ad}_{g^{-1}} X + \mathrm{Ad}_{g^{-1}} \circ \vartheta(R_{g^{-1}*} v - \tilde{X}_b)\end{aligned}$$

Since ω is a connection on B in the ordinary sense, we already know how it induces parallel transport in E; define this to be the parallel transport in E associated with the Cartan connection ϑ on Q.

Definition Let σ be the section $p \mapsto \mathscr{P}(b, o)$, $b \in Q_p$, of the bundle $E = Q \times_H (G/H)$ from 9.28. Let ϑ be a Cartan connection on Q. Given a curve γ in M, with $\gamma(0) = p \in M$, the *development of* γ *into the "tangent homogeneous space"* $E_p \cong G/H$ is the curve c in E_p such that for all t there is a parallel section $\overset{t}{Z}$ of E along γ with $\overset{t}{Z}(0) = c(t) \in E_p$, and $\overset{t}{Z}(t) = \sigma_{\gamma(t)} \in E_{\gamma(t)}$.

For a kinematic description of development, see [N3].

9.32 The following example will make it clear why we did not simply start with the connection ω on B. We will construct a Cartan connection on the linear frame bundle $LM = B(TM)$ of the tangent bundle of M, and then look at the associated parallel transport in TM.

The *canonical 1-form* θ *on* LM is defined by $\theta v := b^{-1}\wp_* v$, $v \in LM_b$; here if $\wp b = p \in M$, then b is considered as an isomorphism from $\mathbb{R}^n$ to M_p.

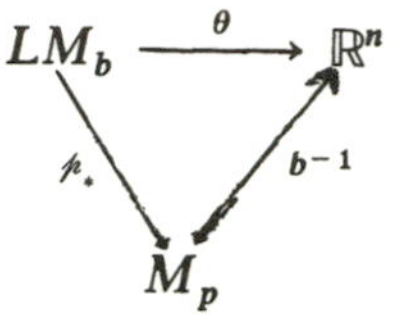

For all $g \in GL(n, \mathbb{R})$,

$$R_g^*\theta(v) = \theta R_{g*} v = (bg)^{-1}\wp_* R_{g*} v = g^{-1}b^{-1}\wp_* v = g^{-1}\theta v$$

Now fix a connection form μ on LM, and set $\vartheta := \theta + \mu$ as a 1-form on LM with values in $\mathbb{R}^n \oplus \mathfrak{gl}(n, \mathbb{R})$; by 2.113, $\mathbb{R}^n \oplus \mathfrak{gl}(n, \mathbb{R})$ is the underlying vector space for the Lie algebra $\mathfrak{a}(n, \mathbb{R})$ of the affine group $A(n, \mathbb{R}) = \mathbb{R}^n \rtimes GL(n, \mathbb{R})$ on $\mathbb{R}^n$. With respect to this decomposition of $\mathfrak{a}(n, \mathbb{R})$, the adjoint action of the subgroup $GL(n, \mathbb{R})$ on $\mathfrak{a}(n, \mathbb{R})$ is

$$\mathrm{Ad}_g(z + X) = gz + \mathrm{Ad}_g X \qquad z \in \mathbb{R}^n,\ X \in \mathfrak{gl}(n, \mathbb{R})$$

Now consider ϑ as a 1-form with values in $\mathfrak{a}(n, \mathbb{R})$; by the equivariance properties of θ and μ, given $b \in LM$, $v \in LM_b$, and $g \in GL(n, \mathbb{R})$,

$$\begin{aligned} R_g^*\vartheta(v) &= (R_g^*\theta + R_g^*\mu)(v) = g^{-1}\theta v + \mathrm{Ad}_{g^{-1}}\mu v \\ &= \mathrm{Ad}_{g^{-1}}(\theta v + \mu v) = \mathrm{Ad}_{g^{-1}} \circ \vartheta(v) \end{aligned}$$

which is one of the requirements for a Cartan connection.

If $X \in \mathfrak{gl}(n, \mathbb{R})$, then $\wp_* X = 0$ by the definition in 9.15, so $\vartheta\tilde{X} = \theta\tilde{X} + \mu\tilde{X} = X$. Finally, if $\vartheta v = 0$, then $\wp_* v = 0$ and $\mu v = 0$, so v is both vertical and horizontal; hence $v = 0$.

Thus ϑ is a Cartan connection on LM. Let ω be the connection on the so-called affine frame bundle $B := LM \times_{GL(n, \mathbb{R})} A(n, \mathbb{R})$ such that $\iota^*\omega = \vartheta$, where ι is the natural embedding of LM into B. Now we determine the associated parallel transport in the associated bundle $B \times_{A(n, \mathbb{R})} \mathbb{R}^n$. As a fiber bundle this is just TM; the only difference between the two is that since $A(n, \mathbb{R})$ does not respect the vector space structure of $\mathbb{R}^n$, we are not allowed to use the vector bundle structure of TM. In particular, parallel transport induced in TM by ω on B will not be linear parallel transport. Instead, the parallel transport defined in TM by ω is just the affine parallel transport in TM (defined in 2.108) associated to the linear connection μ on LM.

To prove this it is sufficient to show that the connection [that is, the horizontal bundle (9.24)] on TM determined by the connection on B is the affine connection $\bar{\mathscr{A}}\mathscr{H}$ associated with the linear connection μ on LM. By 9.4 and 2.108 we must prove the following: fix $(b, \xi) \in B \times \mathbb{R}^n$, with $w = b\xi = \mathscr{P}(b, \xi) \in TM$; if $v \in B_b$ is horizontal with respect to ω, then $\mathscr{P}_*|_{(b,\xi)}(v, 0) \in (TM)_w$ is of the form $z - \mathscr{I}_w \pi_* z$, where $z \in (TM)_w$ is horizontal with respect to the linear connection μ on LM.

Without loss of generality, assume $b \in LM \subset B$; let $v \in \ker(\omega_b) \subset B_b$. There exists a μ-horizontal vector $u \in LM_b$ and an $X \in \mathfrak{a}(n, \mathbb{R})$ such that $v = u + \tilde{X}_b$. By 9.31,

$$\omega v = X + (\theta + \mu)(v - \tilde{X}_b) = X + \theta v + \mu u = X + \theta v$$

Since ωv is zero, $X = -\theta v \in \mathfrak{a}(n, \mathbb{R})$, hence $v = u - \theta u_b$. By 9.4 the vector $z := \mathscr{P}_*|_{(b,\xi)}(u, 0) \in TM_w$, $w = \mathscr{P}(b, \xi)$, is horizontal with respect to the linear connection μ.

Now to calculate $\mathscr{P}_*|_{(b,\xi)}(-\widetilde{\theta u_b}, 0)$. Since $p \circ \mathrm{pr}_1 = \pi \circ \mathscr{P}$ from $B \times \mathbb{R}^n$ to M, $\pi_* z = p_* u$.

$$\begin{array}{ccc} B \times \mathbb{R}^n & \xrightarrow{\ pr_1\ } & B \\ {\scriptstyle \mathscr{P}}\big\downarrow & & \big\downarrow{\scriptstyle p} \\ TM & \xrightarrow[\ \pi\ ]{} & M \end{array}$$

Therefore, $\theta u = b^{-1} p_* u = b^{-1}\pi_* z \in \mathbb{R}^n \subset \mathfrak{a}(n, \mathbb{R})$. By the form of the exponential map of $A(n, \mathbb{R})$ (2.113, exercise) and 9.15,

$$\widetilde{\theta u_b} = \frac{d}{dt}\bigg|_0 b \cdot \exp(t\theta u) = \frac{d}{dt}\bigg|_0 b \cdot (tb^{-1}\pi_* z, I)$$

so

$$\mathscr{P}_*(\widetilde{\theta u_b}, 0_\xi) = \frac{d}{dt}\bigg|_0 \mathscr{P}(b \cdot (tb^{-1}\pi_* z, I), \xi) = \frac{d}{dt}\bigg|_0 b \circ (tb^{-1}\pi_* z, I)(\xi)$$

$$= \frac{d}{dt}\bigg|_0 b(tb^{-1}\pi_* z + \xi) = \frac{d}{dt}\bigg|_0 (w + t\pi_* z)$$

$$= \mathscr{I}_w \pi_* z$$

Hence $\mathscr{P}_*(u - \widetilde{\theta u_b}, 0) = z - \mathscr{I}_w \pi_* z$, where z is μ-horizontal, and therefore $\mathscr{P}_*(u - \widetilde{\theta u_b}, 0)$ is affine horizontal by definition 2.108. *QED.*

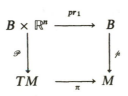

We have now shown that given a linear connection μ on LM, the parallel transport in TM induced by the Cartan connection $\theta + \mu$ on LM is just the affine parallel transport in TM associated to the linear parallel transport determined by μ. By the remark after the proposition in 2.109, the zero section of TM, which is the section of $TM = LM \times_{GL(n,\,\mathbb{R})} \mathbb{R}^n$ determined by the fixed point 0 of $GL(n, \mathbb{R})$ in $\mathbb{R}^n$ as in 9.28, is not parallel with respect to affine parallel transport.

9.33 **Exercises** Check directly that the restriction of $\mathscr{V}TM$ to the zero section of TM is isomorphic to TM (9.30). Show that a curve γ in M, with $p = \gamma(0)$, is a geodesic if and only if its development c into the affine tangent space M_p (9.31) is of the form $c(tv)$ for some $v \in M_p$. Prove the "first structure equation" (cf. 2.103i) $d\theta = \eta - [\theta, \mu]$, where the $\mathbb{R}^n$-valued *torsion form* η on LM is defined by $b \circ \eta(u, v) := T(\not{p}_* u, \not{p}_* v)$ for $u, v \in LM_b$, T being the torsion tensor of the linear connection μ.

9.34 Now we study the geometry of a conformal manifold M in terms of the parallel transport induced by a Cartan connection. Fix a representative Riemannian metric on M.

Let θ be the restriction to the conformal frame bundle $CO(M)$ of the canonical 1-form on LM (9.32); similarly, let μ be the restriction to $CO(M)$ of the connection form on LM which represents the Levi-Civita connection of the metric. The $(\mathbb{R}^n \oplus \mathfrak{co}(n))$-valued 1-form $\theta + \mu$ on $CO(M)$ is a Cartan connection which induces conformal affine parallel transport in the conformal affine tangent bundle; unfortunately, this tells us little beyond what was known in 9.32.

To get something new we complete each conformal affine tangent space to get a conformal sphere, that is, we consider the conformal sphere bundle $E := CO(M) \times_{CO(n)} S^n$ from 9.25c. Each fiber will be the conformal homogeneous space $S^n = O(n+1, 1)/H$ from 5.19, where $O(n+1, 1)$ is the subgroup of $GL(n+2, \mathbb{R})$ which leaves invariant the quadratic form $x^{1^2} + \cdots + x^{n^2} - 2x^0x^{n+1}$, and

$$H = \left\{ \begin{bmatrix} a^{-1} & u & b \\ 0 & h & \zeta \\ 0 & 0 & a \end{bmatrix} \in O(n+1, 1) \right\} \supset \left\{ \begin{bmatrix} a^{-1} & 0 & 0 \\ 0 & h & 0 \\ 0 & 0 & a \end{bmatrix} \right\} \cong CO(n)$$

where $\zeta = ah^t u$ and $b = a\|u\|^2/2$.

Define $G := O(n+1, 1)$; let B be the principal G-bundle $CO(M) \times_{CO(n)} G$ over M, and let Q be the principal H-subbundle $CO(M) \times_{CO(n)} H$ of B; then

$$E := CO(M) \times_{CO(n)} S^n = Q \times_H S^n = B \times_G S^n = B/H$$

Before extending $\theta + \mu$ to a Cartan connection on Q we must look more closely at the Lie algebra structure of $\mathfrak{g} = \mathfrak{o}(n+1, 1)$. Set

$$\mathfrak{g}_{-1} := \left\{ \begin{bmatrix} 0 & 0 & 0 \\ v & 0 & 0 \\ 0 & {}^t v & 0 \end{bmatrix} \middle| v \in \mathbb{R}^n \right\} \qquad \mathfrak{g}_1 := \left\{ \begin{bmatrix} 0 & \zeta & 0 \\ 0 & 0 & {}^t\zeta \\ 0 & 0 & 0 \end{bmatrix} \middle| \zeta \in \mathbb{R}^{n*} \right\}$$

$$\mathfrak{g}_0 := \left\{ \begin{bmatrix} -a & 0 & 0 \\ 0 & X & 0 \\ 0 & 0 & a \end{bmatrix} \middle| X \in \mathfrak{o}(n), a \in \mathbb{R} \right\} \qquad \mathfrak{g}_j := 0 \text{ for } |j| > 1$$

Then $\mathfrak{g} = \mathfrak{g}_{-1} \oplus \mathfrak{g}_0 \oplus \mathfrak{g}_1$ is a graded Lie algebra: $[\mathfrak{g}_i, \mathfrak{g}_j] \subset \mathfrak{g}_{i+j}$. The Lie algebra of H is $\mathfrak{h} = \mathfrak{g}_0 \oplus \mathfrak{g}_1$. With the identification

$$\mathfrak{g}_0 \cong \mathfrak{o}(n) \times \mathbb{R} \cong \mathfrak{co}(n) \qquad \begin{bmatrix} -a & 0 & 0 \\ 0 & X & 0 \\ 0 & 0 & a \end{bmatrix} \leftrightarrow X + aI$$

we obtain an isomorphism $\mathfrak{g} = \mathbb{R}^n \oplus \mathfrak{co}(n) \oplus \mathbb{R}^{n*}$, where $\mathfrak{co}(n)$ (with its usual structure) is a subalgebra, $\mathbb{R}^n$ and $\mathbb{R}^{n*}$ are abelian subalgebras, and given $A \in \mathfrak{co}(n)$, a column vector $v \in \mathbb{R}^n$, and a row vector $\zeta \in \mathbb{R}^{n*}$, $[A, v] = A \cdot v$, $[A, \zeta] = -\zeta \cdot A$, and $[v, \zeta] = v \otimes \zeta - {}^t\zeta \otimes {}^t v + \zeta(v)I \in \mathfrak{co}(n)$.

Suppose that $\vartheta = \vartheta_{-1} + \vartheta_0 + \vartheta_1$ is a Cartan connection on Q, where ϑ_j is $\mathfrak{g}_j$-valued. Let ω be the unique connection form on B such that $\jmath^*\omega = \vartheta$, $\jmath$ being the embedding of Q into B. Conformal parallel transport in $E = B \times_G S^n$ is defined with respect to ω, and in turn defines the development of a curve in M into a fiber of E over a point of the curve.

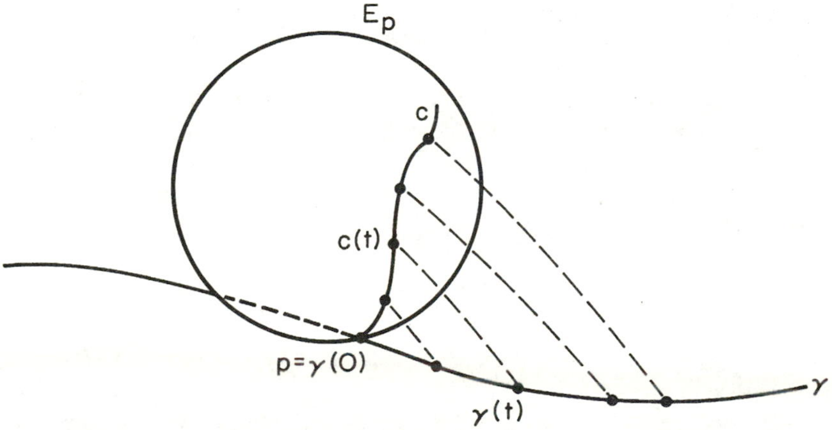

If Ω is the curvature form of ω, define the curvature form of ϑ to be $\Theta := \jmath^*\Omega$. Since Ω is horizontal, so is Θ, that is, $\iota_v\Theta = 0$ if v is vertical. The structure equation $\Omega = d\omega + \frac{1}{2}[\omega, \omega]$ on B implies the structure equation $\Theta = d\vartheta + \frac{1}{2}[\vartheta, \vartheta]$ on Q. Write $\Theta = \Theta_{-1} + \Theta_0 + \Theta_1$, where Θ_j is $\mathfrak{g}_j$-valued. The structure equation then decomposes as

$$\Theta_{-1} = d\vartheta_{-1} + [\vartheta_{-1}, \vartheta_0] \qquad \Theta_1 = d\vartheta_1 + [\vartheta_0, \vartheta_1]$$
$$\Theta_0 = d\vartheta_0 + \tfrac{1}{2}[\vartheta_0, \vartheta_0] + [\vartheta_{-1}, \vartheta_1]$$

9.35 Now we can impose some normalization conditions on ϑ. First assume that $\iota^*\vartheta_{-1} = \theta$ and $\iota^*\vartheta_0 = \mu$, where $\iota: CO(M) \to Q$ is the bundle embedding. The first structure equation (9.33) $d\theta = -[\theta, \mu]$ for the torsion-free connection μ on $CO(M)$ is also valid for ϑ_{-1} and ϑ_0 on Q, that is, $d\vartheta_{-1} = -[\vartheta_{-1}, \vartheta_0]$. Thus $\Theta_{-1} = 0$ (actually, this is true without these assumptions on ϑ_{-1} and ϑ_0 [Ko2: p. 129]).

For the remaining normalization condition, let $\mathscr{H}$ be the subbundle $\ker(\vartheta_0) \cap \ker(\vartheta_1)$ of TQ; observe that $\mathscr{H}$ is trivial, for the map from $\mathscr{H}$ to $Q \times \mathfrak{g}_{-1}$ which takes $v \in \mathscr{H}_b$ to $(b, \vartheta_{-1}v)$ is a bundle isomorphism. The form Θ_0 then induces a tensorial map $\mathscr{H} \wedge \mathscr{H} \otimes \mathscr{H} \to \mathscr{H}$ such that

$$u \wedge v \otimes w \mapsto \vartheta_{-1}|_b^{-1}[\Theta_0(u, v), \vartheta_{-1}w] =: C(u, v)w \qquad u, v, w \in \mathscr{H}_b$$

In terms of the algebra $\mathbb{R}^n \oplus \mathfrak{co}(n) \oplus \mathbb{R}^{n*} \cong \mathfrak{g}$, $[\Theta_0(u, v), \vartheta_{-1}w]$ is just the product of the matrix $\Theta_0(u, v) \in \mathfrak{co}(n)$ and the vector $\vartheta_{-1}w \in \mathbb{R}^n$.

The final normalization condition is that for each $b \in Q$ and v, $w \in \mathscr{H}_b$, the map $\mathscr{H}_b \to \mathscr{H}_b$, $u \mapsto C(u, v)w$, has trace zero. In 9.36 it will be seen how this imposes a condition on ϑ_1.

According to É. Cartan [Ca2] (see also [Ko2: IV.4.2], [Oc]), given the Levi-Civita connection form μ, there exists a unique Cartan connection ϑ on Q, called the *normal conformal connection*, such that $\iota^*\vartheta_{-1} = \theta$, $\iota^*\vartheta_0 = \mu$, and C has trace zero in the first argument.

Weyl's approach to conformal geometry helps to justify the trace-zero condition on C.

9.36 Weyl [Wy*1* and *2*] constructed a generalized curvature tensor ([ST], [*N3*]) on a Riemannian manifold which vanishes whenever the metric is (locally) conformally equivalent to a flat matric; for this reason he called it the *conformal curvature tensor* of the metric. Schouten [Sch] showed that for $n > 3$, the converse is true (a similar tensor exists in dimension 3 [Co] [Ya2]).

To construct Weyl's conformal curvature tensor on (M, g), first set

$$F(X, Y) := \frac{1}{n-2}\left(\frac{\langle X, Y\rangle s}{2n-2} - \mathscr{R}(X, Y)\right) \qquad X, Y \in \mathfrak{X}M$$

where $\mathscr{R}$ is the Ricci tensor of g (3.43) and s is the scalar curvature of g (3.45). For each vector field X on M, let $f(X)$ be the vector field $F(X, \cdot)^*$ dual to the 1-form $F(X, \cdot)$: $\langle f(X), Y\rangle := F(X, Y)$ for all $Y \in \mathfrak{X}M$. The *Weyl conformal curvature tensor* W on M is then defined by

$$W(X, Y)Z := R(X, Y)Z + F(Y, Z)X - F(X, Z)Y + \langle Y, Z\rangle f(X) - \langle X, Z\rangle f(Y)$$

for $X, Y, Z \in \mathfrak{X}M$.

Exercise (See [Ya2].) The Weyl tensor W is invariant under a conformal change of metric, and therefore vanishes if g is locally conformally equivalent to a flat metric.

Direct calculation shows that W satisfies the Jacobi identity $W(X, Y)Z + W(Y, Z)X + W(Z, X)Y = 0$; similarly, the "Ricci tensor" $\mathscr{W}(X, Y) := \sum_j \langle W(E_j, X)Y, E_j\rangle$ is zero (where $\{E_j\}$ is an orthonormal moving frame).

As in 9.20 there exists a $\mathfrak{co}(n)$-valued horizontal 2-form Φ on $CO(M)$ which represents W by the rule

$$W(p_* u, p_* v)bz = b(\Phi(u, v) \cdot z) \qquad u, v \in CO(M)_b,\ z \in \mathbb{R}^n$$

9.37 Let β be the $\mathfrak{g}_1$-valued 1-form on Q such that (i) $\iota^*\beta = 0$, where ι is the embedding of $CO(M)$ into Q, (ii) $\beta\tilde{A} = A$ for all $A \in \mathfrak{g}_1 \subset \mathfrak{h}$, and (iii) β is Ad_H-equivariant: $R_h^*\beta = \mathrm{Ad}_{h^{-1}}\,\beta$ for all $h \in H$. Let τ be a $\mathfrak{g}_1$-valued, Ad_H-equivariant, horizontal 1-form on Q; set $\vartheta_1 := \beta + \tau$, and $\vartheta := \vartheta_{-1} + \vartheta_0 + \vartheta_1$ (with ϑ_{-1} and ϑ_0 as in 9.34). Then ϑ is a Cartan connection on Q. The $\mathfrak{g}_0$-component of its curvature form is

$$\begin{aligned}\Theta_0 &= d\vartheta_0 + \tfrac{1}{2}[\vartheta_0, \vartheta_0] + [\vartheta_{-1}, \vartheta_1]\\ &= d\vartheta_0 + \tfrac{1}{2}[\vartheta_0, \vartheta_0] + [\vartheta_{-1}, \beta] + [\vartheta_{-1}, \tau]\end{aligned}$$

therefore

$$\iota^*\Theta_0 = \iota^*(d\vartheta_0 + \tfrac{1}{2}[\vartheta_0, \vartheta_0]) + [\iota^*\vartheta_{-1}, \iota^*\tau] = d\mu + \tfrac{1}{2}[\mu, \mu] + [\theta, \iota^*\tau]$$

which represents the curvature tensor R plus another tensorial term.

Exercise (See [Ko2: IV.4.2].) If τ is chosen so that $\iota^*\Theta_0 = \Phi$, that is, so that $[\theta, \iota^*\tau]$ represents Weyl's correction tensor $W - R$, then ϑ is the normal conformal connection with respect to the Levi-Civita connection of the metric g.

9.38 Projective differential geometry was also studied by Weyl and Cartan (and many others). Weyl [Wy*1* and *2*] used the approach outlined in chap. 5, and also defined a projective curvature tensor; this tensor had trace zero, and its vanishing characterized a "projectively flat manifold." Cartan [Ca*3*] used a normalized projective connection on an appropriate extension of the frame bundle LM to define parallel transport in a projective space bundle for which the fibers were homogeneous spaces diffeomorphic to the projective completions of the tangent spaces of M. The Ricci zero condition for Weyl's projective curvature tensor is equivalent to the normalization condition for Cartan's normal projective connection. The two viewpoints were brought together by Yano [Ya*1*]. Further work on projective differential geometry from Cartan's viewpoint can be found in [Tan], [KNa], and [Ko*1* and *2*].

SPIN STRUCTURES

The final example of a geometric structure in this book uses the covering group Spin(n) of the special orthogonal group $SO(n)$, $n \geq 2$. Recall (6.49) that the universal covering homomorphism ρ: Spin$(n) \to SO(n)$, $n > 2$, is a 2-fold covering map; similarly, define $\rho: S^1 \to S^1$ by $\rho(e^{it}) := e^{2it}$, and refer to this as the 2-fold covering (not universal) of $SO(2)$ by the group Spin(2).

The covering homomorphism ρ induces a representation of Spin(n) on $\mathbb{R}^n$, but this representation is not faithful [that is, the action of Spin(n) on $\mathbb{R}^n$ is not effective (6.53)]. For geometric purposes we need a vector space large enough for a natural action of Spin(n) to be effective.

9.39 **Definition** Let $\mathbb{F}$ be one of the fields $\mathbb{R}$ or $\mathbb{C}$. A *Clifford algebra of a quadratic form* Q on a vector space $\mathscr{V}$ over $\mathbb{F}$ is an associative algebra C with 1 over $\mathbb{F}$, together with a homomorphism η of $\mathscr{V}$ into C, such that

(i) for each v in $\mathscr{V}$, $\eta(v)^2 = Q(v) \cdot 1$;
(ii) the pair (C, η) is *universal*, that is (cf. [W: 2.2(a)]), given another pair $(C', \mathscr{V} \xrightarrow{\eta'} C')$ which satisfies condition (i), there is a homomorphism ψ from C to C' such that $\eta' = \psi \circ \eta$.

Universality implies that if a Clifford algebra exists for Q, then it is unique up to isomorphism; in the examples we will see two models for the Clifford algebra of Q.

9.40 **Examples** Let Q be a quadratic form on a finite-dimensional vector space $\mathscr{V}$ over $\mathbb{F}$.

(*a*) If $Q = 0$, then C is just the exterior algebra $\Lambda\mathscr{V}$.
(*b*) More generally, let $\mathscr{T} = \bigoplus_{k \geq 0} \otimes^k \mathscr{V}$, and let $\mathscr{I}$ be the ideal in $\mathscr{T}$ generated by the elements $v \otimes v - Q(v) \cdot 1$, $v \in \mathscr{V}$. The algebra $\mathscr{T}/\mathscr{I}$

is a Clifford algebra of Q; thus each Q has a Clifford algebra. *Exercise:* Show that the map η embeds $\mathscr{V}$ into C.

(*c*) Let B be the symmetric bilinear form associated to Q: $2B(u, v) = Q(u+v) - Q(u) - Q(v)$, $u, v \in \mathscr{V}$. For each $v \in \mathscr{V}$, define $\delta_v : \mathscr{V} \to \mathbb{F} \subset \Lambda\mathscr{V}$ by $\delta_v w := B(v, w) \cdot 1$; extend δ_v to a skew-derivation of $\Lambda\mathscr{V}$:

$$\delta_v(y \wedge z) = (\delta_v y) \wedge z + (-1)^{\deg(y)} y \wedge \delta_v z \qquad y, z \in \Lambda\mathscr{V}$$

Let $\cdot$ be the associative multiplication on the vector space $\Lambda\mathscr{V}$ such that $v \cdot z = v \wedge z + \delta_v z$ for $v \in \mathscr{V}$ and $z \in \Lambda\mathscr{V}$. The algebra $(\Lambda\mathscr{V}, \cdot)$ is isomorphic to the Clifford algebra of Q. In particular, as a vector space the Clifford algebra of Q is just $\Lambda\mathscr{V}$.

(*d*) For each finite-dimensional vector space $\mathscr{V}$ with quadratic form Q over $\mathbb{F}$, let $C(\mathscr{V}, Q)$ be the Clifford algebra of Q; given vector spaces $\mathscr{V}_1$ and $\mathscr{V}_2$ over $\mathbb{F}$, each with a quadratic form Q_j, and given a linear map $f: \mathscr{V}_1 \to \mathscr{V}_2$ such that $Q_2 = Q_1 \circ f$, define $C(f)$ to be the algebra homomorphism from $C(\mathscr{V}_1, Q_1)$ to $C(\mathscr{V}_2, Q_2)$ such that

$$(Cf)(v_1 \cdot \ldots \cdot v_k) := f(v_1) \cdot \ldots \cdot f(v_k) \qquad v_i \in \mathscr{V}_1$$

The map C is a covariant C^∞ functor (1.34) from the category of finite-dimensional vector spaces with quadratic forms over $\mathbb{F}$ (and form-preserving linear maps) to the category of algebras over $\mathbb{F}$ (and algebra homomorphisms).

(*e*) Let Q be a quadratic form on a real vector space $\mathscr{V}$. Extend the algebra structure of $C(\mathscr{V}, Q)$ to a complex algebra structure on the vector space $\mathbb{C} \otimes C(\mathscr{V}, Q)$. The algebra $\mathbb{C} \otimes C(\mathscr{V}, Q)$ is isomorphic to the complex Clifford algebra $C(\mathbb{C} \otimes \mathscr{V}, \hat{Q})$, where $\hat{Q}(zv) := z^2 Q(v)$ for $z \in \mathbb{C}$ and $v \in \mathscr{V}$.

9.41 Let C_n be the Clifford algebra for the quadratic form $Q(v) = -\|v\|^2$ on Euclidean space $\mathbb{R}^n$. With the usual orthonormal basis $\{e_j\}$ for $\mathbb{R}^n$, $e_i e_i = -1$, and $e_i e_j = -e_j e_i$ for $i \neq j$. More generally, for all $u, v \in \mathbb{R}^n$, $uv + vu = -2\langle u, v\rangle 1$.

Exercise Show that $C_1 \cong \mathbb{C}$, $C_2 \cong \mathbb{H}$, and $C_3 \cong \mathbb{H} \oplus \mathbb{H}$.

Let $$C_n^+ := \{v_1 \cdot \ldots \cdot v_k \in C_n \mid v_j \in \mathbb{R}^n,\ k \text{ is even}\}$$

and $$C_n^- := \{v_1 \cdot \ldots \cdot v_k \in C_n \mid v_j \in \mathbb{R}^n,\ k \text{ is odd}\}$$

so that $C_n = C_n^+ \oplus C_n^-$. The multiplication table shows that C_n is a $\mathbb{Z}_2$-graded algebra.

	C_n^+	C_n^-
C_n^+	C_n^+	C_n^-
C_n^-	C_n^-	C_n^+

In particular, C_n^+ is a subalgebra of C_n; in fact, the map

$$C_{n-1} \to C_n^+ \subset C_n$$

$$x^+ + x^- \mapsto x^+ + x^- e_n \qquad x^\pm \in C_{n-1}^\pm$$

maps the Clifford algebra C_{n-1} isomorphically to C_n^+. Therefore $\dim C_n^+ = \dim C_n^- = \dim C_{n-1} = 2^{n-1}$.

9.42 Now we will see how Spin(n) lies naturally in C_n.

Define an involution of C_n by $(e_{i_1} \cdots e_{i_k})^* := e_{i_k} \cdots e_{i_1}$. It follows that $(xy)^* = y^* x^*$. Suppose that $x = v_1 \cdots v_k$, where each $v_j \in S^{n-1} \subset \mathbb{R}^n \subset C_n$; if $x \in C_n^+$, then $xx^* = 1$, while $xx^* = -1$ for $x \in C_n^-$.

Definition Let $\text{Pin}(n) := \{v_1 \cdots v_k \in C_n \mid v_j \in S^{n-1}\}$, with the induced multiplication.

9.43 Since Pin(n) is a closed subgroup of $GL(C_n)$, it is a Lie group [W: 3.42]. For all $u, v \in \mathbb{R}^n$, the identity $uv + vu = -2\langle u, v\rangle 1$ implies that if $u \in S^{n-1}$, then $uvu = (-vu - 2\langle u, v\rangle)u = v - 2\langle u, v\rangle u \in \mathbb{R}^n$, that is, uvu is the reflection of v in the hyperplane in $\mathbb{R}^n$ orthogonal to u. More generally, for all $x \in \text{Pin}(n)$, the expression $\rho(x)v := xvx^*$, $v \in \mathbb{R}^n$, is a composition of reflections of $\mathbb{R}^n$ applied to v, and therefore $\rho(x)$ is an orthogonal map of $\mathbb{R}^n$. Thus there is a map ρ: $\text{Pin}(n) \to O(n)$, which is clearly a homomorphism. For $x \in C_n^+ \cap \text{Pin}(n)$, $\rho(x)$ belongs to $SO(n)$ since the product of an even number of reflections of $\mathbb{R}^n$ is a rotation.

Proposition The map ρ is a 2-fold covering map. The subgroup $C_n^+ \cap \text{Pin}(n)$ of Pin(n) is the connected 2-fold covering group of $SO(n)$; in particular, $\text{Spin}(n) \cong C_n^+ \cap \text{Pin}(n)$.

PROOF It suffices to show that the kernel of ρ is $\mathbb{Z}_2$, and that $C_n^+ \cap \text{Pin}(n)$ is connected.

If $x = u_1 \cdots u_k \in \ker \rho$, then $xx^* = \det(\rho u_1) \cdot \ldots \cdot \det(\rho u_k) = \det(\rho x) = 1$, so $x \in C_n^+$. Also, $x = -vxv$ for all $v \in \mathbb{R}^n$ because $xvx^* = v$. But if $x = \sum t_{i_1 \cdots i_n} e_1^{i_1} \cdots e_n^{i_n}$, where each $i_j = 0$ or 1 and $i_1 + \cdots + i_n$ is even, then for all j,

$$x = -e_j x e_j = \sum (-1)^{i_j} t_{i_1 \cdots i_n} e_1^{i_1} \cdots e_n^{i_n}$$

so $t_{i_1 \cdots i_n} = 0$ unless all i_j are zero. Hence x is a multiple of $1 \in C_n$; since $xx^* = 1$, $x = \pm 1$. Therefore the kernel of ρ is $\mathbb{Z}_2$.

The curve

$$\gamma(t) = e_1 \cdot (-\cos(t)e_1 + \sin(t)e_2) = \cos t + \sin(t)e_1e_2$$

$0 \le t \le \pi$, joins the points 1 and -1 of ker ρ, so the 2-fold covering space $C_n^+ \cap \text{Pin}(n)$ of $SO(n)$ is connected, and is therefore the non-trivial 2-fold covering of $SO(n)$.

Exercise (see 6.50) Map $C_3^+ \cong C_2 \to \mathbb{H}$ linearly so that

$$1 \mapsto 1 \qquad e_1e_2 \mapsto i \qquad e_1e_3 \mapsto j \qquad \text{and} \qquad e_2e_3 \mapsto k$$

Show that this is an algebra isomorphism, and that its restriction to Spin(3) is an isomorphism onto S^3. Similarly, exhibit an isomorphism from $\text{Spin}(4) \subset C_4^+$ to $S^3 \times S^3$.

9.44 By 9.43, Spin(n) lies in the closed subgroup $G := \{x \in C_n \mid xvx^* \in \mathbb{R}^n$ for all $v \in \mathbb{R}^n$, and $xx^* = 1\}$ of $GL(C_n)$. As in 6.40, the Lie algebra of G is the subalgebra $\mathfrak{g} := \{x \in C_n \mid [x, v] \in \mathbb{R}^n$ for all $v \in \mathbb{R}^n$, and $x + x^* = 0\}$ of the Lie algebra C_n under commutation. If $i_1 < \cdots < i_k$, then for all $v \in \mathbb{R}^n$, $[e_{i_1} \cdots e_{i_k}, v] \in C_n$ is a sum of products of at least $k - 1$ vectors in $\mathbb{R}^n$; as a result, $\mathfrak{g}$ is contained in the subalgebra $\mathbb{R} \oplus [\mathbb{R}^n, \mathbb{R}^n]$ of the Lie algebra C_n. The condition $x + x^* = 0$, $x \in \mathfrak{g}$, then implies $\mathfrak{g} \subset [\mathbb{R}^n, \mathbb{R}^n]$.

Conversely, the relation $uv + vu = -2\langle u, v\rangle 1$ for $u, v \in \mathbb{R}^n$ implies that for all $u, v, w \in \mathbb{R}^n$ such that $[u, v] \ne 0$,

$$\begin{aligned}
\text{ad}_{[u, v]}\, w &= [[u, v], w] = uvw - vuw - wuv + wvu \\
&= -uwv - wuv - vuw - vwu + wvu + vwu + uwv + uvw \\
&= 2(\langle u, w\rangle v + v\langle u, w\rangle 1 - \langle v, w\rangle u - u\langle v, w\rangle 1) \\
&= 4(\langle u, w\rangle v - \langle v, w\rangle u) \in \mathbb{R}^n
\end{aligned}$$

Similarly, $[u, v] + [u, v]^* = uv - vu + vu - uv = 0$, $u, v \in \mathbb{R}^n$. Therefore $\mathfrak{g}$ is the Lie subalgebra $[\mathbb{R}^n, \mathbb{R}^n]$ of C_n. But as in 3.13, $\langle u, w\rangle v - \langle v, w\rangle u = v \wedge u(w) = -u \wedge v(w)$, where $u \wedge v \in \Lambda^2 \mathbb{R}^n \cong \mathfrak{o}(n)$.

Therefore $\mathfrak{g} \cong \mathfrak{o}(n)$; in fact, the map from $\mathfrak{g}$ to $\mathfrak{o}(n)$ such that $x = [u, v]$, $u, v \in \mathbb{R}^n$, goes to $\text{ad}_{[u, v]} = -4u \wedge v$ is the differential of the map "Ad": $G \to SO(n)$, "$\text{Ad}_x\, v$" $:= xvx^* = xvx^{-1}$, $v \in \mathbb{R}^n$. Since "Ad"$|_{\text{Spin}(n)}$ is exactly the covering homomorphism ρ: Spin(n) $\to SO(n)$, it follows that Spin(n) is the identity component of G, and $\mathfrak{g}$ is the Lie algebra of Spin(n) in C_n.

Exercise Show that $G = \text{Spin}(n)$ and $\text{Pin}(n) = \{x \in C_n | \text{"Ad}_x\, v\text{" is in } \mathbb{R}^n \text{ for all } v \in \mathbb{R}^n, \text{ and } xx^* = \pm 1\}$.

9.45 As a subset of C_n^+, Spin(n) acts linearly by left multiplication on C_n and C_n^+. This action extends naturally to a representation of Spin(n) on the complexification $C_n^{\mathbb{C}} := \mathbb{C} \otimes C_n$ of the Clifford algebra C_n; by 9.40*e*, $C_n^{\mathbb{C}}$ is the Clifford algebra of the extension of $-\|\cdot\|^2$ to a quadratic form on C^n.

Exercises Prove:

$$C_1^{\mathbb{C}} \cong \mathbb{C} \oplus \mathbb{C} \qquad C_2^{\mathbb{C}} \cong \text{End}(\mathbb{C}^2)$$

$$C_3^{\mathbb{C}} \cong \text{End}(\mathbb{C}^2) \oplus \text{End}(\mathbb{C}^2)$$

Using the universal mapping property for Clifford algebras, show that $C_{n+2}^{\mathbb{C}} = C_n^{\mathbb{C}} \otimes C_2^{\mathbb{C}}$, where $(u \otimes x) \cdot (v \otimes y) := uv \otimes xy$.

9.46 Now we will find a representation of Spin(n) on a complex subspace of $C_n^{\mathbb{C}}$. First work just in the even-dimensional case. The usual orthonormal basis $\{e_j\}$ for $\mathbb{R}^{2n}$ is also a basis for the complexification $\mathbb{C} \otimes \mathbb{R}^{2n} = \mathbb{C}^{2n}$ of $\mathbb{R}^{2n}$. Define a new basis for $\mathbb{C}^{2n}$ by

$$\epsilon_j := \tfrac{1}{2}(e_j + ie_{n+j}) \qquad \epsilon_{n+j} := \tfrac{1}{2}(e_j - ie_{n+j}) \qquad j = 1, \ldots, n$$

In $C_{2n}^{\mathbb{C}}$, $\epsilon_j \epsilon_k + \epsilon_k \epsilon_j$ equals 1 if $|j - k| = n$, and equals zero otherwise.

Exercise Show that if $\{\omega^j\}$ is the basis dual to $\{\epsilon_j\}$, then the extension of $\|\cdot\|^2$ on $\mathbb{R}^{2n}$ to $\mathbb{C}^{2n}$ satisfies

$$\|v\|^2 = \sum_{j=1}^{n} \omega^j(v)\omega^{n+j}(v) \qquad v \in \mathbb{C}^{2n}$$

9.47 **Definition** The left ideal

$$S := C_{2n}^{\mathbb{C}} \cdot \epsilon_{n+1} \cdots \epsilon_{2n} = \{v\epsilon_{n+1} \cdots \epsilon_{2n} | v \in C_{2n}^{\mathbb{C}}\}$$

is called the *spinor space of* $\mathbb{R}^{2n}$. The *half-spinor spaces of* $\mathbb{R}^{2n}$ are

$$S^+ := \{v\epsilon_{n+1} \cdots \epsilon_{2n} | v \in \mathbb{C} \otimes C_{2n}^+\}$$

and

$$S^- := \{v\epsilon_{n+1} \cdots \epsilon_{2n} | v \in \mathbb{C} \otimes C_{2n}^-\}$$

Clearly $S = S^+ \oplus S^-$ (the direct sum of complex vector spaces). The commutation relations for the ϵ_j imply that the vectors $\epsilon_{j_1} \cdots \epsilon_{j_k} \epsilon_{n+1} \cdots \epsilon_{2n}$, $0 \leq k \leq n$, $1 \leq j_1 < \cdots < j_k \leq n$, form a basis over $\mathbb{C}$ for S, so $S \cong \Lambda\mathbb{C}^n \cong \mathbb{C}^{2^n}$ as a vector space. Use this isomorphism to fix a Hermitian inner product on S; as in 8.15, its real part is an inner product on the real vector space underlying S.

It must be emphasized that the spinor space S is not canonical; a different orthonormal basis for $\mathbb{R}^{2n}$ will in general yield a different (although isomorphic) left ideal of $C_{2n}^{\mathbb{C}}$.

Exercises Find an element $v \in C_{2n}^{\mathbb{C}}$ so that S^+ and S^- are the (± 1)-eigenspaces of left multiplication by v on S. For each nonzero $w \in \mathbb{C}^{2n}$, show that left multiplication by w on S is an automorphism of order 4 which permutes S^+ and S^-; in particular, S^+ and S^- are isomorphic.

9.48 Left multiplication of $C_{2n}^{\mathbb{C}}$ on S induces an injective homomorphism $C_{2n}^{\mathbb{C}} \to \operatorname{End}(S)$; equality of the dimensions implies that $C_{2n}^{\mathbb{C}}$ and $\operatorname{End}(S)$ are isomorphic complex algebras. Furthermore, the subalgebra $\mathbb{C} \otimes C_{2n}^{+}$ leaves the half-spinor spaces S^+ and S^- invariant.

Exercise Use the identity $C_{k+2}^{\mathbb{C}} \cong C_k^{\mathbb{C}} \otimes C_2^{\mathbb{C}}$ from the exercise in 9.45 to give another proof that $C_{2n}^{\mathbb{C}}$ is the endomorphism algebra of a complex vector space of dimension 2^n. In addition, $C_{2n+1}^{\mathbb{C}} \cong \operatorname{End}(\mathbb{C}^{2^n}) \oplus \operatorname{End}(\mathbb{C}^{2^n})$.

Definition The restriction to $\operatorname{Spin}(2n) \subset C_{2n}^{+} \subset C_{2n} \subset C_{2n}^{\mathbb{C}}$ of the representation $C_{2n}^{\mathbb{C}} \cong \operatorname{End}(S)$ is called the *spin representation of* Spin(2*n*). The induced representations of Spin(2*n*) on the half-spinor spaces S^+ and S^- are called the *half-spin representations of* Spin(2*n*). The representation (9.44) $u \wedge v = -\frac{1}{4}\operatorname{ad}_{[u,v]} \mapsto -\frac{1}{4}[u, v] \in \operatorname{End}(S)$ of the Lie algebra $\mathfrak{o}(2n) \subset C_{2n}$ of Spin(2*n*) on S is called the *spin representation of* $\mathfrak{o}(2n)$, and similarly for the representations on S^+ and S^-.

The action of Spin(2*n*) on S preserves the Hermitian and real inner products on S from 9.47, that is, Spin(2*n*) acts isometrically on S.

Since $C_{2n}^{\mathbb{C}} \cong \operatorname{End}(S)$, the spin representations of Spin(2*n*) and $\mathfrak{o}(2n)$ on S are faithful, as are their representations on S^+ and S^-. The half-spin representations are irreducible and inequivalent [Ja: VII.8].

9.49 Now look at the odd-dimensional case. The inclusion of $\operatorname{Spin}(2n-1) \subset C_{2n-1}^{\mathbb{C}}$ into $\mathbb{C} \otimes C_{2n}^{+}$ embeds Spin(2*n* − 1) into Spin (2*n*) as a Lie subgroup. This commutes with the covering maps onto the respective special orthogonal groups, and induces representations of Spin(2*n* − 1) on the spinor space S of $\mathbb{R}^{2n}$, and also on the half-spinor spaces S^+ and S^-. If α is the isomorphism from S^+ to S^- given by left multiplication by $e_{2n} \in \mathbb{C}^{2n}$, then the representations of Spin$(2n-1)$ on

$$\begin{array}{ccc} \mathrm{Spin}(2n-1) & \longrightarrow & \mathrm{Spin}(2n) \\ \downarrow & & \downarrow \\ \mathrm{SO}(2n-1) & \longrightarrow & SO(2n) \end{array} \qquad \begin{array}{ccc} \mathrm{Spin}(2n-1) \times S^+ & \longrightarrow & S^+ \\ {\scriptstyle \mathrm{id} \times \alpha}\downarrow & & \downarrow{\scriptstyle \alpha} \\ \mathrm{Spin}(2n-1) \times S^- & \longrightarrow & S^- \end{array}$$

S^+ and S^- are equivalent under α.

Definition The *spin representation of* Spin$(2n-1)$ is the induced representation on the half-spinor space S^+ of $\mathbb{R}^{2n}$. The spin representation of $\mathfrak{o}(2n-1)$ is the induced representation on S^+; as in 9.44, $u \wedge v \mapsto -\frac{1}{4}[u, v]$.

Just as in 9.48, Spin$(2n-1)$ acts isometrically on S^+.

For more information on Clifford algebras, Spin(n), and the spinor space, see [ABS], [At], [Che*1*], [Che*2*], [Hch], [Ja], or [Mi*2*].

9.50 Now we are ready to do everything on the bundle level. Let E be a Riemannian vector bundle over M, with associated quadratic form $Q = -\|\cdot\|^2$. Since the Clifford algebra of a quadratic form on a vector space is functorial by 9.40c, associated to Q is a Clifford algebra bundle $C(E, Q)$ as in 1.38; for each $p \in M$, the fiber of $C(E, Q)$ over p is the Clifford algebra $C(E_p, Q_p)$ of the quadratic form $Q_p = -\|\cdot\|_p^2$ on the vector space E_p.

Alternatively, given $W, X \in \Gamma E$, define $\delta_W X := -\langle W, X\rangle 1$, and extend δ_W so that $\delta_W(Y \wedge Z) = (\delta_W Y) \wedge Z + (-1)^{\deg(Y)} Y \wedge \delta_W Z$ for all Y, $Z \in \Gamma\Lambda E$; then $C(E, Q)$ is the vector bundle ΛE with the associative multiplication $\cdot$ such that

$$W \cdot Z = W \wedge Z + \delta_W Z \qquad W \in \Gamma E,\ Z \in \Gamma\Lambda E$$

If B is the orthonormal frame bundle of E, then $C(E, Q) = B \times_{O(m)} C_m$, where C_m is the Clifford algebra of the standard fiber $\mathbb{R}^m$ of E.

Observe that E is a vector subbundle of $C(E, Q)$.

9.51 **Proposition** If ∇ is a Riemannian connection in E, there is a unique extension to a connection in $C(E, Q)$ which respects the algebra bundle structure: $\nabla_V(Y \cdot Z) = (\nabla_V Y) \cdot Z + Y \cdot \nabla_V Z$ for $V \in \mathfrak{X}M$ and Y, $Z \in \Gamma C(E, Q)$. If $Y, Z \in \Gamma C(E, Q)$ are parallel along a curve γ in M, then so is $Y \cdot Z$.

PROOF See 2.14 and 9.27.

9.52 **Definition** Let $\not{p}\colon B \to M$ be the orthonormal frame bundle of an oriented Riemannian vector bundle of rank $m \geq 2$ over M; the group of B is $SO(m)$. A *spin structure on E* is a principal Spin(m)-bundle $\tilde{\not{p}}: \tilde{B} \to M$ together with a principal bundle homomorphism $h: \tilde{B} \to B$, that is, a map such that the diagram

$$\begin{array}{ccccc} \tilde{B} \times \mathrm{Spin}(m) & \xrightarrow{\tilde{\mu}} & \tilde{B} & & \\ \downarrow{\scriptstyle h\times\rho} & & \downarrow{\scriptstyle h} & \searrow & \\ & & & & M \\ & & & \nearrow & \\ B \times SO(m) & \xrightarrow[\mu]{} & B & & \end{array}$$

commutes, where ρ is the group covering map, and $\tilde{\mu}$ and μ are the respective group actions on the bundles. A *spin manifold* is an oriented Riemannian manifold M, together with a spin structure on TM.

It follows that h is a 2-fold covering map.

9.53 Let B be the orthonormal frame bundle of an oriented Riemannian vector bundle E over M, so that $E = B \times_{SO(m)} \mathbb{R}^m$. The spinor group Spin($m$) acts on $\mathbb{R}^m$ via the covering homomorphism ρ from Spin(m) to $SO(m)$; it follows that if $(\tilde{B}, h)$ is a spin structure on E, then $E = \tilde{B} \times_{\mathrm{Spin}(m)} \mathbb{R}^m$. In particular, E admits a spin structure if and only if (see 1.12i and j, 1.49) the $SO(m)$-valued transition functions $f_{\varphi,\psi}$ of E and B can be lifted to Spin(m)-valued functions $\tilde{f}_{\varphi,\psi}$ satisfying the relation $\tilde{f}_{\varphi,\psi}(p) = \tilde{f}_{\varphi,\eta}(p) \cdot \tilde{f}_{\eta,\psi}(p)$ from 1.6. This can be rephrased in terms of some $\mathbb{Z}_2$-characteristic classes of the bundle E, a topic which will be very briefly sketched next.

9.54 Associated with a real vector bundle E over M are characteristic classes $w_i E \in H^i(M; \mathbb{Z}_2)$, called the *Stiefel-Whitney classes* of E (see [Mi*3*: 4] and [S: 38]); unlike the Euler, Chern, and Pontrjagin characteristic classes, the Stiefel-Whitney classes of E are not de Rham cohomology classes of M, and hence are not represented in terms of the curvature operator of a connection in E. As an example of the sort of information carried by the Stiefel-Whitney classes, a vector bundle E is orientable if and only if $w_1 E = 0$ [Mi*3*: 12.4].

An oriented Riemannian vector bundle E over M admits a spin structure if and only if $w_2 E \in H^2(M; \mathbb{Z}_2)$ is zero, (see [BH: 26.5] and [Hae]). For example, $w_2(TCP^n) \neq 0$ for even n [Mi*3*: 11.15], so $\mathbb{C}P^n$ is not a spin manifold for those n; on the other hand, S^n is a spin manifold for $n \geq 2$ because $w_i(TS^n) = 0$ for $i > 0$ [Mi*3*: 4.4.2].

Exercise Show that the Hopf fibration of S^3 over S^2, together with the bundle homomorphism from S^3 to $SO(3)$ from 6.50, is a spin structure on the tangent bundle of S^2.

One proof [Li5] of this result uses sheaf theory. The exact sequence $0 \to \mathbb{Z}_2 \to \mathrm{Spin}(m) \to SO(m) \to 0$ yields a long exact cohomology sequence [W: 5.18]:

$$\cdots \to H^1(M; \mathbb{Z}_2) \to H^1(M; \mathrm{Spin}(m)) \to H^1(M; SO(m)) \to H^2(M; \mathbb{Z}_2) \to \cdots$$

The condition $f_{\varphi,\psi}(p)^{-1} f_{\varphi,\eta}(p) f_{\eta,\psi}(p) = 1$ from 1.6 says that the collection of $SO(m)$-valued transition functions of E is a 1-Čech cocycle on M [W: 5.33]; let $f \in H^1(M; SO(m))$ be the cohomology class of this cocycle. The image of f under α is $w_2 E \in H^2(M; \mathbb{Z}_2)$ (this takes some work). Thus $w_2 E = 0$ if and only if $f = \beta(\tilde{f})$ for some $\tilde{f}$ in $H^1(M; \mathrm{Spin}(m))$; in this case any Čech cocycle representing $\tilde{f}$ is a collection of Spin(m)-valued transition functions for E, and therefore determines a spin structure on E.

This proof implies more than just the result quoted above. Suppose $w_2 E = 0$, so that E admits a spin structure. Identify spin structures $(\tilde{B}, h)$ and $(\tilde{B}_1, h_1)$ on E if there is a principal bundle isomorphism $k: \tilde{B} \to \tilde{B}_1$ such that $h_1 \circ k = h$. The distinct spin structures on E are then in one-to-one correspondence with $H^1(M; \mathbb{Z}_2)$. For example, for $n \geq 2$ the sphere S^n admits a unique spin structure.

9.55 If $(\tilde{B}, h)$ is a spin structure on an oriented Riemannian vector bundle (E, g) over M, then the Clifford algebra bundle of $Q = -\|\cdot\|^2$ can be written $C(E, Q) = B \times_{SO(m)} C_m = \tilde{B} \times_{\mathrm{Spin}(m)} C_m$ by means of the action of Spin(m) on C_m from 9.45. In addition we can use the spin representation of Spin(m) on S to attach a spinor space at each point of M (where S is the spinor space of $\mathbb{R}^m$ if m is even; if $m = 2k - 1$, then by abuse of language, let S denote the half-spinor space S^+ of $\mathbb{R}^{2k}$).

> **Definition** If $(\tilde{B}, h)$ is a spin structure on an oriented Riemannian vector bundle E of rank m, the associated *spinor bundle* is the complex vector bundle $S\tilde{B} := \tilde{B} \times_{\mathrm{Spin}(m)} S$, where S is the spin representation space of Spin(m). The notation $S\tilde{B}$ emphasizes the dependence of the spinor bundle on the choice of spin structure $\tilde{B}$ on E.

Two things are evident from the definition. Since Spin(m) acts isometrically on S (9.48, 9.49), $S\tilde{B}$ is a Hermitian vector bundle (9.5b), 9.7h), and the underlying real vector bundle is Riemannian. Next, the complexified Clifford algebra bundle $C(E, Q)^{\mathbb{C}}$ is the endomorphism bundle End($S\tilde{B}$), and unit vectors in E act isometrically on $S\tilde{B}$.

Given $p \in M$, an orthonormal basis for E_p allows us to identify the spinor space $S\tilde{B}_p$ with a subspace of $C(E, Q)^{\mathbb{C}}$; this identification is not canonical because of the dependence of the spinor space $S \subset C_m^{\mathbb{C}}$ on the choice of orthonormal basis (9.47). In general $S\tilde{B}$ is *not* a subbundle of $C(E, Q)^{\mathbb{C}}$.

Exercise If E is the realification of a holomorphic Hermitian vector bundle over a complex manifold M, is $S\tilde{B}$ a holomorphic vector bundle?

9.56 Let ∇ be a Riemannian connection in an oriented Riemannian vector bundle E over M. If B is the orthonormal frame bundle of E, then by 9.3, 9.4, and 9.14, ∇ is equivalent to a connection form ω on B; since $C(E, Q) = B \times_{SO(m)} C_m$, the connection form ω on B also induces the connection ∇ in $C(E, Q)$ from 9.51.

Now suppose that $(\tilde{B}, h)$ is a spin structure on E. The pullback $h^*\omega =: \tilde{\omega}$ is a connection form on $\tilde{B}$ which is equivalent to the connections in $E = \tilde{B} \times_{\mathrm{Spin}(m)} \mathbb{R}^m$ and $C(E, Q) = \tilde{B} \times_{\mathrm{Spin}(m)} C_m$. In addition, $\tilde{\omega}$ induces a *spinor connection* in the spinor bundle $S\tilde{B}$; it follows that the induced connection on the endomorphism bundle $\mathrm{End}(S\tilde{B})$ from 2.62 is the extension to $C(E, Q)^{\mathbb{C}}$ of the connection on $C(E, Q)$, for $C(E, Q)^{\mathbb{C}} = \tilde{B} \times_{\mathrm{Spin}(m)} C_m^{\mathbb{C}}$ [Ka].

9.57 Given a curve $\tilde{\beta}$ in $\tilde{B}$ and a curve ξ in S, denote by $\tilde{\beta} \cdot \xi$ the curve in $S\tilde{B}$ such that $(\tilde{\beta} \cdot \xi)(t) := \tilde{\beta}(t) \cdot \xi(t)$ (as in 9.5, each point in $\tilde{B}$ is an isomorphism from S to the corresponding fiber of $S\tilde{B}$). By 9.4, each parallel section of $S\tilde{B}$ along a curve γ in M is of the form $\tilde{\beta} \cdot \xi$, where $\tilde{\beta}$ is a horizontal lift of γ to $\tilde{B}$, and ξ is a point in S. But if $\beta := h \circ \tilde{\beta}$, then $\tilde{\omega}(\dot{\tilde{\beta}}) = \omega(\dot{\beta})$, so $\tilde{\beta}$ is a horizontal lift of γ to $\tilde{B}$ if and only if β is a horizontal lift of γ to B.

Now let $\tilde{\beta}$ be an arbitrary lift of γ to $\tilde{B}$; by 9.19, for each $\xi \in S$ the covariant derivative of the section $\tilde{\beta} \cdot \xi$ of $S\tilde{B}$ along γ is $\nabla_D(\tilde{\beta} \cdot \xi) = \tilde{\beta} \cdot \tilde{\omega}(\dot{\tilde{\beta}})\xi$, where for each t, $\tilde{\omega}(\dot{\tilde{\beta}}(t))\xi$ is the product of $\tilde{\omega}(\dot{\tilde{\beta}}(t)) \in \mathfrak{o}(m)$ and $\xi \in S$. If $\beta := h \circ \tilde{\beta}$, then as in 3.13,

$$\tilde{\omega}(\dot{\tilde{\beta}}) = \omega(\dot{\beta}) = [\omega_{ij}(\dot{\beta})] \cong \sum_{i<j} \omega_{ij}(\dot{\beta}) e_i \wedge e_j \in \mathfrak{o}(m)$$

where $\{e_j\}$ is the usual basis for $\mathbb{R}^m$. By 9.48 and 9.49, under the spin representation of $\mathfrak{o}(m)$ on S,

$$\tilde{\omega}(\dot{\tilde{\beta}})\xi = -\frac{1}{4}\sum_{i<j} \omega_{ij}(\dot{\beta})[e_i, e_j]\xi = -\frac{1}{4}\sum_{i,j} \omega_{ij}(\dot{\beta}) e_i e_j \xi \in S$$

Thus

$$\nabla_D(\tilde{\beta} \cdot \xi) = -\frac{1}{4}\tilde{\beta} \cdot \Big(\sum_{i,j} \omega_{ij}(\dot{\beta}) e_i e_j \xi\Big) = -\frac{1}{4}\sum_{i,j} \omega_{ij}(\dot{\beta})\tilde{\beta} \cdot (e_i e_j \xi)$$

For each t, extend the isomorphism $\beta(t)\colon \mathbb{R}^m \to E_{\gamma(t)}$ to an isomorphism $\mathrm{End}(S) \cong \mathrm{End}(S\tilde{B}_{\gamma(t)})$. Given $u \in \mathbb{R}^m$ and $\xi \in S$, the identity $h \circ \tilde{\beta} = \beta$ implies that for all t, $\tilde{\beta}(t)(u\xi) = (\beta(t)u) \cdot (\tilde{\beta}(t)\xi)$. Thus if β is the orthonormal basis field $\{E_j\}$ along γ, then

$$\nabla_D(\tilde{\beta} \cdot \xi) = -\frac{1}{4}\sum_{i,j} \omega_{ij}(\dot{\beta}) E_i E_j \tilde{\beta} \cdot \xi$$

9.58 Similarly we can calculate the curvature tensor R of ∇ in $S\tilde{B}$ in terms of the curvature form $\Omega = d\omega + \frac{1}{2}[\omega, \omega]$ of ω on B. In fact, the curvature form $\tilde{\Omega} = d\tilde{\omega} + \frac{1}{2}[\tilde{\omega}, \tilde{\omega}]$ of $\tilde{\omega}$ on $\tilde{B}$ equals $h^*\Omega$; hence given $u, v \in M_p$, with horizontal lifts $\tilde{u}, \tilde{v} \in \tilde{B}_{\tilde{b}}$ and $\bar{u}, \bar{v} \in B_b$, where $h\tilde{b} = b$ and $\tilde{p}\tilde{b} = pb = p \in M$,

$$\tilde{\Omega}(\tilde{u}, \tilde{v}) = \Omega(\bar{u}, \bar{v}) \cong \sum_{i<j} \Omega_{ij}(\bar{u}, \bar{v}) e_i \wedge e_j \in \mathfrak{o}(m)$$

$$\cong -\frac{1}{4}\sum_{i,j} \Omega_{ij}(\bar{u}, \bar{v}) e_i e_j \in \mathrm{End}(S)$$

Therefore if $be_j = \epsilon_j \in E_p$, and if $\xi \in S$, with $z = \tilde{b}\xi \in S\tilde{B}_p$, then by 9.20,

$$R(u, v)z = \tilde{b} \cdot \tilde{\Omega}(\tilde{u}, \tilde{v})\xi = -\frac{1}{4}\tilde{b}\,(\sum_{i,j} \Omega_{ij}(\bar{u}, \bar{v}) e_i e_j \xi) \in S\tilde{B}_p$$

$$= -\frac{1}{4}\sum_{i,j} \Omega_{ij}(\bar{u}, \bar{v}) \epsilon_i \epsilon_j z$$

By exercise 3 in 9.20, $\Omega_{ij}(\bar{u}, \bar{v}) = \langle R(u, v)\epsilon_j, \epsilon_i \rangle$, so

$$R(u, v)z = \frac{1}{4}\sum_{i,j} \langle R(u, v)\epsilon_i, \epsilon_j \rangle \epsilon_i \epsilon_j z$$

9.59 Now we specialize to the case of a spin manifold, that is, we consider a spin structure $(\tilde{B}, h)$ on the tangent bundle of an oriented Riemannian manifold (M, g), where h is a homomorphism from the principal Spin(n)-bundle $\tilde{B}$ to the orthonormal frame bundle B of M, $n = \dim M \geq 2$. Let ∇ be the spinor connection in $S\tilde{B}$ induced by the Levi-Civita connection of (M, g).

For each *spinor field* $X \in \Gamma S\tilde{B}$, the covariant differential ∇X of X is a section of the bundle $T^*M \otimes S\tilde{B}$; by means of the metric induced isomorphism $\#: T^*M \to TM$ (3.8) we may think of ∇X as a section of $TM \otimes S\tilde{B}$. But TM is a subbundle of $C(TM, Q) \subset \mathrm{End}(S\tilde{B})$, so there is a natural mutiplication map $m: TM \otimes S\tilde{B} \to S\tilde{B}$; thus a section X of the spinor bundle $S\tilde{B}$ determines another section $m \circ \nabla X$ of $S\tilde{B}$.

Definition The *Dirac operator* of the spin structure $\tilde{B}$ on M is the first-order differential operator $D: \Gamma S\tilde{B} \to \Gamma S\tilde{B}$ such that $DX := m \circ \nabla X$.

If $\{\epsilon_j\}$ is an orthonormal basis for M_p, $p \in M$, then

$$(DX)_p = \sum_j \epsilon_j \cdot \nabla_{\epsilon_j} X \qquad X \in \Gamma S\tilde{B}$$

9.60 If M is compact, then integration over M of the Riemannian metric on the spinor bundle $S\tilde{B}$ yields an inner product $\int_M \langle X, Y \rangle$ on $\Gamma S\tilde{B}$ (as a real vector space).

Proposition The Dirac operator of $\tilde{B}$ is self-adjoint with respect to the integration inner product on $\Gamma S\tilde{B}$.

PROOF Fix spinor fields $X, Y \in \Gamma S\tilde{B}$; define $\mu \in A^1 M$ by $\mu(V) := \langle X, V \cdot Y \rangle$, $V \in \mathfrak{X}M \subset \Gamma\ \mathrm{End}(S\tilde{B})$. Fix $p \in M$ and an adapted moving frame $\{E_j\}$ near p, $\epsilon_j := E_j|_p$. Since ∇ is Riemannian and unit vectors in $TM \subset \mathrm{End}(S\tilde{B})$ act isometrically, 9.56 implies

$$\begin{aligned}\delta\mu(p) &= -\sum \epsilon_j \langle X, E_j Y \rangle = -\sum (\langle \nabla_{\epsilon_j} X, \epsilon_j Y \rangle + \langle X, \nabla_{\epsilon_j}(E_j Y) \rangle) \\ &= \sum (\langle \epsilon_j \nabla_{\epsilon_j} X, Y \rangle - \langle X, \epsilon_j \nabla_{\epsilon_j} Y \rangle) \\ &= (\langle DX, Y \rangle - \langle X, DY \rangle)(p)\end{aligned}$$

Now apply Stokes' theorem to the integral over M of the n-form $\star(\langle DX, Y \rangle - \langle X, DY \rangle)$.

Exercise Show that if D is the Dirac operator on a compact spin manifold $(M, \tilde{B})$, then D and D^2 have the same kernel in $\Gamma S\tilde{B}$.

9.61 **Definition** The square D^2 of the Dirac operator D on a spin manifold $(M, \tilde{B})$ is called the *spinor Laplacian.* If $D^2 X = 0$ for a spinor field $X \in \Gamma S\tilde{B}$ on M, then X is called a *harmonic spinor*.

If $\{E_j\}$ is an adapted moving frame near $p \in M$, with $\epsilon_j := E_j|_p$, then for each spinor field X on M, 9.56 implies

$$\begin{aligned}(D^2 X)_p &= \sum_{i,j} \epsilon_i \nabla_{\epsilon_i} E_j \nabla_{E_j} X = \sum_{i,j} \epsilon_i \epsilon_j \nabla_{E_i} \nabla_{E_j} X \\ &= -\sum_i \nabla_{\epsilon_i} \nabla_{E_i} X + \sum_{i<j} \epsilon_i E_j (\nabla_{\epsilon_i} \nabla_{E_j} X - \nabla_{\epsilon_j} \nabla_{E_i} X) \\ &= -\operatorname{div}(\nabla X)_p + \sum_{i<j} \epsilon_i \epsilon_j R(\epsilon_i, \epsilon_j) X\end{aligned}$$

where $\operatorname{div}(\nabla X)$ is defined by analogy with 4.3. This should be compared with the formulas of Weitzenböck (4.22) and Chern (4.25) for Δ on $A(M)$. In particular, if M is flat, then D^2 is formally the Laplacian Δ.

9.62 As in 3.45, let s be the scalar curvature of M,

$$s(p) := \operatorname{tr} \mathscr{R}_p = \sum \mathscr{R}(\epsilon_j, \epsilon_j) = \sum \langle R(\epsilon_i, \epsilon_j)\epsilon_j, \epsilon_i \rangle \qquad p \in M$$

where $\mathscr{R}$ is the Ricci tensor of M, and $\{\epsilon_j\}$ is an orthonormal basis for M_p.

Theorem Lichnerowicz' formula for the spinor Laplacian: If D is the Dirac operator on a spin manifold $(M, \tilde{B})$, then for all $X, Y \in \Gamma S\tilde{B}$,

$$D^2X = -\operatorname{div}(\nabla X) + \frac{s}{4}X$$

$$\langle D^2X, Y \rangle = \frac{s}{4}\langle X, Y \rangle + \langle \nabla X, \nabla Y \rangle - \delta\langle \nabla.X, Y \rangle$$

Here $\langle \nabla.X, Y \rangle$ denotes the 1-form $v \mapsto \langle \nabla_v X, Y \rangle$.

Proof By 9.61 and 9.59, if $\{E_j\}$ is an orthonormal moving frame, then

$$D^2X + \operatorname{div}(\nabla X) = \sum_{i<j} E_i E_j R(E_i, E_j)X$$

$$= \frac{1}{4} \sum_{i<j} \sum_{k,l} \langle R(E_i, E_j)E_k, E_l \rangle E_i E_j E_k E_l X$$

Let us use the abbreviated notation

$$R_{ijkl} := \langle R(E_i, E_j)E_k, E_l \rangle \qquad E_{ijkl} := E_i E_j E_k E_l$$

for all integers i, j, k, and l. The symmetries for R (3.35) are then

$$R_{ijkl} + R_{jikl} = 0 \qquad R_{ijkl} + R_{jkil} + R_{kijl} = 0$$

$$R_{ijkl} + R_{ijlk} = 0 \qquad R_{ijkl} = R_{klij}$$

These and the symmetries for the E_{ijkl} in $C(TM, Q)$ yield (where we suppress the spinor X)

$$D^2 + \operatorname{div} \nabla = \frac{1}{8} \sum_{i,j,k,l} R_{ijkl} E_{ijkl}$$

$$= \frac{1}{24} \sum_{l} \sum_{i,j,k} (R_{ijkl} E_{ijkl} + R_{jkil} E_{jkil} + R_{kijl} E_{kijl})$$

which we write as a sum of terms of five types.

If $i = j = k$, the terms are zero since $R_{iiil} = 0$.

If $i = j \neq k$, then for each l we have

$$\frac{1}{24} \sum_{k \neq i} (R_{iikl} E_{iikl} + R_{ikil} E_{ikil} + R_{kiil} E_{kiil}) = -\frac{1}{12} \sum_{k \neq i} R_{kiil} E_{kl}$$

After reindexing we obtain exactly the same thing for each l in the terms where $j = k \neq i$ and $k = i \neq j$.

Finally, if i, j, and k are distinct, then for each l we have

$$\frac{1}{24} \sum_{\substack{i,\, j,\, k \\ \text{distinct}}} (R_{ijkl} + R_{jkil} + R_{kijl}) E_{ijkl} = 0$$

Thus
$$\begin{aligned}
D^2 + \operatorname{div} \nabla &= -\frac{1}{4} \sum_{l} \sum_{i \neq k} R_{kiil} E_{kl} \\
&= -\frac{1}{4} \sum_{i \neq k} \Big(R_{kiii} E_{ki} + R_{kiik} E_{kk} + \sum_{l \neq i,\, k} R_{kiil} E_{kl} \Big) \\
&= \frac{1}{4} \sum_{i,\, k} \Big(R_{kiik} - \sum_{l<k} (R_{ikli} E_{kl} + R_{ilki} E_{lk}) \Big) \\
&= \frac{1}{4} \sum_{i,\, k} \langle R(E_i, E_k) E_k, E_i \rangle - \frac{1}{4} \sum_{i} \sum_{l<k} (R_{ikli} + R_{liki}) E_{kl} \\
&= \frac{s}{4} + \frac{1}{4} \sum_{i} \sum_{l<k} R_{klii} E_{kl} \\
&= \frac{s}{4}
\end{aligned}$$

Therefore $D^2 X = -\operatorname{div}(\nabla X) + \dfrac{s}{4} X$, $X \in \Gamma S\tilde{B}$.

Finally, if $\{E_j\}$ is an adapted moving frame near p, $\epsilon_j := E_j|_p$, then

$$\begin{aligned}
\langle \operatorname{div}(\nabla X), Y \rangle (p) &= \sum_{i} \langle \nabla_{\epsilon_i} \nabla_{E_i} X, Y \rangle \\
&= \sum_{i} (\epsilon_i \langle \nabla_{E_i} X, Y \rangle - \langle \nabla_{\epsilon_i} X, \nabla_{\epsilon_i} Y \rangle) \\
&= \delta \langle \nabla . X, Y \rangle - \langle \nabla X, \nabla Y \rangle
\end{aligned}$$

9.63 Corollary If M is a compact spin manifold of nonnegative scalar curvature s, then every harmonic spinor on M is parallel; if in addition s is positive somewhere on M, then M admits no nonzero harmonic spinors.

PROOF Use the theorem to integrate

$$\frac{s}{4} \|X\|^2 + \|\nabla X\|^2$$

over M for each harmonic spinor X on M.

For the topological implications of this result via the Atiyah-Singer index theorem, see [Li4]; for a generalization, see [Hch]. Another implication of positive scalar curvature is found in [LY].

9.64 The similarities between D^2 and the usual Laplacian Δ on $A(M)$ led to the question of a possible spinor version of the Hodge theorem: is the dimension of the space of harmonic spinors on a compact spin manifold a topological invariant of the manifold? Hitchin [Hch] showed that although the dimension of the space of harmonic spinors is invariant under a conformal change of metric (this was already known to physicists), it is not invariant under an arbitrary change of metric; therefore the dimension of the space of harmonic spinors is not in general a topological invariant.

BIBLIOGRAPHY

[AM] Abraham, R., and Marsden, J. E.: *Foundations of Mechanics*, 2d ed., Benjamin-Cummings, New York, 1978.

[APS] Ambrose, W., Palais, R. S., and Singer, I. M.: "Sprays," *An. Acad. Bras. Ciênc.*, 32, 1960, pp. 163–178.

[AS] Ambrose, W., and Singer, I. M.: "A theorem on holonomy," *Trans. Am. Math. Soc.*, 75, 1953, pp. 428–443.

[Ad] Arnold, V. I.: *Ordinary Differential Equations*, MIT Press, Cambridge and London, 1978.

[Ar] Artin, E.: *Geometric Algebra*, Interscience, New York, 1957.

[At] Atiyah, M. F.: "Classical groups and classical differential operators on manifolds," *Differential Operators on Manifolds* (*C.I.M.E.*), Edizioni Cremonese, Rome, 1975.

[ABS] Atiyah, M. F., Bott, R., and Shapiro, A.: "Clifford modules," *Topology*, 3, suppl. 1, 1964, pp. 3–38.

[AHS] Atiyah, M. F., Hitchin, N., and Singer, I. M.: "Self-duality in four-dimensional riemannian geometry," *Proc. R. Soc. London Ser. A*, 362, 1978, pp. 425–461.

[AuM] Auslander, L., and MacKenzie, R. E.: *Introduction to Differentiable Manifolds*, McGraw-Hill, New York, 1963; Dover, New York, 1977.

[BM] Beers, B. L., and Millman, R. S.: "The spectra of the Laplace-Beltrami operator on compact, semisimple Lie groups," *Am. J. Math.*, 99, 1977, pp. 801–807.

[Ber] Bérard Bergery, L.: "Les variétés riemanniennes homogènes simplement connexes de dimension impaire à coubure strictement positive," *J. Math. Pures Appl.*, 55, 1976, pp. 47–68.

[B*1*] Berger, M.: "Sur les groupes d'holonomie homogènes des variétés à connexions affine et des variétés riemanniennes," *Bull. Soc. Math. Fr.*, 83, 1955, pp. 279–330.

[B*2*] Berger, M.: "Sur quelques variétés riemanniennes suffisamment pincées," *Bull. Soc. Math. Fr.*, 88, 1960, pp. 57–71.

[B*3*] Berger, M. "Sur les variétés à opérateur de courbure positif," *C. R. Acad. Sci. Ser. A*, 253, 1961, pp. 2832–2834.

[B4] Berger, M.: "On the diameter of some Riemannian manifolds," mimeographed notes, University of California, Berkeley, 1962.

[BE] Berger, M., and Ebin, D.: "Some decompositions of the space of symmetric tensors on a Riemannian manifold," *J. Diff. Geom.*, 3, 1969, pp. 379–392.

[Be] Besse, A. L.: *Manifolds All of Whose Geodesics are Closed*, *Ergeb. Math.*, 93, Springer, New York, 1978.

[BC] Bishop, R. L., and Crittendon, R. J.: *Geometry of Manifolds*, Academic Press, New York, 1964.

[BG] Bishop, R. L., and Goldberg, S. I.: *Tensor Analysis on Manifolds*, Macmillan, New York, 1968.

[BO] Bishop, R. L., and O'Neill, B.: "Manifolds of negative curvature," *Trans. Am. Math. Soc.*, 145, 1969, pp. 1–49.

[Bl] Bleuler, K., et al. (eds.): *Differential Geometric Methods in Mathematical Physics II*, *Lect. Notes Math.*, 676, Springer, New York, 1979.

[Bo*1*] Borel, A.: "Some remarks about Lie groups transitive on spheres and tori," *Bull. Am. Math. Soc.*, 55, 1949, pp. 580–587.

[Bo*2*] Borel, A.: "Le plan projektif des octaves et les sphères commes espaces homogènes," *C. R. Acad. Sci. Ser. A*, 230, 1950, pp. 1378–1380.

[Bo*3*] Borel, A.: "Semi-simple groups and symmetric spaces," mimeographed notes, Tata Institute, 1961.

[BH] Borel, A., and Hirzebruch, F.: "Characteristic classes and homogeneous spaces II," *Am. J. Math.*, 81, 1959, pp. 315–382.

[BoL] Borel, A., and Lichnerowicz, A.: "Groupes d'holonomie des variétés riemanniennes," *C. R. Acad. Sci. Ser. A*, 234, 1952, pp. 1835–1837.

[BS] Borel, A., and Serre, J. P.: "Detérmination des *p*-puissances réduites de Steenrod dans la cohomologie des groupes classiques, Applications," *C. R. Acad. Sci. Ser. A*, 233, 1951, pp. 680–682.

[Bn] Bourguignon, J. P.: Formules de Weitzenböck en dimension 4, *Séminaires A. Besse sur la géométrie riemannienne de dimension 4*, exp. no. XVI, pp. 1–26.

[BB] Bourguignon, J. P., and Bérard Bergery, L.: "Laplacian and Riemannian submersions with totally geodesic fibers," preprint.

[BK] Bourguignon, J. P., and Karcher, H.: "Curvature operators: Pinching estimates and geometric examples," *Ann. Sci. Ec. Norm. Sup.*, ser. 4, vol. 11, 1978, pp. 71–92.

[BL] Bourguignon, J. P., and Lawson, H. B., Jr.: "Yang-Mills theory: Its physical origins and differential geometric aspects," preprint.

[BLS] Bourguignon, J. P., Lawson, H. B., Jr., and Simons, J.: "Stability and gap phenomena for Yang-Mills fields," *Proc. Natl. Acad. Sci. U.S.A.*, 76, 1979, pp. 1550–1553.

[BD] Boyce, W. E., and DiPrima, R. C.: *Elementary Differential Equations and Boundary Value Problems*, 3d ed., Wiley, New York, 1977.

[BrC] Brickell, F., and Clark, R. S.: *Differentiable Manifolds*, Reinhold, London, 1970.

[BrG] Brown, R. B., and Gray, A.: "Riemannian manifolds with holonomy group Spin(9)," *Differential Geometry in honor of K. Yano*, Kirokuniya, Tokyo, 1972, pp. 41–59.

[Cal] Calabi, E.: "On the group of automorphisms of a symplectic manifold," in R. Gunning (ed.), *Problems in Analysis*, Princeton University Press, Princeton, N.J., 1970, pp. 1–26.

[CaE] Calabi, E., and Eckmann, B.: "A class of compact, complex manifolds which are not algebraic," *Ann. Math.*, 58, 1953, pp. 494–500.

[Ca*1*] Cartan, É.: "Sur une généralisation de la notion de courbure de Riemann et les espaces à torsion," *C. R. Acad. Sci. Ser. A*, 174, 1922, pp. 593–597.

[Ca*2*] Cartan, É.: "Sur les variétés à connexion affine et la théorie de la relativité généralisée," *Ann. Ec. Norm. Sup.*, 40, 1923, pp. 325–412.

[Ca3] Cartan, É.: "Les espaces à connexion conforme," *Ann. Soc. Pol. Math.*, 2, 1923, pp. 171–221.

[Ca4] Cartan, É.: "Sur les variétés à connexion projective," *Bull. Soc. Math. Fr.*, 52, 1924, pp. 205–241.

[Ca5] Cartan, É.: "La théorie des groupes et la géométrie," *L'enseignement Math.*, 26, 1927, pp. 200–225.

[Ca6) Cartan, É.: "La topologie des espaces représentatifs des groupes de Lie," *L'enseignement Math.*, 35, 1936, pp. 177–200.

[Car] Cartan, H.: *Elementary Theory of Analytic Functions of One or Several Complex Variables*, Addison-Wesley, Reading, Mass., 1963.

[Cha] Chavel, I.: "A class of Riemannian homogeneous spaces," *J. Diff. Geom.*, 4, 1970, pp. 13–20.

[Cgr] Cheeger, J.: "Some examples of manifolds of nonnegative curvature," *J. Diff. Geom.*, 8, 1973, pp. 623–628.

[CE] Cheeger, J., and Ebin, D.: *Comparison Theorems in Riemannian Geometry*, North Holland, Amsterdam, 1975.

[Ch1] Chern, S-S.: "A simple intrinsic proof of the Gauss-Bonnet formula for closed Riemannian manifolds," *Ann. Math.*, 45, 1944, pp. 747–752.

[Ch2] Chern, S-S.: "On the curvatura integral in a Riemannian manifold," *Ann. Math.*, 46, 1945, pp. 674–684.

[Ch3] Chern, S-S.: "On a generalization of Kähler geometry," *Algebraic Geometry and Topology Symposium in honor of S. Lefschetz*, Princeton University Press, Princeton, N.J., 1957, pp. 103–121.

[Ch4] Chern, S-S.: The geometry of *G*-structures, *Bull. Am. Math. Soc.*, 72, 1966, pp. 167–219.

[Ch5] Chern, S-S.: *Complex Manifolds Without Potential Theory*, 2d ed., Springer, New York, 1979.

[Che1] Chevalley, C.: *Theory of Lie Groups*, Princeton University Press, Princeton, N.J., 1946.

[Che2] Chevalley, C.: *The Algebraic Theory of Spinors*, Columbia University Press, New York, 1954.

[CV] Cohn-Vossen, S.: "Kürzeste Wege und Totalkrümmung auf Flächen," *Compositio Math.*, 2, 1935, pp. 69–133.

[Co] Couty, R.: "Transformations infinitésimales projectives," *C. R. Acad. Sci. Ser. A*, 247, 1958, pp. 804–806.

[DaZ] D'Atri, J. E., and Ziller, W.: Naturally reductive metrics and Einstein metrics on compact Lie groups, *Mem. Am. Math. Soc.*, 18, no. 215, 1970.

[Di] Dieudonné, J. A.: *Foundations of Modern Analysis*, vol. 3, Academic Press, New York, 1972.

[Do1] Dombrowski, P.: "On the geometry of the tangent bundle," *J. reine angew. Math.*, 210, 1962, pp. 73–88.

[Do2] Dombrowski, P.: "Krümmungsgrößen gleichungsdefinierter Untermannigfaltigkeiten Riemannscher Mannigfaltigkeiten," *Math. Nach.*, 38, 1968, pp. 133–180.

[DM] Drechsler, W., and Mayer, M. E.: *Fiber Bundle Techniques in Gauge Theory, Lect. Notes Phys.*, 67, Springer, New York, 1977.

[EK] Eells, J., and Kuiper, N.: "An invariant for certain smooth manifolds," *Ann. Mat. pura appl.*, LX, 1963, pp. 93–110.

[En] Ehresmann, C.: "Les connexions infinitésimales dans un espace fibré différentiable," *Colloq. Topologie*, Bruxelles, 1950.

[Ep] Epstein, D. B. A.: "Natural tensors on Riemannian manifolds," *J. Diff. Geom.*, 10, 1975, pp. 631–645.

[Fi] Fishback, W. T.: *Projective and Euclidean Geometry*, Wiley, New York, 1962.
[Fr] Frazer, W. R.: *Elementary Particles*, Prentice-Hall, Englewood Cliffs, N.J., 1966.
[GaM] Gallot, S., and Meyer, D.: "Opérateur de courbure et Laplacien des formes différentielles d'une variété riemannienne," *J. Math. pures appl.*, 54, 1975, pp. 259–284.
[Gr] Greenberg, M. J.: *Lectures on Algebraic Topology*, Benjamin, New York, 1967.
[GW] Greene, R. E., and Wu, H.: "C^∞ convex functions and manifolds of positive curvature," *Acta Math.*, 137, 1976, pp. 209–245.
[GHV] Greub, W., Halperin, S., and Vanstone, R.: *Connections, Curvature, and Cohomology I*, Academic Press, New York, 1972.
[GKM] Gromoll, D., Klingenberg, W., and Meyer, W.: *Riemannsche Geometrie im Großen*, 2d ed., *Lect. Notes Math.*, 55, Springer, 1975.
[GM] Gromoll, D., and Meyer, W.: "An exotic sphere with nonnegative sectional curvature," *Ann. Math.*, 100, 1974, pp. 401–406.
[GS] Guillemin, V., and Sternberg, S.: "Geometric asymptotics," *Math. Surv. Am. Math. Soc.*, 14, 1977.
[Hae] Haefliger, A.: "Sur l'extension du groupe structural d'un espace fibré," *C. R. Acad. Sci. Ser. A*, 243, 1956, pp. 558–560.
[Ha] Halmos, P. R.: *Finite-Dimensional Vector Spaces*, 2d ed., Van Nostrand, Princeton, N.J., 1958.
[Hn] Hano, J-I.: "On affine transformations of a Riemannian manifold," *Nagoya Math. J.*, 9, 1955, pp. 99–109.
[HM] Hano, J-I., and Morimoto, A.: "Note on the group of affine transformations of an affinely connected manifold," *Nagoya Math. J.*, 8, 1955, pp. 71–81.
[HS] Hausner, M., and Schwartz, J. T.: *Lie Groups · Lie Algebras*, Gordon and Breach, New York, 1968.
[Hz] Heintze, E.: "On homogeneous manifolds of negative curvature," *Math. Ann.*, 211, 1974, pp. 23–34.
[He] Helgason, S.: *Differential Geometry, Lie Groups, and Symmetric Spaces*, Academic Press, New York, 1978.
[Her] Hernandez, H.: "A class of compact manifolds with positive Ricci curvature," *Proc. Symp. Pure Math. A.M.S. XXVII, Differential Geometry*, 1975, part 1, pp. 73–87.
[Hch] Hitchin, N.: "Harmonic spinors," *Adv. Math.*, 14, 1974, pp. 1–55.
[Hi] Hicks, N. J.: *Notes on Differential Geometry*, Van Nostrand, Princeton, N.J., 1965.
[Ho] Hochschild, G.: *The Structure of Lie Groups*, Holden-Day, San Francisco, 1965.
[Hu] Hu, S-T.: *Homotopy Theory*, Academic Press, New York, 1959.
[Hus] Husemoller, D.: *Fibre Bundles*, 2d ed., Springer, New York, 1975.
[Ja] Jacobson, N.: *Lie Algebras*, Interscience, New York, 1962.
[Je*1*] Jensen, G. R.: "The scalar curvature of left-invariant Riemannian metrics," *Ind. Univ. Math. J.*, 20, 1971, pp. 1125–1144.
[Je*2*] Jensen, G. R.: "Einstein metrics on principal fibre bundles," *J. Diff. Geom.*, 8, 1973, pp. 599–614.
[Je*3*] Jensen, G. R.: "Imbeddings of Stiefel manifolds into Grassmannians," *Duke Math. J.*, 42, 1975, pp. 397–407.
[Ka] Karrer, G.: "Einführung von Spinoren auf Riemannschen Mannigfaltigkeiten," *Ann. Acad. Sci. Fenn. Ser. A*, 336/5, 1963, pp. 1–5.
[Ke] Kelley, J. L.: *General Topology*, Van Nostrand, Princeton, 1955.
[Kl*1*] Klingenberg, W.: "Contributions to Riemannian geometry in the large," *Ann. Math.*, 69, 1959, pp. 654–666.
[Kl*2*] Klingenberg, W.: *Lectures on Closed Geodesics*, *Grund. Math. Wiss.*, 230, Springer, New York, 1978.
[Ko*1*] Kobayashi, S.: "On connections of Cartan," *Can. J. Math.*, 8, 1956, pp. 145–156.

[Ko2] Kobayashi, S.: *Transformation Groups in Differential Geometry*, *Ergeb. Math.*, 70, Springer, New York, 1972.

[KNa] Kobayashi, S., and Nagano, T.: "On projective connections," *J. Math. Mech.*, 13, 1964, pp. 215–236.

[KN] Kobayashi, S., and Nomizu, K.: *Foundations of Differential Geometry*, 2 vols, Interscience, New York, 1963, 1969.

[Kr] Kreyszig, E.: *Differential Geometry*, University of Toronto Press, Toronto, 1959.

[Ku] Kulkarni, R.: "On the Bianchi identities," *Math. Ann.*, 99, 1972, pp. 175–204.

[LY] Lawson, H. B., Jr., and Yau, S. T.: "Scalar curvature, non-abelian group actions, and the degree of symmetry of exotic spheres," *Comm. Math. Helv.*, 49, 1974, pp. 232–244.

[Le] Lelong-Ferrand, J.: "Transformations conformes et quasiconformes des variétés riemanniennes; applications à la démonstration d'une conjecture de A. Lichnerowicz," *C. R. Acad. Sci. Ser. A*, 269, 1969, pp. 583–586.

[Lv] Levi-Civita, T.: "Nozione di parallelismo in una varietà qualunque e consequente specificazione geometrica della curvatura Riemanniana," *Rend. Palermo*, 42, 1917, pp. 173–205.

[Li*1*] Lichnerowicz, A.: "Généralisations de la géométrie kählérienne globale," *Colloque de Géométrie Différentielle*, Louvain, 1951, pp. 99–122.

[Li*2*] Lichnerowicz, A.: "Propagateurs et commutateurs en relativité générale," *Inst. Hautes Etud. Sci. Publ. Math.*, 10, 1961, pp. 293–344.

[Li*3*] Lichnerowicz, A.: "Laplacien sur une variété riemannienne et spineurs," *Lincei Rend. Sci. Fis. Mat. Nat.*, XXXIII, 1962, pp. 187–191.

[Li*4*] Lichnerowicz, A.: "Spineurs harmoniques," *C. R. Acad. Sci. Ser. A*, 257, 1963, pp. 7–9.

[Li*5*] Lichnerowicz, A.: "Topics on space-times," in C. M. Dewitt and J. A. Wheeler (eds.), *Battelle Rencontres, Lect. Math. Phys.*, Benjamin, New York, 1968.

[Li*6*] Lichnerowicz, A.: "The Lie algebra of the symplectic automorphisms," preprint.

[Lo] Loos, O.: *Symmetric Spaces*, 2 vols., Benjamin, New York, 1969.

[Ma] Maillot, H.: "Sur les variétés riemanniennes à opérateur de courbure pur," *C. R. Acad. Sci. Ser. A*, 278, 1974, pp. 1127–1130.

[Ms] Matsushima, Y.: "Holomorphic vector fields on compact Kähler manifolds," *Am. Math. Soc.* CBMS 7, 1971.

[Me] Meyer, D.: "Sur les variétés riemanniennes à opérateur de courbure positif," *C. R. Acad. Sci. Ser. A*, 272, 1971, pp. 482–485.

[Mi*1*] Milnor, J.: "On manifolds homeomorphic to the 7-sphere," *Ann. Math.*, 64, 1956, pp. 399–405.

[Mi*2*] Milnor, J.: "The representation rings of some classical groups," mimeographed notes, Princeton University, Princeton, N.J., 1974.

[Mi*3*] Milnor, J.: *Topology from the Differentiable Viewpoint*, University Press of Virginia, Charlottesville, Va., 1965.

[Mi*4*] Milnor, J.: "Curvatures of left invariant metrics on Lie groups," *Adv. Math.*, 21, 1976, pp. 293–329.

[MiS] Milnor, J., and Stasheff, J. D.: *Characteristic Classes*, *Ann. Math. Studies*, 76, Princeton University Press, Princeton, N.J., 1974.

[MTW] Misner, C. W., Thorne, K. S., and Wheeler, J. A.: *Gravitation*, Freeman, San Francisco, 1973.

[MoS] Montgomery, D., and Samelson, H.: "Transformation groups of spheres," *Ann. Math.*, 44, 1943, pp. 454–470.

[MS] Myers, S. B., and Steenrod, N. E.: "The group of isometries of a Riemannian manifold," *Ann. Math.*, 40, 1939, pp. 400–416.

[Na] Nash, J. C.: "Positive Ricci curvature on fibre bundles," *J. Diff. Geom.*, 14, 1979, pp. 241–254.

[NN] Newlander, A., and Nirenberg, L.: "Complex analytic coordinates in almost complex manifolds," *Ann. Math.*, 65, 1957, pp. 391–404.

[N*1*] Nomizu, K.: "On the group of affine transformations of an affinely connected manifold," *Proc. Am. Math. Soc.*, 4, 1953, pp. 816–823.

[N2] Nomizu, K.: "Invariant affine connections on homogeneous spaces," *Am. J. Math.*, 76, 1954, pp. 33–65.

[N*3*] Nomizu, K.: "On the decomposition of generalized curvature tensor fields," *Differential Geometry in Honor of K. Yano*, Kirokuniya, Tokyo, 1972, pp. 335–345.

[N*4*] Nomizu, K.: "Kinematics and differential geometry of submanifolds," *Tôhoku Math. J.*, 2d series, no. 4, 1978, pp. 623–637.

[Ob] Obata, M.: "Conformal transformations of Riemannian manifolds," *J. Diff. Geom.*, 4, 1970, pp. 311–333.

[Oc] Ochiai, T.: "Geometry associated with semi-simple flat homogeneous spaces," *Trans. Am. Math. Soc.*, 152, 1970, pp. 1–33.

[Om] Omori, H.: *Infinite Dimensional Lie Transformation Groups, Lect. Notes Math.*, 427, Springer, New York, 1974.

[Ol] O'Neill, B.: "The fundamental equations of a submersion," *Mich. Math. J.*, 13, 1966, pp. 459–469.

[Pa] Patterson, L-N.: "Connexions and prolongations," *Can. J. Math.*, XXVII, 1975, pp. 766–791.

[Pr*1*] Poor, W. A.: "Some exotic spheres with positive Ricci curvature," *Math. Ann.*, 216, 1975, pp. 245–252.

[Pr2] Poor, W. A.: "A holonomy proof of the positive curvature operator theorem," *Proc. Am. Math. Soc.*, 79, 1980, pp. 454–456.

[Pt] Porteous, I. R.: *Topological Geometry*, Van Nostrand, London, 1969.

[Pn] Preissman, A.: "Quelques propriétés globales des espaces de Riemann, *Comm. Math. Helv.*, 15, 1942–43, pp. 175–216.

[Re] Rebbi, C.: "Solitons," *Sci. Am.*, 240, no. 2, February 1979, pp. 92–116.

[Sn] Sampson, J. H.: "On a theorem of Chern," *Trans. Am. Math. Soc.*, 177, 1973, pp. 141–153.

[Sa] Sasaki, S.: "On the differential geometry of tangent bundles of Riemannian manifolds," *Tôhoku Math. J.*, 10, 1958, pp. 338–345.

[Sch] Schouten, J. A.: "Über die konforme Abbildung *n*-dimensionaler Mannigfaltigkeiten mit quadratischer Maßbestimmung auf eine Mannigfaltigkeit mit euklidischer Maßbestimmung," *Math. Z.*, 11, 1921, pp. 58–88.

[Sha] Shanahan, P.: *The Atiyah-Singer Index Theorem, Lect. Notes Math.*, 638, Springer, 1978.

[Sh] Shimada, N.: "Differentiable structures on the 15-sphere and Pontrjagin classes of certain manifolds," *Nagoya Math. J.*, 12, 1957, pp. 59–69.

[ST*1*] Singer, I. M., and Thorpe, J.: "The curvature of 4-dimensional Einstein spaces," *Global Analysis in Honor of K. Kodaira*, University of Tokyo Press and Princeton University Press, Tokyo, Princeton, N.J., 1969, pp. 355–365.

[ST2] Singer, I. M., and Thorpe, J.: *Lecture Notes on Elementary Topology and Geometry*, Scott-Foresman, Glenview, Ill., 1967.

[Spi] Spivak, M.: *A Comprehensive Introduction to Differential Geometry*, 5 vols., Publish or Perish, Boston, 1970–1975.

[Spr] Springer, G.: *Introduction to Riemann Surfaces*, Addison-Wesley, Reading, Mass., 1957.

[S] Steenrod, N. E.: *The Topology of Fibre Bundles*, Princeton University Press, Princeton, N.J., 1951.

[Sb] Sternberg, S.: *Lectures on Differential Geometry*, Prentice-Hall, Englewood Cliffs, N.J., 1965.

[Sw] Stewart, I. N.: *Lie Algebras, Lect. Notes Math.*, 127, Springer, 1970.

[St] Stredder, P.: "Natural differential operators on Riemannian manifolds and representations of the orthogonal and special orthogonal groups," *J. Diff. Geom.*, 10, 1975, pp. 647–660.

[Tc] Tachibana, S.: "A theorem on Riemannian manifolds of positive curvature operator," *Proc. Japan. Acad.*, 50, 1974, pp. 301–302.

[Tan] Tanaka, N.: "Projective connections and projective transformations," *Nagoya Math. J.*, 11, 1957, pp. 1–24.

['t H] 't Hooft, G.: "Gauge theories of the forces between elementary particles," *Sci. Am.*, 242, 1980, pp. 104–138.

[Ur] Urakawa, H.: "On the least positive eigenvalue of the Laplacian for compact group manifolds," *J. Math. Soc. Japan*, 31, 1979, pp. 209–226.

[Wal] Wallach, N.: "Compact homogeneous Riemannian manifolds with strictly positive curvature," *Ann. Math.*, 96, 1972, pp. 277–295.

[Wa] Walter, R.: "A generalized Allendoerfer-Weil formula and an inequality of the Cohn-Vossen type," *J. Diff. Geom.*, 10, 1975, pp. 167–180.

[W] Warner, F.: *Foundations of Differentiable Manifolds and Lie Groups*, Scott-Foresman, Glenview, Ill., 1971.

[Wl*1*] Weil, A.: *Introduction à l'étude des variétés kählériennes*, Hermann, Paris, 1958.

[Wl2] Weil, A.: "Un théorème fondamental de Chern en géométrie riemannienne," *Sém. Bourbaki*, no. 239, 1961/62.

[We*1*] Weinstein, A.: "Distance spheres in complex projective spaces," *Proc. Am. Math. Soc.*, 39, 1973, pp. 649–650; erratum to "Distance spheres in complex projective spaces," *Proc. Am. Math. Soc.*, 48, 1975, p. 519.

[We2] Weinstein, A.: "On the volume of manifolds all of whose geodesics are closed," *J. Diff. Geom.*, 9, 1974, pp. 513–517.

[We*3*] Weinstein, A.: *Lectures on Symplectic Manifolds, Am. Math. Soc.*, CBMS 29, 1977.

[Ws] Wells, R. O., Jr.: *Differential Analysis on Complex Manifolds*, Prentice-Hall, Englewood Cliffs, N.J., 1973.

[Wy*1*] Weyl, H.: "Reine Infinitesimalgeometrie," *Math. Z.*, 2, 1918, pp. 384–411.

[Wy*2*] Weyl, H.: "Zur Infinitesimalgeometrie: Einordnung der projektiven und der konformen Auffassung," *Göttingen Nachrichten*, 1921, pp. 99–112.

[Wy*3*] Weyl, H.: *The Classical Groups*, Princeton University Press, Princeton, N.J., 1939.

[Wy*4*] Weyl, H.: *Space—Time—Matter*, 4th ed., Dover, New York, 1952.

[Wo] Wolf, J. A.: *Spaces of Constant Curvature*, 2d ed., J. A. Wolf, Berkeley, Ca., 1972.

[Yam] Yamabe, H.: "On an arcwise connected subgroup of a Lie group," *Osaka Math. J.*, 2, 1950, pp. 13–14.

[YM] Yang, C. N., and Mills, R. L.: "Conservation of isotopic spin and isotopic gauge invariance," *Phys. Rev.*, 96, 1954, pp. 191–195.

[Ya*1*] Yano, K.: "Les espaces à connexion projective et la géométrie projective des 'paths,'" *Ann. Sci. Univ. Jassy*, 24, 1938, pp. 395–463.

[Ya*2*] Yano, K.: *Integral Formulas in Riemannian Geometry*, Marcel Dekker, New York, 1970.

[YB] Yano, K., and Bochner, S.: *Curvature and Betti Numbers, Ann. Math. Studies*, 32, Princeton University Press, Princeton, N.J., 1953.

[Y] Yau, S. T.: "Remarks on the group of isometries of a Riemannian manifold," *Topology*, 16, 1977, pp. 239–247.

[Zi*1*] Ziller, W.: "Closed geodesics on homogeneous spaces," *Math. Z.*, 152, 1976, pp. 67–88.

[Zi2] Ziller, W.: "The Jacobi equation on naturally reductive compact Riemannian homogeneous spaces," *Comm. Math. Helv.*, 52, 1977, pp. 573–590.

INDEX OF NOTATION

INDEX